MICROPROPAGATION OF WOODY PLANTS

FORESTRY SCIENCES

Volume 41

The titles published in this series are listed at the end of this volume.

Micropropagation of Woody Plants

edited by

M. R. AHUJA

Institute of Forest Genetics
Federal Research Centre for Forestry and Forest Products
Sieker Landstrasse 2, 2070 Grosshansdorf, Germany

Kluwer Academic Publishers

Dordrecht / Boston / London

Library of Congress Cataloging-in-Publication Data

Micropropagation of woody plants / edited by M.R. Ahuja.
p. cm. -- (Forestry sciences ; v. 41)
Includes bibliographical references and index.
ISBN 0-7923-1807-2 (hb : acid-free paper)
1. Woody plants--Propagation--In vitro. 2. Trees--Propagation--In vitro. I. Ahuja, M. R. II. Series.
SB123.6.M52 1992
634.9'56--dc20 92-14824

ISBN 0-7923-1807-2

Published by Kluwer Academic Publishers,
P.O. Box 17, 3300 AA Dordrecht, The Netherlands.

Kluwer Academic Publishers incorporates the publishing programmes of D. Reidel, Martinus Nijhoff, Dr W. Junk and MTP Press.

Sold and distributed in the U.S.A. and Canada
by Kluwer Academic Publishers,
101 Philip Drive, Norwell, MA 02061, U.S.A.

In all other countries, sold and distributed
by Kluwer Academic Publishers Group,
P.O. Box 322, 3300 AH Dordrecht, The Netherlands.

printed on acid-free paper

Printed in the Netherlands

Contents

Preface

Woody plants have been cultured *in vitro* since the 1930s. Since that time much progress has been made in the culture of tissues, organs, cells and protoplasts of woody plants. That is not to say that all woody plants, including forest trees, can be induced to grow and differentiate *in vitro*. Some do, while others are still recalcitrant. In general juvenile tissues from woody plants are more responsive to *in vitro* manipulations than are mature tissues. Woody plants have long generation cycles. They have an extended vegetative phase, ranging from 10 to 50 years. During the juvenile phase, starting from the embryo and perhaps lasting up to a decade, tissues from woody plants are responsive to *in vitro* conditions. As maturation sets in, tissues from mature trees become less responsive in tissue culture. Other than the age of a tree, the response of an explant/tissue is also determined by the genotype, physiological state of the tissue, time of the year when the explant is cultured, and the composition of the medium. The term 'response' implies that the tissues will be able to grow, differentiate and plantlets can be regenerated *in vitro*. By employing juvenile tissues, clonal propagation has been achieved in a large number of woody plant species. On the other hand, only in a handful of tree species have mature trees been clonally propagated by tissue culture technology.

This book deals with recent developments and limitations in the micropropagation of woody plants. Regeneration, micropropagation, clonal fidelity, and deployment are some of the important aspects of a clonal forestry program. This book has been planned to cover these and other aspects. The material is organized in three sections: fundamentals, micropropagation, and tree improvement.

There is more to micropropagation than just *in vitro* regeneration of plantlets. Before and after micropropagation are just as important as *in vitro* regeneration. With that in mind, innovative thinking and concepts of fundamentals and tree improvement are emphasized in this volume. No effort has been made to cover the micropropagation of all woody plants. Only representative tree species have been selected to demonstrate the feasibility of micropropagation and system development for future research in woody plants.

I am grateful to the contributors to this volume for their commendable jobs on the manuscripts. I also thank Ir. Ad C. Plaizier of Kluwer Academic Publishers for his patience and interest in this book. And finally I am indebted to my wife Ila for her help and understanding during the preparation of this volume.

M. R. Ahuja

Grosshansdorf, Germany

M. R. Ahuja (Ed.), Micropropagation of Woody Plants, xiii–xvii.

Contributors

Abu-Qaoud, H. Department of Horticulture and Forestry, University of Illinois, Urbana-Champaign, IL 61801, USA.

Ahuja, M. R. Institute of Forest Genetics, Federal Research Centre for Forestry and Forest Products, Sieker Landstrasse 2, 2070 Grosshansdorf, FRG.

Bapat, V. A. Plant Biotechnology Section, Bhabha Atomic Research Centre, Trombay, Bombay 400 085, India.

Bonga, J. M. Forestry Canada Maritimes Region, P.O. Box 4000, Fredericton, N.B. E3B 5P7, Canada.

Chun, Y. W. Department of Forestry, College of Forestry, Kookmin University, Seoul, 136–702, Republic of Korea.

Dawkins, M. D. Department of Biology, University of Victoria, Victoria, B.C. V8W 2Y2, Canada.

Dhawan, V. Tata Energy Research Institute, 90 Jor Bagh, New Delhi 110 003, India.

Donnelly, D. J. Plant Science Department, Macdonald College of McGill University, 21, 111 Lakeshore Road, Ste. Anne de Bellevue, Quebec H9X 1CO, Canada.

Durzan, D. J. Department of Environmental Horticulture, University of California, Davis, CA 95616, USA.

Ellis, D. D. Department of Horticulture, 1575 Linden Drive, University of Wisconsin-Madison, Madison, WI 53706, USA.

Ernst, S. G. Department of Forestry, Fisheries and Wildlife, University of Nebraska, Lincoln, NE 68583-0814, USA.

Ganapathi, T. R. Plant Biotechnology Section, Bhabha Atomic Research Centre, Trombay, Bombay 400 085, India.

Gupta, P. K. Technology Centre, Weyerhaeuser Company, Tacoma, WA 98477, USA.

Hackett, W. P. Department of Horticultural Science, University of Minnesota, St. Paul, MN 55108, USA.

Haissig, B. E. USDA Forest Service, North Central Forest Experiment Station, Forestry Science Laboratory, P.O. Box 898, Rhinelander, WI 54501, USA.

Harry, D.E. Department of Horticulture and Forestry, University of Illinois, Urbana-Champaign, IL 61801, USA.

Huang, Y. School of Forestry, Michigan Technological University, Houghton, MI 49931, USA.

Hutzell, M. J. Department of Environmental Horticulture, University of California, Davis, CA 95616, USA.

Jones, O. P. Breeding and Genetics Department, AFRC Institute of Horticultural Research, East Malling, Maidstone, Kent ME19 6Bj, UK.

Karnosky, D. F. School of Forestry, Michigan Technological University, Houghton, MI 49931, USA.

Kendurkar, S. V. Biochemical Sciences Division, National Chemical Laboratory, Pune 411 008, India.

Khuspe, S. S. Biochemical Sciences Division, National Chemical Laboratory, Pune 411 008, India.

Kreitinger, M. Technical Centre, Weyerhaeuser Company, Tacoma, WA 98477, USA.

Kumar, P. P. Plant Physiology Research Group, Department of Biological Sciences, University of Calgary, Calgary, Alberta T2N 1N4, Canada.

Lakshmi Sita, G. Department of Microbiology and Cell Biology, Indian Institute of Sciences, Bangalore 560 012, India.

Libby, W. J. Department of Forestry and Resource Management, University of California, Berkeley, CA 94720, USA.

Mascarenhas, A. F. Biochemical Sciences Division, National Chemical Laboratory, Pune 411 008, India.

Merkle, S. A. School of Forest Resources, University of Georgia, Athens, GA 30605, USA.

Michler, C. H. USDA Forest Service, North Central Forest Experiment Station, Forestry Sciences Laboratory, P.O. Box 898, Rhinelander, WI 54501, USA.

Murray, J. R. Department of Horticultural Science, University of Minnesota, St. Paul, MN 55108, USA.

Nairn, B. J. Forest Nursery and Tree Improvement Centre, Tasman Forestry Ltd., P.O. Box 149, Te Teko, New Zealand.

Ostry, M. E. USDA Forest Service, North Central Forest Experiment Station, St. Paul, MN 55108, USA.

Park, Y. S. Forestry Canada Maritimes Region, P.O. Box 4000, Fredericton, N.B. E3B 5P7, Canada.

Rao, P. S. Plant Biotechnology Section, Bhabha Atomic Research Centre, Trombay, Bombay 400 085, India.

Riemenschneider, D. E. USDA Forest Service, North Central Forest Experiment Station, Forestry Sciences Laboratory, P.O. Box 898, Rhinelander, WI 54501, USA.

Russell, J. A. Cornell University, New York State Agricultural Experiment Station, Department of Horticultural Sciences, Hendrick Hall, Geneva, NY 14456, USA.

Shin, D. I. School of Forestry, Michigan Technological University, Houghton, MI 49931, USA.

Skirvin, R. M. Department of Horticulture and Forestry, University of Illinois, Urbana-Champaign, IL 61801, USA.

Sommer, H. E. School of Forest Resources, University of Georgia, Athens, GA 30605, USA.

Sriskandarjah, S. Department of Horticulture and Forestry, University of Illinois, Urbana-Champaign, IL 61801, USA.

Thorpe, T. A. Plant Physiology Research Group, Department of Biological Sciences, University of Calgary, Calgary, Alberta T2N 1N4, Canada.

Tisdall, L. Plant Science Department, Macdonald College of McGill University, 21, 111 Lakeshore Road, Ste. Anne Bellevue, Quebec H9X 1CO, Canada.

von Aderkas, P. Department of Biology, University of Victoria, Victoria, B.C. V8W 2Y2, Canada.

Webb, D. T. The Institute of Paper Sciences, 575 14th NW, Atlanta, GA 30318, USA.

Welander, M. Department of Horticultural Science, The Swedish University of Agricultural Sciences, S-230 53 Alnarp, Sweden.

Wilde, H. D. School of Forest Resources, University of Georgia, Athens, GA 30605, USA.

Žel, J. Institute of Biology, Askerveca 12, 61000 Ljubljana, Yugoslavia.

SECTION ONE

Fundamentals

1. Micropropagation *à la carte*

M. R. AHUJA
Institute of Forest Genetics, Federal Research Centre for Forestry and Forest Products, Sieker Landstrasse 2, 2070 Grosshansdorf, Germany

1. Introduction

Tissues from woody plants have been cultured since the 1930s [15,16]. Over the next 40 years callus cultures and organ cultures became established in woody plants. A degree of organogenesis was observed, but apparently shoots and roots developed separately on these cultures, or occasionally they may have given the appearance of plantlets [19]. It was not until 1968 that Winton [24] produced the first plantlets in callus cultures of triploid aspen, *Populus tremuloides*. Subsequently, it has been demonstrated that tissue explants from a number of tree species can be grown, and induced to undergo organogenesis and plantlet regeneration *in vitro* [3,5,12–14,20,23,25]. That is not to say that tissues from all woody plants, including forest trees, can be induced to grow and differentiate *in vitro*. Some do, while others are recalcitrant. In addition to the genotypic control of organogenesis, maturation also restricts *in vitro* differentiation. In general juvenile explants, such as embryos, cotyledons, hypocotyls, or bud explants from seedlings are more responsive to *in vitro* regeneration than are tissues, for example bud explants, from mature trees. For this reason, juvenile explants have been extensively employed for the clonal propagation of woody plants. Micropropagation from juvenile explants, although useful for differentiation studies, has a disadvantage in that it is rather difficult to predict how a seedling will perform upon reaching maturity. Progenies from selected trees might provide juvenile material that could be employed in clonal forestry programs. However, this approach has also limitations in that all the progeny will not be superior; there will be gentic variation within families. The next best possibility would be to have biochemical (isozymes) or molecular (restriction fragment length polymorphisms, RFLPs) markers in the juvenile phase that are correlated to the adult yield traits. If availabel, these markers would make juvenile micropropagation more acceptible.

2. Micropropagation

Plant regeneration *in vitro* may be accomplished by employing callus, organ,

M. R. Ahuja (Ed.), Micropropagation of Woody Plants, 3–9.

cell and protoplast cultures. As mentioned earlier, tissue explants from tree species are generally difficult to grow and differentiate *in vitro*. However, the first two types of cultures, callus and organ, have been employed with varying degrees of success for the micropropagation of a number of woody plants [2,4,6,7,11]. Although callus cultures were initially used for plantlet regeneration, organ cultures (embryos, cotyledons, bud meristems) are now mostly routinely employed for micropropagation. Cell suspensions and protoplast cultures have not been successfully employed for large-scale clonal propagation in woody plants. Genetic fidelity of clonal material is important in clonal forestry programs. Long-term callus cultures invariably exhibit genetic variability. Organ cultures, meristem cultures in particular, on the other hand, are relatively more stable and involve minimum genetic risk in clonal propagation.

Plantlets have been regenerated in more than 60 tree species by employing tissue culture technology [14,26]. However, only in a limited number of tree species have the *in vitro* techniques advanced far enough for production of large numbers of plants for testing and deployment under field conditions [3,5,22]. In some angiospermic species both juvenile and mature explants have been employed for micropropagation. On the other hand, in gymnosperms mainly juvenile explants have been cultured for micropropagation, while clonal propagation from tissues of mature conifers is still very difficult by tissue culture [10].

2.1. *Organogenesis and somatic embryogenesis*

Micropropagation may be accomplished by either organogenesis or somatic embryogenesis. Organogenesis involves differentiation of microshoots and roots at different time periods during plantlet development. Usually microshoots are induced on tissues in a cytokinin-enriched medium, and subsequently the microshoots are rooted in an auxin-enriched medium to give rise to plantlets. Organogenesis has been induced on callus, organ, cell, and protoplast cultures in plants. However, organogenesis is greatly influenced by the genotype, physiological state of the explant, age of the explant, and the *in vitro* environment, both the light/temperature regimes and the constitution of the medium, in particular hormone concentrations.

Micropropagation via somatic embryogenesis involves the development of embryos from embryogenically-competent somatic cells *in vitro*. In contrast to organogenesis, where microshoots and roots develop on differnt media, somatic embryogenesis is apparently a one-step process involving the development of embryos having both a shoot and a root pole, as in the zygotic embryos. That is not to say that one does not require different media for the induction, development and maturation phases of somatic embryos *in vitro*. In reality, one does. The developmental program in the embryogenically-competent somatic cells, under the influence of specific phytohormones, proceeds in the same or closely similar manner as it does in a zygote for the development of a zygotic embryo. The underlying difference between the embryogenically-competent cells and cells that undergo organogenesis apparently lies in the differential developmental genetic program that is prone to activation by specific phytohormones. In the same organ or tissue, different cell types may follow different pathways to differentiation of shoots or roots or somatic embryos *in vitro*. In general cells from zygotic embryos are more prone to develop somatic embryos than are cells from other organs, for example bud explants, leaves, stems or roots. It is conceivable that the genetic program for the development of embryos in the cells of immature or even mature zygotic embryos is not fully repressed and can be relatively easily activated as compared to that in the cells from leaf, petiole, or bud meristem. In the conifers, so far the somatic embryos have been mainly derived from the zygotic embryos (see Ref. 8). In the angiospermic tree species, the somatic embryos have been obtained from organs/tissues other than embryos as well.

For the purpose of clonal propagation in woody plants, the organogenesis approach has mostly been employed for micropropagation. The somatic embryogenesis approach is still in the developmental phase: some basic problems need further attention. In addition to the difficulty in initiation of somatic embryos in some genotypes, there are problems with maturation and conversion (germination) of somatic embryos, and the development of viable somatic seedlings in a number of woody plant species. After resolution of these problems the somatic embryogenesis approach should be available for the mass cloning of woody species in the near future. There are several distinct advantages to micropropagation via somatic embryogenesis. For one thing somatic embryos can be grown in a liquid media, and therefore it should be possible to scale-up embryo production in bioreactors. It has been estimated that one litre of conifer embryogenic culture may contain approximately 100 000 somatic embryos [9]. Secondly, the somatic embryos can be encapsulated to produce 'synthetic seed' [21], which can be stored or planted as seed.

3. Clonal fidelity

Regardless of the micropropagation method, it is important to investigate the morphologic and genetic stability of the plants produced via organogenesis or somatic embryogenesis. That may be important in a clonal forestry program. True-to-type propagules are the expected products of vegetative propagation. But do they always occur *in vitro*? The answer is that tissue-culture-associated variation, termed somaclonal variation [18], has frequently been observed in long-term cultures (see Skirvin *et al.*, this volume). While it may be undesirable in a clonal forestry program, somaclonal variation may be useful for the recovery of newer genotypes.

Somaclonal variation has not been extensively investigated in the woody plants [1]. It is not that somaclonal variation does not occur in woody plants, but rather it is veryfying the nature (heritable) of somaclonal variation. Forest trees have long generation cycles, and consequently do not readily lend themselves to genetic analysis and progeny tests. Nevertheless, somaclonal variation once induced and fixed in a genotype, could be maintained by vegetative propagation. The somaclonal selection approach may be used for the isolation of single gene variants, but more likely for the recovery of genotypes with polygenic traits (see Michler, this volume). Most of the economically important traits, for example growth, height, volume, leaf/needle size etc., are controlled by polygenes. Therefore, the somaclonal selection approach may have some value for the selection of economic traits.

4. Acclimatization

Following the regeneration of plantlets *in vitro*, they are acclimatized to greenhouse and field conditions (see Donnelly, this volume). The transition between the *in vitro* environment, with almost 100% relative humidity, to the *ex vitro* environment of field conditions with about 50% relative humidity is rather critical for the survival of plantlets. It is therefore necessary to gradually lower the humidity from 100% to field conditions. The *in vitro* plantlets not only live in a 100% relative humidity milieu, they are also dependent upon the medium for the supply of sugars and other nutrients. Although the leaves/needles of *in vitro* plantlets are green, they are probably not very photosynthetically active. Besides, the leaves/needles of the *in vitro* plantlets have practically no cuticle. As soon as the plantlets come out of the *in vitro* environment, the photosynthesis genes must turn on immediately in order for the plantlets to survive. It is important to have a certain number of leaves and a good root system before transfering the plantlets from an *in vitro* to an *ex vitro* environment; a soilless potting mixture is usually used. Initially, the plants are kept under intermittant mist or under plastic tents in which the relative humidity is nearly 100%. The humidity is gradually lowered during the following few weeks. In the meanwhile, the plants develop an efficient root system, build cuticle on the leaves, and become photosynthetically active. After hardening in the greenhouse, the plants are transplanted to the field for evaluation of morphologic and genetic stability.

5. Biotechnology and tree improvement

Once the technology for regeneration of plants from tissues, cells and protoplasts has been established, it then becomes possible to attempt genetic manipulation for tree improvement. Therefore, *in vitro* regeneration of plants is a prerequisite for genetic engineering. Tissues from woody plant species are somewhat difficult to grow and differentiate *in vitro*. The juvenile tissues are generally more responsive to *in vitro* regeneration as compared to mature tissues in the tree species. Regeneration from protoplasts is rather difficult in most commercially important tree species. Gene transfer technologies based on protoplast regeneration system may therefore not be applicable to all economically important woody plants. On the other hand, genetic transformation mediated by *Agrobacterium* or by employing biolistic microprojectile systems have been demonstrated in a number of woody plants. Although reporter gene expression has been detected, the fate of foreign genes introduced into longlived woody plants, on a long-term basis, needs to be fully investigated. The analysis of genome structure and molecular biology of cell differentiation would considerably enhance our understanding of woody plant biotechnology.

6. Outlook

The main steps in the micropropagation and testing and deployment of micropropagated plants are shown in Figure 1. Although plant regeneration has been accomplished in a large number of woody plants by tissue culture technology, there are still a number of unresolved problems. These relate to growth and differentiation, in particular mature tissues, genetic fidelity of the regenerants, acclimatization and economics of micropropagated plants and their deployment. These and future applications of biotechnology to clonal propagation, genetic modification, and strategic planning of woody plants are discussed in this volume.

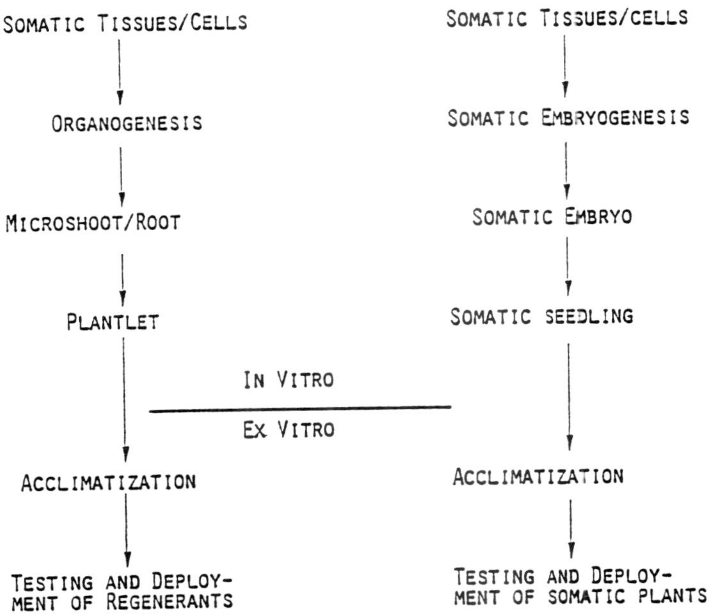

Figure 1. Pathways to micropropagation via organogenesis and somatic embryogenesis. The main *in vitro* and *ex vitro* steps are outlined.

Summary

The components of micropropagation are considered in the framework of physiological and genetic parameters. Organogenesis and somatic embryogenesis are the two differential pathways a somatic cell(s) can follow for *in vitro* regeneration of plants. In addition to the physiological state and genotype of the donor plant, the source of the explant, age of the tree, and *in vitro* conditions also affect the morphogenic response. Following micropropagation, *ex vitro* acclimatization, testing and deployment of clonal material are just as important as *in vitro* regeneration of woody plants.

References

1. Ahuja MR (1987) In vitro propagation of poplar and aspen. In: Bonga JM and Durzan DJ (Eds.), Cell and Tissue Culture in Forestry, Vol. 3. Martinus Nijhoff Publishers, Dordrecht, pp. 207–223.
2. Ahuja MR (Ed.) (1988) Somatic Cell Genetics of Woody Plants, Kluwer Academic Publishers, Dordrecht, pp. 225.
3. Ahuja MR (1991) Biotechnology in forest tress. Plant Res. & Development 23: 106–120.
4. Ahuja MR (Ed.) (1991) Woody Plant Biotechnology. Plenum Press, New York, pp. 373.
5. Ahuja MR and Muhs HJ (1985) In vitro technique in clonal propagation of forest trees. In: Schäfer-Menuhr A (Ed.), In Vitro Techniques – Propagation and Long-Term Storage. Martinus Nijhoff Publishers, Dordrecht, pp. 41–49.
6. Bajaj YPS (Ed.) (1986) Biotechnology in Agriculture and Forestry. I. Trees I. Springer Verlag, Berlin, Heidelberg, New York, pp. 515.
7. Bajaj YPS (Ed.) (1989) Biotechnology in Agriculture and Forestry 5. Trees II. Springer Verlag, Berlin, Heidelberg, New York, pp. 622.
8. Becwar MR (1993) Conifer somatic embryogenesis and clonal forestry. In: Ahuja MR and Libby WJ (Eds.) Clonal Forestry, Vol. 1, Genetics and Biotechnology. Springer Verlag, Berlin, Heidelberg, New York (in press).
9. Becwar MR, Wann SR, Johnson MA, Verhagen SA, Feirer RA, and Nagmani R (1988) Development and characterization of *in vitro* embryogenic systems in conifers. In: Ahuja MR (Ed.) Somatic Cell Genetics of Woody Plants. Kluwer Academic Publishers, Dordrecht, pp. 1–18.
10. Bonga JM (1991) *In vitro* propagation of conifers: fidelity of the clonal offspring. In: Ahuja MR (Ed.) Woody Plant Biotechnology. Plenum Press, New York, pp. 13–21.
11. Bonga JM and Durzan DJ (Eds.) (1987) Cell and Tissue Culture in Forestry, Vol. 3. Martinus Nijhoff Publishers, Dordrecht, pp. 416.
12. Brown CL and Sommer HE (1982) Vegetative propagation of dicotyledonous trees. In: Bonga JM and Durzan DJ (Eds.) Tissue Culture in Forestry. Martinus Nijhoff Publishers, Dordrecht, pp. 109–149.
13. David A (1982) In vitro propagation of gymnosperms. In: Bonga JM and Durzan DJ (Eds.) Tissue Culture in Forestry. Martinus Nijhoff Publishers, Dordrecht, pp. 72–108.
14. Dunstan DI and Thorpe TA (1986) Regeneration in forest trees. In: Vasil IK (Ed.) Cell Culture and Somatic Cell Genetics of Plants, Vol. 3. Academic Press, New York, pp. 223–241.
15. Gautheret RJ (1934) Cultur du tissu cambial. CR Acad Sci (Paris) 198: 2195–2196.
16. Gautheret R (1940) Nouvelles recherches sur le bourgeonnement di tissu cambial d'*Ulmus campestris* Cultive in vitro. CR Acad Sci (Paris) 210: 744–746.
17. Karnosky DF (1981) Potential of forest tree improvement via tissue culture. Bioscience 31: 114–120.

18. Larking PJ and Scowcroft WR (1981) Somaclonal variation – a noval source of variability from cell cultures for plant improvement. Theor Appl Genet 60: 197–214.
19. Mathes MC (1964) The *in vitro* formation of plantlets from isolated aspen tissues. Phyton 21: 137–141.
20. Mott RL (1981) Trees. In: Conger BV (Ed.) Cloning Agricultural Plants via In Vitro Techniques. CRC Press, Florida, pp. 217–254.
21. Redenbaugh K and Ruzin SE (1989) Artificial seed production and forestry. In: Dhawan (Ed.) Application of Biotechnology in Forestry and Horticulture. Plenum Press, New York, pp. 57–71.
22. Talbert CB, Ritchie GA, and Gupta P (1993) Conifer vegetative propagation: an overview from a commercialization perspective. In: Ahuja MR and Libby WJ (Eds.) Clonal Forestry, Vol. 1, Genetics and Biotechnology. Springer Verlag, Heidelberg, Berlin, New York (in press).
23. Thorpe TA and Biondi S (1984) Conifers. In: Sharp WR, Evans DA, Ammirato PV, and Yamada Y (Eds.) Handbook of Plant Cell Culture, Vol. 2. Macmillan Publishing Company, New York, pp. 435–470.
24. Winton LL (1968) Plantlet formation in aspen tissue culture. Science 160: 1234–1235.
25. Winton LL (1978) Morphogenesis in clonal propagation of woody plants. In: Thorpe TA (Ed.) Frontiers of Plant Tissue Culture 1978. University of Calgary Offset Printing Service, Calgary, pp. 419–426.
26. Zimmerman RH (1986) Regeneration in woody ornamentals and fruit trees. In: Vasil IK (Ed.) Cell Culture and Somatic Cell Genetics of Plants, Vol. 3. Academic Press, New York, pp. 243–258.

2. Cellular control of morphogenesis

TREVOR A. THORPE and PRAKASH P. KUMAR[1]
Plant Physiology Research Group, Department of Biological Sciences, University of Calgary, Calgary, Alberta T2N 1N4, Canada
[1] *Present address: Botany Department, National University of Singapore, Singapore*

1. Introduction

Our current knowledge in the vast area covered by the title of this chapter is fragmentary. In addition to the multitude of intrinsic factors, the complex form of plants is affected by environmental factors. One of the major problems faced by plant morphogeneticists is to elucidate the mechanism(s) by which the derivatives of meristematic or parenchymatous cells become specialized into distinct cell and tissue types at successive stages of development [85]. The occurrence of differential growth has enabled higher plants to assume complex three-dimensional forms with a greater possibility of establishing gradients than if the organism were of simple spherical or filamentous structure [4]. Morphogenesis, or the origin of form, can be studied at different levels of organization, e.g., cell, tissue, organ and whole plant level [85]. It is conceivable that control points exist at each level. Because, in their absence the progressive specialization would only lead to a mass of cells that are different in structure and function, but without integration of their activities [84].

The study of problems associated with growth and differentiation can be performed at the structural, physiological, biochemical and macro-molecular levels. The use of cell, tissue and organ cultures, as will be discussed in this chapter, has provided considerable information on the structural as well as physiological aspects of differentiation. In this chapter, we will use the terms morphogenesis and differentiation interchangeably for the process which appears to have several distinct phases or stages.

2. Phases in differentiation

The early events in *de novo* organogenesis can be divided into different phases based on the responses of the explants in culture [18,19,25,68]. They are: (1) attainment of competence or pre-induction phase; (2) induction or determination phase; and (3) expression phase or post-initiation phase. The attainment of competence by the cells of the explant is comparable to dedifferentiation; however, it does not necessarily involve proliferation of cells to produce unorganized callus tissue [18]. Once the cells become competent,

M. R. Ahuja (Ed.), Micropropagation of Woody Plants, 11–29.

they can be induced to follow a determined morphogenic pathway. This induction is achieved by external stimuli such as phytohormones in the medium and/or a balance between the exogenous and endogenous levels of phytohormones. At least three requirements have been recognized, namely: (a) cell dedifferentiation; (b) cell interactions; and (c) reaction to specific signals [72]. The states of competence and induction (or determination) are not always reflected by morphological changes of the explant, and as a result their occurrence can only be verified by experimental manipulations [18,50].

If one assumes that there is an initial lag period, namely, attainment of competence phase, before the explanted tissue becomes induced by a specific stimulus to undergo specific morphogenesis, then growing the tissue on a non-inductive medium during this period should not adversely affect the subsequent induction [18]. In other words, growing an explant that is to be used for shoot induction on medium for callus or root induction or on basal medium up to the beginning of the induction phase should not alter the capacity to form shoots. Therefore, the exact duration of the attainment of competence can be determined by preculturing explants for various lengths of time before exposing them to the induction medium [19]. One of the disadvantages of this approach is that one also assumes that factors such as phytohormones, especially the ones required for induction, are not needed for the attainment of competence by the responding cells. It has been shown that excised cotyledons of radiata pine require cytokinin from the start of culture and fail to form adventitious shoot buds if they are treated with BA after the first 24 hours *in vitro* [5,80]. Also, adventitious shoot bud formation occurred on embryos of *Picea abies* if they were cultured on hormone free medium after they were treated with 250 μM BA for as little as 1 min; their shoot-forming response decreased if the exposure to BA was delayed until after 2 days of initiation of culture [83].

At the end of the induction phase a determined tissue should express morphogenesis, e.g., shoot bud differentiation, even if it is transferred from the shoot induction medium to basal medium. Again, the exact length of the induction period can be determined by transferring explants from induction medium to basal medium at various intervals and observing the morphogenetic response. It has been shown that leaf explants of *Convolvulus* spend discrete amounts of time in attaining competence and being induced to give rise to shoot buds *in vitro* [19]. Also, *de novo* shoot formation in tobacco callus requires the exposure of the tissue to shoot-forming (SF) conditions for a period between 10–15 days prior to subjecting the callus to non-shoot-forming (NSF) conditions [35].

The response of tissues to growth active substances may be different in each phase. Thus, while indolebutyric acid (IBA) was required during the pre-initiative and post-initiative phases of adventitious root initiation in hypocotyls of *Pinus radiata*, kinetin was inhibitory during the former phase [68,69]. On the other hand, the presence of gibberellic acid was inhibitory to the pre-initiative phase and the phase after the establishment of the meristemoids, but it can enhance rooting if applied coincident with the first observable stage of root initiation. It is evident, therefore, that several factors play roles in the regulation of morphogenesis. These will now be reviewed.

3. Physiological gradients

The establishment of gradients of nutrients and phytohormones within tobacco callus tissues has been demonstrated [62]. That the point of origin of shoot buds might be determined, in part, by such gradients is supported by the observations from manipulation of the orientation of tobacco callus pieces cultured *in vitro* [62,74]. Orientation of the explants on the medium in tissue cultures is known to influence morphogenesis. For example, cotyledons in culture respond differently depending on whether their adaxial or abaxial surface is in contact with the medium, as well as on the genotype [36,52]. Also, when 0.5–2.5 mm long segments of hypocotyl tissue from *Eucalyptus* were cultured *in vitro*, it was found that the uppermost portion of the hypocotyl produced the maximum number of nodular tissue and subsequently, shoots [56]. Furthermore, if the explants were placed with their basal ends pointing upward, no shoots were differentiated; up to 20% of the explants formed shoots when they were placed horizontally or with their apical end up.

3.1. *Electrical polarity*

Physiological gradients may play an important role in the establishment and maintenance of polarity, of which the latter is the first requirement for organized development to commence (see Ref. 20). While the establishment of polarity in an isolated cell may lead to cytodifferentiation, e.g., differentiation of tracheids (see Ref. 27), in multicellular systems it can lead to the spatial organization and pattern formation characteristic of the species [20]. Also, it has been hypothesized that self-generated electric currents may be responsible for establishing polarity in the zygotes of *Fucus*, as it is among the earliest events noted (see Ref. 20).

The naturally occurring electrical polarity of somatic embryos of wild carrot (*Daucus carota* L) was measured by vibrating a platinum electrode back and forth, 30 mm from the surface of the embryo at closest approach [13]. Inward currents could be detected at the future shoot apical region and outward currents at the future radicle pole in globular embryos. Although this kind of electrical polarity remained as the embryos developed into a plantlet, the current density increased with maturity of the embryo. The presence of inward current at the point of growth has been reported in other systems as well, and it is probably involved in the uptake of nutrients from the medium at the points of rapid growth (references in Ref. 13). In support of this hypothesis, it was found that although the outward current was strongest at the elongation zone of the developing radicles such current was not detectable at the tip of the radicle, which was becoming a root tip [13]. It was speculated that the developing root tip probably represents a second region of inward current being differentiated. No current could be detected on the surface of the torpedo stage embryos. There was a positive correlation between current density and growth. Also, indication that H^+ is a major component of the inward current was obtained by altering the pH of the medium. Thus, when the pH was increased from 5 to 4, inward current declined and when it was lowered, current increased. Treatment of wild carrot embryos with 3mM IAA resulted in an almost immediate decline in the current density [13]. This decline was reversible when the exogenous IAA was removed. Incubation with micromolar quantities of 2,4-D, kinetin and zeatin failed to affect the current. The reduction in current caused by IAA might be due to its ability to induce proton efflux from the tissues (see Ref. 13). Treatment with IAA also resulted in disorganization of the embryos.

When electric current of the order of 1 to 2 microamperes (μA) was passed through tobacco callus, it not only increased the average number of shoots formed per culture by up to five-fold, but in some cases it enabled the callus to produce shoots which otherwise would not have produced them [59]. The first formed shoot buds arose from the most negative region of the culture when 2 μA current was used. Thus, the buds became visible first near the electrode in the callus when it was negative and near the medium when the latter was negative. Root formation from wheat callus also responded in a similar fashion, in which case, differentiation of some shoot buds was also made possible by this treatment [59]. Along with the stimulation of shoot bud differentiation, the applied electric current could also enhance growth of the callus by about 70% over the controls [31]. However, the stimulation did not occur in the absence of a polarly transportable auxin, such as IAA, in the medium. Thus, when 2,4-D, which is not polarly transported, is the auxin of choice in the medium, or if an inhibitor of polar transport such as 2,3,5-tri-iodobenzoic acid was used along with IAA, the electrical stimulation did not occur. These observations further the idea that the observed stimulation of growth and differentiation by the applied current may be mediated by a rearrangement of the polarity of the cells of the callus.

Since the applied electrical current was small and the cell membranes have high resistance the current could not have penetrated the cells [59]. Instead, it could have polarized the cells of the callus in a manner analogous to that occurring in fucoid zygotes [38,39,61]. That charged molecules such as proteins (including hormone receptors and ion channels) could be rearranged by the applied electrical field according to their charge was demonstrated by Woodruff and Telfer [87] using oocytes of *Cecropia* injected with fluorescent proteins of various charges. Goldsworthy [31] pointed out that the cells of the unorganized callus may not share a common polarity, and the applied current can cause a rearrangement of charged proteins and enzymes in the fluid-mosaic plasmamembrane, thereby orientating the cellular polarities along a common axis within a callus piece. This would increase the chances of development of organized meristems (shoot or root).

One of the recently developed techniques for the direct introduction of nucleic acids into protoplasts, namely, electroporation, involves subjecting the protoplasts to micro- or milli-seconds bursts of high voltages. Electroporation enhanced division and plant regeneration in many species [30,55,60]. When protoplasts of colt cherry (*Prunus avium X pseudocerasus*) and *Solanum dulcamara* L. were subjected to electroporation (250 V/cm for 87 msec and 750 V/cm for 29 msec, respectively) the onset of subsequent cell divisions was speeded up [60]. The DNA synthesis was increased significantly in these protoplasts, which could be correlated with the observed early onset of cell division. Ochatt *et al.* [55] also showed that the capacity for enhanced DNA synthesis and early cell division acquired by *Prunus* protoplasts following electroporation was retained by the second and subsequent generations of protoplasts. In other words, freshly isolated, 'unelectroporated' protoplasts derived from long-term (even after 16 subcultures) cell suspension cultures, which, in turn, were established from electroporated protoplasts, exhibited the enhanced growth and regeneration capacity.

Electromagnetic fields enhanced growth of animal cells by increasing the activities of enzymes such as ornithine decarboxylase [15]. Electroporation itself enhanced the uptake of ions and nucleic acids from the surrounding medium in animal [40] and plant cells [26,65,77]. Therefore, it is not unlikely that such physiological changes might have been introduced permanently during electroporation and transmitted to the cells from these protoplasts [55].

3.2. *Physiological state of the explants*

In woody plants, juvenile tissues – stump sprouts, sprouts from pruned trees, zygotic embryos or seedling parts (epicotyl, hypocotyl and cotyledon) – are the best explants if regeneration of plants is to be achieved [1,2,8]. Selection of the appropriate explant is critical if optimum number of rootable shoots is to be achieved [1,2]. Bonga [9] observed that bud regeneration from immature ovulate strobili of *Larix* sp. occurred only if they were cultured not later than one week after meiosis in the developing ovules. This study and others suggesting that tissues arising from recent meiosis (e.g., anthers and seedling parts) responded better with respect to *in vitro* regeneration led to the speculation that meiosis probably removes certain epigenetic constraints on regeneration that might exist in the adult plant tissues [10,34]. Meiosis, which might render cells responsive to stimuli inducing regeneration, also caused a high rate of turnover of ribosomes and other organelles, coupled with an overall reduction in the number and complexity of cytoplasmic organelles (see Ref. 8). Hence, it has been suggested that conditions associated with meiosis could be mimicked by manipulating physical and chemical factors in order to rejuvenate adult trees [8]. Adventitious bud formation on internodal segments of *Populus* spp. cultured in the absence of exogenous phytohormones was observed at the physiological apex of the explants [22]. Also, internodes close to the shoot apex failed to produce shoot buds under these condtions and the capacity to form buds increased with increasing distance from the apex.

While the explant state is very important for ease in manipulating organized development, the medium and the culture conditions are equally important. As a matter of fact there is an interplay between these factors [72]. One key component is the composition of the medium [28]. Different mineral salt formulations and different concentrations of the minerals (e.g., one-half or double strength media) may bring about different responses *in vitro*. The levels and types of the carbohydrate and reduced nitrogen supplements in the medium can also influence morphogenesis. As discussed below, phytohormones have a profound influence in regulating organized development *in vitro*.

3.3. *Phytohormones*

The controlling influence of phytohormones on differentiation has long been demonstrated; their exact role in this process, however, is not clear cut, because each class of phytohormone can elicit a wide array of responses in different parts of different plants [85]. The processes determining the specifity of the varied developmental responses are little known when compared to the biochemistry and hormone-induced effects.

Proof that organ differentiation from callus tissue occurs in response to the appropriate auxin/cytokinin ratio as proposed by Skoog and Miller [66] has recently been obtained using modified transfer-DNA (T-DNA) of *Agrobacterium tumefaciens* [37,64]. Inactivation by transposon insertions of T-DNA genes 1 and 2, which code for the biosynthesis of auxin, causes shoot formation, presumably by decreasing the auxin/cytokinin ratio within the tissue. Also, inactivation of the gene responsible for cytokinin biosynthesis results in root formation, again by altering the phytohormone balance [3,29]. Nevertheless, there are numerous cases where the observed type of morphogenesis was opposite to that which was expected of the phytohormone balance to which the explants were exposed to (see Refs. 34, 72). Although data are lacking at present, it is plausible that the endogenous balance of phytohormones in such tissues might conform to the Skoog and Miller dogma.

Pulse treatment of the explants with high concentrations of phytohormones at the beginning of culture was adequate to bring about shoot bud regeneration in some plants. Presoaking fir (*Abies balsamea*) needles in a solution containing 1000 mg/l indolebutryc acid (IBA) permitted organogenesis to occur when they were cultured on normal media, which otherwise would not support it [7]. Also, cotyledons and embryos of *Picea abies* were capable of regenerating shoot buds after pulse treatment with high concentration of BA and subsequent transfer to basal medium [12,83].

Plantlets could be regenerated directly from white pine cotyledons, but when the cotyledon was induced to form callus the cells appeared to have lost their capacity to regenerate shoot buds [53]. Indication that such apparent loss of morphogenetic capacity might be due to physiological or epigenetic constraints, and not due to irreversible genetic changes, was provided by the observation that under appropriate conditions shoot buds and somatic embryos could be regenerated from callus of Engelmann spruce ([48], and Sharma and Thorpe, unpublished).

The effect of a phytohormone may be related to the developmental state of the tissue, which could alter the 'sensitivity' of the tissue to the applied phytohormone [79]. As mentioned earlier, IBA was essential for the induction of root primordia on seedling-derived cuttings of radiata pine between days 2 and 8, but kinetin was inhibitory to this process [69]. Gibberellic acid was inhibitory to root formation, except between days 4–6 when it promoted root initiation.

Shoot bud formation from cultured micro-cross sections (MCS) of poplar leaves that were up to 500 mm thick required the presence of NAA, an auxin, along with BA [47]. But, if 1 cm long explants were used, shoot formation occurred in the absence of NAA. It is likely that any endogenous auxin that might be present in the MCS explants might leach out of the tissue into the liquid culture medium, thus making an external auxin supply essential for regeneration [47].

Wetmore and Rier [86] found that shoot buds grafted onto *Syringa* callus caused differentation of vascular tissue, namely, xylem and phloem, in the callus. They also observed that auxin and sucrose could produce a similar effect in the absence of the grafted buds. The sucrose concentration determined the nature and amount of tissue formed. Thus, a low sucrose concentration favored xylem differentiation, a high concentration favored phloem formation and intermediate levels of the sugar resulted in the differentiation of both xylem and phloem tissue. At present, it is known that various factors such as carbohydrates, amino acids, light, temperature and osmotic pressure influence the differentiation of tracheary elements [27]. However, their modes of action are not understood.

Excised cotyledons of radiata pine cultured under shoot-forming (SF) conditions produced significant levels of CO_2 and ethylene [42]. Under non-shoot-forming (NSF) conditions the accumulation of these gases was very low. To establish the role of these gases on morphogenesis the composition of the headspace gas was manipulated by: (1) allowing the gases to accumulate within the flasks by sealing them with gastight serum caps on various days, and replacing the caps with foam plugs on various days during incubation; (2) absorbing the gases by setting up chemical traps singly and in combination; and (3) incubating the cultures under a continous flow of gas mixtures of known composition [42]. It was observed that both CO_2 and ethylene, which build up during the initial 10 to 15 days of culture promote shoot bud differentiation even after they are released from within the tissues. The optimum number of buds per explant was obtained when the cultures had accumulated about 5 to 8 μl l⁻ ethylene and about 10% (v/v) CO_2 in the headspace during the first 10 to 15 days *in vitro*. Excessive accumulation of these gases beyond this stage caused slight dedifferentiation, which inhibited the elongation of shoot buds in the shoot elongation medium subsequent to the induction. The cut ends of the cotyledons in such flasks turned necrotic. A similar browning of the cut ends was observed in the cotyledons incubated under a continuous flow of ethylene+O_2+N_2, indicating that this might be a result of exposure to high concentrations of ethylene. When both gases were eliminated from the headspace of the culture vessels, shoot bud induction and growth were inhibited.

The role of ethylene in influencing bud differentiation in excised cotyledons of radiata pine may be at least two-fold [42]. It can influence the orientation of cellulose microfibrils in plant cells [46]. The first ten days in culture, when ethylene and CO_2 have a positive influence on shoot bud differentiation, coincided with a period of intense cell divisions leading to the formation of meristemoids under the SF conditions. Therefore, it is conceivable that the role of ethylene during this stage may be in regulating the cell division as it has been known to induce partial synchrony in cell suspension cultures [21]. In addition, ethylene may have an entirely independent, as yet unidentified role in influencing the process of organogenesis. CO_2, on the other hand, may be required during shoot formation to modify the biosynthesis and action of ethylene at various stages [42]. Also, CO_2 may be important because it is required for the maintainance of high activity of the non-photosynthetic C-fixing enzyme PEPC, which is high in these cotyledons during the early stages of shoot bud induction [43].

Embryogenic and non-embryogenic cell suspension cultures of white spruce also produced significant levels of ethylene and CO_2 [45]. Allowing the gases to accumulate within the flasks results in an inhibition of growth and multiplication of the cells and proembryos in the suspension cultures within four days after sealing the flasks. The high levels of CO_2 accumulated within the flasks suggests that the cultures may be under anoxic or hypoxic conditions. However, incubation of embryogenic cell suspension cultures under a continuous flow of ethylene and CO_2, at levels comparable to those in the headspaces of day 4 serum-capped cultures, in the presence of 20% O_2 also resulted in a 2–3-fold reduction of growth as measured by the packed cell volume, fresh and dry weights. This indicates that the observed inhibition of growth is due to CO_2 and ethylene accumulated in the headspace, and not due to O_2 starvation.

The contrasting observations on the influence of ethylene and CO_2 on *de novo* organogenesis and somatic embryogenesis raise the possibility that these two unique pathways of *in vitro* morphogenesis are controlled differently. Also, the data suggest that the effect of ethylene and CO_2 on morphogenesis may vary between species.

Polyamines, mainly putrescine, spermidine and spermine, have been reported to be required for growth and differentiation in many plant species [67]. Spermidine was found to be the major polyamine in cultured cotyledons of radiata pine [6,44]. However, putrescine, spermidine or spermine could not replace the requirement for the cytokinin, BA in the medium (Kumar and Thorpe, unpublished). When ^{14}C-putrescine was provided in 2 h pulses to the excised cotyledons of radiata pine on different days of culture, it was metabolized primarily to γ-aminobutryc acid, aspartate and glutamate [44]. Only traces of ^{14}C were found to enter the higher polyamines, spermidine and spermine. However, if the cotyledons were cultured for extended periods in the presence of ^{14}C-putrescine, the higher polyamines were also labeled to appreciable levels after two days in culture [6]. During the pulse-feeding period, $^{14}CO_2$ was also released by the cotyledons indicating that the derivatives of putrescine catabolism enter the tricarboxylic acid cycle [44]. Therefore, putrescine, and possibly other polyamines, may serve as reserves of C and N during *de novo* organogenesis. The activity of arginine decarboxylase, an enzyme involved in the biosynsthesis of putrescine, was maintained at 3–5 fold higher levels in the SF tissues than in the NSF cotyledons early in culture, suggesting the possible importance of polyamines in *de novo* organogenesis (Kumar and Thorpe, unpublished).

The techniques of histochemistry and autoradiography have proven useful for analyses of changes occurring in differentiating systems as a result of the above manipulative treatments [72]. The net result of these changes can be observed structurally, which is the usual marker of differentiation in plant systems, as we currently lack the biochemical and biophysical markers used in animal systems.

4. Histochemical and autoradiographic studies

Autoradiographic studies of incorporation of radioactive precursors into protein (^{3}H-leucine), RNA (^{3}H-uridine) and DNA (^{3}H-thymidine) by excised cotyledons of radiata pine cultured under SF and NSF conditions revealed that during the initial two days in culture there were no significant differences in the patterns of incorporation [81]. But by days 3 to 5 labeling became concentrated in the epidermal and subepidermal cells of the SF (BA-treated) cotyledons, which give rise to adventitous shoot buds later in culture. In contrast, the cells of the NSF cotyledons incorporated very little of the labeled precursors. ^{3}H-thymidine labeling of the plastids was found in the latter tissues, in which plastids matured earlier than in the SF cotyledons [81]. These observations were confirmed by histochemical studies, where increased staining for RNA, DNA and nuclear and cytoplasmic proteins in the SF cotyledons as compared to their NSF counterparts was observed [57].

Increased staining for several enzymes including lipase, acid phosphatase, ATPase and succinate dehydrogenase was observed when the SF cotyledons of radiata pine were examined histochemically [57]. Although lipase was distributed evenly in the cotyledons at the time of inoculation, the activity became restricted to the SF cell layers by day 3 in culture. Interestingly, during the later stages of bud differentiation, lipase activity was highest in the mesophyll cells internal to the SF layer, indicating mobilization of lipids from the mesophyll tissue towards the SF cells [57], which are in need of an increased energy supply [72]. The cotyledons contained starch grains at the time of inoculation. When cultured in the absence of BA (NSF), starch grains disappeared from the cells of the cotyledon within 3 days in culture. In contrast, cells of the SF layer in the BA-treated cotyledons of the same age retained abundant starch grains, although starch from the rest of the cells in these cotyledons was depleted. When the organogenic centers were formed as the SF process progressed, these starch grains were slowly used up [57].

The major reserves stored in the cotyledons of radiata pine are lipids and proteins, which decline during *de novo* shoot formation [23]. Analyses indicated that the most dramatic change in lipid content of SF cotyledons was the rapid degradation of triglycerides, as in the cotyledons during germination. Apart from this quantitative change there were qualitative changes in the lipids as well. Thus, linolenic acid content of the polar lipids increased and stigmasterol, a component of plant membranes, was synthesized at a higher level during the early stages of shoot initiation. However, the overall changes of the lipid profile of the cultured cotyledons did not differ notably from those that occur during germination and seedling growth, suggesting that lipid mobilization serves primarily for supplying energy in the cultured cotyledons [23].

5. Structural aspects

Much of the information regarding the early structural and physiological events associated with *de novo* organogenesis comes from studies carried out with tobacco callus and radiata pine cultures (see Refs. 73, 74). The early events could be traced to single cells within the explants. However, the entire primordium does not necessarily arise from the products of such activated cells, because some of the adjoining cells may become incorporated in the developing meristemoids and primordia at various stages [68,73,82].

5.1. *Promeristemoids and meristemoids*

De novo organogenesis in plant tissues cultured *in vitro* is often associated with localized cell divisions in specific directions and the formation of well defined centers of meristematic activity at the future sites of bud development.

The initial response to root-inducing stimulus, namely, treatment with IBA, in hypocotyls of radiata pine could be traced to individual cells in the cortex [68]. Four days after exposure to the auxin these cells and their nuclei exhibited slight swelling. The cells contained many plastids concentrated around their nuclei. They could be identified by their darkening response to treatment with dihydroxyphenylalanine, indicating the presence of polyphenol oxidase or tyrosinase in them. The neighboring cells did not show any change in color, suggesting that the cells destined to give rise to root primordia had undergone biochemical changes at this stage. There were certain cytoplasmic changes also associated with these newly formed meristematic cells; the most obvious being the presence of numerous microvacuoles as opposed to a large central vacuole in the rest of the cortical parenchyma cells. Between days 5 and 7 the cytoplasm of one or more of these neighboring cortical cells migrated towards the nearest meristematic locus and their large vacuoles were replaced by numerous microvacuoles as in the meristematic cell. Concurrent with these changes asymmetric cell divisions also occurred in such neighboring cells [68]. Continued divisions of these cells and their derivatives led to the formation of meristemoids, which are well defined groups of meristematic cells, by day 10. Periclinal division of cells at the outer surface of these meristemoids gave rise to a layer which eventually became the root cap. Clearly identifiable root primordia were differentiated by day 12 from the meristemoids. Thus, the authors identified the period prior to the specialization of the single meristematic cells as preinitiation phase, formation of the meristematic locus and specialization of its surrounding cells as the initiation phase and the subsequent period of development of the meristemoid and root primordium as the post-initiative phase [68].

The excised cotyledons of radiata pine, used as explants for *de novo* shoot bud initiation, exhibited numerous mitotic figures scattered throughout the cotyledons [89]. The plane of such cell divisions was predominantly anticlinal (with respect to the long axis of the cotyledon). The cells of such explants were rich in reserve substances, had large nuclei and dense cytoplasm. When cultured under NSF conditions the cells developed large vacuoles and most of the reserve substances were depleted and the plastids enlarged [81,82]. The chloroplasts and stomatal complexes were fully developed by days 3 and 5, respectively.

Although the cell divisions in the SF cotyledons were random at first, they became localized in the epidermal and hypodermal regions of the surface in contact with the BA-containing medium within three days of incubation [82,89]. These early divisions led to the formation of the first organized structures called promeristemoids. The origin of these promeristemoids could be traced to single cells immediately below the epidermis. The first division of such mother cells was periclinal. Subsequently, both periclinal and anticlinal divisions took place leading to the formation of 6–8-celled, spherical or oval structures, namely, the promeriostemoids, by days 5 [82,89].

The cells within each promeristemoid had large nuclei and dense cytoplasm [82]. The daughter cells had prominent plasmodesmatal connections between each other. However, plasmodesmata were absent between different promeristemoids, suggesting that the individual promeristemoids probably functioned independently of each other. The depletion of reserve nutrients was slower in the cells of the SF cotyledons than in their NSF counterparts. Chloroplasts were abundant and randomly distributed within the cells. Concomitant with the depletion of the stored lipids, starch grains appeared within the chloroplasts.

The cells of the promeristemoids continued to divide in the presence of BA and light to give rise to meristematic nodules or meristemoids by day 10 *in vitro* [82]. Comparative histological examination of adventitious shoot formation from excised embryos, epicotyls or hypocotyl explants in three species of *Pinus* and three species of *Picea* indicated that although the timing and the sites of initiation of buds differed among the species examined, the patterns of shoot initiation showed remarkable similarity [76]. In all the cases examined, bud differentiation was preceded by the formation of meristemoids and bud primordia. There was no intermediate callus stage in this sequence of development.

The formation of meristemoids beneath the epidermis after the tenth day of culture imparted a nodular appearance to the cotyledons of radiata pine [82]. These meristemoids gave rise to bud primordia, which were visible by day 21. Scanning electron microscopic examination of such developing bud primordia indicated that the epidermis of the cotyledon explants formed part of the protoderm of the newly formed buds [82]. The developing primordia in the SF cotyledons of Douglas fir appeared to emerge from the subepidermal regions by rupturing the cotyledonary epidermis [41], suggesting that these buds are formed entirely from either single hypodermal cells or from a few cells in that region.

In SF tobacco callus cultures, zones of preferential cell division activity were observed by eight days of culture in the lower half, from which meristemoids and bud primordia developed eventually [49,62,75]. The cells of the meristemoids were small, isodiametric with dense cytoplasm and prominent nuclei. More recently, the formation of similar meristemoids has been reported to occur during shoot formation in excised cotyledons of cashew nut (*Anacardium occidentale*) cultured *in vitro* [58]. Furthermore, in various conifers, specific target regions such as cells towards the periphery of the explants and between the bases of the cotyledons have been identified as the most likely sites of development of meristemoids [63,76].

Adventitous shoot but development could be induced on 200 μm to 500 μm thick sections of hybrid poplar leaves with 1 to 2 mm lamina tissue on each side of the midvein (MCS) [47]. Thin sections of the lamina alone or of the midvein without attached lamina tissue failed to regenerate shoots. Bud formation occurred from the junction area where the lamina and midvein meet. It was possible to observe the developmental sequence directly under the microscope because of the use of thin explants. The responding cells consisted of groups of about 40 to 50 cells per junction area with numerous chloroplasts and they produced green callus (0.5 to 1.0 mm in diameter) about two weeks after inoculation. Adventitious buds were visible on such callus. The ability to form buds from MCS was influenced by the thickness of the explants as well as by the presence of lamina tissue attached to the midvein [47]. Thus, 400–500 μm thick sections containing 10–16 cell layers formed as many shoots buds as 1 cm explants, while 100 μm sections, which contained only 2–4 cell layers, failed to regenerate buds. The shoots regenerated from 300–500 μm thick sections appeared uniform and normal green, but those that originated from 200 μm thick sections exhibited phenotypic variability, which included variegated, chlorotic and normal green shoots from different sections. Therefore, such thin sections could be used as the starting material if it is necessary to harness the naturally occurring (chimeric) variability and in genetic engineering [47].

From the above discussion it is evident that the sequence of events leading to the formation of shoot buds are comparable in both herbaceous and woody angiosperms as well as in gymnospermous trees. These events are preceded by biochemical and biophysical changes resulting from selective gene activation. While patterns from studies with herbaceous species are beginning to emerge [14,72,73], much less is known about woody species.

6. Metabolic and molecular aspects

Using excised cotyledons of *Pinus radiata* as the experimental system, it was demonstrated that non-autotrophic CO_2 fixation was probably more important than autotrophic CO_2 fixation during *de novo* shoot bud formation [43]. Histochemical, precursor incorporation and autoradiographic studies showed that the synthesis of macromolecules such as DNA, RNA and proteins occurred preferentially in the epidermal and subepidermal cell layers, from where the shoot buds originate (see Ref. 74). Feeding the tissue with radioactive glucose and acetate indicated that these compounds were metabolized rapidly with the release of labeled CO_2. When these tissues were extracted and analyzed, it was found that most of the label went into the ethanol soluble fraction, which includes lipids, amino acids, organic acids and sugars. The patterns of incorporation of label into specific amino acids and organic acids were different between the SF and NSF cotyledons (see Ref. 74). Similar observations had been made in SF tobacco callus cultures as well, indicating that the basic biochemical processes underlying *de novo* shoot bud differentiation may be comparable in diverse plant species [73,74].

Exogenously applied phytohormones modulate gene expression in various plants [51,70,90]. Since *de novo* organogenesis and somatic embryogenesis require the presence of exogenous phytohormones in the majority of the cases it is likely that one of the modes of action of these phytohormones in controlling morphogenesis is by selective gene activation.

Adventitious bud formation on Douglas fir cotyledon explants exposed to NAA and BA was associated with specific changes in the cytoplasmic soluble proteins [88]. Synthesis of low molecular weight (16 to 20 kD) proteins was increased in SF cotyledons between days 2 and 4 in culture. If the SF cotyledons were transferred to callus-forming medium on day 2 the synthesis of these low molecular weight proteins was suppressed, concomitant with the loss of SF capacity. Similarly, SF cotyledons of *Pinus ponderosa* exhibited the appearance of a new 4 kD peptide and a reduced level of expression of a 52 kD polypeptide in comparison to the protein profile of the NSF cotyledons; the latter profile was similar to that of seedling cotyledons [24].

Limited but important progress has been made in understanding the molecular events occurring early during somatic embryogenesis in carrot cell cultures (see Refs. 11, 17, 71). Although the rate of cell division was lowered, there was active turnover of RNA and new species of mRNA were being synthesized in the globular stage embryos compared to the single cells and embryogenic cell clusters in carrot [54]. Also, polarized synthesis of DNA and RNA along with localization of free Ca^{2+} ions was observed in the embryogenic cell clusters. Distinct changes in metabolism associated with the various phases of somatic embryogenesis have been described in these cultures (see Ref. 54). The protein profiles of embryogenic and non-embryogenic carrot cultures following two-dimensional polyacrylamide gel electrophoresis of the steady-state and newly synthezized polypeptides exhibited few differences [16]. Less than 2% of the resolved polypeptides were different in these two cultures. Also, there were limited differences in the mRNA expressed in these cultures (see Ref. 71). Hybridization experiments performed using randomly selected cDNA recombinants representing different mRNA populations isolated from various stages of embryogenesis indicated few differences between the patterns of transcription in the proembryogenic mass of cells (PEMs) and the heart-shaped embryos [71]. A significant number of cDNA recombinants demonstrated either qualitative or quantitative differences in their representation in embryo and leaf RNA populations. Although some cDNAs that were embryo-specific could be detected in the PEMs, they were absent from mature leaves. Some species of RNAs were transcribed at a high level in the PEMs and to a lesser extent in the torpedo stage embryos. These transcripts were undetectable in the cotyledon, hypocotyl, callus or mature leaves. Also, they were absent from the nonembryogenic cell population within the same cultures. These results indicate that the embryo-specific gene expression, or the causal events of somatic embryogenesis, might begin in target cells prior to the withdrawal of auxins from the medium, and much prior to any visible changes [71].

7. Conclusions and future perspectives

Many of the problems associated with development (differentiation) appear to be common to both plants and animals, especially at the molecular and sub-cellular levels [84]. Thus, a student of plant morphogenesis can benefit by being familiar with the progress in this area in animal science. One should be aware, however, that because of the non-motile nature of plants, the presence of rigid cell walls surrounding their cells and the continued presence of embryonic cells (meristems) in the mature plant body, the patterns of differentiation in plants and animals at higher levels of organization are unique to each group. Notwithstanding the above, morphogenesis in plants and animals can be regarded as the result of programmed sequential gene expression [84]. Consequently, understanding the mechanism of activation of the right genes at the right time in the right cells is a basic requirement in order to fully comprehend the problem of differentiation. At a higher level of organization, the influence of certain biophysical factors on morphogenesis cannot be neglected. For example, the reinforcing hoops or rings of cellulose on the transverse walls of the cells in cylindrical organs have been suggested to be one of the important biophysical factors controlling the directionality of growth [32]. The stress pattern at the surface, determined in part by the hoop-reinforcement, could influence the rate and direction of localized growth by altering the polarities. Hence, as pointed out by Green [32], the surface layers of the new-forming organs are very important, perhaps even more important than the traditional view of the central core as the primary biophysical factor controlling the form of plants. This hypothesis is supported by thin cell layers which can give rise to full-fledged organs under appropriate nutrient and hormonal conditions *in vitro* [78]. Examination of the changes in the structure at the cellular, tissue and organ levels complement our knowledge of the biochemical, physiological and sub-cellular changes associated with morphogenesis. Therfore, much more concerted effort needs to be made in these areas, and only by integrating the information gained in these diverse yet inter-related fields of study can one begin to understand morphogenesis.

8. Summary

The complex sequence of events associated with morphogenesis, or the origin of form, in plants appear to be controlled at different levels of organization. Hence, the process has been studied at the structural, metabolic and molecular levels. Plant cell and tissue cultures coupled with analytical techniques such as histochemistry, autoradiography, physiology and biochemistry have proven to be extremely useful in studying morphogenesis. One of its important contribitions has been the unequivocal proof that phytohormones are essential for the process. The establishment of polarity and distinct physiological gradients have been shown to be among the earliest events observed. Studies have been limited in woody plants, but the available information suggests that the process is similar in herbaceous and woody species. Also, at present, a major effort is being directed at understanding the molecular basis of organized development, particularly in herbaceous species. Nevertheless, from the current trends in research, we can confidently expect significant new information on the process of morphogenesis during the 1990s.

References

1. Aitken J, Horgan KJ and Thorpe TA (1981) Influence of explant selection on the shoot-forming capacity of juvenile tissue of *Pinus radiata*. Can J For Res 11: 112–117.
2. Aitken-Christie J, Singh AP, Horgan KJ and Thorpe TA (1985) Explant developmental state and shoot formation in *Pinus radiata* cotyledons. Bot Gaz 146: 196–203.
3. Akiyoshi DE, Klee H, Amasino RM, Nester EW and Gordon MP (1984) T-DNA of *Agrobacterium tumefaciens* codes an enzyme of cytokinin biosynthesis. Proc Natl Acad Sci USA 81: 5994–5998.
4. Barlow PW (1989) Differential growth in plants – A phenomenon that occurs at all levels of organization. Env Expt Bot 29: 1–5.
5. Biondi S and Thorpe TA (1982) Growth regulator effects, metabolite changes, and respiration during shoot initiation in cultured cotyledon explants of *Pinus radiata*. Bot Gaz 143: 20–25.
6. Biondi S, Torrigiani P, Sansovini A and Bagni N (1988) Inhibition of polyamine biosynthesis by dicyclohexylamine in cultured cotyledons of *Pinus radiata*. Physiol Plant 72: 471–476.
7. Bonga JM (1977) Organogenesis in *in vitro* cultures of embryonic shoots of *Abies balsamea* (Balsam fir). In Vitro 13: 41–48.
8. Bonga JM (1982) Vegetative propagation in relation to juvenility, maturity, and rejuvenation. In: Bonga JM and Durzan DJ (Eds.) Tissue Culture in Forestry pp. 387–412. Martinus Nijhoff/Dr. W. Junk Publ., The Hague.
9. Bonga JM (1984) Adventitous shoot formation in cultures of immature female strobili of *Larix decidua*. Physiol Plant 62: 416–421.
10. Bonga JM (1987) Clonal propagation of mature trees: Problems and possible solutions. In: Bonga JM and Durzan DJ (Eds.) Cell and Tissue Culture in Forestry, Vol. 1, pp. 249–271. Martinus Nijhoff, Dordrecht.
11. Borkird C, Choi JH, Jin Z-H, Franz G, Hatzopolous P, Chorneau R, Bonas U, Pelegri F and Sung ZR (1988) Developmental regulation of embryonic genes in plants. Proc Natl Acad Sci USA, 85: 6399–6403.
12. Bornmann CH (1983) Possibilities and constraints in the regeneration of trees from cootyledonary needles of *Picea abies in vitro*. Physiol Plant 57: 5–16.
13. Brawley SH, Wetherell DF and Robinson KR (1984) Electrical polarity in embryos of wild carrot precedes cotyledon differentiation. Proc Natl Acad Sci USA, 81: 6064–6067.
14. Brown DCW and Thorpe TA (1986) Plant regeneration by organogenesis. In: Vasil IK (Ed.) Cell Culture and Somatic Cell Genetics of Plants, Vol. 3, pp. 49–65. Acad Press, Inc., New York.
15. Byus CV, Pieper SE and Adey WR (1987) The effects of low energy 60-Hz environmental electromagnetic fields upon the growth related enzyme ornithine decarboxylase. Carcinogenesis 8: 1385–1389.
16. Choi JH and Sung ZR (1984) Two-dimensional gel analysis of carrot somatic embryonic proteins. Plant Mol Biol Reporter 2: 19–23.
17. Choi JH, Liu L-S, Borkird C and Sung ZR (1987) Cloning of genes developmentally regulated during plant embryogenesis. Proc Natl Acad Sci USA 84: 1906–1910.
18. Christianson ML (1987) Causal events in morphogenesis. In: Green CE, Somers DA, Hacket WO and Biesboer DD (Eds.), Plant Tissue and Cell Culture pp. 45–55. Alan R. Liss, Inc., New York.
19. Christianson ML and Warnick DA (1983) Competence and determination in the process of *in vitro* shoot organogenesis. Dev Biol 95: 288–293.
20. Coleman WK and Thorpe TA (1985) Polarity. In: Pharis RP and Reid DM (Eds.) Encyclopedia of Plant Physiology. New Series. Vol. 11, Hormonal Regulation of Development pp. 116–138. Springer-Verlag, Berlin.
21. Constabel F, Kurz WGW, Chatson KB and Kirkpatrick JW (1977) Partial synchrony in soybean cell suspension cultures induced by ethylene. J Cell Res 105: 263–268.

22. Douglas GC (1984) Formation of adventitous buds in stem internodes of *Populus* spp. cultured *in vitro* on basal medium: Influence of endogenous properties of explants. J Plant Physiol 116: 313–321.
23. Douglas TJ, Villalobos VM, Thompson MR and Thorpe TA (1982) Lipid and pigment changes during shoot initiation in cultured explants of *Pinus radiata*. Physiol Plant 55: 470–477.
24. Ellis DE and Judd RC (1987) SDS-PAGE analysis of bud-forming cotyledons of *Pinus ponderosa*. Plant Cell Tissue Organ Culture 11: 57–65.
25. Everett N (1982) The determination phase of differentiation. In: Fujiwara A (Ed.) Plant Tissue Culture 1982 (Proc. 5th Intl. Cong. Plant Tissue Cell Culture) pp. 93–94. Jap. Assoc. Plant Tissue Culture, Tokyo.
26. Fromm ME, Taylor LP and Walbot V (1986) Stable transformation of maize after gene transfer by electroporation. Nature 319: 791–793.
27. Fukuda H and Komamine A (1985) Cytodifferentiation. In: Vasil IK (Ed.) Cell Culture and Somatic Cell Genetics of Plants, Vol. 2 pp. 149–212. Academic Press, New York.
28. Gamborg OL, Murashige T, Thorpe TA and Vasil IK (1976) Plant tissue culture media. In Vitro 12: 473–478.
29. Garfinkel OJ, Simpson RB, Ream LW, White FF, Gordon MP, and Nester EW (1981) Genetic analysis of crown gall: Fine structure map of the T-DNA by site-directed mutagenesis. Cell 27: 143–153.
30. Gill R, Mishra KP and Rao PS (1987) Stimulation of shoot regeneration of *Vigna aconitifolia* by electrical control. Ann Bot 60: 399–403.
31. Goldsworthy A (1986) The electric compass of plants. New Scientist 2 Jan 86: 22–23.
32. Green PB (1980) Organogenesis – A Biophysical View. Ann Rev Plant Physiol 31: 51–82.
33. Halperin W (1969) Morphogenesis in cell cultures. Ann Rev Plant Physiol 20: 395–418.
34. Halperin W (1986) Attainment and retention of morphogenic capacity *in vitro*. In: Vasil IK (Ed.) Cell Culture and Somatic Cell Genetics of Plants, Vol. 3, pp. 3–47. Acad Press, Inc., New York.
35. Hammersley-Straw DH and Thorpe TA (1987) Modulation of shoot formation in tobacco callus by tissue transfer between permissive and inhibitory media. In Vitro Cellular Developmental Biol 23: 867–872.
36. Hartweck LM, Lazzeri PA, Cui D, Collins GB and Williams EG (1988) Auxin orientation effects on somatic embryogenesis from immature soybean cotyledons. In Vitro Cellular Develop Biol 24: 821–828.
37. Inze D, Follin A, Van Lijsebettens M, Simoens C, Genetello C, Van Montagu M, and Schell J (1984) Genetic analysis of the individual T-DNA genes of *Agrobacterium tumefaciens* further evidence that two genes are involved in indole-3-acetic acid synthesis. Mol Gen Genet 194: 265–274.
38. Jaffe LF (1966) Electrical currents through the developing *Fucus* egg. Proc Natl Acad Sci USA 56: 1102–1129.
39. Jaffe LF (1977) Electrophoresis along cell membranes. Nature 265: 600–602.
40. Kinosita K and Tsong TY (1977) Voltage-induced pore formation and hemolysis of human erythrocytes. Biochem Biophys Acta 471: 227–242.
41. Kirby EG and Schalk ME (1982) Surface structural analysis of cultured cotyledons of Douglas fir. Can J Bot 60: 2729–2733.
42. Kumar PP, Reid DM, and Thorpe TA (1987) The role of ethylene and carbon dioxide in differentiation of shoot buds in excised cotyledons of *Pinus radiata in vitro*. Physiol Plant 69: 244–252.
43. Kumar PP, Bender L, and Thorpe TA (1988) Activities of ribulose bisphosphate carboxylase and phosphoenolpyruvate carboxylase and ^{14}C-bicarbonate fixation during *in vitro* culture of *Pinus radiata* cotyledons. Plant Physiol 87: 675–679.
44. Kumar PP and Thorpe TA (1989) Putrescine metabolism in excised cotyledons of *Pinus radiata* cultured *in vitro*. Physiol Plant 76: 521–526.
45. Kumar PP, Joy IV RW, and Thorpe TA (1989) Ethylene and carbon dioxide accumulation,

and growth of cell suspension cultures of white spruce (*Picea glauca*). J Plant Physiol 135: 592–596.
46. Lang JM, Eisinger WR, and Green PB (1982) Effects of ethylene on the orientation of microtubules and cellulose microfibrils of pea epicotyl cells with polylamellate cell walls. Protoplasma 110: 5–14.
47. Lee-Stadelmann OKY, Lee SW, Hackett WP and Read PE (1989) The formation of adventitious bids *in vitro* on micro-cross sections of hybrid *Populus* leaf midveins. Plant Sci 61: 263–272.
48. Lu C-Y and Thorpe TA (1988) Shoot-bud regeneration in subcultured callus of Engelmann spruce. In Vitro Cellular Develop Biol 24: 239–241.
49. Maeda E and Thorpe TA (1979) Shoot histogenesis in tobacco callus cultures. In Vitro 15: 415–424.
50. Meins F and Binns AN (1979) Cell determination in plant development. Bioscience 29: 221–225.
51. Memelink J, Hoge JHC and Schilperoort RA (1987) Cytokinin stress changes the developmental regulation of several defence-related genes in tobacco. EMBO J 6: 3579–3583.
52. Merkle SA, Wetzstein HY, and Sommer HE (1987) Somatic embryogenesis in tissue cultures of pecan. HortSci 22: 128–130.
53. Minocha SC (1980) Callus and adventitious shoot formation in excised embryos of white pine (*Pinus strobus*). Can J Bot 58: 366–370.
54. Nomura K and Komamine A (1986) Somatic embryogenesis in carrot cells. Dev Growth Differ 28: 511–517.
55. Ochatt SJ, Rech, EL, Davey MR, and Power JB (1988) Long-term effect of electroporation on enhancement of growth and plant regeneration of colt cherry (*Prunus avium* × *pseudocerasus*) protoplasts. Plant Cell Rep 7: 393–395.
56. Oka S, Yeung EC, and Thorpe TA (1982) Shoot formation in *Eucalyptus globulus* hypocotyl explants. New Zealand J For Sci 12: 501–509.
57. Patel KR and Thorpe TA (1984) Histochemical examination of shoot initiation in cultured cotyledon explants of radiata pine. Bot Gaz 145: 312–322.
58. Philip VJ (1984) *In vitro* organogenesis and plantlet formation in cashew (*Anacardium occidentale* L.) Ann Bot 54: 149–152.
59. Rathore KS and Goldsworthy A (1985) Electrical control of shoot regeneration in plant tissue cultures. Bio/Technology 3: 1107–1109.
60. Rech EL, Ochatt SJ, Chand PK, Davey MR, Mulligan BJ, and Power JB (1988) Electroporation increases DNA synthesis in cultured plant protoplasts. Bio/Technology 6: 1091–1093.
61. Robinson KR and Jaffe LF (1975) Polarizing Fucoid eggs drive a calcium current through themselves. Science 187: 70–72.
62. Ross MK, Thorpe TA, and Costerton JW (1973) Physiological gradients and shoot initiation in tobacco callus cultures. Plant Cell Physiol 14: 473–480.
63. Rumary C, Patel KR and Thorpe TA (1986) Plantlet formation in white and black spruce. II. Histological analysis of adventitious shoot formation *in vitro*. Can J Bot 64: 997–1002.
64. Schroder G, Waffenschmidt S, Weiler EW, and Schroder J (1984) The T-region of Ti plasmids codes for an enzyme synthesizing indole-3-acetic acid. Eur J Biochem 138: 387–391.
65. Shillito RD, Saul MW, Paszkowski J, Muller M, and Potrykus I (1985) High efficiency direct gene transfer to plants. Bio/Technology 3: 1099–1103.
66. Skoog and Miller CO (19577) Chemical regulation of growth and organ formation in plant tissue cultured *in vitro*. Symp Soc Exp Biol 11: 118–130.
67. Slocum RD, Kaur-Sawhney R, and Galston AW (1984) The physiology and biochemistry of polyamines in plants. Arch Biochem Biophys 235: 283–303.
68. Smith DR and Thorpe TA (1975) Root initiation in cuttings of *Pinus radiata* seedlings: I. Developmental sequence. J Exp Bot 26: 184–192.
69. Smith DR and Thorpe TA (1975) Root initiation in cuttings of *Pinus radiata* seedlings: II. Growth regulator interactions. J Exp Bot 26: 193–202.
70. Theologis A (1986) Rapid gene regulation by auxin. Ann Rev Plant Physiol 37: 407–438.

71. Thomas TL and Wilde D (1987) Analysis of carrot somatic embryo gene expression programs. In: Green CE, Somers DA, Hackett WP and Biesboer DD (Eds.) pp. 83–93. Alan R. Liss, Inc., New York.
72. Thorpe TA (1980) Organogenesis *in vitro*: Structural, physiological, and biochemical aspects. Int Rev Cytol Suppl 11A: 71–111.
73. Thorpe TA (1983) Morphogenesis and regeneration in tissue culture. In: Owens LD (Ed.) Genetic Engineering: Applications to Agriculture, Beltsville Symp. 7 pp. 285–303. Rowman & Allanheld, Totowa, USA.
74. Thorpe TA (1988) Physiology of bud induction in conifers *in vitro*. In: Hanover JW and Keathley DE (Eds.) Genetic Manipulation of Woody Plants, pp. 167–183. Plenum Publ., New York.
75. Thorpe TA and Murashige T (1970) Some histochemical changes underlying shoot initiation in tobacco callus culture. Can J Bot 48: 277–285.
76. Thorpe TA and Patel KR (1986) Comparative morpho-histological studies on the sites of shoot initiation in various conifer explants. New Zealand J For Sci 16: 257–268.
77. Toriyama K, Arimoto Y, Uchimiya H, and Hinata K (1988). Transgenic rice plants after direct gene transfer into protoplast. Bio/Technology 6: 1072–1074.
78. Tran Thanh Van M, Thi Dien N and Chlyah A (1974) Regulation of organogenesis in small explants of superficial tissue of *Nicotiana tabacum* L. Planta 119: 149–159.
79. Trewavas AJ and Cleland RE (1983) Is plant development regulated by changes in the concentration of growth substances or by changes in the sensitivity to growth substances? TIBS 8: 354–357.
80. Villalobos VM, Leung DWM and Thorpe TA (1984) Light-cytokinin interaction in shoot formation in cultured cotyledon explants of radiata pine. Physiol Plant 61: 497–504.
81. Villalobos VM, Oliver MJ, Yeung EC, and Thorpe TA (1984) Cytokinin-induced switch in development in excised cotyledons of radiata pine cultured *in vitro*. Physiol Plant 61: 483–489.
82. Villalobos VM, Yeung EC and Thorpe TA (1985) Origin of adventitious shoots in excised radiata pine cotyledons cultured *in vitro*. Can J Bot 63: 2172–2176.
83. Von Arnold S and Eriksson T (1985) Initial stages in the course of adventitious bud formation on embryos of *Picea abies*. Physiol Plant 64: 41–47.
84. Wareing PF and Graham CF (1984) Introduction-Problems of Development. In: Warring PF and Graham CF (Eds.) Developmental Control in Animals and Plants, 2nd Edition, pp. 3–4. Blackwell Scientific Publications, London, UK.
85. Wareing PF and Phillips IDJ (1981) Growth and Differentiation in Plants, 3rd Edition, Pergamon Press, New York, USA.
86. Wetmore RH and Rier JP (1963) Experimental induction of vascular tissues in callus of angiosperms. Am J Bot 50: 418–430.
87. Woodruff RI and Telfer WH (1980) Electrophoresis of proteins in intercellular bridges. Nature 286: 84–86.
88. Yasuda T, Hasegawa PM and Cheng T-Y (1980) Analysis of newly synthesized proteins during differentiation of cultured Douglas fir cotyledons. Physiol Plant 48: 83–87.
89. Yeung EC, Aitken J, Biondi S, and Thorpe TA (1981) Shoot histogenesis in cotyledon explants of radiata pine. Bot Gaz 142: 494–501.
90. Zurfluh LL and Guilfoyle TJ (1982) Auxin- and ethylene-induced changes in the population of translatable messenger RNA in basal sections and intact soybean hypocotyl. Plant Physiol 69: 338–340.

3. Light regimes used in conifer tissue culture

DAVID D. ELLIS[1] and DAVID T. WEBB[2]

[1] *University of Wisconsin-Madison, Department of Horticulture, 1575 Linden Drive, Madison, WI 53706, U.S.A.*

[2] *Department of Botany, University of Hawaii, 3190 Maile Way Honolulu, HI 96822, U.S.A.*

1. Introduction

Light has a major influence on the growth, development and morphogenesis of plants. In an *in vitro* environment, where conditions are manipulated to optimize a given response, careful consideration should be given to light quantity and intensity as well as the photoperiod. Unfortunately, all too often the light regime used for a given microculture situation is based either on (1) incomplete information of the optimum *in vitro* requirements for the species and tissue with which one is working or (2) the conditions that currently exist or are available in a particular research setting. Often a compromise must be also made when choosing a light regime to satisfy a wide range of species within a single growth area.

As scientists, we are fortunate that plants are tolerant of a wide range of environmental, particularly light, conditions. Due in part to this tolerance, there is a lack of information on light responses for individual species. Therefore, optimizing light conditions for a particular morphogenetic response or growth condition is often not done. In this chapter, we review some of the light regimes which have been used for the *in vitro* manipulation of conifers, and highlight studies which have investigated the effect of varying the *in vitro* light regime on conifer morphogenesis. In addition, since temperature is often altered with photoperiod, a brief discussion of its effects is also included. Finally in Table 1, we include a compilation of some of the light regimes that have been used for different conifers species.

The *in vitro* culture of plants provides an artificial environment where many substances necessary for sustained life of the plant are available in luxury amounts. It is assumed that the requirement for light in an *in vitro* environment is not the same as that required in a natural situation. Photosynthesis for example, is not an absolute necessity *in vitro* because carbohydrates, such as surcrose, are added to the culture medium as a carbon source [103]. Despite this, plants and tissues retain their photoautotrophic nature and are dependent on light for overall survival. Light also continues to play a very important role in the regulation and maintenance of many developmental and morphogenetic processes.

M. R. Ahuja (Ed.), Micropropagation of Woody Plants, 31–55.

Table 1. An example of some light regimes used in recent literature on conifer tissue culture. It is not intended as a complete survey, instead it is intended to be used as a rough guideline from where to start. Most reports do not give complete information, hence the abundance of missing data.

Abbreviations:

Explants

Emb	=	embryonic tissue including seedling material
bud	=	buds or shoots from mature trees
F-gam	=	mega-gametophyte
f-cone	=	female cone
sus cell	=	cell suspension
emb call	=	embryogenic callus

Induced

buds	=	adventitious buds
ax	=	axillary buds
emb call	=	embryogenic callus
elong	=	elongation of the explant
proto	=	protoplast
m-strob	=	male strobilus

Light Quality

2.8, 0.5, 2.0 = μWatts cm^{-2} nm^{-1} in the blue, far-red, red spectra respectively

Light Intensity

fc	=	foot candles
μE	=	μEinsteins m^{-2} s^{-1}
μM	=	μ m^{-2} s^{-1}
ergs	=	ergs cm^{-2} s^{-1}
W	=	Watts m^{-2}

Source

flu	=	fluorescent
cw	=	cool-white fluorescent
ww	=	warm-white fluorescent
inc	=	incandescent
gl	=	gro–lux
tung	=	tungsten

Temperature

21/18 = day temp/night temp °C

Explant	Induced tissue/ organ	Dura-tion hours	Quality	Intensity	Source	°C Temp.	Refer-ence
Abies balsamea							
bud	buds	16		540lux	flu	21	19
Biota orientalis							
emb	buds	16		500lux	flu	25	126
Chamaecyparis nootkatensis							
emb	buds	16		25μE	cw	23	85
	root	16		20μE	cw	23	85
Chamaecyparis obtusa							
emb	buds	16		18μE	cw	25	70
Cryptomeria japonica							
emb	buds	14		1900lux	flu	25	71
Cunninghamia lanceolata							
bud	ax bud	16		20W	gl	26/22	17
emb	buds	16		20W	gl	26/22	17
Cupressus lusitanica							
emb	buds	16		60μE	gl	25	57
Larix decidua							
f-cone	buds	16		50μE	gl	21	23,24
emb	buds	24		1000lux	flu+inc	20	77
f-cone	buds	16		500lux	gl	21	22,20
f-gam	callus	16		50μM	gl	21	104
bud	buds	24		80μE	cw+inc	20	43
bud	elong	24		20μE	cw	22	51

Table 1 (continued)

Explant	Induced tissue/ organ	Duration hours	Quality	Intensity	Source	°C Temp.	Reference
Larix × eurolepis							
bud	callus	16	400–700nm	30μE	flu	22/20	87
Larix occidentalis							
emb	buds	16	300–800nm	80μE	gl	24	67
Larix sibirica							
emb	buds	24		2140lux		20	96
Picea abies							
emb	buds	24				20	135
bud	buds	24				20	135
needle	buds	24				20	135
emb	buds	16		27W	flu	24/18	25
emb	emb call	16		10–40μE	ww	25	63
emb	buds	14		250fc+75fc	cw+inc	25/21	115
emb	buds	24	400–700nm	20W	flu	20	134
bud	buds	24	400–700nm	20W	flu	20	135
emb	buds	16	400–700nm	1–5W	flu	20	137
bud	buds	16	400–700nm	2W	flu	20	138
emb	buds	14	400–700nm	2–3W	flu	20	139
emb	emb call	16		15–50μE	cw+inc	23	105 143
emb	emb call	16		15–50μE	cw+inc	23	12
emb call	culture	16		10–40μE	ww	25	63
emb call	emb	16		10–40μE	ww	25	62
som emb	devel.	24	5.0,2.0,0.5			23	27
emb	buds	16		180μE		20	140
micro-spore	callus/ roots/buds	18		1000lux	flu		118
Picea engelmanii							
emb	buds	16	300–800nm	80–100μM	gl	27	108
	elong	20		80–100μM	flu	22	
	roots	16		60–80μM	flu	20	
emb	emb call	16		20–40μE	cw	25	147
Picea glauca							
12yr bud	elong	16				24	79
emb	buds	13.5	210,110,240			20/7	29
emb	buds	24				20	30
	roots	24				20	30
emb	buds	24	210,110,240			20	30
emb	sus cell	24	6.2,2.7,5.5			22.5	46
emb	buds	16		80μE		27	116
emb	emb call	16		20–40μE	cw	25	147
emb	emb call	16		15–50μE	cw+inc	23	105
sus emb call	culture	24		15–20μE		25	45
emb call	emb	16	300–800nm	80μM		26	89
emb call	emb sus	24		$1.6 \times 10^{-4} Wcm^{-2}$	cw	28	65
emb	buds	24		20μE	cw	27	52

Table 1 (continued)

Explant	Induced tissue/ organ	Duration hours	Quality	Intensity	Source	°C Temp.	Reference
emb	emb call	24		20μE	cw	27	Ellis, unpubl.
Picea mariana							
emb	buds	16		80μE		27	116
emb	emb call	12				25	64
Picea pungens							
bud	buds	16				22/18	36
bud	buds	16		1000lux	gl	26	93
Picea sitchensis							
emb	buds	16		10.5W		21	73
emb	buds	13.5				20	74
emb	buds	13.5		1.25×10 ergs		20	148
bud	elong+bud	16		2.2–260μM	cw+inc	20	117
Pinus banksiana							
emb	buds	24		1000lux	flu+inc	20	77
Pinus brutia							
emb	buds	16	400–700nm	80μM	flu+inc	23	1,2
Pinus caribaea							
emb	buds	24		75μE	cw	25	16
emb	buds	12		1×10^{-4}Wcm^{-2}	cw	29/25	145
emb	ax	16		70μE	flu+tung	25	11
Pinus contorta							
emb	buds	16		80μE	gl	27	107
emb	elong	16		50μE	gl	22	107
emb	buds	24	400–700nm	2W	flu	20	136
emb	buds	16		1.25×10 ergs		25	148
Pinus coulteri							
emb	buds	24		1076lux	cw	25	106
Pinus elliottii							
emb	callus	24		150–200fc		23	10
emb	buds	24		40–50μE	cw	25	110
Pinus eldarica							
emb	org. callus	16		150μE	cw	25	59
Pinus monticola							
plantlet	roots	24		1500lux	cw	22	5
emb	buds	24		17W	flu	22	99
Pinus mugo							
emb	bud	16	400–700nm	80–100μM	cw	24	102
Pinus nigra							
emb	buds	16		700–1500lux	flu	26	72
emb	buds	16		700–1500lux	flu	26	82
emb	buds	16		1250–1500lux	flu		83
Pinus oocarpa							
emb	buds	16		60μE	gl	25	57
proto	callus	16		3W	gl	21	86
emb	ax	16		70μE	flu+tung	25	11
Pinus palustris							
emb	buds	16		8608lux	cw	25	123

Table 1 (continued)

Explant	Induced tissue/ organ	Duration hours	Quality	Intensity	Source	°C Temp.	Reference
Pinus pinaster							
emb+bud	buds	24		1200lux		26	41
	elong	16		10000lux		26/21	41
	root	16		5000lux		26/21	41
emb	proto	16		0.6W	gl	25/21	42
bud	buds	24		1200lux		26	56
	elong	16		8000lux		26/21	56
	root	16		4000lux		26/21	56
emb	buds	16		2000lux	flu	20	111
emb	root	16		15W	flu	21	112
Pinus pithyusa							
emb	buds	24		2140lux		20	97
Pinus ponderosa							
emb	buds	24		2140lux		20	96
emb	buds	24		5W	cw	25	48
Pinus radiata							
emb	buds	16		80μE		28/24	3
	elong	16		80μE		24/20	3
plantlets	roots	16		120μE		25/20	4
emb	callus	24	2.0,0.5,1.4		cw	22	9
emb	buds	16		80μE		28	44
emb	buds	16	400–700nm	17.2W	cw	25/20	59
emb	root	12		3000lux	flu+inc	25/20	34
emb	buds				cw	25	113
emb	root	16		2000lux	flu+inc	27	119
emb	buds			55–90μE		25	120
emb	buds	16	380–800nm	80μE	gl	28	131,133
callus	different	24	2.0,0.5,1.4		cw	22	144
emb	buds	16		80μE		28/24	150
Pinus rigida							
emb	buds	16		80μE	gl	27	109
Pinus strobus							
emb	buds	24		400–500W/cm^2	cw	26	78
	root	16		150–200W/cm^2		17	
emb	buds	12		15μE	cw	25	55
emb	buds	12		15μE	cw+ww	25	54
fasicles	elong	24		120μE	cw+inc	22	124
emb	buds	12		15μE	cw	27	146
Pinus strobus							
emb	buds	12		2000lux	flu	25/20	92
Pinus sylvestris							
emb	buds	16		500lux		25	129
emb	buds	16		12.5W		20	129
Pinus taeda							
emb	buds	24		2200+800lux	cw+inc	23	7
	elong	24		2700+200lux	cw+inc		7
		16		3200+200lux	cw+inc		7

Table 1 (continued)

Explant	Induced tissue/ organ	Duration hours	Quality	Intensity	Source	°C Temp.	Reference
	root	24		1830+320lux	cw+inc	25	7
emb	buds	24		9000fc	flu	21/7	91
	root	24		9000fc	flu	22	91
emb	buds	24		1000–2000lux	cw+inc	21	100
				300–400lux			100
	root	24		1000–2000lux	ww	23	100
emb	buds			1250/750lux	cw+inc		101
callus		24		75μE		25	101
sus cell	proto	16		300lux	flu	23	125
emb	sus cell	16		3000lux	cw	23	125
Pinus tecunumanii							
emb	ax	16		70μE	flu+tung	25	11
Pseudotsuga menziesii							
som emb	devl	24	2.8,2.0,0.5		cw	23	47
emb	buds	15		500fc	cw	25	28
bud	ax bud	24	5.0,2.0,0.5		cw	23	61
emb	buds			25μE	cw	23	90
bud	buds	16		20W	flu+inc	20	127
bud	buds	16	2.8,0.5,2.0		cw	22	60
emb	buds	18		200fc	flu+inc	25	31,33
	roots					19	34
bud	buds			22W			53
emb	buds	16	660nm	0.42mW/day	flu	26	76
emb	buds	18		1800lux	cw	25	81
emb	buds	16		2000lux	cw	24/18	149
emb	buds	16		67.6μM	cw	25	95
Sequoia sempervirens							
bud	elong	18		150fc	flu+inc		8
bud	elong	16		14W	flu	20	98
Sequiadendron giganteum							
bud	buds	16		200lux	flu	21	26
Taxus brevifolia							
bud	callus/ elong	24		20μE	cw	27	Ellis unpubl.
Taxus cuspidata							
bud	callus/ elong	24		20μE	cw	27	Ellis unpubl.
Thuja occidentalis							
emb	buds	16		80μM	gl	27	66
bud	elong	24		20μE	cw	27	Ellis unpubl.
Thuja plicata							
emb+bud	buds	24		2000lux	cw	25	37
bud	m-strob	24		2000lux	flu	25	38
Tsuga heterophylla							
emb	buds	18		1000fc		25/19	32

Plants respond to light in three general ways: (1) photoperiodism, the response to duration and timing of day and night; (2) phototropism, growth based on the direction of the light source; and (3) photomorphogenesis, the influence of light on the development of the plant. The level of response can depend on the way in which light is presented to the plant, that is the light quality, intensity and duration of exposure. To confound the problem of studying light effects on plants, the response may depend on the way light is perceived and the particular phase of the life cycle of the plant. Seedlings for example, may respond to a light stimulus that has no measurable effect on mature tissues, such as the requirement of red light for some varieties of lettuce seed germination. In addition, other external stimuli, such as cold (i.e. vernalization) may also effect how a plant might respond to light.

2. Light quality

Light quality is the spectral distribution of photons [121] and is usually measured in wavelengths. The visible radiation which we can perceive with our eyes, the visible spectrum, is in the range of 400–700 nm [68,121] (Figure 1). Although the very important process of photosynthesis occurs within this range, the visible spectrum accounts for less than half of the light energies within the biological window [68]. Studies of plant photoresponses have lead to the generalization that there are two classes of signal transducing photoreceptors. One class of photoreceptor is phytochrome, which absorbs mainly in the 600–800 nm region and exists in plants in two forms; *Pr* with an absorption maximum of ca. 665 nm and *Pfr* with an absorption maximum of ca. 730 nm [122]. The second, a blue-absorbing photoreceptor absorbs in the 300–500 nm region [121]. Therefore, depending on the desired morphogenetic response, published spectra of light sources limited to the visible spectrum

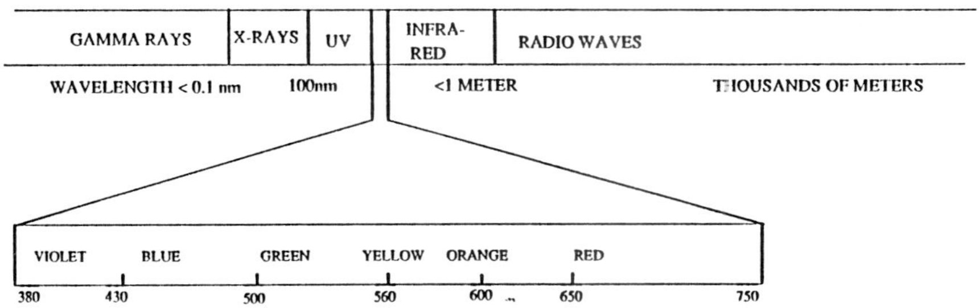

Figure 1. Wavelengths in the electromagnetic spectrum which correspond to each color. The further to the left on the spectrum, the shorter the wavelengths (the distance between one peak and the next) and the greater the energy. Redrawn from Raven, Evert and Curtis, *Biology of Plants*, p. 69.

SUNLIGHT AND SKYLIGHT

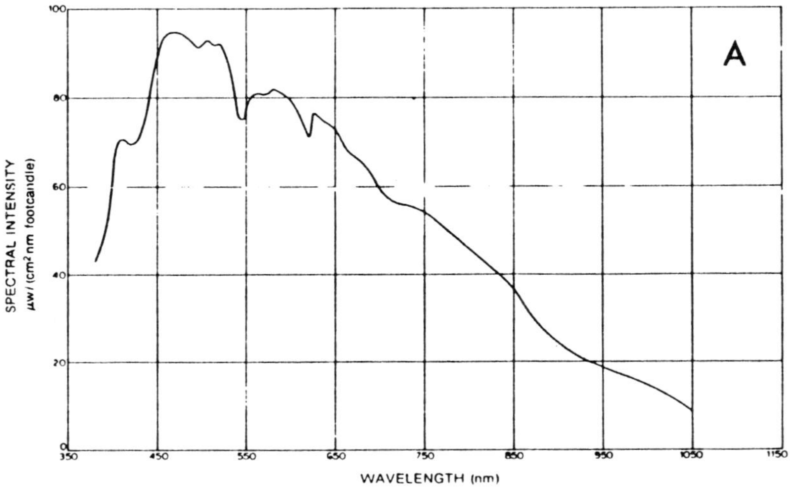

**WARM WHITE FLUORESCENT LAMP

40 WATT, LUMINOUS EFFICACY: 81.2 LUMENS/WATT

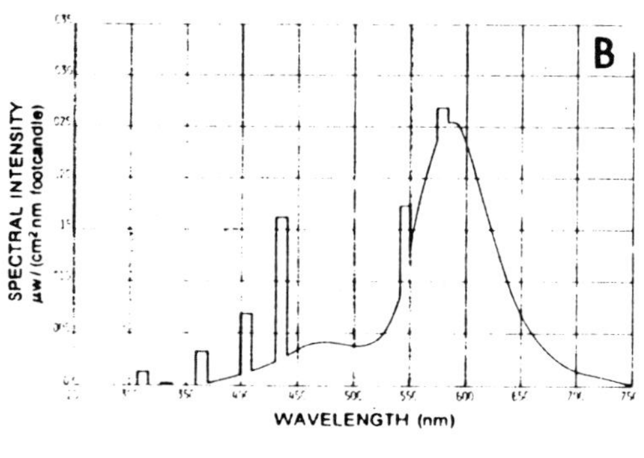

*METAL HALIDE VAPOR LAMP

400 WATT; LUMINOUS EFFICACY: 75.0 LUMENS/WATT

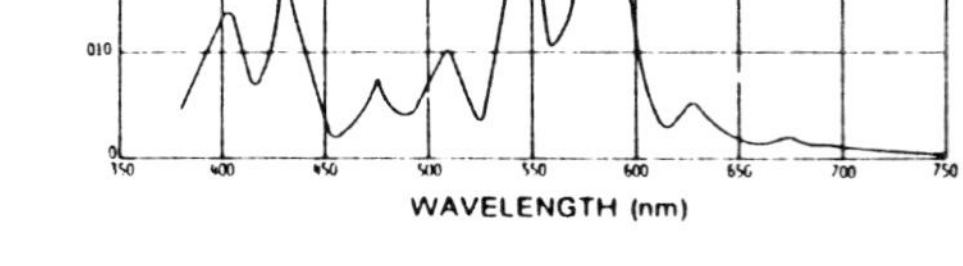

*HIGH PRESSURE SODIUM VAPOR LAMP

400 WATT; LUMINOUS EFFICACY: 107 LUMENS/WATT

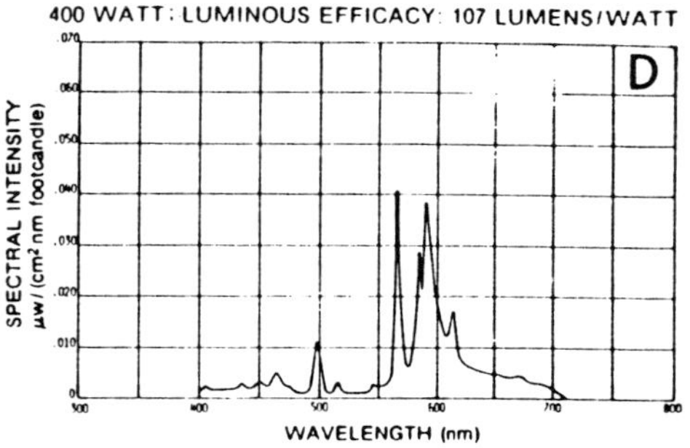

**DAYLIGHT FLUORESCENT LAMP

LAMP TYPE	WATTAGE	LUMINOUS EFF.
STANDARD	40	63.8
HIGH OUTPUT	60	54.0
VERY HIGH OURPUT OR POWER GROVE (PG)	110	55.9

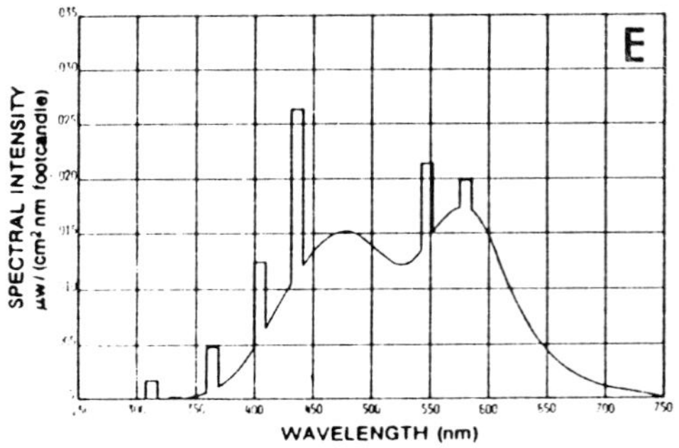

***100 WATT TUNGSTEN INCANDESCENT LAMP**

100 WATT: LUMINOUS EFFICACY 17.5 LUMENS/WATT

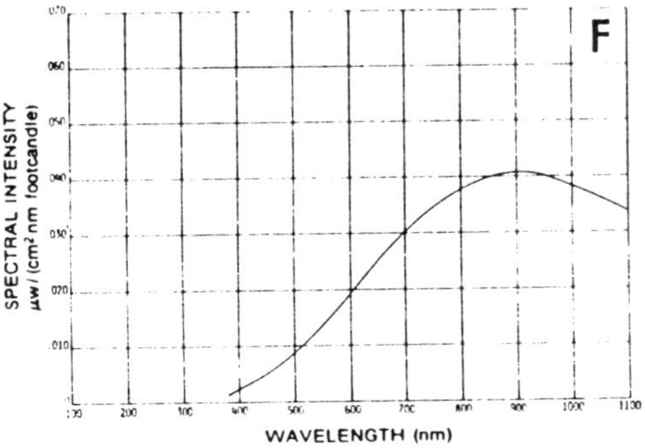

***25 WATT TUNGSTEN INCANDESCENT LAMP**

25 WATT: LUMINOUS EFFICACY 10.5 LUMENS/WATT

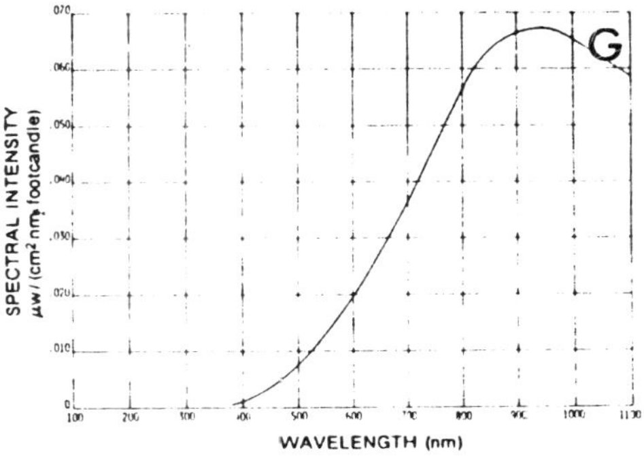

****COOL WHITE FLUORESCENT LAMP**

LAMP TYPE	WATTAGE	LUMINOUS EFF
STANDARD	40	79.0
HIGH OUTPUT	60	66.0
VERY HIGH OUTPUT OR POWER GROVE (PG)	110	62.5

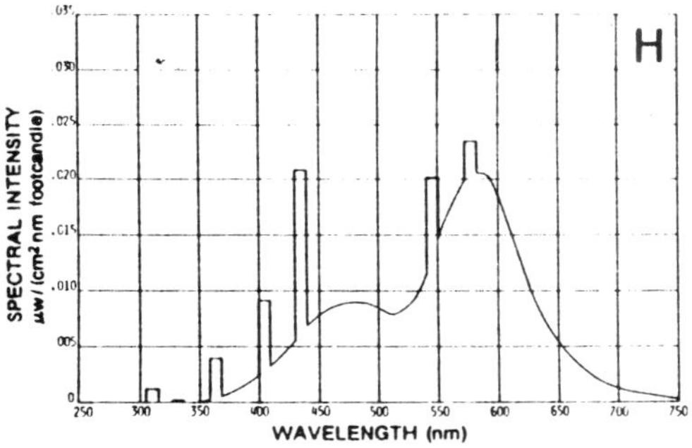

****SYLVANIA GRO-LUX WIDE SPECTRUM LAMP**

40 WATT: LUMINOUS EFFICACY: 35.8 LUMENS/WATT

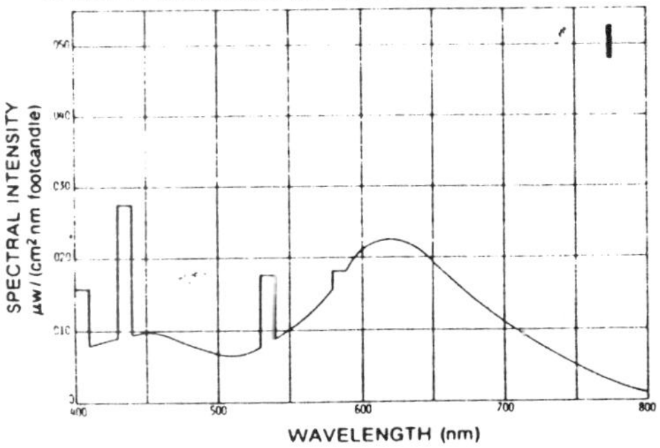

****SYLVANIA STANDARD GRO-LUX LAMP**

40 WATT: LUMINOUS EFFICACY: 23 LUMENS/WATT

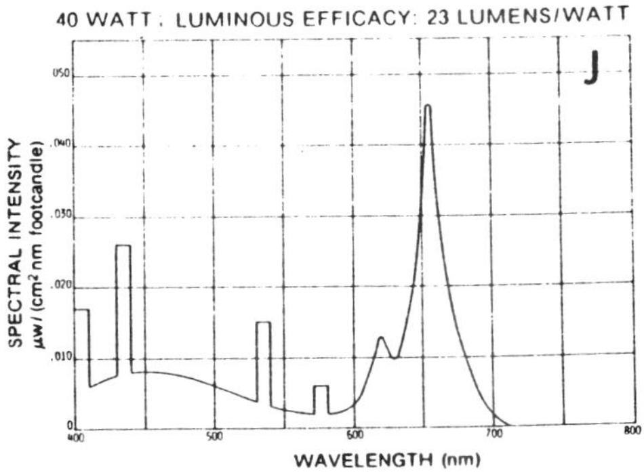

Figure 2. Spectra emitted from various sources of light. Reprinted with permission from Isco Tables (ISCO Inc., Lincoln, Nebraska) and General Electric Company (Cleveland, OH).

are useful, but may only be of limited value. The spectra emitted from various sources of artificial light are included in Figure 2.

In nature, sunlight provides the energy for photosynthesis and photomorphogenesis. Sunlight provides maximal irradiation in the blue region (450–500 nm) with slightly less illumination at longer wavelengths in the visible and near-infrared regions of the spectrum (Figure 2A). Incandescent bulbs are a rich source of far-red and red light with little energy in the blue region (Figures 2F, 2G). Fluorescent lights provide energy throughout the visible spectrum and generally have two sharp peaks in the blue region but vary in terms of their emission in the yellow, red and far-red regions (Figures 2B, 2E, 2H).

Fluorescent lights generally provide the only source of illumination for *in vitro* experiments and cool white tubes are most frequently used. Cool white tubes have a broad emission peak in the yellow-red region of the spectrum (Figure 2H). Warm white tubes have a similar emission spectrum with slightly more output in the red and less in the blue (Figure 2B). However, they both provide little energy in the red region corresponding with the action spectrum of photosynthesis [121] and phytochrome. Consequently, the light from cool white tubes probably satisfies some of the photomorphogenic requirements of cultured materials while providing a minimal level of energy for photosynthesis.

With the advent of gro-lux tubes which have a large emission peak in the red, corresponding with the absorption peak of chlorophyll a (Figure 2J), scientists gained a new light source which could effectively drive photosynthesis while also providing an energy for phytochrome. Furthermore, gro-lux wide spectrum tubes have an extended emission peak in the red-yellow region and in the near far-red (Figure 1I), thus providing a complete range of wavelengths for photosynthesis and photomorphogenesis. However, these latter two kinds of illumination are not extensively used for *in vitro* work with either conifers or angiosperms.

In most cases, the choice of cool white over the other types of fluorescent tubes is probably a custom derived from prior work rather than a decision related to the light quality requirements of the cultures. In addition, gro-lux and gro-lux wide-spectrum tubes are more expensive than cool white tubes and unless one is trying to specifically stimulate photosynthesis, there may be no need to have the additional wavelengths present. In some cases, a combination of fluorescent and tungsten lights are used to give a more balanced spectrum. However, tungsten bulbs radiate large quantities of heat and unless there is a perceived need for red and far-red light, they are usually avoided.

Most artificial sources also emit some light in the near ultra violet (UV) 300–400 μm range. Although near UV light can effect biological systems detrimentally [68], in cultured cells these are probably minor concerns with the levels emitted. In addition, these deleterious effects may be offset by potential beneficial photomorphogenic effects due to the blue-light receptor.

Near UV light has however been demonstrated to inhibit *Pseudotsuga menziesii* (Douglas fir) callus growth [76] as well as the growth of *Gingko* pollen suspension cultures. The *Gingko* cultures were also inhibited by light in the green/yellow (500–600 nm) range [21].

It should be noted, that a problem with the *in vitro* determination of the effect of a given wavelength of light to a particular tissue is that it is very difficult to vary one region of the spectrum, without affecting other light dependent responses such as photosynthetic rates or other photoreceptors [121]. For conifers the most widely reported light source used is fluorescent light in the visible (400–700 nm) range.

3. Light intensity

Light intensity is generally treated as the amount of light present at the culture level. However, the term light intensity correctly describes the light flux emitted by the source rather than the flux intercepted per unit area [121]. The term irradiance (watts m^{-2}, Joules m^{-2} sec^{-1}) indicates the light flux intercepted per unit area while quantum (photon) flux density (Einsteins cm^{-2} sec^{-1}) expresses the intercepted flux in quantum units. Photometric units like footcandles and lux are prevalent in the tissue culture literature. These can be used to duplicate experimental procedures as long as the emission spectrum of the lights and the absoprtion characteristics of the light meter are known. In general, photometric light meters are designed to match the sensitively of the human eye to color and are most responsive to green light which is generally of little consequence to plants.

The use of photometric terminology should be avoided in preference to irradiation and quantum flux density. However, the latter require the use of a quantum radiometer which can be equally sensitive to visible and near-infra-red light. While such instruments exist, they may not be readily available. Data for converting photometric units into quantum or irradiance units have been published [58,80] and can be used to *approximately* determine the amount of energy at the culture level (Table 2). Another way to deal with this problem is to borrow a radiometer and calibrate a foot-candle meter under the prevailing light conditions. The foot-candle meter can be used to measure light levels thereafter. However, periodic recalibrations should be done to insure that the meter's sensitivity hasn't changed. Further, the correlation between irradiance and photometric units is different for different light sources, therefore calibration must be done for each different light source used.

For the purpose of this report, we will report light intensity values as originally stated in a particular reference to avoid any miscalculation in the conversion from one unit to another unless conversion into another unit will facilitate comparison of light regimes within the text. In these cases, the calculated values will be in parenthesis following the reported values.

Table 2. Conversion factors of foot-candles to Wcm^{-2} for different light sources (Reprinted with permission from Klein [80]).

Radiation source	$\mu W\ cm^{-2}$ per foot-candle (400–800 nm)
Sunlight	6.50
Incandescent lamps	
25 W (2720 K)	7.19
40 W (2780 K)	6.00
60 W (2820 K)	5.69
100 W (2890 K)	5.42
300 W (2930 K)	5.25
500 W (3000 K)	5.14
1000 W (3050 K)	5.09
Tungsten-halogen lamps	
500 W (3000 K)	5.10
Mercury lamps	
H33–1CD	3.77
H33–1GL/C	3.58
H33–1GL/W	3.70
Metalarc lamps	
400 W (3000 K)	5.92
Fluorescent lamps	
Cool White	2.69
Cool White Deluxe	3.18
Warm White	2.64
Warm White Deluxe	3.01
Daylight	3.61
Blacklight BL	4.44
Blue	6.61
Green	1.51
Gold	1.39
Pink	4.50
Red	8.55
Gro-lux (or equal)	8.09
Gro-lux WS (or equal)	4.92

Light intensity plays a role in the morphogentic response and is easily adjusted within a given range by moving the light source or the cultures up or down relative to each other. It should be noted however that compared to sunlight, the amount of energy emitted per unit area from fluorescent lights is relatively low. This could have a direct effect on the tissues, as lower light intensities are associated with changes in chloroplast size, morphology and distribution, changes in leaf structure and changes in photosynthetic and respiratory metabolism [88,121]. Although these effects may not have a direct affect on the morphogentic response desired, they could present additional complications during the acclimation process when moving plants from the *in vitro* environment.

4. Photoperiod

Perhaps more than the other light parameters, the effect of the length of exposure to light over a diurnal cycle has a profound effect on plant morphogenesis. It has been known for over fifty years that day-length can be used in many plants to control flowering. It is known that even a short duration of light during the dark period, in the range of 660 nm corresponding to phytochrome *Pr*, can alter the effect of day-length. Other biological processes influenced by photoperiod, such as the evolution of a circadian rhythm or biological clock, are well known in both plants and animals [75]. In plants, circadian rhythms are known to influence leaf movement, CO_2 metabolism, stomatal movement and numerous biochemical processes including chloroplast enzymes (Rubisco and Cab) and enzymes involved in nitrogen metabolism. Although little is known about the mechanism involved in the control over these processes, these fluctuations may be required for daily chloroplast maintenance [39].

It has not been shown that conifers respond to day-length by flowering, yet daylength does play a major role in the initiation of bud set in the fall, as well as flushing of the buds in the spring. The effect of short-day treatment on bud set may also have a phytochrome responsive element, as short-day followed by a red light pulse did not induce bud set in *Pinus mariana* [40]. Many temperate conifers will display an indeterminent growth pattern as seedlings and will retain such a characteristic with long day-lengths. However, with a reduction in the number of hours of light in a day, bud set will occur. Once the terminal bud is set, the tree is determined in respect to growth and will have a flushing or episotic type growth from then on. Such growth is one factor which complicates the manipulation of mature conifers in culture. For cultures of seedling origin, it should be noted that the avoidance of bud set *in vitro* may be a factor when determining an optimal photoperiod. Further, this may have to be determined for each genotype or seed source. For example, bud set in *Picea glauca* seedlings from the northern extremes of its range will occur with a greater photoperiod than ecotypes from southern ranges which require a shorter day before bud set is initiated (J. Russell, personal communication).

5. *In vitro* response of conifers to light

The organogenic response in many conifers is light dependent [7, 71, 101, 131,148]; however, the dependence on the timing of the light is variable. Bud induction in *Pinus taeda* (loblolly pine) requires light for cell division during the first 28 days in culture. Light is also required for root formation and elongation in loblolly pine hypocotyl cuttings, yet light is not required for the pre-root cell division phase [101]. In *Pinus radiata* (radiata pine), a minimum of three days exposure to light is required for the formation of meristematic tissue, yet light can be delayed for up to ten days and the tissue will still remain competent to form buds. It is speculated that with radiata pine, cytokinin is required for bud initiation, while light is required for shoot formation and development [131,132]. In contrast, *Pinus ponderosa* (ponderosa pine) will form meristematic tissue and shoots even in 28 days of darkness [50], as will *Picea sitchensis* [148], and bud formation is similar in *Picea abies* whether or not the tissue is exposed to light during the first twenty days of culture [139].

Although it appears that light during the induction phase is not detrimental and in some instances is required; with other species, a brief dark period has been found to be advantageous. A 12 day dark pre-treatment was beneficial to bud production in *Picea sitchensis* [74] and *Pinus contorta* [148] and a one-week pre-treatment in the dark stimulated bud formation in *Pseudotsuga menziesii* (Douglas fir) [149].

The rational of a light requirement for bud formation however has not been adequately researched. Ponderosa pine cotyledons cultured in the dark showed a similar SDS-PAGE peptide pattern when compared to light grown cotyledons, with the exception of a reduced amount of a 52000 dalton peptide in the dark grown cotyledons. Although meristematic tissue forms in the dark with this species, it does indicate that no major protein changes occur between light and dark grown cotyledons. Further, the molecular weight of this peptide corresponds with that of the larger subunit of ribulose-1,5-bisphosphate carboxylase and is therefore probably correlated with a reduction in the number of chloroplasts [49]. In *Pinus banksiana* cells, there is a decrease in free amino acids when they are cultured in the dark relative to the light [21]. How this decrease in free amino acids might relate to bud formation is not known.

With embryogenic systems, light is thought to play an important role in the initiation of embryogenic callus [130,147]. However, there is no detailed work published on this and different workers have used a variety of light regimes for initiation. Furthermore, the methods used to measure light levels have been diverse and difficult to understand. Another inherent problem in most of this work revolves around the use of immature embryos which are removed from cones at diverse stages of maturation. Although the date of collection correlates strongly with the maturation stage of the embryo [14,147], other factors affecting the induction of embryogenic callus other than embryo maturation are largely unknown. Therefore, the role of light in the induction of embryogenic callus may be different for different maturation stages of zygotic embryos. Consequently, the role of light in the initiation of embryogenic callus remains an open question. In addition, most work on the maturation of conifer somatic embryos has been done in the light, but no definitive study has been done regarding the need for light, or the effects of light quality or photoperiod.

In the induction of embryogenic callus, 30% of *Picea abies* embryos formed embryogenic callus in the dark, while less than 1% formed embryogenic callus in the light [142]. Although Hakman and Fowke [65] initiated embryogenic callus from *Picea glauca* in the light, they also found that initiation of embryogenic callus from *Picea abies* and *Picea mariana* occurred in both the light and dark [64]. However, no information on the relative frequency of initiation was provided. Becwar *et al.* [13,15] initiated embryogenic callus in the light from *Picea abies*, and achieved a maximum frequency of 62% of the embryos forming embryogenic callus [14]. Embryogenic callus was also initiated in the light in *Picea glauca* and *Picea engelmanii* [147].

The maintenance and culture of the embryogenic callus is usually done in the dark. For example, after embryogenic callus initiation in *Picea glauca*, the callus is separated from the explant and cultured in the dark [147]. One exception is the work by Becwar *et al.* [13,15] where cultures were maintained in the light. In high light, red (anthocyanin) spots associated with the somatic embryos have been observed in *Picea abies* (Webb, unpublished) and *P. glauca* (Flinn, personal communication). Although the callus will maintain its embryogenic potential and morphology in the light, it is uncertain whether light has any affect on callus growth. Cultures are generally transferred to the light for later stages of somatic embryo development. One report of light effects on somatic embryo development of conifers indicates that pretreating the embryos in the light improved root emergence [114].

6. Light quality effects

Kadkade and Jopson [76] found that bud production in Douglas fir is stimulated by light in the 550–660 nm range and that light in the 660 nm range is optimal. Light in the 660 nm range stimulated bud production during the 3rd–5th week. Whether this effect was due to phytochrome was not investigated. In *Pinus sylvestris*, axillary bud formation was stimulated by far-red light (690–720 nm) [151]. Further, the interaction of intensity and quality may be important. Whereas blue light (420 and 467 nm) at irradiances from 100–710 $\mu W\ cm^{-2}$ had no effect on Douglas fir bud formation, red light (600 nm) at 420 $\mu W\ cm^{-2}$ increased bud formation. At irradiances higher than 420 $\mu W\ cm^{-2}$ red light promoted callus growth. Near UV (371 nm) inhibited callus growth [76]. These studies investigated the effect of light in narrow wavelengths and not over a multi-wavelength spectrum. It should also be noted that these studies are of particular interest since in other non-coniferous plants, like tobacco, shoot initiation is stimulated by blue light (450–500 nm) and root initiation is stimulated by red (620–700) [103].

Over a broader wavelength range, rooting of loblolly pine shoots was optimal when only incandescent lights were used [7]. During root growth, however, incandescent light should be supplemented with fluorescent light, as incandescent alone promoted shoot chlorosis [6]. In microspore callus cultures of *Picea abies*, root development was poor in the dark and red light was most effective at initiating roots [118].

7. Light intensity effects

Vegetative buds of Douglas fir vary in their optimum light requirement for growth and morphogenesis depending on the time of year they were excised. Optimum light intensities for buds collected in January, March and April were 36.4 W m^{-2}, 43.0 W m^{-2} and 24.5 W m^{-2} (169 μm m^{-2}, 200 μm m^{-2} s^{-2} and 114 μm m^{-2} s^{-1}) respectively, with an overall average light intensity optimum of 22 W m^{-2} (102 μm m^{-2} s^{-1}) [53]. This indicates that although higher light intensity optima were obtained for specific collection dates, these higher light intensities could be detrimental to buds collected throughout the year.

The percentage of embryonic explants forming buds was greater in *Pinus contorta* with an increased light intensity of 12.5 μm m^{-2} s^{-1} versus 4.5 μm m^{-2} s^{-1}, while there was no difference in the percent of explants forming buds in *Picea sitchensis* under these two light intensities [148]. A light intensity of 50 μm m^{-2} s^{-1} was too high for adventitious shoots of Douglas fir, as at these high light intensities chlorosis occurred [90].

In most cases with *in vitro* rooting, the base of the stem is exposed to light which is generally inhibitory to root elongation, therefore the use of *ex vitro* rooting in potting mix may be better than *in vitro* rooting [94]. One possible cause of this could be the elimination of light from the root initiation zone.

In general, the optimal amount of light required for rooting is less than that required for bud initiation or elongation [7,41,56,99]. *Picea abies* buds however need no light to root [138] while radiata pine buds root under an increased light intensity when compared with bud induction and elongation [4].

8. Photoperiod effects

There have been more reports on the effect of photoperiod, or the duration of light exposure per 24 hour period, on *in vitro* growth of conifer tissues than other aspects of light. In cell suspension cultures of *Picea glauca*, continuous illumination gave a more organized growth than a 14 hour photoperiod. In addition, an increase in tannins and aldehydes were noted when a photoperiod was used rather than continuous illumination [46]. Campbell and Durzan [30] also report a higher percentage of explants forming buds on *Picea glauca* hypocotyls under continuous light. In *Picea abies*, bud formation was greatest with continuous light and when the light period was interrupted, explants became discolored [134]. Further resultes indicated however that in the long term (3 months), a 16 hour photoperiod for *Picea abies* was superior to a longer (20 or 24 hour) photoperiod [137]. In radiata pine, it was found that a 16 hour photoperiod was necessary for optimum meristematic tissue formation and that interrupting this 16 hour cycle with a period of darkness decreased bud formation [131]. A stimulation in bud formation by a photoperiod has also been noted in *Picea abies* [136,140], *Pinus rigida* [109] and *Pinus brutia* [*1*]. In the latter case, bud production was decreased with continuous light due to the production of callus and anthocyanin.

Rooting of conifer tissue also seems to be enhanced by a reduced photoperiod. More *Pinus radiata* needle fasicles rooted with a 12 hour photoperiod than with either an 18 hour photoperiod or continuous light. Rooting was also enhanced by a three month short day (8 hours of light) pre-treatment [84]. *Picea abies* responds favorably to a month-long short day pre-treament prior to rooting in the dark [138]. In contrast, a higher percentage of *Picea glauca* buds rooted under constant light [29].

9. Temperature

A temperature in the range of 20–27 °C is usually adequate for most conifer cultures. Many researchers will vary the temperature with the photoperiod, a decrease in temperature corresponding to the dark cycle. Often a variation in the thermoperiod is a direct result of the photoperiod. Even with air conditioning or proper air circulation, there is generally a localized temperature build up near the lights. Therefore, the cultures are usually exposed to temperatures 2–4 °C higher than the ambient temperature measured by the sensors in the room. Consequently the temperature drops for the cultures during the dark cycle.

It is conceivable that a decrease in temperature is beneficial in that a lower temperature could decrease general cellular metabolism and thus reduce losses due to respiration, assuming that the photosynthetic apparatus is functioning in the light. Although this decrease in night temperature is what occurs *in vitro*, the need for the decrease has not been established. Further, the decrease is often by only a few degrees and is still well within an optimal temperature range. Unfortunately, arbitrary decisions are often all that we can use, as few studies are available even today to guide our decisions. It is important to bear in mind that temperature does play a very important role in developmental processes in conifers, such as controlling bud flushing in *Picea* [18].

Smith [120] reported no difference in bud initiation in *Pinus radiata* over a temperature range of 23–27 °C but that above 27 °C callus formed. In contrast, roots initiation was optimal below 23 °C however root formation was progressively slower as the temperature dropped from 23 °C. In *Picea abies* bud formation is optimal at 20 °C [134]. No bud development occurs below 15 °C and temperatures above 30 °C are lethal [135]. In both *Pinus contorta* [107] and *Pinus rigida* [109], the percentage of embryos forming buds was greater at 27 °C compared to 20 °C. Further, at 27 °C, buds formed from *Pinus contorta* embryos within 15 days while 25–30 days was required for bud formation at 20 °C. When using a BA-pulse treament in *Picea abies*, bud formation is not affected by the temperature during the pulse treatment [139], indicating perhaps that BA uptake is not a limiting factor in temperature dependency.

To further complicate the situation, with a 16 hour photoperiod, a decrease in temperature at night had no observable effect in *Picea abies*, while at other photoperiods, a temperature decrease was necessary. At lower night temperatures there was an increase in the browning of the embryos suggesting an increased production of phenolic compounds [137]. This study shows that under certain conditions, there may be an interaction of temperature and photoperiod.

With *Picea glauca*, rooting was greater with a 20/18 °C temperature regime compared with 24/18 °C,. 20/15 °C or 25/25 °C [116]. In Douglas fir, 19 °C gave increased rooting and normal plants, while at 24 °C few roots formed and at the root junction callus formed causing a discontinuity in the vascular system [35]. In contrast, with *Thuja occidentalis*, at 25 °C 60% of the shoots rooted while only 10% rooted at 20 °C [66]. It has been speculated that light and temperature affect rooting by controlling auxin activity [94].

10. Summary

The micropropagation of conifers requires culture conditions which are flexible in photoperiod, light intensity, light quality and temperature. The given parameters that are required for optimal response will depend not only on the species and explant, but also on the morphogenetic response desired. For example, a decrease in the photoperiod may aid elongation and rooting while a decrease in light intensity and temperature may be beneficial to rooting alone. It is not unreasonable to expect that for optimal culture response, one set of condtions (light and temperature) may be needed for each phase (bud initiation, bud elongation, root initiation, root elongation) of micropropagation. However, for most purposes, a general multi-spectrum light source such as cool-white fluorescent, at 10–20 W m^{-2}, with a 16–24 hour photoperiod and a constant temperature of 25 °C is adequate for most general conifer micropropagation. Optimization of a given system could then proceed from this starting point.

References

1. Abdullah A, Yeoman MM, and Grace J (1985) *In vitro* adventitious shoot formation from embryonic and cotyledonary tissues of *Pinus brutia* Ten. Plant Cell, Tissue Organ Culture 5:35–44.
2. Abdullah A and Grace J (1986) Rapid micropropagation of Calabrian pine from primary and secondary buds on shoot explants. Can J For Res 16:637–641.
3. Aitken J, Horgan KJ, and Thorpe TA (1981) Influence of explant selection on the shoot-forming capacity of juvenile tissue of *Pinus radiata.* Can J For Res 11:112–117.
4. Aitken-Christie, J, Jones C, and Bond S (1985) Wet and waxy shoots in radiata pine micropropagation. Acta Hort 116:93–100.
5. Amerson HV and Mott RL (1982) Improved rooting of western white pine shoots from tissue cultures. Forest Sci 28(4):822–825.
6. Amerson HV (1983) Loblolly pine *in vitro* propagation from juvenile materials: methods and regulatory features. Proc ICTCWG, Tacoma, Washington.
7. Amerson HV, Frampton LJ, McKeand SE, Mott RL, and Weir, RJ (1984) Loblolly pine tissue culture: laboratory greenhouse, and field studies. In: Henke RR, Hughes KW, and Constantin MJ (Eds.) Tissue Culture in Forestry and Agriculture. Plenum Press, New York, pp. 271–288.
8. Ball EA, Morris DM, and Rydelius, JA (1978) Cloning of *Sequoia sempervirens* from mature trees through tissue culture. Proc Multi *In vitro* d'Especes Ligneuses Gembloux, Belgium, pp. 181–226.
9. Barnett JR (1978) Fine structure of parenchymatous and differentiated *Pinus radiata* callus. Ann Bot 42:367–373.
10. Baur PS and Walkinshaw CH (1974) Fine structure of tannin accumulations in callus cultures of *Pinus elliotti* (slash pine). Can J Bot 52:615–619.
11. Baxter R, Brown SN, England NF, Ludlow CHM, Taylor SL, and Womack RW 1989) Production of clonal plantlets of tropical pine in tissue culture via axillary shoot activation. Can J For Res 19:1338–1342.
12. Becwar MR, Noland TL, and Wann SR (1987) Somatic embryo development and plant regeneration from embryogenic Norway spruce callus. Tappi, April:155–160.
13. Becwar MR, Verhagen SA, and Wann SR (1987) The frequency of plant regeneration from Norway spruce somatic embryos. In: Proc 19th Southern Forest Tree Improvement Conf, College Station, Texas, pp. 92–100.
14. Becwar MR, Wann SR, Johnson MA, Verhagen SA, Feirer RP, and Nagmani R (1988) Development and characterization of *in vitro* embryogenic systems in conifers. In: Ahuja RM (Ed.) Proc Int Union of Foresty Research Organization Workshop. Somatic Cell Genetics of Woody Plants. Martinus Nijhoff Publ, Dordrecht.
15. Becwar MR, Noland TL, and Wycokoff JL (1989) Maturation, germination, and conversion of Norway spruce (*Picea abies* L.) somatic embryos to plants. In Vitro Cell & Devel Biol 25(6):575–580.
16. Berlyn GP, Anoruo AO, Beck RC, and Cheng J (1987) DNA content polymorphism and tissue culture regeneration in carribean pine. Can J Bot 65:954–961.
17. Bigot C and Engelman F (1987) Vegetative propagation *in vitro* of *Cunninghamia lanceolata* (Lamb) Hook. In: Bonga JM and Durzan DJ (eds.) Cell and Tissue Culture in Forestry, Martinus Nijhoff Publishers, Dordrecht, pp. 114–127.
18. Blum BM (1988) Variation in the phenology of bud flushing in white and red spruce. Can J For Res 18:315–319.
19. Bonga JM (1977) Organogenesis in *in vitro* cultures of embryonic shoots of *Abies balsamea.* In Vitro 13(1):41–48.
20. Bonga JM (1981) Organogenesis *in vitro* of tissues from mature conifers. In Vitro 17(6): 511–518.

21. Bonga JM (1982a) Tissue culture techniques. In: Bonga J and Durzan D (Eds.) Tissue Culture in Forestry. Martinus Nijhoff Publishers, The Hague, pp. 4–35.
22. Bonga JM (1982b). Shoot formation in callus from the stalks of young female strobili of *Larix decidua*. Can J Bot 60:1357–1359.
23. Bonga JM (1984a) Adventitious shoot formation in cultures of immature female strobili of *Larix decidua*. Physiol Plant 62:416–421.
24. Bonga JM (1984b) Adventitious shoot and root formation in tissue cultures of mature *Larix decidua*. Proc Inter Sym Recent Adv for Biotech, Michigan, pp. 64–68.
25. Bornman CH (1983) Possibilities and constraints in the regeneration of trees from cotyledonary needles of *Picea abies in vitro*. Physiol Plant 57:5–16.
26. Bose A and Clay LKC (1981) Vegetative propagation of redwood. In: Colloque Inter Sur la Culture *in vitro* des Essences Forestieres, France, pp. 197–200.
27. Boulay MP, Gupta PK, Krogstrup, P, and Durzan DJ (1988) Development of somatic embryos from cell suspension cultures of Norway spruce (*Picea abies* Karts.). Plant Cell Reports 7:134–137.
28. Brown CL and Sommer HE (1977) Bud and root differentiation in conifer cultures. Tappi 60(6):72–73.
29. Campbell R and Durzan D (1975) Induction of multiple buds and needles in tissue culture of *Picea glauca*. Can J Bot 53(16):1652–1657.
30. Campbell RA and Durzan DJ (1976) Vegetative propagation of *Picea glauca* by tissue culture. Can J For Res 6(2):240–243.
31. Cheng T-Y (1975) Adventitious bud formation in culture of Douglas fir (*Pseudotsuga menziesii* (Mirb) Franco). Plant Sci Lett 5:97–102.
32. Cheng T-Y (1976) Vegetative propagation of western hemlock (*Tsuga heterophylla*) through tissue culture. Plant Cell Physiol 17:1347–1350.
33. Cheng T-Y (1977a) Factors effecting adventitious bud formation of cotyledon culture of Douglas fir. Plant Sci Lett 9:179–187.
34. Cheng T-Y (1977b) Regenration of Douglas fir plantlets through tissue culture. Science 198:306–307.
35. Cheng T-Y (1978) Clonal propagation of woody plant species through tissue culture techniques. Proc Intern Plant Propagators' Soc 28:139–155.
36. Chretien L and Vieth J (1984) Neoformation de bourgeons adventifs sur des portions de bourgeons vegetatifs de *Picea pungesn* cv. Foliis aureis cultivees *in vitro*. Can J Bot 62:185–187.
37. Coleman WK and Thorpe TA (1977) *In vitro* culture of western redcedar (*Thuja plicata* Donn). I. Plantlet formation. Bot Gaz 138(3):298–304.
38. Coleman WK and Thorpe TA (1978) *In vitro* culture of western redcedar (*Thuja plicata*). II. Induction of male strobili from vegetative shoot tips. Can J Bot 56:557–564.
39. Cuozzo M, Kay SA, and Chua N-H (1988) Regulatory circuits of light-response genes. In: Verma DPS and Goldberg RB (Eds.) Temporal and Spatial Regulation of Plant Genes. Springer-Verlag, New York, pp. 131–153.
40. D'Aoust AL and Hubac C (1986) Phytochrome action and frost hardening in black spruce seedlings. Physiol Plant 67:141–144.
41. David H, Isemukali K, and David A (1978) Obtention de plants de Pin maritime (*Pinus pinaster* Sol.) a partir de brachyblastes ou d'apex caulinaires de tres jeunes sujets cultives *in vitro*. CR Acad Sc Paris 287:245–248.
42. David H, Jarlet E, and David A (1984) Effects of nitrogen source, calcium concentration and osmotic stress on protoplasts and protoplast-derived cell cultures of *Pinus pinaster* cotyledons. Physiol Plant 61:477–482.
43. Diner AM (1990) Clonal propagation of mature *Larix*. New Forests 4:63–66.
44. Douglas TJ, Villalobos VM, Thompson MR, and Thorpe TA (1982) Lipid and pigment changes during shoot initiation in cultured explants of *Pinus radiata*. Physiol Plant 55:470–477.

45. Dunstan DI, Bekkaous F, Pilon M, Fowke LC, and Abrams SR (1988) Effects of abscisic acid and analogous on the maturation of white spruce (*Picea glauca*) somatic embryos. Plant Sci 58:77–84.
46. Durzan D and Lopushanski M (1973) Effects of environmental changes on sugars, tannins and organized growth in cell suspension cultures of white spruce. Planta (Berl) 113:241–249.
47. Durzan DJ and Gupta PK (1987) Somatic embryogenesis and polyembryogenesis in Douglas-fir cell suspension cultures. Plant Sci 52:229–235.
48. Ellis DD and Bilderback DE (1984) Multiple bud formation by cultured embryos of *Pinus ponderosa*. J Plant Physiol 115:201–204.
49. Ellis DD and Judd RC (1987) SDS-PAGE analysis of bud-forming cotyledons of *Pinus ponderosa*. Plant Cell, Tissue Organ Culture 11:57–65.
50. Ellis DD and Bilderback (1991) Ponderosa pine (*Pinus ponderosa* Laws.). In: Bajaj YPS (Ed.) Biotechnology in Forestry, Trees III, Springer-Verlag, Berlin. 339–357.
51. Ellis DD, McCown B, Skilling D, Barker M, Serres R, and Ostry M (1991) Establishment of a shoot culture system for *Larix decidua*. In: Ahuja MR (Ed.) Woody Plant Biotechnology, Plenum Publishing, New York, 335–338.
52. Ellis DD, Barczynska H, McCown BH, and Nelson N (1991) A comparison of BA, zeatin and thidiazuron for adventitious bud formation from *Picea glauca* embryos and epicotyl explants. Plant Cell Tissue Organ Culture, 281–287.
53. Evers PW (1985) Growth and morphogenesis of shoot initials of Douglas fir, *Pseudotsuga menziesii* (Mirb.) Franco, *in vitro*. Neth J Agri Sci 33:179–181.
54. Flinn BS, Webb DT, and Georgis W (1986) *In vitro* control of caulogenesis by growth regulators and media components in embryogenic explants of eastern white pine (*Pinus strobus*). Can J Bot 64:1948–1956.
55. Flinn BS, Webb DT, and Newcomb W (1988) The role of cell clusters and promeristemoids in determination and competence for caulogenesis of *Pinus strobus* cotyledons *in vitro* Can J Bot 66:1556–1565.
56. Franclet A, David A, David H, and Boulay M (1980) Premiere mise en evidence morphologique d' un rejeunissement de meristemes primaires caulinaires de Pin maritime age (*Pinus pinaster* Sol.). CR Acad Sc Paris 290:927–930.
57. Franco EO and Schwarz OJ (1984) Micropropagation of two tropical conifers: *Pinus oocarpa* Schiede and *Cupressus lusitanica* Miller. In: Henke RR, Hughes KW, and Hollaender A. (Eds.) Tissue Culture in Forestry and Agriculture. Plenum Press, New York, pp. 195–214.
58. George EF and Sherrington PD (1984) Plant propagation by tissue culture. Exergenetics Ltd., Eversley, England. pp. 709.
59. Gladfelter HJ and Phillips GC (1987) *De novo* shoot organogenesis of *Pinus eldarica* Medw. *in vitro* I. Reproducible regeneration from long-term callus cultures. Plant Cell Rep 6:163–166.
60. Gupta PK and Durzan DJ (1985) Shoot multiplication from mature trees of Douglas-fir (*Pseudotsuga menziesii*) and sugar pine (*Pinus lambertiana*). Plant Cell Rep 4:177–179.
61. Gupta PK and Durzan DJ (1987) Micropropagation and phase specifity in mature, elite Douglas fir. J Amer Soc Hort Sci 112(6):969–971.
62. Hakman L, Fowke LC, von Arnold S, and Eriksson T (1985) The development of somatic embryos in tissue cultures initiated from immature embryos of *Picea abies* (Norway spruce). Plant Sci 38:53–59.
63. Hakman I and von Arnold S (1985) Plantlet regeneration through somatic embryogenesis in *Picea abies* (Norway spruce). J Plant Physiol 121:149–158.
64. Hakman I and Fowke LC (1987a) Somatic embryogenesis in *Picea glauca* (white spruce) and *Picea mariana* (black spruce). Can J Bot 65:656–659.
65. Hakman I and Fowke LC (1987b) An embryogenic cell suspension culture of *Picea glauca* (White spruce). Plant Cell Rep 6:20–22.
66. Harry IS, Thompson MR, Lu C-Y, and Thorpe TA (1987) *In vitro* plantlet formation from embryonic explants of eastern white cedar (*Thuja occidentalis* L.) Tree Physiol 3:273–283.

67. Harry IS, Thompson MR, and Thorpe TA (1991) Regeneration of plantlets from mature embryos of Western larch. In Vitro Cell Dev Biol 27P:89–98.
68. Hart JW (1987) Light and plant growth. Allen & Unwin Publ., London. pp. 204.
69. Horgan K and Aitken J (1981) Reliable plantlet formation from embryos and seedling shoot tips of radiata pine. Physiol Plant 53:170–175.
70. Ishii K (1986) *In vitro* plantlet formation from adventitious buds on juvenile seedlings of Hinoki cypress (*Chamaecyparis obtusa*). Plant Cell, Tissue Organ Culture 7:247–255.
71. Isikawa H (1974) *In vitro* formation of adventitious buds and roots on the hypocotyl of *Cryptomeria japonica*. Bot Mag 87:73–77.
72. Jelaska S, Kolevska-Pletikapic B, and Vidakovic M (1981) Bud regeneration in *Pinus nigra* embryo and seedling tissue culture. In: Colloque Inter. Sur La Culture *in vitro* des Essences Forestieres, France, pp. 159–166.
73. John A and Murray BW (1981) Micropropagation of sitka spruce (*Picea sitchensis* (Bong.) Carr.). In: Colloque Inter. Sur. La Culture *in vitro* des Essences Forestieres, France, pp. 65–70.
74. John A and Webb KJ (1987) Sitka spruce (*Picea sitchensis* (Bong.) Carr.). In: Bonga JM and Durzan DJ (Eds.) Cell and Tissue Culture in Forestry, Martinus Nijhoff Publishers, Dordrecht, pp. 30–41.
75. Johnson CH and Hastings JW (1986) The elusive mechanism of the circadian clock. American Scientist 74:29–36.
76. Kadkade PG and Jopson H (1978) Influence of light on organogenesis from the embryo-derived callus of Douglas fir (*Pseudotsuga menziesii*). Plant Sci Lett 13:67–73.
77. Karnosky DF, Diner AM, and Mickler RA (1984) Micropropagation of juvenile jack pine and European larch. In: Proc Int Symp Recent Adv in Forest Biotech, Traverse City, Michigan, pp. 69–74.
78. Kaul K (1987) Plant regeneration from cotyledon-hypocotyl explants of *Pinus strobus* L. Plant Cell Rep 6:5–7.
79. Keathley DE (1984) Micropropagation of mature spruce. Proc Int Symp Recent Advv in Forest Biotech, Traverse City, Michigan, pp. 58–63.
80. Klein RM (1973) Determining radiant energy in different wavelengths present in white light. HortScience 8:210–211.
81. Kirby EG and Schalk ME (1982) Surface structural analysis of cultured cotyledons of Douglas-fir. Can J Bot 60(12):2729–2733.
82. Kolevska-Pletikapic B, Jelaska S, Berljak J, and Vidakovic M (1983) Bud and shoot formation in juvenile tissue culture of *Pinus nigra*. Silvae Genetica 32:115–119.
83. Kolevska-Pletikapic B and Krsnik-Raol B (1984) Analysis of adventitious bud formation in *Pinus nigra* Arn. embryo culture. Acta Bot 43:49–57.
84. Kummerow J (1966) Vegetative propagation of *Pinus radiata* by means of needle fascicles. For Sci 12:391–398.
85. Kurtz ML, Webb DT, and Vidaver WE (1989) Micropropragation of yellow cedar (*Chamaecyparis nootkatensis*). Plant Cell Tissue Organ Culture 18:297–312.
86. Laine E, David H, and David A (1988) Callus formation from cotyledon protoplasts of *Pinus oocarpa* and *Pinus patula*. Physiol Plant 72:374–378.
87. Laliberte S and Lalonde M (1988) Sustained caulogenesis in callus cultures of *Larix* × *eurolepis* initiated from short shoot buds of a 12-year-old tree. Amer J Bot 75(6):767–777.
88. Larcher W (1980) Physiological Plant Ecology. 2nd Ed. Springer-Verlag, Berlin 303 pp.
89. Lu C-Y and Thorpe TA (1987) Somatic embryogenesis and plantlet regeneration in cultured immature embryos of *Picea glauca*. J Plant Physiol 128:297–302.
90. Mapes MO, Young PM, and Zaerr JB (1981) Multiplication *in vitro* du Douglas (*Pseudotsuga menziesii*) par induction precoce d'un bourgeonnement adventif et axillaire. In: Colloque Inter. Sur la Culture in vitro des Essences Forestieres, France, pp. 109–114.
91. Mehra-Palta A, Smeltzer RH, and Mott RL (1978) Hormonal control of induced organogenesis. Experiments with excised plant parts of loblolly pine. Tappi 61(1):37–40.

92. Minocha SC (1980) Callus and adventitious shoot formation in excised embryos of white pine (*Pinus strobus*). Can J Bot 58:366–370.
93. Misson JP, Coumans M, Giot-Wirgot P, and Gaspar T (1982) Induction of adventitious buds on buds of *Picea pungens* grown *in vitro*. Z Pflanzenphysiol Bd 107:161–167.
94. Mohammed GH and Vidaver WE (1988) Root production and plantlet development in tissue-cultured conifers. Plant Cell, Tissue Organ Culture 14:137–160.
95. Mohammed GH and Patel KR (1989) Tissue culture propagation of Douglas-fir. New Forests 3:125–139.
96. Momot TS and Smirnov AM (1979) Organogenesis from various organs of the Siberian and Dahur larch (*Larix sibirica* Maxim. and *L. dahurica* Turcz.) and yellow pine (*Pinus ponderosa* Dougl.) cultured *in vitro*. Sov Plant Physiol 78:758–761.
97. Momot TS (1981) Differentiation of pinceana pinyon pine (*Pinus pithyusa* Stev.) plants in culture *in vitro*. Tsitol i Gen 15(6):54–57.
98. Monteuuis O (1987) *In vitro* meristem culture of juvenile and mature *Sequoiadendron giganteum*. Tree Physiol 3:265–272.
99. Mott RL and Amerson HV (1981a) Tissue culture plantlets produced from *Pinus monticola* embryonic materials. For Sci 27(2):299–304.
100. Mott RL and Amerson HV (1981b) A tissue culture process for the clonal production of loblolly pine plantlets. NC Ag Res Serv Tech Bull 271:1–14.
101. Mott RL and Amerson RH (1984) Role of tissue culture in loblolly pine improvement. Proc Int Sym Adv Forest Biotech, Traverse City, Michigan, pp. 24–36.
102. Mudge KW (1986) Micropropagation of mugo pine from embryonic and seedling explants. HortScience 21(2):298–299.
103. Murashige T (1974) Plant propagation through tissue cultures. Ann Rev Plant Physiol 25: 135–166.
104. Nagmani R and Bonga J (1986) Embryogenesis in subcultered callus of *Larix decidua*. Can J For Res 1088–1091.
105. Nagmani R, Becwar MR, and Wann SR (1987) Single-cell origin and development and somatic embryos in *Picea abies* (L.) Karst. (Norway spruce) and *P. glauca* (Moench) Voss (white spruce). Plant Cell Rep 6:1257–159.
106. Patel KR and Berlyn GP (1982) Genetic instability of multiple buds of *Pinus coulteri* regenerated from tissue culture. Can J For Res 12:93–101.
107. Patel KR and Thorpe TA (1984) *In vitro* differentiation of plantlets from embryonic explants of lodgepole pine (*Pinus contorta* Dougl. ex Loud.). Plant Cell, Tissue Organ Culture 3:131–142.
108. Patel KR and Thorpe TA (1986) *In vitro* regeneration of plantlets from embryonic and seedling explants of Engelmann spruce (*Picea engelmannii* Parry). Tree Physiol. 1:289–301.
109. Patel KR, Kim H-R, and Thorpe TA (1986) Plantlet formation in pitch pine (*Pinus rigida* Mill.) by tissue culture methods. For Ecol Manage 15:147–160.
110. Perez-Bermudez P and Sommer HE (1987) Factors affecting adventitious bud induction in *Pinus elliottii* (Engelm.) embryos cultured *in vitro*. Plant Cell, Tissue Organ Culture 11:25–335.
111. Rancillac M (1981) Perspective d' application des cultures d'organes *in vitro* a la multiplication vegetative du Pin maritime, *Pinus pinaster* Sol. Ann Sci Forest 38(1):55–70.
112. Rancillac M, Faye M, and David A (1982) *In vitro* rooting of cloned shoots in *Pinus pinaster*. Physiol Plant 56:97–101.
113. Reilly K and Washer J (1977) Vegetative propagation of radiata pine by tissue culture: plantlet formation from embryonic tissue. NZ J For Sci 7(2):199–206.
114. Roberts DR, Sutton BCS, and Flinn BS (1990) Synchronous and high frequency germination of interior spruce somatic embryos following partial drying at high relative humidity. Can J Bot 68:1086–1090.
115. Romberger JA, Varnell RJ, and Tabor CA (1970) Culture of apical meristems and embryonic shoots of *Picea abies* – approach and techniques. USDA FS Tech Bull 1409:1–30.

116. Rumary C and Thorpe TA (1984) Plantlet formation in black and white spruce. I. *In vitro* techniques. Can J For Res 14:10–16.
117. Selby C and Harvey BMR (1985) The influence of natural and *in vitro* bud flushing on adventitious bud production in sitka spruce (*Picea sitchensis* (Bong.) Carr.) bud and needles cultures. New Phytol 100:549–562.
118. Simola LK and Huhtinen O (1986) Growth, differentiation, and ultrastructure of microspore callus of *Picea abies* as affected by nitrogenous supplements and light. NZ J For Sci 16:357–368.
119. Smith DR and Thorpe TA (1975) Root Initiation of cuttings of *Pinus radiata* seedlings. J Exp Bot 26(91):184–192.
120. Smith DR (1986) Radiata pine (*Pinus radiata* D. Don). In: YPS Bajaj (Ed.) Biotechnology in Agriculture and Forestry. Vol. 1: Trees I, Springer-Verlag, Berlin, pp. 274–291.
121. Smith H (1982) Light quality, photoperception, and plant strategy. Ann Rev Plant Physiol 33:481–518.
122. Smith H and Whitelam GC (1990) Phytochrome, a family of photoreceptors with multiple roles. Plant Cell and Environ 13:695–707.
123. Sommer H, Brown CL, and Kormanik PP (1975) Differentiation of plantlets in longleaf pine (Pinus palustris Mill.) tissue cultured *in vitro*. Bot Gaz 136(2):196–200.
124. Stiff CM, Wenny DL, Roberts LW, Olson JR, and LeTourneau DJ (1988) Micropropagation of western white pine using needle fascicles. Proc Tenth North American Forest Biology Workshop, BC, Canada. pp. 133–139.
125. Teasdalee RD and Rugini E (1983) Preparation of viable protoplasts from suspension-cultured loblolly pine (*Pinus taeda*) cells and subsequent regeneration to callus. Plant Cell, Tissue Organ Culture 2:253–261.
126. Thomas M, Duhoux E, and Vazart J (1977) *In vitro* organ initiation in tissue culture of *Biota* (*Thuja*) *orientalis*. Plant Sci Letters 8:395–400.
127. Thompson DG and Zaerr JB (1981) Induction of adventitious buds on cultured shoot tips of Douglas-fir (*Pseudotsuga menziesii* (Mirb.) Franco). In: Colloque Inter. Sur la Culture *in vitro* des Essences Forestieres, France, pp. 167–174.
128. Torbio M and Pardos JA (1989) Scots pine (*Pinus sylvestris* L.) In: Bajaj YPS (Ed.) Biotechnology in Agriculture and Forestry Trees II. pp. 479–506.
129. Tranvan H and Thomas M (1981) Influence hormonale sur l'organogenese *in vitro* des cotyledons dur *Pinus sylvestris*. In: Colloque Int Sur la Culture *in vitro* des Essences Forestieres, France, pp. 153–154.
130. Verhagen SA and Wann SR (1989) Norway spruce somatic embryogenesis: high-frequency initiation from light cultured mature embryos. Plant Cell Tissue Organ Cult 16:103–111.
131. Villalobos VM, Leung DWM, and Thorpe TA (1984a) Light-cytokinin interaction in shoot formation in cultured cotyledon explants of radiata pine. Physiol Plant 61:497–504.
132. Villalobos VM, Oliver MJ, Yeung EC, and Thorpe TA (1984b) Cytokinin-induced switch in development in excised cotyledons of radiata pine cultured *in vitro*. Physiol Plant 61:483–489.
133. Villalobos VM, Yeung EC and Thorpe TA (1985) Origin of adventitious shoots in excised radiata pine cotyledons cultured *in vitro*. Can J Bot 63:2172–2176.
134. von Arnold S and Eriksson T (19788) Induction of adventitious buds on embryos of Norway spruce *in vitro*. Physiol Plant 44:283–287.
135. von Arnold S and Eriksson T (1979) Induction of adventitious buds on buds of Norway spruce (*Picea abies*) gronw *in vitro*. Physiol Plant 45:29–34.
136. von Arnold S and Eriksson T (1981) *In vitro* studies of adventitious shoot formation in *Pinus contorta*. Can J Bot 59:870–874.
137. von Arnold S (1982) Factors influencing formation, development and rooting of adventitious shoots from embryos of *Picea abies* (L.) Karst. Plant Sci Lett 27:275–287.
138. von Arnold S (1984) Importance of genotype on the potential for *in vitro* adventitious bud production of *Picea abies*. Forest Sci 30(2):314–318.

139. von Arnold S and Eriksson T (1985) Initial stages in the course of adventitious bud formation on embryos of *Picea abies.* Physiol Plant 64:41–47.
140. von Arnold S and Gronroos R (1986) Anatomical changes and peroxidase activity after cytokinin treatments inducing adventitious bud formation on embryos of *Picea abies.* Bot Gaz 147:425–431.
141. von Arnold S and Eriksson T (1986) Norway spruce (*Picea abies* L.) In: Bajaj YPS (Ed.) Biotechnology in Agriculture and Forestry, Vol. 1: Trees I, Springer-Verlag Berlin, pp. 291–310.
142. von Arnold S (1987) Improved efficiency of somatic embryogenesis in mature embryos of *Picea abies* (L.) Karst. J Plant Physiol 128:233–244.
143. Wann Sr, Johnson MA, Noland TL, and Carlson JA (1987) Biochemical difference between embryogenic and non-embryogenic callus of *Picea abies* (L.) Karts. Plant Cell Rep 6:39–42.
144. Washer J, Reilly KJ, and Barnett JR (1977) Differentiation in *Pinus radiata* callus culture: The effect of nutrients. NZJ For Sci 7(3):321–328.
145. Webb DT and Santiago OD (1983) Cytokinin induced bud formation on carribean pine (*Pinus caribaea* Morlet) embryos *in vitro.* Plant Sci Lett 32:17–21.
146. Webb DT, Flinn BS, and Georgis W (1988) Micropropagation of eastern white pine (*Pinus strobus* L.) Can J Bot 18:1570–1580.
147. Webb DT, Webster F, Flinn BS, Roberts DR, and Ellis DD (1989) Factors influencing the induction of embryogenic and caulogenic callus from embryos of *Picea glauca* and *P. engelmanii.* Can J For Res 19:1303–1308.
148. Webb KJ and Street HE (1977) Morphogenesis *in vitro* of *Pinus* and *Picea.* Acta Hort 78:259–269.
149. Winton LL and Verhagen SA (1977) Shoots from Douglas-fir cultures. Can J Bot 55: 1246–1250.
150. Yeung EC, Aitken J, Biondi S, and Thorpe TA (1981) Shoot histogenesis in cotyledon explants of radiata pine. Bot Gaz 142(4):494–501.
151. Zel J, Gogala N, and Camloh M (1988) Micropropagation of *Pinus sylvestris.* Plant Cell, Tissue Organ Culture 14:169–175.

4. Haploid embryogenesis in trees

P. VON ADERKAS and M. D. DAWKINS
Department of Biology, University of Victoria, Victoria, British Columbia, V8W 242, Canada

1. Introduction

Plant breeders have long sought haploid and double-haploid plants [22]. Haploids have been produced in numerous genera, most of them agricultural crops. Because haploid plants do not exhibit phenomena such as dominance and segregation they reveal cryptic and recessive genetic variation. Such plants are useful for producing homozygous diploid (dihaploid) lines for controlled hybridization. Traditionally, homozygosity for certain traits was produced by continuous inbreeding, but for trees this is somewhat problematic. This review concentrates on tree species of conifers (including Taxaceae) as well as angiosperms. Breeding of tree species requires longer periods of time and thus inbreeding is impractical, and in certain cases impossible. For example, inbreeding in dioecious trees is restricted to sib-mating, never selfing. Although haploids could be used to speed up bredeing programs, to study inheritance of qualitative traits and to establish linkage maps, there are relatively few tree species which have been amenable to induction of haploids.

There are a number of ways in which haploid plants can be obtained (see review, Ref. 6). They may occur naturally, they may be parthenogenetic in origin, or the result of embryo rescue from interspecific crosses. They may also arise by chromosome elimination [33]. The most frequent route and the focus of this review, is via androgenesis or gynogenesis. Gynogenesis is plantlet regeneration from cultured female reproductive tissues, and androgenesis is plantlet regeneration from microspores *in vitro*. Since there are a number of recent reviews on haploids of tree species [6,8,28,32], we have decided to concentrate on the embryogenic aspect of androgenesis and gynogenesis.

2. Gynogenesis

2.1. *Angiosperms*

Culturing female tissue of angiosperms is difficult because of the inaccesibility of reproductive tissue, and the complex nature of the embryo sac. The reasons for doing so are predominantly connected with dioecious or functionally dioecious breeding stock in which a desirable characteristic could be captured

M. R. Ahuja (Ed.), Micropropagation of Woody Plants, 57–65.

in haploid plants. For example, several cold-tolerant forms of *Hevea brasiliensis* have stamens which degenerate early in development thus eliminating the possibility of anther culture. Thus, unpollinated ovules were cultured which either produced embryoids directly, or callus, which in turn produced embryoids and finally plantlets [10]. In dioecious species such as poplar [10] and monoecious species such as *Robinia pseudoacacia* [46] unpollinated ovule culture was also used to produce embryoids and plantlets. This is a labour intensive technique which requires dissection of as much somatic tissue as possible prior to culture. Explants are subjected to a series of MS-type media in the following sequence: dedifferentiation, differentiation and plantlet-forming stage [10]. Since whole ovules are cultured, ploidy levels of the plantlets which are derived from the embryogenic process must be established. Generally, this is a low yielding method used when androgenesis is not otherwise possible.

2.2. *Conifers*

The establishment of gynogenic cultures of conifers is also labour-intensive. However, in conifers the female tissue is easily dissected from surrounding somatic tissue. In some genera megagametophytes only develop if pollination occurs. In *Picea*, megagametophytes fail to develop in unpollinated cones resulting in flat seed. However, in other genera, such as *Larix* and *Pseudotsuga*, failure to pollinate does not affect megagametophyte development until after the time when fertilization would occur. Thus unpollinated cones are an excellent source of haploid explants. Embryos found in pollinated megagametophytes are easily removed by dissection or isolated by further partitioning the explant. It is nevertheless necessary to do chromosome counts, since there are reports in the literature that tissues which may develop from megagametophytes of *Taxus* [48], *Picea* [34], *Pinus* [7] and *Sequoia* [3] have been diploid, although Hakman and colleagues [15] found that nuclear content of cytofluorometrically-analyzed protoplasts of megagametophytes of *Picea* were consistently haploid.

Only one genus, *Larix*, has successfully been induced to form plantlets via embryogenesis from cultured megagametophytes [24]. This has been achieved from both pollinated [24,43] and unpollinated megagametophytes [43]. Plantlets have been produced from organogenic cultures of *Picea abies* callus [35]. Incomplete embryogeny have been achieved in this same species by Huhtinen [16] and Simola and Santanen [36].

The initial response of isolated megagametophyte explants is elongation of cells near the corrosion cavity or near a cut surface. Both genotype and environment influence induction from cultured megagametophytes of *Pseudotsuga menziesii* [13], *Larix decidua* [40] and *Picea abies* [17]. Although the tissue may continue to produce long vacuolate cells indefinitely, most tissue dies within a few months of induction, in spite of transfer to hormone-free media, or other types of media [41].

The next stage is the development of a population of short cells either through polar division of the long vacuolate cells or cellularization of large coenocytes [41]. These short cells are capable of producing long and short cells in turn. These types of culture have been reported from *Picea* [16] and *Larix* [41]. This repetitive growth may continue for years and is no guarantee of further embryogeny as many cultures may remain in this state. Attempts to alter development by altering concentrations of media components and growth regulating substances have failed. It is suspected, though not proven, that this is genetic (von Aderkas and Bonga, unpubl.).

In embryogenic cultures, the next stage of embryogenesis involves the development of an axial structure with a domed embryonal mass at one end and continuous suspensor formation at the other. This recognizable polar structure is comparable to an embryonal mass and suspensor seen *in vivo* [44]. The gynogenic system in larch is highly regenerable, and suspensor cells of the embryonal mass will regenerate the long and short cell populations necessary to renew the cycle [29]. Recently, plantlets have been regenerated from protoplasts of haploid cultures.

Media composition can very strongly influence this development. *Larix* haploid cultures are raised with high concentrations of organic supplements such as casein hydrolysate and glutamine [42]. There is a general problem of maturing somatic embryogenic lines which have arrested in a stage prior to late embryogeny. ABA has been successfully used on somatic embryos of *Picea* [45], but this only appears to work in certain lines of haploid *Larix* (von Aderkas and Grimm, unpublished). The number of *Larix* lines which go on to form green embryos is a very low percentage of the total. Presently, there are two-year old seedlings from only one line of *Larix decidua* (Bonga and von Aderkas unpublished) originally cultured by von Aderkas and colleagues [43].

Gynogenesis in conifers is similar in many respects to somatic embryogenesis. These two phenomena have been compared in *Larix* species in which both haploid and diploid lines have been developed [43]. Stages are only similar once an embryonal mass has formed. Somatic embryogenesis generally appears to be a form of polyembryogenesis, or repeated cleavage of an embryonal mass [12] and consequently there are no earlier developmental stages of embryogeny at all. Embryogenic paths unique to *in vitro* diploid embryogenic tissue occur following protoplasting [1,14,20]. Coenocytes play a part in some of these systems, as do polar divisions. Klimaszewska [20] did not find that coenocytic protoplast of *Larix* showed any further embryogeny at all, but were developmentally terminal. It should be stated that no one has yet found regeneration which recapitulates megagametophyte development.

The competence of megagametophyte tissue to be induced is dependent on genotype, seasonal environmental factors affecting tree physiology and megagametophyte development, as well as environmental conditions *in vitro*.

Table 1. Reports of haploid embryoids in trees including angiosperms and conifers where the haploid source tissue is indicated.

Species	Haploid tissue	Ref.
Aesculus	pollen	26
Caesalpinia (Poincinia)	pollen	2
Carica	pollen	37
Cassia	pollen	2
Chimonanthus	pollen	27
Citrus	pollen	8
Fagus	pollen	19
Hevea	pollen	9
Hevea	ovule	10
Jacaranda	pollen	2
Larix	megagametophyte	24
Litchi	pollen	8
Euphorbia	pollen	8
Lycium	pollen	47
Malus	pollen	8
Picea	megagametophyte	36
Poncirus	pollen	8
Populus	ovule	10
Prunus	pollen	49
Pyrus	pollen	18
Robinia	ovule	46

3. Androgenesis

3.1. *Angiosperms*

The most successful overall method for woody plant haploid cultures has been via androgenesis. The list of successfully induced androgenetic haploid genera is, nevertheless, not a very long one (Table 1).

Pollen development in angiosperms is relatively simple. Following meiosis a further mitosis results in a tube cell and a generative cell. The generative cell eventually divides to form the two male gametes. Most androgenic protocols have involved careful choice of the appropriate pollen stage. There is not one stage which is universally amenable. In some genera such as *Quaercus* [19] and *Lycium* [47] early uninucleate pollen responded best. In others, pollen at the mid-late uninucleate stage was optimal (*Hevea*, [8]).

The number of media used is quite limited. Murishige and Skoog [23] and its variants have been used as woody plant media: MS [10,11,19,26], Linsmaier and Skoog [47], RT [8], MB [8]. Occasionally other media have been used such as Nitsch [18], or low ionic strength media such as Woody Plant Medium [19].

Androgenesis has a number of different developmental pathways. In some instances embryoids are produced directly from the pollen. In most others a callus develops which in turn produces embryoids [8]. The callus may otherwise produce shoots organogenically which are then rooted. Some systems have well-developed protocols for establishment of plantlets from anther culture. Alteration involves changing from a medium with growth-regulating substances suitable for induction (a cytokinin and auxin [usually 2,4-D] combination) to a medium free of growth regulating substances or with a different auxin (IAA, NAA or IBA) and in some instances no cytokinin. Further development involves optimizing root establishment and plantlet development prior to greenhouse transfer. The nature of the protocols is very dependent on the type of androgenic pathway followed.

The number of successfully induced woody angiosperms spans a wide range of genera (see Table 1). Most of the success has been in China, and it is apparent that the most elaborate protocols have also been developed there [8]. Embryo maturation has not required ABA, but rather an alteration of sucrose concentration, as well as types and concentrations of auxins and cytokinins. Supplements are usually organic nitrogen in the form of casein hydrolysate, glutamine, myoinositol and even royal jelly.

3.2. *Conifers*

Androgenesis has not yet been achieved in confiers. The number of published reports on attempts at conifer androgenesis does not reflect the effort made by researchers. Rare organogenic events such as shoot initiation and root formation have been reported in callus from microspore culture of *Picea* [34]. Regeneration of cell cultures similar to some stages found in gynogenic *Larix* (composed of small isodiametric as well as elongated cells) have been reported for *Taxus* [38]. More developmentally advanced structures have been reported for *Pinus* [4]. Coenocytes were also reported from a number of genera: *Torreya* [39], *Pinus* [5] and *Taxus* [21]. The coenocytes which gies rise to callus are thought to originate via interruption of linear polar development in the microspore [5]. This may be induced by pretreating microsporangia with a cold treatment as has been reported for *Pinus* [4].

A major problem in research on conifer androgenesis is the seasonal availability of suitable explant material. It would be desirable to have clones of trees which produce pollen more than once a year. Fortunately, the ability to do this in diverse conifers has come about largely because of success in establishing potted breeding orchards during the 70s and 80s. It is now possible to get stock from rooted cuttings, or grafted stock, and within two or three years, to induced cones on these trees by cultural treatments, including application of gibberellins, root pruning, or girdling [30]. This could allow three or four pollen-one crops per year, depending on the species. Such small trees fit easily into growth chambers and each tree will produce hundreds of cones. Thus experimental avenues are opened, and such things as the effect of hot and cold pretreatment on induction may be explored.

The predominant factor influencing androgenesis is genotype. Obviously, it is important to work with a species in which there is some ongoing genetic program and known genotypes are available. In conifers such programs have only recently been established for most of the economically important species, and programs using potted breeding orchards are even fewer in number. Very few published studies on tissue culture of conifers can identify their seed sources to a deliberate cross, and this has a very direct bearing on the repeatability one can expect with haploid experimentation. For example, one may wish to study the influence that genetic load has on induction of androgenesis, but how many present breeding programs can provide three trees of known high, intermediate or low load levels?

The second major consideration is the developmental stage of the pollen when induction is atempted. Most plants induced to date are only amenable during early pollen development, usually up to the late-uninucleate microspore stage. It seems an accepted fact that one starch accumulation occurs during pollen development androgenesis is inhibited [31].

Environmental pretreatment of trees or branches, such as low temperature, are known to reduce the amount of starch formed in the pollen [11,31]. Some conifers do not build up appreciable starch reserves during pollen development, and these may be easier to induce. For many conifers microspore formation and development are either not considered or not described, and it is a very short stage in the development of the complex microgametophyte. Pinaceous pollen is four- or five-celled at maturity. However, a number of conifer genera have much simpler pollen, which is uni- or binucleate when mature, such as that of the Cupressaceae [25]. Thus, this family may be more amenable to androgenesis.

4. Conclusions

The two most common methods for generating haploid plants have been gynogenesis and androgenesis. Gynogenesis has generally been very limited in its success and application. The inherent difficulties with this system – labour-intensive dissection, low explant number and low yield – augur badly for any widespread usefulness. Androgenesis has been much more successful to date, but only in angiosperms. Most of the work has used anther culture but it is foreseeable that there will be a change over to microspore culture as has occurred in herrbaceous angiosperms. The difficulties with conifers may be surmountable by close examination of the developmental stage of the pollen and manipulation of potted breeding orchards to provide appropriate stages of pollen at more frequent intervals.

5. Summary

Haploid embryogenesis from trees may be induced from either microspores and pollen, or from female tissues such as embryo sac or megagametophyte. This review considers these two different types: androgenesis (embryos from male tissue) and gynogenesis (embryos from female tissues). Angiosperms are considered separately from conifers, as conifers have presented more technical difficulties. To improve experimental systems for conifers is a goal which may be attainable using cone induction in potted orchards.

References

1. Attree SM, Bekkaoui F, Dunstand DI, and Fowke LC (1987) Regeneration of somatic embryos from protoplasts isolated from an embryogenic suspension culture of white spruce (*Picea glauca*). Plant Cell Rep 6:480–483.
2. Bajaj YPS and Dhanju MS (1983) Pollen embryogenesis in three ornamental trees *Cassia fistula, Jacaranda acutifolia*, and *Poinciana regia*. J Tree Sci 2:16–19.
3. Ball EA (1987) Tissue culture multiplication of *Sequoia*. In: Bonga JM, Durzan DJ (Eds.) Cell and Tissue Culture in Forestry, Vol. 3, Martinus Nijhoff Publishers, Dordrecht, The Netherlands, pp. 146–158.
4. Bonga JM (1974) *In vitro* culture of microsporophylls and megagametophyte tissue of *Pinus*. In Vitro 9:270–277.
5. Bonga JM (1981) Haploid culture and cytology of conifers. *In* Colloque International sur la Culture 'In vitro' des Essences Forestieres, IUFRO Section 52015 AFOCEL 77370 Nagis France, pp. 283–293.
6. Bonga JM, Von Aderkas P, and James D (1988) Potential application of haploid cultures of tree species. *In*: Hanover JW and Keathley DE (Eds.) Genetic Manipulation of Woody Plants, Plenum Publishing, New York, pp. 57–77.
7. Borchert R (1968) Spontane Diploidisierung in Gewebe Kulturen des Megagametophyten von *Pinus lambertiana*. Z Pflanzenphysiol 59:389–392.
8. Chen Z (1987) Induction of androgenesis in hardwood trees. *In*: Ahuja MR (Ed.) Somatic Cell Genetics of Woody Plants, Kluwer Academic Publishers, Dordrecht, the Netherlands, pp. 247–268.
9. Chen Z, Chen F, Chien C, Wang C, Chang S, Hsu H, Ho Y, and Lu T (1978) A process of obtaining pollen plants of *Hevea brasiliensis* Muell-Arg. Sci Sinica 22:81–90.
10. Chen Z, Li W, Zhang L, Xu X, and Zhang S (1988) Production of haploid plantlets in cultures of unpollinated ovules of *Hevea brasiliensis* Muell.-Arg. *In*: Ahuja (Ed.) Somatic Cell Genetics of Woody Plants, Kluwer Academic Publishers, Dordrecht, The Netherlands, pp. 39–44.
11. Chen Z, Yao Y, and Zhang L (1988) Studies on embryogenesis of woody plants in China. *In*: Ahuja MR (Ed.) Somatic Cell Genetics of Woody Plants, Kluwer Academic Publishers, Dordrecht, The Netherlands, pp. 19–25.
12. Durzan DJ (1988) Somatic polyembryogenesis for the multiplication of tree crops. Biotechnology Genetic Eng Rev 6:341–378.
13. Glock H, Hattemer HH, and Steinhauer A (1988) Untersuchungen überlebensfähigkeit von Endospermkulturen der Douglasie (*Pseudotsuga menziesii*) *in vitro*. Bot Acta 101:240–245.
14. Gupta PK and Durzan DJ (1987) Biotechnology of somatic embryogenesis and plantlet regeneration in loblolly pine. Bio/Technology 5:147–151.
15. Hakman I, Von Arnold S, and Fellner-Feldegg H (1986) Isolation and DNA analysis of protoplasts from developing female gametophytes of *Picea abies* (Norway spruce). Can J

Bot 64:108–112.
16. Huhtinen O (1976) *In vitro* culture of haploid tissue of trees. *In* XVI IUFRO World Congress, Division II, Forest Plants and Forest Production, Norwegian Forest Research Institute N-1432 AS-NHL, Norway, pp. 28–30.
17. Huhtinen O, Honkanen J, and Simola LK (1982) Effects of genotype and nutrient media on callus production and differentiation of Norway spruce endosperms cultured in vitro. *In* Colloque International sur la Culture 'in vitro' des Essences Forestieres. IUFRO, Fontainebleau, pp. 159–166.
18. Jordan M (1975) *In vitro* Kultur von *Prunus*, *Pyrus* und *Ribes*-Antheren. Planta Med (suppl) 59–65.
19. Jörgenson J (1988) Embryogenesis in *Quercus petraea* and *Fagus sylvatica*. J Plant Physiol 132:638–640.
20. Klimaszewska K (1989) Recovery of somatic embryos and plantlets from protoplast cultures of *Larix* × *eurolepis*. Pl Cell Rep 8:440–444.
21. Larue CD (1954) Studies on growth and regeneration in gametophytes and sporophytes of gymnosperms. Brookhaven Symp Biol 6:187–208.
22. Maheshwari SC, Tyagi AK, and Malhotra K (1980) Induction of haploidy from pollen grains in angiosperms. Theor Appl Genet 58:193–206.
23. Murashige T and Skoog F (1962) A revised medium for rapid growth and bio-assays with tobacco tissue cultures. Physiol Plant 15:437–497.
24. Nagmani R and Bonga JM (1985) Embryogenesis in subcultured callus of *Larix decidua*. Can J For Res 15:1088–1091.
25. Owens JN (1982) Pollen magement through phenological control of development. *In*: Edwards DFW and Yeatman CW (Eds.) Proceedings 18th Meeting of the Canadian Tree Improvement Association, Part 2, Duncan BC, pp. 39–48.
26. Radojevic L (1978) *In vitro* induction of androgenic plantlets in *Aesculus hippocastanum*. Protoplasma 96:369–374.
27. Radojevic L (1980) Haploid embryos, plantlets and callus formation in woody species. *In*: Davies DR and Hopwood DA (Eds.) The Plant Genome pp. 259.
28. Rohr R (1987) Haploids (Gymnosperms). *In*: Bonga JM and Durzan DJ (Eds.)) Cell and Tissue Culture in Forestry Vol. 2, Martinus Nijhoff Publishers, Dordrecht, The Netherlands, pp. 230–245.
29. Rohr R, Von Aderkas P, and Bonga JM (1989) Ultrastructural changes in haploid embryoids of *Larix decidua* during early embryogenesis. Amer J Bot 76:1460–1467.
30. Ross SD, Pharis RP, and Binder WD (1983) Growth regulators and conifers: their physiology and potential uses in forestry. *In*: Nickell LG (Ed.) Plant Growth Regulating Chemicals, Vol. II, CRC Press, Boca Raton, pp. 35–78.
31. Sangwan RS and Sangwan-Norreel BS (1987) Ultrastructural cytology of plastids in pollen grains of certain androgenic and nonandrogenic plants. Protoplasma 138:11–22..
32. Sangwan-Norreel BS and Duhoux E (1982) Les conditions de la gynogenèse et de la androgenèse *in vitro* chez les arbres. Rev Cytol Biol végét Bot 5:171–187.
33. Schlarbaum SE (1987) Cytogenic manipulations in forest trees through tissue culture. *In*: Bonga JM and Durzan DJ (Eds.) Cell and Tissue Culture in Forestry, Vol. 1, Martinus Nijhoff Publishers, Dordrecht, The Netherlands, pp. 330–352.
34. Simola LK (1987) Structure of cell organelles and cell wall in tissue cultures of trees. *In*: Bonga JM and Durzan DJ (Eds.) Cell and Tissue Culture in Forestry, Vol. 1, Martinus Nijhoff Publishers, Dordrecht, pp. 389–418.
35. Simola LK and Honkanen J (1983) Organogenesis and fine structure in megagametophyte callus lines of *Picea abies*. Physiol Plant 59:551–561.
36. Simola LK and Santanan A (1990) Improvement of nutrient medium for growth and embryogenesis of megagametophyte and embryo callus lines of *Picea abies*. Physiol Plant 80:27–35.
37. Tsay HS and Cu CY (1985) Anther culture of papaya (*Carica papaya* L). Plant Cell Rep

4:28–30.
38. Tulecke W (1959) The pollen cultures of CD LaRue: A tissue from the pollen of *Taxus*. Bull Torrey Bot Club 86:283–289.
39. Tulecke W and Sehgal N (1963) Cell proliferation from the pollen of *Torreya nucifera*. Contr Boyce Thompson Inst 22:153–163.
40. Von Aderkas P and Bonga JM (1988) Morphological definition of phenocritical period for initiation of haploid embryogenic tissue from explants of *Larix deciaua*. *In*: Ahuja MR (Ed.) Somatic Cell Genetics of Woody Plants, Kluwer Academic Publishers, Dordrecht, The Netherlands, pp. 29–38.
41. Von Aderkas P and Bonga JM (1988) Formation of haploid embryoids of *Larix decidua*: early embryogenesis. Amer J Bot 75:690–700.
42. Von Aderkas P, Bonga JM, and Nagmani R (1987) Promotion of embryogenesis in cultured megagametophytes of *Larix decidua*. Can J For Res 17:1293–1296.
43. Von Aderkas P, Klimaszewska K, and Bonga JM (1990) Diploid and haploid embryogenesis in *Larix leptolepis, L. deciadua*, and their reciprocal hybrids. Can J For Res 20:9–14.
44. Von Aderkas P, Bonga JM, Klimaszewska K, and Owens JN (in press) Comparison of larch embryogeny *in vivo* and *in vitro*. *In*: Ahuja MR (Ed.) Woody Plant Biotechnology, Kluwer Academic Publishers, Dordrecht, The Netherlands.
45. Von Arnold S and Hakman I (1988) Regulation of somatic embryo development in *Picea abies* by abscisic acid (ABA). J Plant Physiol 132:164–169.
46. Wang QZ, Wang ZX, and Zhang XH (1982) Initial success in the culture of monoploid plants from the unpollinated ovary of *Robinia pseudoacacia*. For Abstr 43:845.
47. Zenkteler M (1972) Development of embryos and seedlings from pollen grains in *Lycium halimifolium* Mill. in the *in vitro* culture. Biol Plant 14:420–422.
48. Zenkteler MA and Guzowska I (1970) Cytological studies on the regenerating mature female gametophyte of *Taxus baccata* L. and mature endosperm of *Tilia platyphyllos* Scop. in *in vitro* culture. Acta Soc Bot Poloniea 39:161–173.
49. Zenkteler M, Misiura E, and Ponitka A (1975) Induction of androgentic embryoids in *in vitro* cultured anthers of several species. Experientia 31:289–291.

5. Advances in the protoplast culture of woody plants

JULIE A. RUSSELL
Cornell University, New York State Agricultural Experiment Station, Department of Horticultural Sciences, Hedrick Hall, Geneva, NY 14456, U.S.A.

1. Introduction

During the past few years, several laboratories have reported the successful culture of protoplasts from woody species, a group which has long been considered recalcitrant. A review by McCown and Russell [41] covering the literature through May 1986, listed four woody genera (*Citrus*, *Populus*, *Santalum*, and *Ulmus*) in which protoplasts had been cultured through whole plant regeneration. Now plants have been regenerated from protoplasts of 46 woody genotypes representing 32 species, 18 genera, and 12 families (Table 1). In many of these cases the protoplast source tissue originated from non-seedling (mature) plants (Table 1). Thus, it will be possible to use protoplast-based genetic engineering to add one or a few desirable traits to already proven varieties so that the development time for a saleable product can be reduced.

It is especially important to note the breakthrough with gymnosperm trees. Until recently, only calli had been recovered from protoplasts of this group. Somatic embryos have now been recovered from protoplasts of *Picea glauca* [2], *Pinus taeda* [18,19], and *Pseudotsuga menziesii* [*18*]. The author is aware of at least two laboratories which have recovered complete plants from protoplast-derived embryos of gymnosperms: (1) *Larix* × *eurolepis* (K. Klimaszewska, personal communication) and *Picea glauca* (S. Attree, personal communication).

Protoplasts of many other previously recalcitrant or untested woody genotypes have also been cultured to the callus stage. Those reported since the previous review by McCown and Russell [41] include *Acer pseudoplatanus* [5], *Betula papyrifera* [81], *Fagus sylvatica* [33], *Hibiscus syriacus* [8], *Malus domestica* 'Greensleeves' [10], *Olive europaea* 'Tanche' [7], *Pinus oocarpa* [32], *Prunus persica* 'Hakuho' [37], *Thuja occidentalis* 'Woodwardii' [66], *Vitis rotundifolia* 'Summit' [35], and *Vitis vinifera* 'Cabernet Sauvignon' [35]. Limited organogenesis has also been reported from protoplast-derived calli of *Pyrus communis* 'Conference' [49] and *Malus* × *domestica* 'MM.106' [54].

Protoplast culture is not yet easy for all genotypes; however, the recent successes provide the foundation for protoplast-based genetic engineering in woody crops. Protoplasts are the basis of numerous gene transfer methods,

M. R. Ahuja (Ed.), Micropropagation of Woody Plants, 67–91.

Table 1. Woody species from which whole plants have been regenerated from protoplast.

Species	Variety	Source Tissue	Sdl[1]	Non-Sdl[1]	Reference
Actinidiaceae					
Actinidia chinensis (kiwi)	Chinensis	Leaf callus		X	82
Fabaceae					
Crotalaria juncea (sunhemp)	PLC-71	Cotyledons	X		57
Pithecellobium dulce (huamuchil, opiuma, or manila tamarind)		Shoot cultures	X		68
Magnoliaceae					
Liriodendron tulipifera (yellow poplar)		Embryogenic cell suspension	X		38
Moraceae					
Broussonetia kazinoki (paper mulberry)		Shoot cultures	X		53
Pinaceae					
Larix × eurolepis (larch)		Embryogenic cell suspension	X		K. Klima-szewska pers.comm.
Picea glauca (White spruce)		Embryogenic cell suspension	X		2 S. Attree pers.comm.
Rosaceae					
Malus × domestica	M.9	Shoot cultures		X	54
(apple)	'Spartan'	Shoot cultures		X	
Prunus avium	'Colt'	Shoot cultures		X	47,48,51
× *P. pseudocerasus* (cherry)		Cell suspension		X	
Prunus cerasus	CAB4D	Shoot cultures		X	50
(sour cherry)	CAB5H	Shoot cultures		X	
Pyrus communis (wild pear)	Pyraster	Shoot cultures		X	46
Rubiaceae					
Coffea canephora (coffee)	No. 3481	Somatic embryos		X	70
Rutaceae					
Citrus aurantium (sour orange)		Nucellar callus		X	88
Citrus aurantium × (somatic cybrid)		Nucellar callus		X	85
C. limon	'Villafranca'	Nucellar callus		X	

Table 1 (continued)

Species	Variety	Source Tissue	Sdl[1]	Non-Sdl[1]	Reference
Rutaceae (continued)					
Citrus limon (lemon)	'Villafranca'	Nucellar callus'		X	88
Citrus mitis (lime)		Embryogenic cell suspension[2]		X	75
Citrus paradisi (grapefruit)	'Duncan'	Nucellar callus		X	88
Citrus reticulata (mandarin)	'Murcott'	Nucellar callus		X	88
	'Dancy'	Nucellar callus		X	
	'Ponkan'	Nucellar callus,		X	
		Hypocotyl callus	X		22
Citrus sinensis (orange)	'Shamouti'	Nucellar callus		X	12,84
	'Trovita'	Nucellar callus,		X	30
		Embryogenic cell suspension[2]		X	29
	'Nucellar Shamouti'	Nucellar callus		X	88
	'Shamouti Landau'	Nucellar callus		X	
	'Washington Navel'	Hypocotyl callus	X		22
Citrus sinensis × (somatic hybrid)	'Hamlin'	Embryogenic cell suspension[2]		X	15
Poncirus trifoliata (trifoliate orange)	'Flying Dragon'	Seedling leaves	X		
Citrus sinensis × (somatic hybrid)	'Trovita'	Embryogenic cell suspension[2]		X	52
Poncirus trifoliata (trifoliate orange)		Seedling leaves	X		
Citrus sinensis × (somatic hybrid)	'Hamlin'	Emgryogenic cell suspension[2]		X	16
Severina disticha (philippine box orange)		Epicotyl callus	X		
Citrus yuko		Hypocotyl callus	X		22
Microcitrus sp. (australian wild lime)		Embryo callus	X		87
Poorman × *Poncirus trifoliata* × (somatic cybrid)		Nucellar callus		X	85
Citrus limon	'Villafranca'	Nucellar callus		X	
Poorman × *Poncirus trifoliatata* × (somatic cybrid)		Nucellar callus		X	85
Citrus aurantium		Nucellar callus		X	
Salicaceae					
Populus alba × P. grandidentata (hybrid poplar)	'Crandon'/NC-5339	Shoot cultures		X	63,64

Table 1 (continued)

Species	Variety	Source Tissue	Sdl[1]	Non-Sdl[1]	Reference
Salicaceae (continued)					
Populus maximowiczii × *P.* trichocarpa (hybrid poplar)	NC-11390	Shoot cultures		X	unpubl.
Populus nigra (black poplar)		Petiole callus		X	34
Populus nigra × *P. laurifolia* (hybrid poplar)	NC-5272	Shoot cultures		X	unpubl.
Populus nigra × *P. trichocarpa* (hybrid poplar)	NC-5331	Shoot cultures		X	64
Populus tremula (aspen)	'Erecta'	Shoot cultures		X	64
Populus tremula	706–8	Shoot cultures		X	unpubl.
× *P. alba*	709–21	Shoot cultures		X	
(hybrid poplar)	710–23	Shoot cultures		X	
Santalaceae					
Santalum album (sandalwood)		Embryogenic cell suspension		X	58
Solanaceae					
Duboisia myoporoides		Cell suspension	X		28
Ulmaceae					
Ulmus X (hybrid elm)	'Pioneer'	Leaf callus		X	78

[1] Status of Source Plant Material:
Sdl = Seedling
Non-Sdl =Non-Seedling
[2] Nucellar Origin

though they are no longer strictly required. Nuclear and/or cytoplasmic genomes of different genotypes can be combined via protoplast fusion [31], protoplasts can be stimulated to take up foreign DNA by polyethylene glycol (PEG) treatment or electroporation [56,73], and genes or whole chromosomes can be inserted into protoplasts via microinjection [9,14].

The remainder of this review will cover the specific components of a protoplast-based genetic engineering system for woody crops. The primary focus will be to identify the important parameters affecting protoplast isolation, culture, and regeneration from the viewpoint of the author who is at the completion of a 5 year project on *Populus* protoplasts. The few actual cases of protoplast fusion and protoplast-based direct gene transfer in woody species will also be discussed. Though at first there appear to be endless combinations of parameters to test, it is hoped that this review will aid in making intelligent choices at each step, thus reducing the amount of empirical testing required for each new genotype.

2. Components of an 'ideal' protoplast-based gene transfer system

For the purpose of this review, we will take the approach of developing a protoplast system to support genetic engineering research. It will be assumed that the primary goal is to recover a large number of stably transformed plants so that variations in the transformation event (position effect, copy number, etc.) can be studied and the best transformant(s) can be selected for further propagation. The general requirements of such a system are detailed below.

2.1. *Genetically defined source tissue*

With woody crops it is especially important to take advantage of materials from established breeding and selection programs because the long generation times make it difficult to breed traits back into commercial varieties. Ideally, one or a few desired traits would be added to an already existing, elite variety. Successful protoplast culture of some genotypes (eg. most gymnosperms), however, requires the use of seedling donor tissue. For such genotypes, the selection of seeds from a well maintained seed orchard is the best choice of starting material.

2.2. *Isolation of viable protoplasts*

The isolation of viable protoplasts is the foundation of successful protoplast culture. The exact yields that are required depend upon the method of gene transfer. For example, 20 million or more protoplasts may be required for electroporation experiments in order to test the many combinations of electrical, media, and plasmid parameters. Conversely, it is unlikely that one would want to hand microinject that quantity of protoplasts in a single sitting. Protoplast populations are usually heterogeneous and a greater number of transformants can be recovered if transformation is focused only on those protoplasts that are capable of developing into plants. Thus, the ability to select developmentally competent protoplasts is desirable.

2.3. *Stable transformation of the protoplasts*

Gene transfer is theoretically possible using any of the methods that have been developed for herbaceous crops. A selection scheme for transformants is also required and varies with the transformation method. For example, heterokaryons may be selected mechanically with micromanipulators or via flow cytometry immediately after fusion. A common method is to select for transformants at the callus stage, based upon their growth on selective media.

2.4. *Culture through sustained division*

After transformation, the protoplasts must be cultured in an environment capable of promoting cell wall regeneration and sustained division. Since transformation frequencies are commonly low (less than 1%), high plating efficiencies (dividing protoplasts) increase the probability of recovering transformants.

2.5. *Differentiation*

The dividing protoplast-derived cells must be induced to differentiate into organized structures. This can be either via somatic embryogenesis or via organogenesis from protoplast-derived calli. Again, a high frequency of regeneration is desired to recover the maximum number of transformants.

2.6. *Plants Ex Vitro*

The regenerated embryos or shoots/plantlets must be acclimated to greenhouse and/or field conditions for further evaluation. Standard micropropagation and acclimation methods are sufficient [39].

3. Evaluation of specific protoplast parameters

Numerous parameters must be evaluated throughout each of the above outlined steps. The most important of those parameters are discussed below. Examples are primarily taken from the literature on woody species to highlight important variables to be considered in establishing a protoplast system for a previously untested genotype. Readers are referred to a recent excellent review by Rashid [59] for details of other parameters which can influence protoplast culture.

3.1. *Genotype*

Virtually every parameter throughout protoplast isolation and culture depends upon the specific genotype, and thus there is no single magical formula for success. Even different clones within the same species can have highly specific requirements. For example, in *Malus* × *domestica*, protoplasts of 'M9' and 'Spartan' were regenerated into whole plants, whereas protoplasts of 'M106' and 'Bramley's Seedling' could be cultured only to the callus stage, even though numerous parameters were tested [54]. The only loose generalization that can be made is that a plant that is highly regenerative in other culture systems is likely to have a greater chance of success in protoplast culture. For example, in *Liriodendron* [38], plants could be regenerated only if the protoplasts were isolated from a cell suspension which was characterized by high frequency embryogenesis. A highly regenerative genotype, however, is not a guarantee for successful protoplast culture. Examples of the influence of genotype will be evident throughout the remainder of the text.

3.2. *Source tissue*

Protoplasts can be isolated from almost every type of plant tissue. For woody plants, the choice of source tissue is especially critical. Plant regeneration from protoplasts of woody species has been most successful when either shoot, callus, or cell suspension cultures were used as the protoplast source (Tables 1 and 2). With callus and cell suspension cultures, embryogenic lines appear to be best (Table 2). It is not clear whether this is because of the availability of a previously established regeneration system or because of the physiological state of the source tissue itself. In woody species there is an apparent correlation between successful protoplast culture and source tissue that is in a embryonic or juvenile state [40]. Embryogenic callus and cell suspension cultures often originate from embryonic tissues.

Table 2. The number of reports in woody species which cite varying degrees of success in the culture of protoplasts from different donor tissues. Each notation represents the most advanced report for a given genotype × source tissue. The majority of the references are cited either in this review or in the previous review by McCown and Russell [41].

Source Tissue	Isolation/ preliminary division	Sustained division (callus)	Embryos or organs	Whole plants	Total
Greenhouse/field plts	43	12	0	0	55
In vitro plants	22	10	1	16	49
Cotyledons	5	7	0	1	13
Somatic embryos	0	0	0	1	1
Callus					
Embryogenic:	0	0	3	19	22
Non-embryogenic:	11	9	2	3	25
Cell suspensions					
Embryogenic:	0	0	3	4	10
Non-embryogenic:	9	13	1	1	24
Roots	1	0	0	0	1
Meiotic cells	5	0	0	0	5

Shoot cultures provide a ready source of genetically uniform protoplast donor tissue at any time of the year. Protoplast isolation from shoot cultures is more reproducible and higher yielding than when greenhouse and field grown plants are used [46,76]. Plants have not been regenerated from protoplasts of a greenhouse or field grown woody plant, except when leaf-derived protoplasts of seedling trees were fused with embryogenic cell suspension-derived protoplasts [15,52].

Callus and cell suspensions are also an excellent source of protoplasts for woody crops. These are especially important options for genotypes which are difficult to grow as shoot cultures, for example, most slow growing, episodic hardwoods and conifers [39]. Additionally, protoplasts from rapidly dividing cell suspensions begin cell wall formation and division sooner and have a higher incidence of division than do leaf-derived protoplasts. This is presumably because of less time needed for dedifferentiation. However, with callus and cell suspensions there is a higher chance of mutations, which may cause a loss in the ability of the protoplasts to regenerate or which may cause variation in the protoplast-derived plants. In *Ulmus* [78], protoplast yield progressively declined from 2×10^5 per gram fresh weight in 10 week old callus cultures to 0.002×10^5 per gram fresh weight in 80 week old cultures (after approximately 20 subcultures). The greatest frequency of protoplast division occurred with protoplasts from 30 wk old cultures and the ability to divide was lost when the callus cultures were between 40 and 75 weeks old. In contrast, several lines of *Citrus* callus and cell suspension cultures maintained their embryogenic potential for more than a year [22,30,75,88]. One explanation for the low mutation rate in *Citrus* may be the lack of a requirement for hormones, especially 2,4-D, in the culture medium. Vardi *et al.* [88] reported that seven out of eight nucellar callus lines of *Citrus* maintained a diploid chromosome number after one year in culture. The only cultivar which deviated from diploidy was also the only one for which 2,4-D had been required for culture initiation and early maintenance. The high mutation rate common in callus and cell suspensions may indeed be influenced by hormones, but is also inherent to the unorganized state and rapid division of the cells.

The quality of the tissue can be as critical as the type of tissue used for protoplast isolation. In *Populus*, high quality, succulent, non-stressed leaves yield the best protoplasts. Whenever necrotic, chlorotic, stunted, or vitreous shoots are used, the yield of protoplasts is low. One of the main causes of stress in *Populus* is high cytokinin in the shoot culture growth medium. Thus, a two-step system of first multiplying the shoots on a medium with 0.4 μM benzyladenine (BA) and then growing individual shoot tips on a low cytokinin medium (0.01 μM BA) to produce large, healthy leaves is routinely employed [63]. In *Populus*, it is also important to harvest the tissue before the leaves become thick and pubescent. The environmental conditions under which the shoot cultures are grown can be important [68], but have not been studied in most cases.

For callus and suspension cultures, the growth medium and environment

[38] are undoubtedly important, but generalizations are difficult to make. One universally important parameter, however, is the time since the last subculture. For callus cultures of *Ulmus*, a subculture interval of 4 weeks or less was required; otherwise the yield of protoplasts was substantially reduced and no divisions were obtained [78]. The optimal time for protoplast release is especially critical with cell suspension cultures. In *Betula* [81], maximal protoplast yields were obtained from 4–5 day old cell suspensions which were in the log phase of growth. This period corresponded to a rapidly increasing mitotic index. In *Citrus*, protoplast isolation was best with 8 day old cultures, which were also at the log phase of growth [75].

3.3. *Digestion*

When shoot cultures are used as the protoplast source, the leaves are usually chopped or sliced for better penetration of the enzymes. With *Populus*, however, even when the tissue is chopped the enzymes do not readily penetrate the leaves and only the cells at the cut surfaces are readily attacked [62]. The result is that only a small percentage of the cells receive the optimal enzyme exposure because protoplasts first released from the cut surfaces suffer from overdigestion while those protected within the leaf are underdigested. Exposure of the cells to the enzymes is more uniform when the leaves are first processed for 3–5 seconds in an Omni-Mixer [62]. Success with this technique, however, depends on the structure of the leaf tissue. For example, when shoots of *Thuja* are processed in the Omni-Mixer, the needles are merely stripped from the stems and are not cut up. The timing of the treatment is also critical and damage can result with the addition of only a few seconds.

Preincubation treatments are sometimes employed prior to enzyme exposure. Butt [6] reported that chopped leaves release water soluble compounds (presumably hydroxycinnamic acids) which modify the cell wall structure, rendering the cells resistant to enzymatic digestion. Protoplasts of several greenhouse and field-grown trees could only be released if the chopped leaf tissue was thoroughly washed prior to the digestive enzyme addition. It is not known whether the release of such compounds is limited only to greenhouse/field-grown sources.

For *Pithecellobium*, incubating the leaves overnight at 10 °C in a sucrose-free, liquid MS medium containing 2.5 μM 2iP was necessary to obtain stable protoplasts which were otherwise prone to disruption during purification [68].

Placing the source tissue in a plasmolyzing solution prior to enzyme exposure has also been reported to be beneficial in some cases [48,50,54,70], but is not a universal requirement. Preplasmolysis presumably helps to protect the plasmamembrane from damage by toxins present in the cell wall degrading enzymes [59].

The type and concentration of the cell wall degrading enzymes are perhaps the most critical factors affecting protoplast isolation, other than the source

Table 3. Enzyme treatments which were used in woody plant protoplast systems from which embryos or whole plants were regenerated. Listings are arranged by the protoplast source tissue.

Plant	Enzyme formulation	Duration	Reference
Shoot cultures			
Broussonetia sp.	1.0% Cellulase Onozuka R-10 0.05% Pectolyase Y-23	3 h	53
Pithecellobium sp.	1.0% Cellulase RS 0.05% Macerozyme R-10 0.25% Pectolyase Y-23 0.5% Rhozyme HP-150	7–9 h	68
Populus sp.	0.5% Worthington Cellulase 0.1% Macerase	4 h	63,64
Prunus sp. Colt, CAB4D	1.0% Cellulase Onozuka R-10 0.2% Macerozyme R-10 0.1% Driselase	17–18 h	48,50
Prunus sp. CAB5H & *Malus* sp.	1.0% Cellulase R-10 0.1% Pectolyase Y-23 1.0% Hemicellulase	17–18 h	50,54
Pyrus sp.	0.5% Cellulase Onozuka R-10 0.1% Macerozyme R-10 0.1% Driselase	8 h	46
Callus			
Actinidia sp.	2.0% Cellulase Onozuka R-10 0.5% Macerozyme R-10	4–5 h	82
Citrus sp. Shamouti, Villafranca	0.2% Cellulase Onozuka R-10 0.3% Pectinase 0.1% Driselase	overnight	12,84,88
Nucellar Shamouti, Shamouti Landou, Murcott, Trovita	0.2% Cellulase Onozuka R-10 0.3% Macerozyme 0.1% Driselase	overnight	30,88
Citrus sp. Dancy, Duncan	0.2% Cellulase Onozuka R-10 0.3% Macerozyme	overnight	88
Washington Navel, Yuko, Ponkan	0.33% Cellulase Onozuka RS 0.01% Pectolyase Y-23	15 h	22

Table 3 (continued)

Plant	Enzyme formulation	Duration	Reference
Callus (continued)			
Microcitrus sp.	0.3% Cellulase R-10 0.2% Macerozyme 0.1% Driselase	overnight	87
Populus sp.	2.0% Cellulase Onozuka R-10 0.5% Macerozyme R-10 0.5% Hemicellulase	6 h	34
Ulmus sp.	1.5% Cellulysin 0.5% Macerase	23–25 h	78
Cell suspensions			
Citrus sp.	0.3% Cellulase Onozuka R-10 0.3% Macerozyme Onozuka R-10 0.1% Driselase	16 h	29,75
Duboisia sp.	0.2% Pectolyase Y-23 2.0% Cellulase R-10	16 h	28
Liriodendron sp.	2.0% Cellulysin 1.0% Macerase	12 h	38
Picea sp.	1.0% Cellulase RS 0.25% Rhozyme HP-150 0.25% Pectinase 0.25% Driselase	3.5 h	2
Pinus sp. & *Pseudotsuga* sp.	2.0% Cellulysin 1.0% Macerase 2.0% Pectinase	7–8 h	18,19
Prunus sp.	2.0% Meicelase 2.0% Rhozyme HP-150 0.03% Macerozyme R-10	17 h	47,48,51
Santalum sp.	1.0% Cellulase Onozuka RS 1.0% Macerozyme Onozuka R-10 0.5% Driselase	4–5 h	58

tissue. The optimal enzyme treatment can only be determined empirically. The enzyme treatments used in cases of successful plant regeneration from protoplasts of woody species are listed in Table 3. Cellulase R-10 and RS in conjunction with Macerozyme have been the most commonly used. Driselase has been used successfully in a few cases (Table 3), but has been reported harmful to subsequent protoplast division in others [68,88]. Our laboratory has found Worthington cellulase and Macerase to be a good general combination for protoplast isolation from shoot cultures.

The optimal type and concentration of cell wall degrading enzymes varies not only with the plant genotype but also with the source tissue. Of ten different enzyme concentrations tested for *Prunus* [48], leaf protoplast yield was highest (2.3×10^6 viable protoplasts per g fresh weight) with a combination of 1.0% Onozuka R-10 cellulase, 0.2% Macerozyme, and 0.1% Driselase. Using the same enzyme treatment, cell suspensions yielded 33×10^6 viable protoplasts. However, using a combination of 2.0% Meicelase, 2.0% Rhozyme HP-150, and 0.03% Macerozyme R-10 the yield of cell suspension protoplasts increased to 153×10^6 per g fresh weight. With this combination, however, the yield of leaf protoplasts decreased to 0.25×10^6.

The optimum enzyme treatment cannot be based on protoplast yield alone. For example, with *Betula* [81] a mixture of 1% Cellulase and 0.03–0.5% Macerase yielded 2.75×10^6 protoplasts which had a plating efficiency of 20–25%. In contrast, a mixture of 1% Cellulase, 0.5% Rhozyme, and 0.5% Macerase yielded 4.8×10^6 protoplasts (highest yield), but all of the protoplasts subsequently died in culture. Similarly, in *Citrus* the source of pectinase used had a strong effect on the plating efficiency [86]. Damage from the cell wall degrading enzymes often is the result of contaminants such as proteases and lipases. Additionally, the cells themselves may be a source of toxic factors produced during enzymatic cell wall digestion [20].

The protoplast isolation procedures are further complicated by media and incubation conditions. In addition to the enzymes, the isolation medium must contain an osmoticum. Some media also contain salts, hormones, organics, buffers such as MES, PVP to reduce browning, or bovine serum albumin as a carrier. I would recommend starting out with a simple medium and then adding complexities later if needed.

The only universally critical factor of the isolation medium is that the osmolality be near the isotonic point of the cell to avoid bursting or collapse of the protoplasts. Medium osmolality cannot be based simply on sugar molarities, but is best measured with an osmometer because salts and other medium components can have a significant effect. The isotonic point of the cells can be determined by sequentially subjecting cells or thin viable leaf sections to gradually increasing concentrations of osmoticum until the point of plasmolysis is visualized [77].

Some genotypes exhibit a sensitivity to the type of osmoticum used. For example, in *Pyrus* [46], sucrose is important for the isolation medium. Though the yields with either mannitol or sorbitol are similar to when sucrose is used, the protoplasts die within a few days. Sucrose, glucose, and mannitol are the most common osmotica and have been successfully used in numerous cases.

3.4. *Protoplast purification*

The protoplasts are commonly purified from debris and undigested cells by a combination of filtration, pelleting, and/or flotation. Our laboratory prefers the sucrose flotation method which readily separates live protoplasts from undigested cells as well as from any dead or injured protoplasts. In some cases, however, the flotation method will not work. For example, *Liriodendron* [38] cell suspension-derived protoplasts could not be purified by flotation even in solutions of up to 35% sucrose and 10% ficoll. The protoplasts did float on a 40% Percoll solution but so did the undigested cells. In this case, purified protoplasts could only be obtained by a combination of long digestion times and filtration.

The initial viability of protoplast preparations is commonly assayed by fluorescein diacetate (FDA) staining [89]. This method, however, relies on the accumulation of fluorescein molecules which have been cleaved from the diacetate residues within the cell and merely indicates if the plasmamembrane is intact. To recover transformed plants the protoplasts must not only be viable, but must also be developmentally competent. Most protoplast preparations contain heterogeneous mixtures of protoplasts and the ability to select competent protoplasts could increase recovery of transformants. In *Populus*, using three sucrose flotations instead of two flotations for protoplast purification significantly increased the percentage of dividing protoplasts [64]. Since this treatment decreased the yields by half, it would appear that weak cells were selected against. Density gradient separation can increase the number of dividing protoplasts slightly (see Table 4 below) but more selective methods are needed.

3.5. *Preculture treatments*

The application of three successive electrical pulses between 250 and 1000 V to protoplasts of *Prunus*, *Pyrus*, and three herbaceous species promoted cell division and enhanced the plating efficiency and the number of recoverable colonies [61]. Calli recovered from the treated protoplasts of *Prunus* also have an enhanced capacity for shoot regeneration. Additionally, in *Prunus*, there is a long term carry-over effect of the electrical pretreatment in successive protoplast isolations [47,51]. In *Pennesitum*, the stimulatory effect of electroporation is mimicked by heat shock [17]. Heat shock can also significantly increase protoplast plating efficiency in *Oryza* [80], but the effect on woody genotypes is not known. At present, the physiological action of electroporation and heat shock remains speculative.

3.6. *Protoplast culture*

3.6.1. *Method*

The choice of culture method and vessel depends both on the preference of the individual researcher and on the genotype involved. The ease of handling and the ease of data collection should also be considered. Protoplasts are commonly plated in thin layers of liquid or agarose in petri dishes. With this method, however, data collection may be difficult. Furthermore, when the protoplasts are embedded in agar they are difficult to observe due to diffraction of the light, and the medium cannot be changed. The agarose bead method [74] which involves surrounding agarose embedded protoplasts with a liquid medium, enables medium change and has been succesfully used for a wide range of genotypes. The less purified types of agar may contain contaminants which are toxic to protoplasts.

Liquid culture provides the best control over medium replenishment and change, but a commonly cited problem is aggregation of the protoplasts which subsequently inhibits both observation and growth. Protoplast aggregation can be avoided by culturing the protoplasts in a monolayer together with a polyester screen disc [63]. Additionally, the screen discs provide a grid for data collection, facilitate medium change, and the developing protoplast-derived cells can be transferred as a unit to regeneration medium. Limitations of the floating disc method include the requirement for protoplasts to be able to float and to grow on sucrose medium, as well as variations in some lots of screen material causing the discs to sink.

3.6.2. *Media*

Protoplast culture media are highly genotype dependent and even gross generalizations are difficult to make. As suggested for protoplast isolation, I would recommend starting with a simple medium and then adding or subtracting components based on the response of the protoplasts. If the protoplasts degrade rapidly within the first week of culture as evidenced by budding, clearing, or collapse, then there are probably inhibitory compounds in the medium which need to be eliminated. On the other hand, if the protoplasts sit for several weeks without much change and then gradually degrade, the focus should be on adding components that might stimulate growth. Below is a discussion of factors which are either beneficial or inhibitory to the culture of protoplasts from woody species.

3.6.2.1. *Basal salts.* The mineral salts used for protoplast culture are generally formulations of media developed for other tissue cultures. Common formulations which have been used successfully for woody species are MS [44], MT [45], KM [24], WPM [36], B5 [13], and V47 [4]. A good starting point is to use the preferred medium on which other types of cultures such as shoot, callus, or cell suspensions of that genotype grow best.

The presence of ammonium in the culture medium is detrimental to some genotypes [46,63], but is beneficial [48] or has no effect on others [50,54]. In *Brousonettia* [53], MS salts which contain 20.6 mM NH_4NO_3 support a low level of protoplast division but colonies cannot be recovered. Growth of *Brousonettia* protoplast-derived colonies is best when the concentration of NH_4 NO_3 in the medium is reduced, but not eliminated [53]. In *Populus* [63], the detrimental effect of NH_4NO_3 is characterized by excessive budding of the protoplasts. Once *Populus* protoplasts develop into calli, however, the sensitivity to ammonium is lost.

3.6.2.2. *Osmoticum.* To avoid osmotic shock, the measured osmolality of the culture medium is initially balanced with that of the isolation medium. The best osmotica to use depends on the plant genotype. Sucrose, mannitol, and glucose are the most commonly used osmotic stabilizers either alone or in combination, but a variety of other sugars and sugar alcohols have been tested [59]. The osmolality of the medium is usually reduced after the initial period of culture because high osmolarities can inhibit division [43,72]. The timing and amount of the osmotic reduction are critical. Protoplasts are most tolerant when they have undergone one or more divisions. Reductions in steps of 100 to 250 mOsm at any given time are common. It may be beneficial to first measure the medium in which the protoplasts are growing because culture vessels, especially petri dishes, are prone to desiccation and there may be a significant increase in the medium osmolality. This can be especially important when protoplasts are cultured in multiwell plates because osmotic micro-environments are known to form in different well positions [65].

3.6.2.3. *Growth regulators.* Growth regulators are an important factor in the protoplast medium, but requirements again are genotype dependent. NAA, 2,4-D, and BA are the most commonly employed for initial growth of the protoplasts. Zeatin appears to be an absolute requirement in some cases [50]. *Citrus*, notably, does not require growth regulators for protoplast culture [88].

Growth regulator requirements are not only genotype specific, but may also depend on the source of the protoplasts. For example, in *Prunus* [48] the best plating efficiency of cell suspension-derived protoplasts was in a medium containing 5.38 μM NAA, 1.11 μM BA, and 2.28 μM zeatin. For leaf protoplasts, 4 different growth regulator combinations supported initial division but only a combination of 10.75 μM NAA, 2.22 μM BA, and 2.28 μM zeatin supported good colony growth. The difference between shoot culture-derived and cell suspension-derived protoplasts could be due to a carry-over effect of the culture medium or to differences in hormone receptivity.

3.6.2.4. *Medium supplements.* A variety of amino acids or complex organic supplements are stimulatory to protoplast development in some genotypes. Such compounds, however, can also be detrimental to protoplast culture. In

Populus, a combination of coconut water and casein hydrolysate reduces protoplast survivability, though for some genotypes such supplements are critical for the induction of division [64]. Most gymnosperms appear to have a strong requirement for organic medium supplements. A general rule to follow is not to add such compounds unless their essentiality has been verified. This is especially true of undefined compounds such as coconut water and casein hydrolysate because their composition can vary with different sources.

An interest in the addition of polyamines to protoplast culture medium was stimulated by reports of reduced protoplast lysis combined with induced DNA synthesis and mitosis in leaf protoplasts of *Avena* [1,25,26]. Indeed, in woody species putrescine was reported to be beneficial in *Prunus dulcis* [91], and either ornithine or putrescine were required for division in *Alnus* [23]. A general requirement for polyamines, however, has yet to be established.

3.6.3. *Plating density*

The plating density is critical to successful protoplast culture. In general, the effect of density on plating efficiencies follows a bell-shaped curve. The plating density response was well documented with *Ulmus* [79] in which an optimum of 63% division occurred when the protoplasts were plated at 0.5×10^5 per ml. Raising or lowering the plating density by 40–50 000 protoplasts per ml decreased the incidence of division to 10–15%. When the protoplasts were plated at densities of either 100 000 per ml above or below the optimum, the incidence of division was only 2%.

The optimal plating density may also vary with the culture method used. Protoplasts of *Rosa* were best plated at $0.01–1 \times 10^5$ per ml in liquid culture, whereas in agar the protoplasts could be plated as high as 5×10^5 per ml [55]. As suggested by the authors, this was apparently due to a locally high density within the floating layer of protoplasts in the liquid medium. Insufficient testing of plating density may be one reason for an apparent preference of a particular plating method for a given genotype.

In *Populus*, those clones which require organic medium supplements also have a strong requirement for low plating densities [64]. A combination of high density and organic supplements is always detrimental. Similarly, in *Vicia* [24] plating density could be lowered if undefined components such as coconut water were added to the medium. The authors concluded that the lack of division of protoplasts at low densities was due to the diffusion of necessary metabolites.

Plating density optima can also vary with the type or the physiology of the cells. Protoplasts from 9 week old callus cultures of *Ulmus* were best plated at a density of 0.5×10^4, whereas protoplasts from 30 day old cultures could be plated at densities between $0.05–1.0 \times 10^4$.

Furthermore, in protoplast populations which originate from a heterogeneous mixture of cell types, there may be subpopulations which have different plating density requirements. This was demonstrated with *Populus* protoplasts

Table 4. The influence of plating density on the plating efficiency of *Populus* NC-5339 leaf protoplasts separated by density gradient centrifugation. Methods were modified from Harms and Potrykus [21]. For details see: J. A. Russell (1989) Culture and Manipulation of *Populus* Protoplasts as a Foundation for Genetic Engineering Research, Ph.D. Thesis, University of Wisconsin-Madison.

		Percentage dividing protoplasts			
		Plating Density (protoplasts per ml)			
Fraction	Specific[1] density (g/cm^3)	5 000	10 000	25 000	50 000
A	1.039	0.0 ± 0.0	0.0 ± 0.0	14.2 ± 3.1	13.1 ± 1.9
B	1.047	0.0 ± 0.0	1.3 ± 0.4	7.7 ± 1.9	13.6 ± 0.9
C	1.054	0.3 ± 0.3	4.2 ± 1.3	17.1 ± 3.4	21.8 ± 3.5
D	1.061	1.5 ± 0.7	22.4 ± 3.2	27.1 ± 2.2	18.5 ± 2.4
E	1.069	nt[2]	nt	nt	14.7 ± 3.8
Unseparated	–	0.0 ± 0.0	2.6 ± 1.2	22.0 ± 5.5	9.8 ± 1.8

[1] Density at 20 °C
[2] nt = not tested.

separated into five subpopulations using density gradient centrifugation. A population of small, dense, chloroplast-laden cells could be plated at significantly lower densities (Table 4). Optimal plating densities may depend on whether the protoplasts are isolated from cell suspension, callus, or shoot cultures, but to my knowledge a good comparison between such sources within the same genotype has not been reported.

In addition to the effect on protoplast division, plating density has also been related to subsequent differentiation. For *Citrus*, high plating densities can inhibit the formation of embryoids [22,29]. Protoplasts at high densities exhibited earlier and higher rates of division but formed unorganized structures only.

3.6.4. *Culture conditions*

The environmental conditions under which the protoplasts are grown undoubtedly have some effect on culture, but this has not been well documented in woody species. The light intensity and duration are probably the most significant factors. For *Populus*, dark is important for initial culture [63], however in other species, culture in the light is best. In *Pyrus* [46], plating efficiency was highest (28%) under a 16 hour photoperiod, was intermediate (12.5%) in continuous light, and was the worst (6%) in darkness. The testing of simple light vs. dark culture regimes is warranted when working with a new genotype.

3.6.5. *Regeneration*

The conditions required to achieve differentiation from the dividing protoplast-derived cells are highly genotype specific. One differentiation scheme involves the induction of organogenesis from the protoplast-derived calli. In woody species, shoot organogenesis usually proceeds through an intermediary nodular callus stage [40,42]. The shoots can then be rooted as microcuttings. In some cases, a higher amount of regeneration is achieved by first inducing rhizogenesis from the protoplast-derived calli and subsequently to induce shoot organogenesis from the regenerated roots [49]. The alternative plant differentiation pathway is via somatic embryogenesis either direct or through an intermediary callus stage. Standard micropropagation methods can be used to establish recovered plants in the greenhouse and field.

One reason for the lack of differentiation from protoplasts of some genotypes may be the failure to accomodate changing hormone requirements throughout the culture period. For example, in *Pithecellobium* [68], 2,4-D was required for the initiation of protoplast division. Shoot differentiation, however, required that the 2,4-D be removed from the medium after only two weeks of culture. Thus, hormones used to initiate cell division may be inhibitory to morphogenesis from the protoplast-derived calli.

Another reason for the lack of differentiation in some genotypes is that the protoplast-derived calli do not respond to the common adenine-type cytokinins. In *Populus* NC-5339, differentiation could be obtained only if the substituted phenylurea, thidiazuron was included in the medium [63,64]. Thidiazuron is known to stimulate proliferation in shoot cultures of other woody species [27,83] and thus, may stimulate shoot regeneration from the protoplast-derived calli of other genotypes in addition to *Populus*.

4. Reports of protoplast-based genetic engineering in woody species

4.1. *Protoplast fusion*

One of the earliest reports of protoplast fusion involving tree species was between a herbaceous and two tree genotypes of *Datura* [69]. Hybrids were selected based on complementation of the ability of chlorophyll deficient mutants of *D. innoxia* to divide verses the inability of green wild type protoplasts of *D. sanguinea* and *D. candida* to divide. The hybrid nature of the regenerates was confirmed by amylase isozyme analysis. Chromosome numbers varied from aneuploidy to octaploidy and the hybrid expressed several growth abnormalities.

Hybrid calli of *Acer* were obtained by electrical fusion of protoplasts of a bromodeoxyuridine resistant mutant and an azaguanine resistant mutant [5]. Fused hybrids were selected from the homofusion products and the non-fused protoplasts by growth on hypoxanthine-aminopterin-thymidine (HAT) medium.

The most advanced reports in protoplast fusion are in the Rutaceae, perhaps reflective of early success in protoplast culture. Initially, model fusion systems were developed using sexually compatible genotypes to minimize the possible interference of somatic incompatibility. Oghawara *et al.* [52] employed a hybrid selection scheme based on complementation of the regenerative ability of *Citrus sinensis* 'Trovita' protoplasts with the subsequent expression of the trifoliate leaf character of *Poncirus trifoliata*. The regenerated plants had trifoliate leaves, intermediate morphology, and amphidiploid chromosome numbers. Restriction analysis of rDNA by ECORI showed the presence of a 11.5 kb fragment from *C. sinensis* and an 8.9 kb fragment from *P. trifoliata* confirming the hybrid nature of the regenerates. A similar hybrid was obtained between *C. sinensis* 'Hamlin' and *P. trifoliata* 'Flying Dragon' [15]. In this case, the hybrid nature of the regenerates was confirmed by intermediate morphology, chromosome counts, and malate dehydrogenase isozyme patterns.

Recently, allotetraploid somatic hybrid plants have been recovered after the fusion of protoplasts from two sexually incompatible species [16]. Plant regeneration was a function of complimentation combining the capacity for somatic embryogenesis of *C. sinensis* 'Hamlin' with the organogenic capability of *Severnia disticha*. Regenerated plants demonstrated intermediate morphology, had amphidiploid chromosome numbers, and contained the unique parental bands in phosphoglucose mutase and malate dehydrogenase zymograms.

Cybrids have also been produced in *Citrus* [85]. The cybridization scheme involved arresting the nuclear division of donor plants via gamma irradiation, coupled with transient inhibition of the metabolism of unfused recipient protoplasts by iodoacetate. Plants from the following donor-recipient fusion combinations were obtained: (1) Poorman × *P. trifoliata* × *C. limon* 'Villafranca'; (2) Poorman × *P. trifoliata* × *C. aurantium*; and (3) *C. aurantium* × *C. limon* 'Villafranca'. Regenerated plants from all combinations were recovered, each containing the nuclear encoded morphology of the recipient fusion partner. Mitochondrial DNA restriction profiles revealed novel mitochondrial genomes.

4.2. *Protoplast-based direct gene transfer*

Protoplast-based direct gene transfer has been recently applied to a few woody genotypes. The following cases of transient gene expression in protoplasts after electroporation or PEG-mediated DNA transfer have been reported: (1) the luciferase gene in *Pseudotsuga* and *Pinus* [18]; (2) the CAT (chloramphenicol acetyl transferase) gene in *Picea* [3,90]; and (3) the GUS (β-glucuronidase) gene in *Alnus* [71] and in *Picea* [90].

5. Conclusions and perspectives

Woody plants can no longer be considered generally recalcitrant to protoplast culture. Though many species may remain difficult or impossible to culture, numerous additional reports of success will likely emerge in the next few years. Protoplast isolation, culture, and regeneration techniques can not be generalized, but empirical testing can be reduced by making intelligent choices at each stage. The most critical factors are the protoplast source tissues, the type and concentration of cell wall degrading enzymes, the presence or absence of ammonium, organics, and growth regulators in the plating medium, and the plating density.

Though important contributions will continue to be made in the culture of woody protoplasts, the challenge now is to prove their utility for genetic engineering. Already, protoplast fusion has advanced to high levels in *Citrus* and should prove useful in future breeding programs by allowing the creation of new hybrids from sexually incompatible species. The application of such methods can now be extended to other woody genotypes in which protoplast regeneration has been achieved.

Protoplast-based direct gene transfer is still in its infancy in woody crops. The few reports of transient expression of reporter/marker genes suggest that the techniques developed for herbaceous plants are also applicable to woody species. The exact role of protoplast-based gene transfer methods in the future improvement of woody crops is difficult to predict. Thus far, such methods have resulted in only transient gene expression in trees, whereas *Agrobacterium*-mediated transformation was used to produce transgenic *Populus* trees [11] and the new biolistics technology [67] has already resulted in stably transformed calli and plants of *Populus* (B. McCown, personal communication). The advantages and limitations of most gene transfer methods have not yet been fully explored. Thus, in order to provide maximum flexibility to future genetic engineering programs research using a variety of techniques should continue.

6. Summary

Protoplast technology in woody species has advanced rapidly during the past few years. Woody plants can no longer be considered generally recalcitrant to protoplast culture. Whole plants have been regenerated from protoplasts of 46 woody genotypes representing 12 different families. Two of the successful cases are with gymnosperms.

Perhaps the most critical factor in successful protoplast culture of woody plants is the proper selection of donor cells. Both *in vitro* grown shoot cultures and embryogenic cell suspensions have proven successful for several genotypes. No woody plant has been regenerated from protoplasts isolated from the leaves of field or greenhouse-grown plant material. Other factors which have been commonly reported as important to woody plant protoplast culture include the type and concentration of the cell wall degrading enzymes; the presence or absence of ammonium, organics, and plant growth regulators in the culture medium; and the plating density.

The challenge now for those interested in woody plant protoplasts is to apply genetic engineering technology. Already, protoplast fusion has resulted in somatic hybrids between sexually incompatible species and in cybrid plants. Protoplast-based direct gene transfer has thus far resulted in transient expression of the CAT, GUS, and luciferase genes.

References

1. Altman A, Kaur-Sawhney R, and Galston AW (1977) Stabilization of oat leaf protoplasts through polyamine-mediated inhibition of senescence. Plant Physiol 50:570–574.
2. Attree SM, Bekkaoui F, Dunstan DI, and Fowke LC (1987) Regeneration of somatic mebryos from protoplasts isolated from an embryogenic suspension culture of white spruce (*Picea glauca*). Plant Cell Rep 6:480–483.
3. Bekkaoui F, Pilon M, Laine E, Raju DSS, Crosby WL, and Dunstan DI (1988) Transient gene expression in electroporated *Picea glauca* protoplasts. Plant Cell Rep 7:481–484.
4. Binding H (1974) Regeneration von haploiden und diploiden Pflanzen aus Protoplasten von *Petunia hybrida* L. Z. Pflanzenphysiol 74:327–356.
5. Burlion N and Northcote DH (1987) Electrical fusion of protoplasts from sycamore mutants hybrid selection on hypoxanthine-amimopterin-thymidine (HAT) medium. Protoplasma 138:23–31.
6. Butt AD (1985) A general method for the high-yield isolation of mesophyll protoplasts from deciduous tree species. Plant Sci 42:55–59.
7. Canas LA, Wyssmann AM, and Benbadis MC (1987) Isolation, culture and division of olive (*Olea europaea* L.) protoplasts. Plant Cell Rep 6:369–371.
8. Chung JD and Jo JD (1987) Callus formation from callus derived protoplasts of *Hibiscus syriacus*. Korean J Plant Tissue Culture 14:21–29.
9. Crossway A, Hauptli H, Houck CM, Irvine JM, Oakes JV, and Perani LA (1986). Micromanipulation techniques in plant manipulation. Biotechniques 4:320–334.
10. Doughty S and Power JB (1988) Callus formation from leaf mesophyll protoplasts of Malus × domestica Borkh. cv. Greensleeves. Plant Cell Rep 7:200–201.
11. Fillatti JJ, Sellmer J, McCown B, Haissig B, and Comai L (1987) Agrobacterium mediated transformation and regeneration of *Populus*. Mol Gen Genet 206:192–199.
12. Galun E, Aviv D, Raveh, D, Vardi A, and Zelcher A (1977) Protoplasts in studies of cell genetics and morphogenesis. *In*: Reinhard E and Alfermann AW (Eds.) Proceedings in Life Science, Springer, Berlin, pp. 301–312.
13. Gamborg OL, Miller RA, and Ojima K (1968) Nutrient requirements of suspension cultures of soybean root cells. Exp Cell Res 50:151–158.
14. Griesbach RJ (1987) Chromosome-mediated transformation via microinjection. Plant Sci 50:69–77.
15. Grosser JW, Gmitter Jr FG, and Chandler JL (1988) Intergeneric somatic hybrid plants of *Citrus sinensis* cv. Hamlin and *Poncirus trifoliata* cv. Flying Dragon. Plant Cell Rep 7:5–8.
16. Grosser JW, Gmitter Jr FG, and Chandler JL (1988) Intergeneric somatic hybrid plants from sexually incompatible woody species: *Citrus sinesis* and *Severinia disticha*. Theor Appl Genet 75:397–401.
17. Gupta HS, Rech EL, Cocking EC, and Davey MR (1988) Electroporation and heat shock stimulate division of protoplasts of *Pennisetum squamulatum*. J Plant Physiol 133:457–459.
18. Gupta PK, Dandekar AM, and Durzan DJ (1988) Somatic proembryc formation and transient expression of a luciferase gene in Douglas Fir and Loblolly Pine protoplasts. Plant Sci 58:85–92.

19. Gupta PK and Durzan DJ (1987) Somatic embryos from protoplasts of Loblolly Pine proembryonal cells. Bio/technology 5:710–712.
20. Hahne G and Lorz H (1988) Release of phytotoxic factors from plant cell walls during protoplast isolation. J Plant Physiol 132:345–350.
21. Harms CT and Potrykus I (1978) Fractionation of plant protoplast types by iso-osmotic density gradient centrifugation. Theor Appl Genet 53:57–63.
22. Hidaka T and Kajiura I (1988) Plantlet differentiation from callus protoplasts induced from *Citrus* embryo. Sci Hortic 34:85–92.
23. Huhtinen O, Honkanen J, and Simola, LK (1982/83) Ornithine- and putrescine-supported divisions and cell colony formation in leaf protoplasts of alders (*Alnus glutinosa* and *A. incana*). Plant Sci Lett 28:3–9.
24. Kao KN and Michayluk M (1975) Nutritional requirements for growth of *Vicia hajastana* cells and protoplasts at very low population density in liquid media. Planta 126:105–110.
25. Kaur-Sawhney R, Adams Jr WR, Tsang J, and Galston AW (1977) Leaf pretreatment with senescence retardants as a basis for oat protoplast improvement. Plant & Cell Physiol 18:1309–1317.
26. Kaur-Sawhney R, Flores HE, and Galston AW (1980) Polyamine-induced DNA synthesis and mitosis in oat leaf protoplasts. Plant Physiol 65:368–371.
27. Kerns HR and Meyer Jr MM (1986) Tissue culture propagation of *Acer* × *freemanii* using thidiazuron to stimulate shoot tip proliferation. HortSci 21:1209–1210.
28. Kitamura Y, Morkawa T, and Miura H (1989) Isolation and culture of protoplasts from cell suspension cultures of *Duboisia myoporoides* with subsequent plant regeneration. Plant Sci 60:245–250.
29. Kobayashi S, Ikeda I, and Uchimiya H (1985) Conditions for high frequency embryogenesis from orange (*Citrus sinensis* Osb.) protoplasts. Plant Cell Tissue Organ Culture 4:249–259.
30. Kobayashi S, Uchimiya H, and Ikeda I (1983) Plant regeneration from 'Trovita' orange protoplasts. Japan J Breed 33:119–122.
31. Kumar A and Cocking EC (1987) Protoplast fusion: a novel approach to organelle genetics in higher plants. Am J Bot 74:1289–1303.
32. Laine E, David H, and David A (1988) Callus formation from cotyledon protoplasts of *Pinus oocarpa* and *Pinus patula*. Physiol Plantarum 72:374–378.
33. Lang H and Kohlenbach HW (1988) Callus formation from mesophyll protoplasts of *Fagus sylvatica* L. Plant Cell Rep 7:485–488.
34. Lee JS, Lee SK, Jang SS, and Lee JJ (1987) Plantlet regeneration from callus protoplasts of *Populus nigra*. Res Rep Inst For Gen Korea 23:143–148.
35. Lee N and Wetzstein HY (1988) Protoplast isolation and callus production from leaves of tissue-cultured *Vitis* spp. Plant Cell Rep 7:531–534.
36. Lloyd G and McCown B (1980) Commercially-feasible micropropagation of mountain laurel, *Kalmia latifolia*, by use of shoot-tip culture. Proc Int Plant Prop Soc 30:421–427.
37. Matsuta N, Hirabayashi T, Akihama T, and Kozaki I (1986) Callus formation from protoplasts of peach cell suspension culture. Sci Hortic 28:59–64.
38. Merkle SA and Sommer HE (1987) Regeneration of *Liriodendron tulipifera* (family Magnoliaceae) from protoplast culture. Amer J Bot 74:1317–1321.
39. McCown BH (1986) Woody ornamentals, shade trees, and conifers. In: Zimmerman RH *et al.* (Eds.) Tissue Culture as a Plant Production System for Horticultural Crops, Martinus Nijhoff Publishers, Dordrecht, the Netherlands, pp. 333–342.
40. McCown BH (1988) Recent advances in protoplast culture of horticultural crops: Ornamental trees and shrubs. Sci Hortic 37:257–265.
41. McCown BH and Russell JA (1987) Protoplast culture of hardwoods. In: Bonga JM and Durzan DJ (Eds.) Cell and Tissue Culture in Forestry, Vol. 2: Specific Principles and Methods: Growth and Development, Martinus Nijhoff Publishers, Dordrecht, the Netherlands, pp. 16–30.
42. McCown BH, Zeldin EL, Pinkalla HA, and Dedolph RR (1988) Nodule culture: A

developmental pathway with high potential for regeneration, automated micropropagation, and plant metabolite production from woody plants. In: Hanover JW and Keathley DE (Eds.) Genetic Manipulation of Woody Plants, Plenum Press, New York, pp. 149–166.

43. Mii M, Seeni S, Fowke LC, and King J (1985) The isolation and cultivation of protoplasts from cell suspensions of a pantothenate-requiring auxotroph of *Datura*. Can J Bot 63:779–783.
44. Murashige T and Skoog F (1962) A revised medium for rapid growth and bio-assays with tobacco tissue cultures. Physiol Plant 15:473–497.
45. Murashige T and Tucker DPH (1969) Growth factor requirements of *Citrus* tissue culture. In: Chapman HD (Ed.) Proc. First Int. Citrus Symp., Vol. 3, University of California, Riverside, pp. 1155–1161.
46. Ochatt SJ and Caso OH (1986) Shoot regeneration from leaf mesophyll protoplasts of wild pear (*Pyrus communis* var. Pyraster L.) J Plant Physiol 122:243–249.
47. Ochatt SJ, Chand PK, Rech EL, Davey MR, and Power JB (1988) Electroporation-mediated improvement of plant regeneration from Colt cherry (*Prunus avium* × *pseudocerasus*) protoplasts. Plant Sci 54:165–169.
48. Ochatt SJ, Cocking EC, and Power JB (1987) Isolation, culture and plant regeneration of Colt cherry (*Prunus avium* × *pseudocerasus*) protoplasts. Plant Sci 50:139–143.
49. Ochatt SJ and Power JB (1988) Rhizogenesis in callus from Conference pear (*Pyrus communis* L.) protoplasts. Plant Cell Tissue Organ Culture 13:159–164.
50. Ochatt SJ and Power JB (1988) An alternative approach to plant regeneration from protoplasts of sour cherry (*Prunus cerasus* L.). Plant Sci 56:75–79.
51. Ochatt SJ, Rech EL, Davey MR, and Power JB (1988) Long-term effect of electroporation on enhancement of growth and plant regeneration of colt cherry (*Prunus avium* × *pseudocerasus*) protoplasts. Plant Cell Rep 7:393–395.
52. Ohgawara T, Kobayashi S, Oghawara E, Uchimiya H, and Ishii S (1985) Somatic hybrid plants obtained by protoplast fusion between *Citrus sinensis* and *Poncirus trifoliata*. Theor Appl Genet 71:1–4.
53. Oka S and Ohyama K (1985) Plant regeneration from leaf mesophyll protoplasts of *Broussonetia kasinoki* Sieb. (Paper Mulberry). J Plant Physiol 119:455–460.
54. Patat-Ochatt EM, Ochatt SJ, and Power JB (1988) Plant regeneration from protoplasts of apple rootstocks and scion varieties (*Malus* × *domestica* Borkh.) J Plant Physiol 133:460–465.
55. Pearce RS and Cocking EC (1973) Behaviour and culture of isolated protoplasts from 'Paul's Scarlet' rose suspension culture cells. Protoplasma 77:165–180.
56. Potrykus I, Shillito RD, Saul MW, and Paszkowski J (1985) Direct gene transfer-state of the art and future potential. Plant Mol Biol Rep 3:117–128.
57. Rao IVR, Mehta U, and Rom HYM (1982) Whole plant regeneration from cotyledonary protoplasts of *Crotalaria juncea*. In: Fujiwara A (Ed.) Plant Tissue Culture, 1982. Proc. 5th Int. Congress Plant Tissue and Cell Culture. Jpn. Assoc. Plant Tissue Culture, Tokyo, Japan, pp. 595–596.
58. Rao PS and Ozias-Akins P (1985) Plant regeneration through somatic embryogenesis in protoplast cultures of sandalwood (*Santalum album* L.). Protoplasma 124:80–86.
59. Rashid A (1988) Protoplast – isolation and cell regeneration. In: Rashid A (Ed.) Cell Physiology and Genetics of Higher Plants, Vol. II, CRC Press, Boca Raton, Florida, pp. 1–46.
60. Rech EL, Ochatt SJ, Chand PK, Davey MR, Mulligan BJ, and Power JB (1988) Electroporation increases DNA synthesis in cultured plant protoplasts. Bio/technology 6:1091–1093.
61. Rech EL, Ochatt SJ, Chand PK, Power JB, and Davey MR (1987) Electro-enhancement of division of plant protoplast-derived cells. Protoplasma 141:169–176.
62. Russell JA and McCown BH (1986) Techniques for enhanced release of leaf protoplasts in *Populus*. Plant Cell Rep 5:284–287.
63. Russell JA and McCown BH (1986) Culture and regeneration of *Populus* leaf protoplasts isolated from non-seedling tissue. Plant Sci 46:133–142.

64. Russell JA and McCown BH (1988) Recovery of plants from leaf protoplasts of hybrid-poplar and aspen clones. Plant Cell Rep 7:59–62.
65. Russell JA and McCown BH (1988) Osmotic microenvironments within tissue culture well plates are an important cause of variability in plant protoplast cultures. *In vitro* Cellular and Developmental Biology 24:605–608.
66. Russell JA and McCown BH (1988) Callus regeneration from protoplasts of a nonseedling gymnosperm tree. In: Hanover JW and Keathley DE (Eds.) Genetic Manipulation of Woody Plants, Plenum Press, New York, pp. 485 (abstract).
67. Sanford JC (1988) The biolistic process – a new concept in gene transfer and biological delivery. Trends in Biotechnology 6:229–302.
68. Saxena PK and Gill R (1987) Plant regeneration from mesophyll protoplasts of the tree legume *Pithecellobium dulce* Benth. Plant Sci 53:257–262.
69. Schieder O (1980) Somatic hybrids between a herbaceous and two tree *Datura* species. Z Pflanzenphysiol 98:119–127.
70. Schopke C, Muller LE, and Kohlenbach HW (1987) Somatic embryogenesis and regeneration of plantlets in protoplast cultures from somatic embryos of coffee (*Coffea canephora* P. ex Fr.). Plant Cell Tissue Organ Culture 8:243–248.
71. Seguin A and Lalonde M (1988) Gene transfer by electroporation in Betulaceae protoplasts: *Alnus incana*. Plant Cell Rep 7:367–370.
72. Shekhawat NS and Galston AW (1983) Mesophyll protoplasts of fenugreek (*Trigonella foenumgraecum*): Isolation, culture and shoot regeneration. Plant Cell Rep 2:119–121.
73. Shigekawa K and Dower WJ (1988) Electroporation of eukaryotes and prokaryotes: A general approach to the introduction of macromolecules into cells. BioTechniques 6:742–751.
74. Shillito RD, Paszkowski J, and Potrykus I (1983) Agarose plating and a bead type culture technique enable and stimulate development of protoplast-derived colonies in a number of plant species. Plant Cell Rep 2:244–247.
75. Sim GE, Loh CS, and Goh CJ (1988) Direct somatic embryogenesis from protoplasts of *Citrus mitis* Blanco. Plant Cell Rep 7:418–420.
76. Smith MAL and McCown BH (1982/83) A comparison of source tissue for protoplast isolation from three woody species. Plant Sci Lett 28:149–156.
77. Smith MAL, Palta JP, and McCown BH (1984) The measurement of isotonicity and maintenance of osmotic balance in plant protoplast manipulations. Plant Sci Lett 33:249–258.
78. Sticklen MB, Domir SC, and Lineberger RD (1986) Shoot regeneration from protoplasts of *Ulmus* × 'Pioneer'. Plant Sci 47:29–34.
79. Sticklen MB, Lineberger RD, and Domir SC (1985) Isolation and culture of protoplasts of *Ulmus* × 'Homestead'. Plant Sci 41:117–120.
80. Thompson JA, Abdullah R, and Cocking EC (1986) Enhanced protoplast division in rice (*Oryza sativa* L.) following heat shock treatment. J Plant Physiol 127:367–370.
81. Tremblay FM (1988) Callus formation from protoplasts of *Betula papyrifera* March. cell suspension culture. J Plant Physiol 133:247–251.
82. Tsai CK (1988) Plant regeneration from leaf callus protoplasts of *Actinidia chinensis* Planch. var. Chinensis. Plant Sci 54:231–235.
83. Van Nieuwkerk JP, Zimmerman RH, and Fordham I (1986) Thidiazuron stimulation of apple shoot proliferation *in vitro*. HortSci 21:516–518.
84. Vardi A (1977) Isolation of protoplasts in *Citrus*. Proc Int Soc Citriculture 2:167–174.
85. Vardi A, Breiman A, and Galun E (1987) *Citrus* cybrids: production by donor-recipient protoplast-fusion and verification by mitochondrial-DNA restriction profiles. Theor Appl Genet 75:51–58.
86. Vardi A and Galun E (1988) Recent advances in protoplast culture of horticultural crops: *Citrus*. Sci Hortic 37:217–230.
87. Vardi A, Hutchinson DJ, and Galun E (1986) A protoplast-to-tree system in *Microcitrus* based on protoplasts derived from a sustained embryogenic callus. Plant Cell Rep 5:412–414.

88. Vardi A, Speigel-Roy P, and Galun E (1982) Plant regeneration from *Citrus* protoplasts: Variability in methodological requirements among cultivars and species. Theor Appl Genet 62:171–176.
89. Widholm JM (1972) The use of fluorescein diacetate and phenosafranine for determining viability of cultured plant cells. Stain Tech 47:189–194.
90. Wilson SM, Thorpe TA, and Moloney MM (1989) PEG-mediated expression of GUS and CAT genes in protoplasts from embryogenic suspension cultures of *Picea glauca*. Plant Cell Rep 7:704–707.
91. Wu SC and Kuniyuki AH (1985) Isolation and culture of almond protoplasts. Plant Sci 41:55–60.

6. Maturation and rejuvenation in woody species

WESLEY P. HACKETT and JOHN R. MURRAY
Department of Horticultural Science, University of Minnesota, St. Paul, MN 55108, U.S.A.

1. Introduction

In the development of all woody plants from seed there is a so-called juvenile phase lasting up to 30–40 years in centain forest trees, during which flowering does not occur and cannot be induced by the normal flower-initiating treatment or conditions. In time, however, the ability to flower is achieved and maintained under natural conditions; at this stage, the tree is usually considered to have attained the adult or mature condition. The length of the juvenile period can be influenced by environmental and genetic factors [20]. Maturation is distinct from aging as used by Wareing [47] to describe changes such as reduced growth rate and type of branching due to increased size and complexity of the tree, which disappear when a scion is grafted onto a young rootstock or when a stem cutting is rooted.

The transition from the juvenile to the mature phase has been referred to as phase change by Brink [7], ontogenetic aging by Fortanier and Jonkers [12], or meristem aging (cyclophysis) by Seeliger [45] and Oleson [39]. Associated with this transition are progressive changes in morphological and developmental attributes including leaf cuticular characteristics [13] and chemical composition [24]; bark characteristics [40]; leaf plastochron [10], shape and thickness [44]; phyllotaxis, thorniness, and shoot orientation [44]; branch number and branching pattern [32]; tracheid width [40] and length [43]; shoot growth vigor [16,17,46]; seasonal leaf retention and stem pigmentation [10]; leaf chlorophyll content [19,3]; ability to form adventitious roots and buds [5,21]; partitioning of photosynthates into main stem, branches, or needles [32]; disease resistance [49]; insect resistance [30]; and cold resistance [28]. Changes in such characteristics during development vary from species to species and most change gradually during the period preceding the mature phase, resulting in transitional forms. Usually no distinct change in any one characteristic is apparent at the time the ability to flower is attained. Because attainment of the ability or potential to flower is the most consistent and often a late characteristic to change during development of woody species, it has usually been used to define maturation [20].

Phase change is of considerable theoretical importance relative to morphogenetic control, determination and differentiation in plant development. It has practical significance for the following reasons:

M. R. Ahuja (Ed.), Micropropagation of Woody Plants, 93–105.

1. The length of the juvenile period is inversely related to the breeding efficiency of woody perennials and to the selection of improved cultivars [23].
2. The ease of cuttage propagation and *in vitro* morphogenesis and somatic embryogenesis for all types of woody perennials is strongly affected by ontogenetic age [18,25].
3. The quantity and quality of productivity of a forest tree species is related to its degree of maturity[25].

Maturation and rejuvenation have been reviewed comprehensively several times in recent years [5,18,20]. The objective of this chapter isn't to present a review but rather to present a modified perspective of the phenomenon and thereby contribute to building a conceptual framework that is theoretically and practically useful in understanding and manipulating maturation related characteristics.

2. Maturation as determination and differentiation processes

Because of the observations described above, a question can be raised as to whether the changes in characteristics that occur in a particular species during maturation are mechanistically related through the control of their expression [6], or whether they are developmental processes that occur independently during the transition from juvenile to mature phenotypes. (I.e. occur independently with chronological or developmental age.) Systematic efforts to document the time course of changes in characteristics are sparse even though phase change has been described in many woody plant species. Greenwood *et al.* [19] have assessed the time course of change of six vegetative characteristics and flowering in *Larix laricina* by observing the level of expression of the characteristics in the subsequent growth of young scions that were removed from donor trees that ranged in age from 1–74 years and grafted on two year old containerized seedling rootstocks. The most pronounced change in the level of expression of the six vegetative characteristics was observed for scions from donor trees between 1 and 5 years of age. However, they found that the vegetative changes exhibited two distinct time course patterns when plotted as a function of the $\log_{10}$ of donor tree age. Orthotropic growth and the number of secondary branches per cm of scion shoot both decreased abruptly between donor tree ages of 1 and 5 years and showed little change after that. In contrast, height, stem diameter, chlorophyll content and rooting ability of the grafted scions were a linear function of the $\log_{10}$ of donor tree age, although the relative rates of change differed. Greenwood *et al.* [19] conclude that although there appears to be two distinct types of time courses, a question remains whether some of these vegetative maturation changes vary independently of one another. With regard to flowering, scions from juvenile (young) donor trees produced more strobili per tree, 3 years after grafting, than did scions from mature trees. Greenwood *et al.* [19] conclude from these results that achievement of a minimum size is more important than maturation state for flowering by *Larix*. This suggests that the mechanism for acquisition of flowering potential is not closely related to mechanisms for changes in the vegetative maturation characteristics studied. That flowering ability may not be closely related to changes in vegetative maturation characteristics is suggested by observations of some eucalyptus species in which flowering occurs before change in foliage characteristics. Similarly, in *Hedera helix* L. (English ivy) the juvenile characteristics of aerial stem root formation and anthocyanin accumulation can occur under certain conditions on plants that have the ability to flower.

Such observations indicate that there probably is not a single switch (regulatory event) that controls the acquisition of maturation-related characteristics. These observations also suggest that maturation is not a single phenomenon but a composite of processes. The unexpected production of greater numbers of strobili on scions from younger versus older donor trees of *Larix* and the expression of some juvenile characteristics on *Hedera* plants with flowering ability further suggests that there is not a rigid hierarchy of maturation-related characteristics. Based on these observations, we should consider alternatives to a single switch mechanism. For example, there may be several switches in parallel, one for each character or separate switches for sets of characters, or perhaps there are switches in series such that change in one character is a prerequisite for change in one or more characteristics so that there might be two stage or multi-stage switches or even a cascade.

These latter possibilities are suggested by results of experiments involving GA_3-induced rejuvenation of mature ivy plants [42]. Such GA_3-treated mature plants demonstrate a differential sensitivity of various physiological and morphological characteristics to GA_3 dosage such that a very low dose represses flowering ability, a slightly higher dose represses flowering ability and induces some rooting ability, an even higher dose represses flowering ability, induces greater rooting ability and induces some anthocyanin accumulation and so on until a saturating GA_3 dosage induces maximum expression of a complete complement of juvenile characteristics. These results don't prove that a change in one characteristic is required for a change in another but certainly are suggestive of such a possibility and strongly indicate that all charateristics are not controlled by a single switch or mechanism.

No matter what the mechanisms for or the timing of changes in phenotypic characteristics during development, they accumulate in what we term the mature condition. They accumulate because the characteristics are quite stable and are transmitted developmentally through cell division from one somatic cell generation to the next. Such stability is characteristic of a developmental state referred to as determination. Determination is operationally defined [33] as the state in which a cell or group of cells exhibit the same competence to respond to environmental signals and the same developmental (differentiation) fate whether grown *in situ*, in isolation (excised), or in a new place in or on the organism (grafted). This definition of determination certainly fits the observations for maturation related characteristics when assessed by the stability of shoot apical meristems for expression of charateristics or response to environmental signals. However, such stability or determination doesn't mean that maturation-related characteristics are permanent and non-reversible. There are many examples which indicate that they are reversible or manipulable. Such reversibility in animal systems is referred to as trans-determination. However, all of the maturation-related characteristics may not have the same ease of reversibility and in addition, ease of reversibility of any one characteristic may change over developmental time.

What is most interesting about the above observations from both a practical and theoretical biological standpoint is that these maturation-related characteristics are stable but reversible. This gives us the opportunity to study their control one at a time to investigate if and/or how one characteristic is related to another and to be able to determine if there is a single switch. To be able to manipulate and maintain each characteristic independently is also very attractive from a practical standpoint for forestry and horticulture.

What, then, is a useful way to think about maturation in woody species from both a practical and theoretical standpoint? From what has been outlined above it seems that a reasonable way to think about maturation is that it is a very dramatic and protracted example of determination and differentiation. The integrated expression of characters during development, which results in a recognizable phenotype, is a manifestation of the processes of determination followed by differentiation. Determination and differentiation, leading to structural and functional specialization of cell, tissue and organ types, rusults from the differential expression of genes [15] during particular phases of development, such as the juvenile and mature phases and is referred to as epigenetic variation [31,37]. Epigenetic control of determination and differentiation not only results in the difference in phenotypic characters observed in the juvenile and mature phases, but also determines the competency of the tissue to undergo subsequent cell, tissue, and organ differentiation. For example, stem cuttings from the juvenile phase of woody plants form adventitious roots easily, while cuttings from mature phase tissue form them slowly or not at all [21].

English ivy, a woody perennial with a juvenile phase lasting 10 or more years, exhibits a number of distinct differences in phenotypic characters between the juvenile and mature phases [22]. The types of characters that are obviously different between the two phases of ivy and probably the developmental bases of these characters are quite diverse. However, there is much in common between the juvenile and mature phases of ivy. The anatomy of the cell types and the structural organization of the leaf laminae, petioles and stems of the distinct phase phenotypes are similar. This similarity in cell types and structural organization but marked difference in the phenotypes of the two phases suggests there may be at least two levels of control of determination and differentiation. The control of primary differentiation, which determines the cell types and structural organization of the tissues, is fundamental and similar in both phases. Superimposed on primary differentiation, epigenetic control of determination and differentiation also leads to the unique expression of the phase-specific characters. In *H. helix*, the epigenetic control of determination and differentiation in a relatively small percentage of the cells in the plant body results in rather large differences in phenotype. For example, red stems and leaf petioles in juvenile plants as compared to green stems and petioles in mature plants is the result of differential expression of anthocyanin accumulation in four or five collenchyma cell layers in the hypodermis [36]. Likewise, the presence of aerial stem roots in the juvenile and the lack of them in the mature stem appears to be due to a difference in response to auxin in a very small discrete area of cells in the exterior part of the phloem and adjacent cortex [14].

The apical or eumeristem and its derivatives in primordial leaves and the subapical meristem region of the shoot apex are the sites of origin of different paths of determination and cellular, tissue and organ differentiation. This fact leads logically to the conclusion that these same sites are involved in the origin of maturation-related changes in development. It is well documented that the apical and subapical meristems of juvenile and mature *H.helix* are very different in size, configuration, and cell division activity and these differences in characteristics can be related to differences in some whole plant phenotypic characteristics such as phyllotaxis and internode length [22]. Changes in cells in meristematic regions that account for differences in anthocyanin accumulation could be very useful for analyzing the basis for differentiation of a maturation-related characteristic. Recently our laboratory has demonstrated that the lack of anthocyanin accumulation in the hypodermis of mature *H. helix* stems and petioles is due to very low expression of activity for dihydroflavonol reductase (DFR), an enzyme catalyzing a late step in the flavonoid biosynthetic pathway leading to production of anthocyanin [36]. This is the only example we know of in which expression of a phase-related characteristic is due to the differential activity of a single polypeptide. This work, along with that of Hutchison *et al.* [29] with chlorophyll a/b binding protein expression in larch, suggests that differential gene expression is involved in control of maturation-related characteristics and provides a unique opportunity to investigate the molecular basis for epigenetic control of maturation-related determination and differentiation in shoot apical meristems.

What mechanisms are involved in the epigenetic control of maturation-related determination and differentiation in shoot meristems? This is a very big unknown but as indicated above probably involves differential gene expression. It has recently been shown that maize transposable element expression, and timing and frequency of element transposition are developmentally regulated with more active expression in the juvenile portion of the plant than the mature portion [11]. Perhaps an understanding of the mechanisms of regulation of transposable element expression during development can be useful as a model for the mechanism for epigenetic control of maturation related characteristics.

Federoff and Banks [11] have recently studied the genetic mechanism responsible for the developmental control of suppressor-mutator (Spm) transposable element expression in maize. They note that the striking correlation between position on the plant and Spm expression points to the possibility that the element is regulated by the same genetic mechanism that determines the development al fate of apical and lateral meristems. Their discussion of the developmental implications of the mechanism that controls the Spm element is in many ways pertinent to an analysis of the mechanism of control of heritable, epigenetically controlled, maturation related characteristics if each maturation related characteristic is considered as a separate transposable element. We quote:

> 'The mechanism that controls the SPM element comprises components that correspond to the two distinct types of genetic events that are involved in implementing a developmental program: determination and differentiation. What we have called the SPM's *program* determines in a heritable but reversible fashion, how and when it will be expressed. The actual change in phase of activity corresponds to differentiation, the expression of the determined state and depends on the element's position in the plant as well as its ability to count developmental time. The genetic properties of the Spm's developmental programs constrain the choice of the responsible molecular mechanisms to those that are reversible (epigenetic). But the heritability of the element's developmental programming also implicates a type of epigenetic modification that is replicated with fidelity. Finally, the variety of relatively labile developmental expression patterns (programs) and the incremental transition from a stably active to a stably inactive state suggest the involvement of multiple epigenetic modification, each combination of which has a characteristic stability and genetic consequence.'

The Spm control system they describe is remarkably reminiscent of the characteristics of the phase-related changes in woody species described in preceding pages.

Epigenetic control of differentiation may be mediated through the methylation and demethylation of genes. This hypothesis is basad on the widely observed inverse correlation between the presence of 5-methyl-cytosine residues at a specific locus and the level of expression of that gene [9,26]. In plants, an inverse correlation between methylation and expression has been demonstrated for T-DNA genes of crown gall tumors [1], ribosomal genes [48], storage protein genes [4] and the activity of maize transposable elements including Spm [1].

DNA methylation and demethylation at specific sites is an attractive mechanism for long-term stable activity or inactivity of phase-dependent genes[12]. If this hypothesis is true, it is likely there are key regulatory events that establish a new pattern of methylation at specific loci during phase change or rejuvenation, but once established, methylation could maintain the long-term, stable activity or inactivity of phase-dependent genes. This is because a hyper- or hypomethylation pattern of a gene would be propagated through many cell cycles by the presence of a high-fidelity maintenance methyltransferase [26]. Maintenance methyltransferases recognize the hemimethylated state, methylated template strands and unmethylated nascent strands that arise with semiconservative DNA replication. Although the maintenence methyltransferases would have to exhibit a high degree of fidelity, they would only have to recognize the relatively non-specific sequences of hemimethylated DNA in order to self-propagate the phase-specific, stable state over extended periods of time. At or near meiosis, an abrupt but directed change in methylation pattern of phase related genes would be required to account for the sudden and complete return to juvenile phase characteristics of the zygote and seedling.

Recently, Banks *et al.* [2] have used three forms of the maize Spm transposable element (see above), stably active, stably inactive or cryptic, and programmable (can be either active or inactive and be reset and reprogrammed), to study the molecular mechanism for developmental regulation of the element. They found that element expression is related to the pattern and degree of cytosine methylation of the DNA immediately upstream and downstream from the element's transciptional start site. An active Spm element in a genome with a cryptic element promotes its partial demethylation but *not* its transcriptional activation. In contrast, a trans-acting Spm promotes extensive demethylation and transcriptional activation of an inactive programmable element. Banks *et al.* [2] conclude that these observations define the molecular components of the Spm element's developmental regulatory mechanism.

With maturation related characteristics similar variation in states of activity are available to use for analysis of the molecular basis for their developmental control. For example, in English ivy anthocyanin accumulation and DFR activity are stably expressed in juvenile stems and petioles, inducible in juvenile leaf lamina, not expressed in mature stems and leaves [36] but expressed after rejuvenation of mature shoots with GA_3. Thus, in genotypically identical plants, it is possible to have tissue for which the phase dependent DFR enzyme is; (1) stably determined and expressed; (2) stably determined, not expressed but inducible with sucrose; (3) stably *not* determined or expressed but reversible with GA_3. These various states of determination, expression, and inducibility suggest the involvement of more than one component of control at the molecular level.

3. Rejuvenation

Procedures to obtain juvenile material from mature plants is of considerable importance for the micropropagation of woody plants because establishment of explants *in vitro* and the induction of adventive embryos, and adventitious shoots and roots is greatly influenced by the maturation state of the tissue used for the primary explant [5,18,21]. Obtaining juvenile material from mature plants can be approached in two ways:

1. Use of juvenile parts of mature plants
2. Rejuvenation of mature parts of plants.

Procedures relating to both approaches have been recently reviewed critically and in detail by Hackett [20] and those relating to rejuvenation by Greenwood [18] and therefore comprehensive and detailed coverage will not be given here. However, procedures for rejuvenation of mature parts of plants is controversial and therefore, deserves further discussion here.

The term rejuvenation implies a reversal of the maturation process. The concept of rejuvenation is based on the general conclusion discussed earlier that although maturation related chracteristics are relatively stable, their expression is reversible or manipulable under certain conditions. If the premise presented earlier that maturation in woody species is a dramatic and protracted example of the processes of determination and differentiation is correct, then rejuvenation requires that these processes be interfered with or manipulated whether via a single switch or several switches in parallel or series. In this sense it may be more accurate to think of rejuvenation in terms of manipulation of determination and differentiation. As foresters, horticulturists, or developmental physiologists, what we want to do is manipulate the phenotypic characteristics either transiently or stably. When rejuvenation is thought of in these terms, it is not surprising that some maturation related characteristics appear to be more easily rejuvenated than others, that the ease of rejuvenation of any one characteristic may change over developmental time or that a particular treatment or treatment intensity or duration may influence one or some characteristics but not all of them. In this sense rejuvenation is a relative term and not an absolute one.

When maturation and rejuvenation are thought of in terms of the determination and differentiation processes, attention is directed to meristems in general, but particularly the shoot apical and subapical meristem and their early cellular antecedents where the determination and biochemical and structural differentiation processes are occurring. It follows then that manipulation of the determination and differentiation processes (rejuvenation) should focus on meristems because the cells in them are in a more plastic, less determined state.

There are several examples indicating that mature shoot apical meristems and early antecedents of them (particularly if isolated from the mature mother plant) as well as exogenously induced adventitious shoot meristems in mature tissue are susceptible to rejuvenation in varying degrees [20]. Monteuuis [34] limits this generalization, citing evidence that mature meristems are most susceptible to rejuvenation during a very short period corresponding mainly to time of budbreak.

The most striking example of meristem susceptibility to rejuvenation procedures is the work of Monteuuis [34]. He demonstrated that *in vitro* culture of an apical meristem removed during budbreak from a 100-year-old ortet of *Sequoiadendron giganteum* resulted in stabilized juvenile characteristics *in vitro* and in outdoor conditions without any evidence of losing them over a three year period. There are numerous examples [20] in which *in vitro* multiplication of shoots from mature shoot tips and axillary buds through several subcultures results in restoration of some juvenile characteristics, especially rooting potential and leaf shape. Long term observation of the resulting plants is seldom reported but Broome and Zimmerman [8] found that *in vitro* shoot tip derived blackberry plants displayed mature leaf shape when established in the greenhouse. In addition, there was no delay in flowering of *in vitro* derived blackberries [50] or apples [35,50]. As noted earlier, these observations indicate that some characteristics are more susceptible than others to rejuvenation, particularly when correlative influences of tissues subjacent to the apical meristem are present.

Grafting mature meristems onto juvenile seedling root stocks *in vitro* has also been shown to induce rejuvenation. Monteuuis (personal communication) has been successful in rejuvenating *Pinus pinnaster* and *Pseudotsuga menziesii* by using mature meristems consisting of the apical dome and two or three primordia and minimizing the amount of subjacent tissue remaining with the meristem. Plants derived from mature meristems by such procedures have been transferred to the greenhouse and show juvenile foliage characteristics, orthotropic growth habit and ease of rooting and establishment *in vitro* comparable to seedling derived material. However, grafting mature meristems of *Sequoiadendron giganteum* onto juvenile seedling rootstocks *in vitro* only results in transient juvenile characteristics in the grafted scion. Monteuuis (personal communication) attributes the transient nature of the rejuvenation in this case to the large amount of subjacent tissue that must be transferred with the apical meristem in order to obtain a successful graft union and shoot growth.

Pliego-Alfaro and Murashige [41] have demonstrated rejuvenation of adult avocado scions by graftage onto juvenile rootstocks *in vitro*. In this case lateral buds from greenhouse-grown flowering-age plants were used as scions. About 50% of shoots that emerged from such scions grafted onto juvenile rootstocks showed restored rooting competence. The shoots continued to show high rooting potential when the terminals of rooted shoots were severed and retested. The grafted adult stems elongated more rapidly than those of ungrafted adults and the growth characteristics resembled more nearly those of juvenile shoots. The authors conclude that the restored rooting competence reflects a phase reversal from adult to juvenile rather than a simple transfer of rooting cofactors from rootstock to scion.

Contrary to the above cited micro-grafting results, the observations of Navarro *et al.* [38] and Navarro (personal communication) indicate that the apical meristems of citrus grafted to seedling rootstocks *in vitro* are very stable. Resulting plants show no juvenile characteristics and flower as soon as cutting grown or conventionally grafted plants.

4. Conclusions

1. Maturation is probably not a single phenomenon but a composite of processes.

2. A useful way to think about maturation is that it is a very dramatic and protracted example of determination and differentiation.
3. Maturation related chracteristics are stable but reversible. However, all of the maturation related characteristics may not have the same ease of reversibility and the ease of reversibility of any one characteristic may change over developmental time.
4. Rejuvenation is a relative term and not an absolute one.

5. Summary

Modified concepts of maturation and rejuvenation are developed. It is concluded that maturation is probably not a single phenomenon but a composite of processes. This tentative conclusion suggests there is not a single switch (regulatory event) that controls the acquisition of maturation-related characteristics. Therefore, we should consider alternatives to a single switch mechanism.

No matter what the mechanisms for or the timing of changes in phenotypic characteristics during development, they accumulate in what we term the mature condition. They accumulate because the characteristics are quite stable and are transmitted developmentally through cell division from one somatic cell generation to the next. Such stability is characteristic of a developmental state referred to as determination. This stability or determination doesn't mean that maturation-related characteristics are permanent and non-reversible. There are many examples which indicate that they are reversible or manipulable. The stable but reversible nature of maturation-related characteristics gives us opportunity to study their control one at a time, to investigate if and/or how one characteristic is related to another, to be able to determine if there is a single switch, and, from a practical standpoint, to be able to manipulate and maintain each characteristic independently.

The integrated expression of characters during development, which results in a recognizable phenotype, is a manifestation of the processes of determination followed by differentiation. It seems that a useful way to think about maturation is that it is a very dramatic and protracted example of determination and differentiation.

The term rejuvenation implies a reversal of the maturation process. The concept of rejuvenation is based on the general conclusion that although maturation related characteristics are relatively stable, their expression is reversible or manipulable under certain conditions. If maturation in woody species is a dramatic and protracted example of the processes of determination of determination and differentiation, then rejuuvenation requires that these processes be interfered with or manipulated whether via a single switch or several switches in parallel or series. In this sense, it may be more accurate and useful to think of rejuvenation in terms of manipulation of determination and differentiation. As foresters, horticulturists, or developmental physiologists, what we want to do is manipulate the phenotypic characteristics either transiently or stably. When rejuvenation is thought of in these terms, it is not surprising that some maturation related characteristics appear to be more easily rejuvenated than others, that the ease of rejuvenation of any one characteristic may change over developmental time or that a particular treatment or treatment intensity or duration may influence one or some characteristics but not all of them. In this sense rejuvenation is a relative term and not an absolute one.

References

1. Amasino RM, Powell ALT, and Gordon MP (1984) Changes in T-DNA methylation and expression are associated with phenotypic variation and plant regeneration in a crown gall tumor line. Mol Gen Genet 197:437–446.
2. Banks JA, Masson P, and Federoff N (1988) Molecular mechanisms in the developmental regulation of the maize suppressor-mutator transporable element. Genes and Development 2:1364–1380.
3. Bauer H and Bauer U (1980) Photosynthesis in the leaves of the juvenile and adult phase of ivy (*Hedera helix*). Physiol Plant 49:366–372.
4. Bianchi MW and Viotti A (1988) DNA methylation and tissue-specific transcription of the storage protein genes of maize. Plant Mol Biol 11:203–214.
5. Bonga JM (1982) Vegetative propagation in relation to juvenility, maturity, and rejuvenation. In: Bonga JM and Durzan DJ (Eds.) Tissue Culture in Forestry, Martinus Nijhoff/Dr. W. Junk, Publishers. pp. 387–412.
6. Borchert R (1976) The concept of juvenility in woody plants. Acta Hort 56:21–36.
7. Brink RA (1962) Phase change in higher plants, and somatic cell heredity. Quart Rev Biol 37:1–22.
8. Broome OC and Zimmerman RH (1978) *In vitro* propagation of blackberry. HortScience 13:151–153.
9. Cedar H (1988) DNA methylation and gene activity. Cell 53:3–4.
10. Cordero R, Murray J, and Hackett WP (1985) Plastochron indices for juvenile and mature forms of *Hedera helix* L. (Araliaceae). Amer J Bot 72:324–327.
11. Federoff NV and Banks JA (1988) Is the suppressor-mutator element controlled by a basic developmental regulatory mechanism? Genetics 120:559–577.
12. Fortanier EJ and Jonkers HH (1976) Juvenility and maturity of plants as influenced by their ontogenetic and physiological aging. Acta Hort 56:37–44.
13. Franich RA, Wells LG, and Barnett JR (1977) Variation with tree age of needle cuticle topography and stomatal structure in *Pinus radiata* D. Don. Ann Bot 41:621–626.
14. Geneve RL, Hackett WP, and Swanson BT (1988) Adventitious root initiation in debladed petioles from juvenile and mature phases of English ivy. J Amer Soc Hort Sci 113:630–635.
15. Goldberg RB (1987) Emerging patterns of plant development. Cell 49:298–300.
16. Goodin JR (1964) Shoot growth rates as a factor in growth phase transitions in *Hedera*. Proc Amer Soc Hort Sci 84:600–605.
17. Greenwood MS (1984) Phase change in loblolly pine: shoot development as a function of age. Physiol Plant 61:518–522.
18. Greenwood MS (1987) Rejuvenation of forest trees. Plant Growth Regulation 6:1–12.
19. Greenwood MS, Hooper CA, and Hutchinson KW (1989) Maturation in larch I. Effects of age on shoot growth foliar characteristics and DNA methylation. Plant Physiol 90:406–412.
20. Hackett WP (1985) Juvenility, maturation, and rejuvenation in woody plants. Hortic Rev 7:109–155.

21. Hackett WP (1988) Donor plant maturation and adventitious root formation. In: Davis TD, Hassig BE, and Shankhla N (Eds.) Dioscorides Press, Portland, OR.
22. Hackett WP, Cordero RE, and Srinivasan C (1987) Apical meristem characteristics and activity in relation to juvenility in *Hedera*. In: Atherton JG (Ed.) Manipulation of Flowering, Butterworths, London.
23. Hansche PE and Beres W (1980) Genetic remodling of fruit and nut trees to facilitate cultivar development. HortScience 15:710–715.
24. Hayman AR, Perry NB, and Weavers RT (1986) Juvenile-adult chemical dimorphism in foliage of *Dacrydium biforme*. Phytochem 25:649–653.
25. Heybroek HH and Visser T (1976) Juvenility in fruit growing and forestry. Acta Hort 56: 71–80.
26. Holliday R (1987) The inheritance of epigenetic defects. Science 238:163–170.
27. Hood JV and Libby WJ (1979) Continuing effects of maturation state in radiata pine and a general maturation model. In: Hughes KW, Henke R, and Contatin M. (Eds.) Proc Symposium on Propagation of Higher Plants Through Tissue Culture. University of Tennessee, Knoxville, TN.
28. Hood JV and Libby Jr WJ (1980) A clonal study of intraspecific variability in radiata pine. I. Cold and animal damage. Austral For Res 10:9–20.
29. Hutchinson KW, Singer PB, and Greenwood MS (1988) Gene expression during maturation in eastern larch. In: Halgren J-E (Ed.) Proc. Frans Kempe Symp. 8:101–114.
30. Kearsley MJC and Whitman TG (1989) Developmental changes in resistance to herbivory: Implications for individuals and populations. Ecol 70:422–434.
31. Lewin B (1983) Genes. John Wiley and Sons. New York. pp. 354–355.
32. Libby Jr. WJ and Hood JV (1976) Juvenility in hedged radiata pine. Acta Hort 56: 91–98.
33. McDaniel C (1982) Shoot meristem development. In; Barlow P and Carr DJ (Eds.) Positional Controls in Plant Development. Cambridge University Press.
34. Monteuuis O (1988) Maturation concept and possible rejuvenation of arborescent species. Limits and promises of shoot apical meristems to ensure successful cloning. In Union Forest Research Organizations Meeting on Breeding Tropical Trees. Pattayai, Thailand, Now–Dec 1988.
35. Mullins MG (1985) Regulation of adventitious root formation in micorcuttings. Acta Hort 166:53–60.
36. Murray JR (1988) Flavonoid metabolism in relation to differential anthocyanin accumulation in juvenile and mature phase *Hedera helix* L. Ph D Thesis, University of Minnesota.
37. Nanney DL (1958) Epigenetic control systems. Proc Nat Acad Sci 44:712–717.
38. Navarro L, Rorstacher CN, and Miurashige T (1975) Improvement of shoot tip grafting *in vitro* for virus-free *Citrus*. J Amer Soc Hort Sci 100:471–479.
39. Oleson PO (1978) On cyclophysis and topophysis. Silvae Genetica 27:173–178.
40. Oleson PO (1982) The effect of cyclophysis on tracheid width and basic density in Norway spruce. For Tree Improvement Arbor (Horsholm) 15:1–80.
41. Pliego-Alfaro F and Murashige T (1987) Possible rejuvenation of adult avocado by graftage onto juvenile rootstocks *in vitro*. HortScience 22:1321–1324.
42. Rogler CE and Hackett WP (1975) Phase change in *Hedera helix*: induction of the mature to juvenile phase change by gibberellin A_3. Physiol Plant 34:141–147.
43. Rumball W (1963) Wood structure in relation to heteroblastism. Phytomorphoplogy 13: 206–214.
44. Schaffalitzky de Muckadell (1959) Investigations on aging of apical meristems in woody plants and its importance in silviculture. Forstl Forsgsv Danm 25:310–455.
45. Seeliger R (1924) Topophysis und zyklophysis pflanzlicker organe und ihre bedeutung für die Pflanzenkultur. Angew Bot 6:191–200.
46. Sweet GB and Wells LG (1974) Comparison of the growth of vegetative propagules and seedlings of *Pinus radiata*. New Zealand J For Sci 4:399–409.

47. Wareing PF (1959) Problems of juvenility and flowering in trees. J Linn Soc Lond Bot 56:282–289.
48. Watson JC, Kaufman LS, and Thompson WF (1987) Developmental regulation of cytosine methylation in the nuclear ribosomal RNA genes of *Pisum sativum*. J Mol Biol 193: 15–26.
49. Zagory D and Libby WJ (1985) Maturation-related resistance of *Pinus radiata* to western gall rust. Phytopath 75:1443–1447.
50. Zimmerman RH (1981) Micropropagation of fruit plants. Acta Hort 120:217–222.

7. Synthetic seeds in forest trees

PRAMOD K. GUPTA and MARY KREITINGER
Technology Center, Weyerhaeuser Company, Tacoma, WA 98477, U.S.A.

1. Introduction

Artificial seed consists of somatic embryos surrounded by a protective coating. The coating around the somatic embryo provides the necessary protection during storage, handling and mechanical planting. The artificial seed may contain nutrients to provide the energy required for germination which is normally supplied by endosperm or female gametophyte tissue. In this case either a mechanical barrier or antimicrobial compounds are needed in the seed to prevent contamination by microorganisms. Hormones may also be included to aid development during germination.

The first report of encapsulation was by Keto and Janic (1985) [37,38] who successfully germinated carrot somatic embryos coated in a wafer disc made of 2.5% polyoxyethylene. Each wafer contained many embryos. Desiccation of these encapsulated embryos resulted in only as low percentage germination, however. Higher percentage of germination have been obtained from desiccated somatic embryos without encapsulation such as soybean [61]d, orchid grass, grape [22] and wheat [12].

Redenbaugh *et al.* (1986) [56] reported encapsulation of alfalfa and celery somatic embryos in a hydrated gel. They tested different types of gels (see Table 1) and found the most suitable gel to be 2% sodium-alginate complexed with calcium. Sodium-alginate was dropped from a separatory funnel into a 0.1M solution of $Ca(NO_3)_2$ or $CaCl_2$. Somatic embryos were inserted into the drop of sodium-alginate as it was released from the tip of the funnel. After a 15–30 minute complexing period, the capsules are rinsed with water to remove excess calcium and sodium nitrate. These encapsulated somatic embryos retained viability when stored in an airtight container at 4 °C for extended periods of time. Somatic embryos of alfalfa, celery, carrot, lettuce, Brassica, cotton and corn have also been encapsulated in sodium-alginate and converted into plants *in vitro* [57]. Maximum success has been achieved with alfalfa with a 10% germination in soil [58].

In order to produce artificial seed on a commercial scale, it is first necessary to establish somatic embryogenesis in suspension culture. Liquid culture systems are amenable to automation, making them economically feasible relative to solid culture systems. In addition, embryo development may be controlled much more readily in liquid, especially if a bioreactor is utilized.

M. R. Ahuja (Ed.), Micropropagation of Woody Plants, 107–119.

Table 1. Gels for encapsulation of somatic embryos [18].

Gel	Concentration (%)	Complexing agents	Concentration (mM)
sodium or potassium alginate	0.5–5.0	$CaCl_2$	30–100
		$LaCl_2$	
		$CoCl_2$	
		$FeCl_2$	
		$Ca(NO_3)_2$	
		$Ca(OH)_2$	
sodium alginate	2.0		
with gelatin	5.0	$CaCl_2$	300–100
carrageenan	0.2–0.8	KCl	500
with locust bean gum	0.4–1.0	NH_4Cl	500
guar gum	2.0	sodium tetraborate	100–120
agar	5–8	tannic seed	100
gelrite	0.15–0.25	none	–
sodium pectate	2.0	$CaCl_2$	100
		$CaSO_4$	100

To maximize efficiency, the suspension culture would contain a large proportion of somatic embryos developing in synchrony. Maturation of somatic embryos should be comparable to that of zygotic embryos in order to optimize germination. A desiccation treatment may be needed to complete embryo maturation. Even if this were not the case, it would be advantageous to be able to desiccate embryos for ease of storage.

Somatic embryos must be singulated and encapsulated. Manually, this is a very labor intensive process and makes the cost of artificial seed prohibitive. Success has been achieved in automating the encapsulation procedure [19]. The cost of artificial seed containing alfalfa somatic embryos was calculated to be 0.026 cents each. This price is still significantly higher than that of true seed (0.00066 cents each) but gains in yield realized with clonal populations of superior genotypes will outweight the extra cost. This cost includes labor, materials and mechanical encapsulation [59]. Artifical seed production will significantly reduce the cost of somatic seedlings since the embryos can go directly from an *in vitro* environment to the soil.

2. Application to forest trees

Most forest trees are propagated through seeds. Seeds of superior genotypes are produced in seed orchards. It takes approximately 20–30 years to produce large quantities of seeds for reforestation. Somatic embryogenesis has the potential to reduce this time dramatically. Millions of somatic embryos can be produced in the lab all year round. Embryonal masses can be stored in liquid nitrogen until clones have been field tested, which will take 7–10 years. Plantlets have been regenerated from embryonal masses of Norway spruce, loblolly pine [23] and white spruce [36] which have been stored in liquid nitrogen. After field testing is completed, embryonal masses corresponding to the superior genetotypes can be thawed and large numbers of somatic embryos can be produced in a relatively short amount of time. This technology is also very important for propagation of genetically engineered trees. Transformed embryonal masses can be cryo-stored until a genetically engineered clone has been tested and proven successful in the field.

3. Somatic embryogenesis

3.1. *Conifers*

Since 1985 excellent success has been achieved on somatic embryogenesis in conifers. Hakman *et al.* (1985) [30] reported somatic embryogenesis in Norway spruce and Gupta and Durzan (1986) [24] reported somatic polyembryogenesis in sugar pine. Now there are more than fifty papers published on somatic embryogenesis in different species of conifers (Table 2). The most successful plant regeneration has been with Norway spruce and white spruce, but somatic seedlings have also been regenerated with Douglas-fir and loblolly pine. Most of the reports on somatic embryogenesis in conifers have been on work with cultures initiated from immature or mature seed. The exception is Norway spruce where somatic embryogenesis has been initiated from cotyledons. The highest rate of somatic embryogenesis has been obtained from immature seed 4–5 weeks after fertilization [25,30]. There have been no reports of somatic embryogenesis from mature plant tissue.

Embryonal suspensor masses (ESM) were obtained by culturing immature zygotic embryos on modified MS [24] or DCR [26] basal medium supplemented 2,4-D, kinetin and BAP, and with reduced sources of nitrogen such as casein hydrolysate, glutamine, asparagine and arginine. ESM were maintained by subculturing at 12–15 day intervals. ESM can also be maintained in suspension culture by subculturing at 6–7 day intervals [27]. Somatic embryos were multipled by true-to-type conifer cleavage polyembryogenesis [15].

ABA has been found to be very important in the maturation of somatic embryos [1,2]. ABA inhibits the cleavage polyembryony which allows for the development and maturation of individual somatic embryos. This effect was first reported by Durzan and Gupta (1987) [15] with Douglas-fir followed by several other reports with Norway spruce [5,9] and white spruce [14,31,67].

Although there are numerous reports on successful somatic embryogenesis in conifers, there are very few on conversion and establishment of somatic seedlings. However, at the present time, there are several thousand Norway spruce and small numbers of both Douglas-fir and loblolly pine established in soil at Weyerhaeuser Company, Washington [28].

Table 2. Somatic embryogenesis in conifers.

Species	Explants	Response	Converted	Reference
Norway Spruce	IME	ESM, PL	No	[30]
Picea abies			No	[32]
			Yes	[5,6,7]
	ME	ESM, PL	Yes	[68]
			Yes	[24]
			Yes	[69]
			No	[67]
			No	[35]
			Yes	[66]
	cotyledon	ESM, PL	No	[40]
			Yes	[44]
	ME	ESM, PL	Yes	[28]
White Spruce				
Picea glauca	IME	ESM, PL	No	[33]
			Yes	[45]
			Yes	[14]
			Yes	[71]
			No	[51]
			No	[64]
			Yes	[31]
Picea engelmannii	IME	ESM	No	[71]
Black Spruce				
Picea mariana	IME	ESM, PL	No	[34]
Sitka Spruce				
Picea sitchensis	IME, ME	ESM, PL	No	[70]
			Yes	[41]
Larch				
Larix decidua	FG	ESM, PL	No	[52]
Hybrid Larch				
decidua × *leptolepsis*	IME	ESM, PL	No	[39]
Sugar Pine				
Pinus lumbertiana	IME, ME	ESM, PL	No	[25]
Loblolly Pine				
Pinus taeda	IME	ESM, PL	Yes	[27]
			Yes	[16]
			Yes	[29]
			Yes	[28]
	ESM		No	[53]
Pond Pine				
Pinus serotiana	IME	ESM	No	[6]
White Pine				
Pinus strobus	IME	ESM	No	[6]
			No	[17]
Hudrid Pine				
taeda × *resinosa*	IME	ESM	No	[8]
Radiata Pine				
Pinus radiata	IME	ESM, PL	Yes	[62]
Douglas-fir				
Pseudotsuga menziesii	IME	ESM, PL	Yes	[15,16]
			Yes	[28]

Table 2 (continued).

Species	Explants	Response	Converted	Reference
Redwood				
Sequoia sempervirens	IME	ESM, PL	No	[11]
White Fir				
Albies alba	IME	ESM	No	[60]

ME = Mature Embryo
IME = Immature Embryo
FG = Female Gametophyte
ESM = Embryonal Suspensor Mass
PL = Plantlet
SS = Somatic Seedling

3.2. *Angiosperms*

Somatic embryogenesis has been achieved in many ansiosperm trees (Table 3). Most of the reports indicate initiation of cultures from mature seeds or seedling explants. There are a few reports on somatic embryogenesis from mature trees such as sandalwood [42] and populus spp. [13]. However, there is not much data available on quality of somatic embryos, conversion or establishment in soil. There are no reports on field performance of somatic seedlings.

Table 3. Somatic embryogenesis in forest trees (Angiosperms).

Species	Explants	Response	Converted	Reference
Albizia lebbeck	hypocotyl	SE, PL	Yes	[20]
Bamboosa arundinacea	ME	SE, PL	Yes	[47]
Fraximes americana	ME	SE, PL	No	[54]
Juglan regia	cotyledon	SE, PL	Yes	[65]
Eucalyptus citriodora	cotyledon	SE, PL	Yes	[50]
E. gunii	hypocotyl	SE		[10]
Liquidambar styraciflua	cotyledon	SE, PL	No	[63]
Liriodendron tulipifera	IME	SE	No	[48]
Paulownia tomentosa	ME	SE, PL	No	[55]
Populus ciliata	leaves of mature trees	SE	No	[13]
P. nigra × *P. maximowiczii*	leaves of 1–2 old trees	SE, PL	Yes	[72]
Quercus rubra	IME	SE, PL	Yes	[21]
Santalum album	hypocotyl, mature trees, endosperm	SE, PL	Yes	[3] [43] [42]
Robinia pseudoacacia	IME	SE, PL	Yes	[49]

Table 4. Artificial seed production in forest trees.

Species	Explants	Response	Converted	Reference
Angiosperm				
Eucalyptus citriodora	sodium alginate + $Ca(NO_3)_2$	60	4	[46]
Sandalwood *Santalum album*	sodium alginate + $CaCl_2$	10	0	[4]
Conifers				
Sugarpine *Pinus lamberciana*	sodium alginate + $Ca(NO_3)_2$	0	0	[24]
Loblolly pine *Pinus taeda*	sodium alginate + $Ca(NO_3)_2$	0	0	[27]
Norway spruce *Picae abies*	sodium alginate + $Ca(NO_3)_2$	4	0	[16]

4. Studies on artificial seed production in forest trees

4.1. *Conifers*

Somatic embryos of sugar pine, loblolly pine and Norway spruce have been encapsulated in sodium-alginate gel [24,27]. The best results were obtained with 1.5–2% sodium-alginate and 0.1M $Ca(NNO_3)_2$ in half-strength MS basal medium containing 1% sucrose (Fig. 1, 2, 3). Four percent of encapsulated Norway spruce somatic embryos germinated *in vitro* [16] (Fig. 4). Encapsulated embryos remained viable after storage for four months at 4 °C. Encapsulated embryos of loblolly pine and sugar pine did not germinate.

4.2. *Angiosperms*

Somatic embryos of *Eucalyptus citirodora* [46] and sandalwood [4] have been encapsulated in sodium-alginate gel and stored at 4 °C. Somatic embryos of *E. citirodora* were encapsulated in 1.5% sodium-alginate and 0.1M $Ca(No_3)_2$ in MS basal medium. *In vitro* germination on MS medium with 2% sucrose was 60%. Germination in the soil was only 4% [46]. However, survival of *in vitro* germinated encapsulated embryos transferred to the soil was 50% in contrast to a survival rate of 20% for that of *in vitro* germinated nonencapsulated embryos. Bapat and Rao (1988) [4] tested different concentrations of sodium-alginate and $CaCl_2$ in MS basal medium for encapsulation of sandalwood somatic embryos. They obtained an *in vitro* germination of 10% when using a 3% sodium-alginate gel. Encapsulated embryos were germinated on a filter paper bridge with MS liquid medium. They were also able to regenerate embryos from encapsulated ESM which had been stored at 4 °C for 45 days. No data are available for survival of encapsulated somatic embryos planted in soil.

Figure 1. Somatic embryos of Norway spruce developing on solid media.

Figure 2. Encapsulated somatic embryos of Norway spruce in calcium alginate.

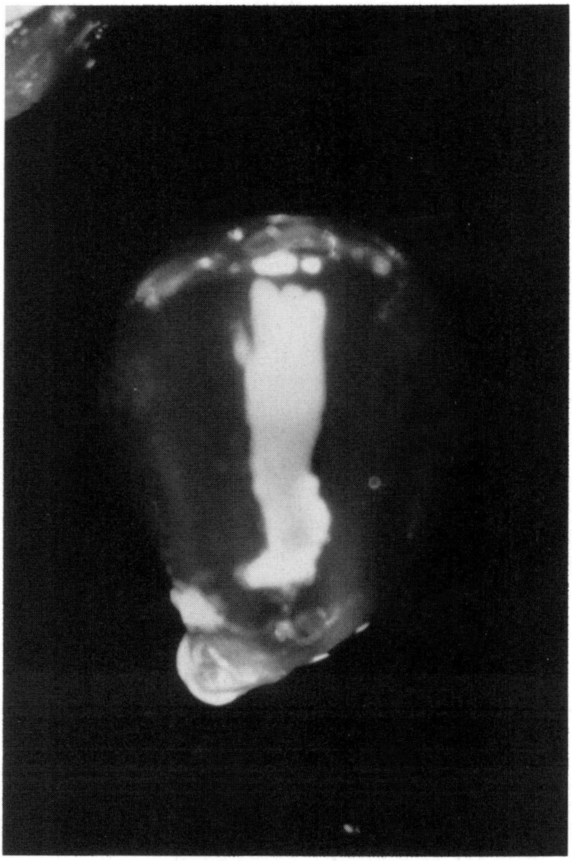

Figure 3. Close view of synthetic seed of Norway spruce (somatic embryo in alginate gel).

5. Conclusion

Tremendous success has been achieved in somatic embryogenesis in coniferous species, but further work is needed to improve embryos quality in order to obtain a rate of germination closer to that of zygotic embryos. To this date only low numbers of encapsulated somatic and zygotic embryos have been germinated. Even with herbaceous plants such as alfalfa, where the quality of somatic embryos is much closer to that of zygotic embryos, only 10% germination could be obtained.

Artificial seed technology still needs substantial research before it can be made commercially feasible.Efforts are expected to accelerate, however, due to the immense potential this technology holds for making clonal forestry a viable option for the future.

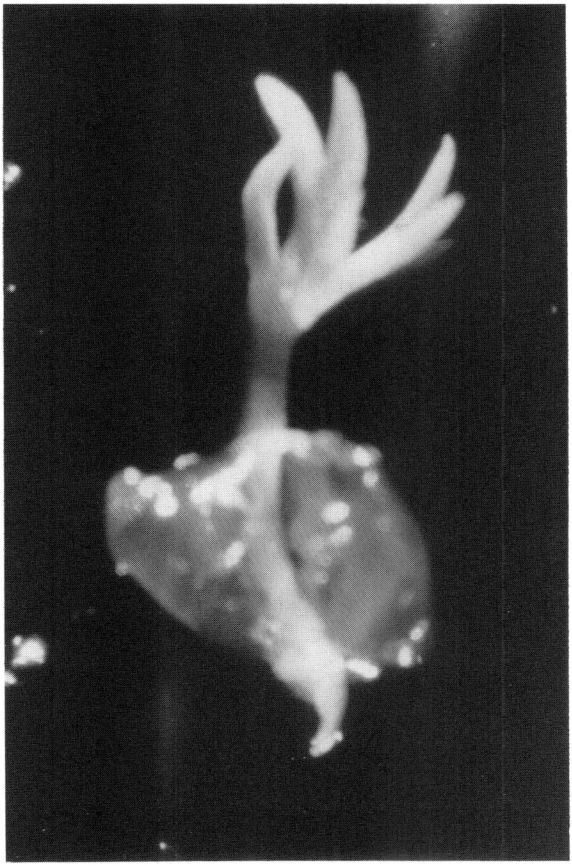

Figure 4. Germinating synthetic seed of Norway spruce in soil *in vitro*.

6. Summary

There are several reports on somatic embryo development with forest trees (Table 2, 3). However, there are not many reports on delivery of somatic embryos into the soil. Germination of somatic embryos in vitro and then transfer into the soil is labor intensive. Delivery of somatic embryos into the soil via synthetic seeds will reduce the cost of somatic seedlings dramatically. Encapsulation will protect somatic embryos from injury during storage, handling, mechanical plpanting and also provide nutrients during germination.

Despite the fact that synthetic seed technology holds great promise [56–59], there have been no reports to date on successful application for any plant species. However, there has been great progress in the production of good quality mature somatic embryos [21,22,28,50]. It is clear that additional research and development are needed towards synthetic seed production in order to determine the commercial feasibility of this technology.

References

1. Ammirato PV (1974) The effects of abscisic acid on the development of somatic embryos of caraway (Carum carvi L.) Bot Gaz 135:328–337.
2. Ammirato PV (1983) The regulation of somatic embryo development in plant cell culture: suspension culture techniques and hormone requirements. Bio/Tech 1:68–74.
3. Bapat V and Rao P (1979) Somatic embryogenesis and plantlet formation in tissue cultures of sandalwood. Ann Bot 44:629–630.
4. Bapat VA and Rao PS (11988) Sandalwood plantlets from 'synthetic seeds'. Plant Cell Rep 7:434–436.
5. Becwar MR, Wann SR, and Noland TL (1987) Somatic embryo development and plant regeneration from embryogeneic Norway spruce callus. Tappi J 70:155–160.
6. Becwar MR, Wann SR, Johnson MA, Verhagen VA, Feirer RP, and Nagmani R (1988) Development and characterization of *in vitro* embryogenic systems in conifers. In: Ahuja MR (Ed.) Somatic cell genetics of woody plants. Kluwer Academic Publ., Dordrecht, the Netherlands, pp. 1–18.
7. Becwar MR, Noland TL, and Wyckoff JL (1989) Maturation, germination and conversion on Norway spruce (*Picea abies* L.) somatic embvryos to plants. In Vitro Cell Dev Biol (in press).
8. Becwar MR, Wann SR, and Nagmani R (1989) A survey of initiation frequency of embryogenic callus from immature embryos on ten clones of loblolly pine (*Pinus taeda* L.) Can J For Res (submitted).
9. Boulay M, Gupta PK, Krogstrup P, and Durzan DJ (1988) Development of somatic embryos from cell suspension cultures of Norway spruce (*Picea abies* Karts.) Plant Cell Rep 7:134–137.
10. Boulay M (1987) *In vitro propagation of tree species.* In: Green CE, Somers DA, Hackett WP, Biesboer DD (Eds.) Plant Tissue and Cell Culture, Alan R. Liss Inc., N.Y. pp. 367–382.
11. Bourgkard F and Favre JM (1988) Somatic embryogenesis from callus of *Sequoia sempervirens* Pl Cell Rep 7:445–448.
12. Carman J, Jefferson N. and Campbell W (1987) ABA, O_2, temperature and relative humidity effect on embryoid maturation in *Triticum aestivum* L. calli. Ann Agron Meeting, Atlanta, Georgia, November 29–December 4, p. 148.
13. Cheema GS (1989) Somatic embryogenesis and plantlet regeneration from cell suspension and tissue culture of mature himalayan poplar (*Populus ciliata*). Pl Cell Rep 8:124–127.
14. Dunstan DI, Bekkaoui F, Pilon M, Fowke LC, and Abrams SR (1988) Effects of abscisic acid and analogous on the maturation of white spruce (*Picea glauca*2) somatic embryos. Plant Sci 58:77–84.
15. Durzan DJ and Gupta PK (1987) Somatic embryogenesis and polyembryogenesis in Douglas-fir cell suspension cultures. Plant Sci 52:229–235.
16. Druzan DJ and Gupta PK (1988) Somatic embryogenesis and polyembryogenesis in conifer. Adv Biotech Process 9:53–81.
17. Finer JJ, Kriebel H, and Becwar MR (1989) Initiation of embryogenic callus and suspension cultures of white pine (*Pinus strobus* L.). (submitted)
18. Fuji JA, Slade DT, Redenbaugh K, and Walker KA (1987) Artificial seeds for plant propagation. TIBTCH, 5:335–3359.
19. Gautz L, Upadhaya and Garrelt (1987) Vibratory separation of gel encapsulated plants propagules. Trans ASAE 30:652–656.
20. Gharyal PK and Maheshware SC (1981) *In vitro* differentiation of somatic embryos in a leguminous tree (*Alhizzia lebbeck*). Naturwissenschaften 68:379–380.
21. Gingas VM and Lineberger RD (1989) Asexual embryogenesis and plant regeneration in Quercus. Pl Cell Tissue Organ Cult 17:191–203.
22. Gray D (1987) Quiescence in monocotyledons and dicotyledonous somatic embryos induced by dehydraton. Hort Sci 22:810–814.

23. Gupta PK, Durzan DJ, and Finkle BJ (1987) Somatic polyembryogenesis in embryogeneic cell masses of *Picea abies* (Norway spruce) and *Pinus taeda* (loblolly pine) after thawing from liquid nitrogen. Can J For Res 17:1130–1133.
24. Gupta PK and Durzan DJ (1986) Plantlet regeneration via somatic embryogenesis from subcultured callus of mature embryos of *Picea abies* (Norway spruce). *In Vitro* Cell Dev Biol 22:685–688.
25. Gupta PK and Durzan DJ (1986) Somatic polyembryogenesis from callus of mature sugar pine embryos. Bio/Tech 4:643–645.
26. Gupta PK and Durzan DJ (1985) Shoot multiplication from mature trees of Douglas-fir (*Pseudotsuga menziesii*) and sugar pine (*Pinus lambertiana*). Plant Cell Rep 4:177–179.
27. Gupta PK and Durzan DJ (1987) Biotechnology of somatic polyembryogenesis and plantlet regeneration in loblolly pine. Bio/Tech 5:147–151.
28. Gupta PK, Pullman GS, Kreitinger ME, and Roger T (1989) Somatic embryogenesis in conifers. Abst *In vitro* cellular and development biology. 25(3).12A.
29. Gupta PK and Durzan DJ (1989) Loblolly pine. In: Bajaj YPS (Ed.) Biotechnology of Agriculture and Forestry, Vol. 12, Springer Verlag Berlin (in press).
30. Hakman I, Fowke LC, von Arnold D, and Eriksson T (1985) The development of somatic embryos in tissue cultures initiated from immature embryos of *Picea abies* (Norway spruce). Plant Sci 38:53–59.
31. Hakman I and von Arnold S (1988) Somatic embryogenesis and plant regeneration from suspension cultures of *Picea glauca* (white spruce). Physiol Plantarum 72:579–587.
32. Hakman I and von Arnold S (1985) Plantlet regeneration through somatic embryogenesis in *Picea abies* (Norway spruce). J Plant Physiol 121:149–158.
33. Hakman I and Fowke LC (1987) An embryogenic cell suspension culture of *Picea glauca* (white spruce). Plant Cell Rep 6:20–22.
34. Hakman I and Fowke LC (1987) Somatic embryogenesis in *Picea glauca* (white spruce) and *P. amriana* (balck spruce). Can J Bot 65:656–659.
35. Jain SM, Newton RJ, and Solta BJ (1988) Enhancement of somatic embryogenesis in Norway spruce (*Picea abies*). Theor Appl Genet 76:501–506.
36. Kartha KK, Fowke IC, Leung NL, Caswell KL, and Hakman I (1988) Induction of somatic embryos and plantlets from cryopreserved cell cultures of white spruce (*Picea glauca*). J Plant Physiol 132:529–539.
37. Kitto S and Janick J (1985) Production of synthetic seeds by encapsulating asexual embryos of carrot. J Amer Soc Hort Sci 110:227–282.
38. Kitto S and Janick J (1985) Hardening treatments increase survival of synthetically coated asexual embryos of carrot. J Amer Soc Hort Sci 110:283–286.
39. Klimaszewska J (1989) Somatic embryogenesis and plantlet regeneration from immature zygotic embryos of hybrid larch. Can J For Res (in press).
40. Krogstrup P (1986) Embryo-like structures from cotyledons and ripe embryos of Norway spruce (*Picea abies*). Can J For Res 16:644–668.
41. Krogstrup P, Eriksen EN, Moller JD, and Roulund H (1988) Somatic embryogenesis in Sitka spruce (*Picea sitchensis* (Bong.) Carr.). Plant Cell Rep 7:594–597.
42. Lakshimisita G (1986) Sandalwood. In: Bajaj YPS (Ed.) Biotechnology in Agriculture and Forestry, Vol. I, Springer Verlag Berlin, pp. 363–374.
43. Lakshimisita G, Raghava Ram NV, and Vaidyanathan CS (1979) Differentiation of embryoids and plantlets from shoot culture of sandalwood. Pl Sci Lett 15:265–270.
44. Lelu MA, Boulay M, and Arnaud Y (1987) Formation of embryogenic calli from cotyledons of *Picea abies* (L.) Karst. collected from 3 to 7 days old seedings. C R Acad Sci, Paris 305:105–107.
45. Lu CY and Thorpe TA (1987) Somatic embryogenesis and plantlet regeneration in cultured immature embryos of *Picea glauca*. J Plant Physiol 128:297–302.
46. Mascarenhas A, Nadgauda R, Khuspe SS, Gupta KK, Khau B, and Muralidharau M (1989) Biotechnological application of plant tissue culture to forestry species in India. In: Dhaven V and Mott R (Eds.) Application of Biotechnology in Forestry and Horticulture, Plenum Press, N.Y. (in press).

47. Mehta V, Ramanuja Rao IV, and Mohan Ram HY (1982) Somatic embryogenesis in bamboo. In: Fujiwara A (Ed.) Plant Tissue and Cell Culture. Tokyo, Japan, pp. 109–110.
48. Merkle SA and Sommer HE (1986) Embryogenesis in *Liriodendron tulipfera* tissue culture. Can J For Res 16:420–422.
49. Merkle SA and Wiecko AT (1989) Regeneration of *Robina pseudoacacia* via somatic embryogenesis. Can J For Res 19:285–288.
50. Muralidharan EM, Gupta PK, and Mascarenhas AF (1989) Plantlet production through high-frequency embryogenesis in long-term cultures of *E. citriodora*. Pl Cell Rep 8:41–43.
51. Nagmani R, Becwar MR, and Wann SR (1987) Single-cell origin and development of somatic embryos in *Picea abies* (L.) Karts. (Norway spruce) and *P. glauca* (Moench) Voss (white spruce). Plant Cell Rep 6:157–159.
52. Nagmani R and Bonga JM (1985) Embryogenesis in subcultured callus of *Larix decidua* Can J For Res 15:1088–1091.
53. Nagmani R and Becwar MR (1988) Factors affecting somatic embryo development in loblolly pine (*Pinus taeda*). Abst Proc 4th Intl Conifer Tissue Culture working group 8–12 Aug, Sakatoon, Canada, p. 18.
54. Preece JE, Zhao Jl, and Kung FH (1989) Callus production and somatic embryogenesis from white ash. Hort Sci 24(2):377–380.
55. Radojevic L (1979) Somatic embryogenesis and plantlets from callus culture of Paulownia tomentosa. Pflanzenphysical 91:57–62.
56. Redenbaugh K, Paasch BD, Nichol JW, Kossler ME, Viss PR, and Walker KA (1986) Somatic seeds: encapsulation of asexual plant embryos. Bio/Tech 4:797–801.
57. Redenbaugh K, Viss P, Slade D, and Fuji JA (1987) Scale up: Artificial seeds. In: Green CE, Sommers DA, Hackett WP, and Biesboer DD (Eds.) Plant Tissue and Cell Culture. Alan R. Liss Inc., New York, pp. 473–493.
58. Redenbaugh K (1988) Artificial seed application to tropical crops. Hort Sci (in press).
59. Redenbaugh K, Slade D, Viss P, and Fuji J (1987) Encapsulation of somatic embryos in synthetic seed coat. Hort Sci 22:803–809.
60. Schuller A, Reuther G, and Geier T (1988) Somatic embryogenesis from seed explants of *Abies alba*. P Cell Tissue Organ Cult 17:53–58.
61. Slawinska J and Obendorf R (1987) Soybean somatic embryo composition, respiration and water relation during maturation to a desiccation tolerant state. Ann Agron Meeting, Atlanta, Georgia, Nov. 29–Dec. 4, p. 153.
62. Smith DR, Jones C, and Grace L (1989) Somatic embryogenesis in *Pinus radiata* progress toward artificial seed. Can J For Res (in press).
63. Sommer HE and Brown CL (1980) Embryogenesis in tissue culture of sweetgum. For Sci 26:257–260.
64. Tremblay FM (1988) Somatic embryogenesis and plantlet regeneration from embryos isolated from stored seeds of *Picea glauca*. Abst In Proc 4th Intl Conifer Tissue Culture Working Group 8–12 Aug, Saskatoon, Canada, p. 20.
65. Tulecke W and McGranahan G (1985) Somatic embryogenesis and plantlet regeneration from cdotyledons of walnut, *Juglans regia*. Pl Sci 40:57–63.
66. Verhagen S.A. and Wann SR (1989) Norway spruce somatic embryogenesis: high-frequency initiation from light-cultivated mature embryos. Pl Cell Tissue Organ Cult 16:130–111.
67. von Arnold S and Hakman I (1988) Regulation of somatic embryo development in *Picea abies* by abscisic acid (ABA). J Plant Physiol 132:164–169.
68. von Arnold S and Hakman I (1986) Effect of surcose on initiation of embryogenic callus culture from mature zygotic embryos of *Picea abies* (L.) Karts (Norway spruce). J Plant Physiol 122:261–265.
69. von Arnold S (1987) Improved efficicency of somatic embryogenesis in mature embryos of *Picea abies*. J Plant Physiol 128:233–244.
70. von Arnold S and Woodward S (1988) Organogenesis and embryogenesis in mature zygotic embryos of *Picea sitchensis*. Tree Physiol 4:291–300.
71. Webb DT, Webster F, Flinn, BS, Roberts DR, and Ellis DE (1989) Factors influencing the

induction of embryogenic and caulogenic callus from embryos of *Picea glauca* and *Ficea engelmannii*. Can J For Res (in press).

72. Young GP and Sung HS (1988) *In vitro* organogenesis and somatic embryogenesis from punctured leaf of *Populus nigra* × *P. maximowiczii*. Pl Cell Tissue Oran Cult 15:95–105.

8. Genetics of micropropagated woody plants

R. M. SKIRVIN[1], H. ABU-QAOUD[2], S. SRISKANDARAJAH[3], and D. E. HARRY[1]

[1] *University of Illinois, Department of Horticulture and Forestry, Urbana-Champaign, IL 61801, U.S.A.*
[2] *University of Jordan, Department of Plant Production, Amman, Jordan*
[2] *University of Sydney, School of Crop Science, Sydney, NSW, Australia*

1. Introduction

To perform genetic studies with any organism, it is imperative that a source of variation be available. In the past variation has come from wide crosses, germplasm collection and utilization, mutagens, and sports. With the advent of tissue culture protocols, variation has been ubiquitious. The variability associated with tissue culture has provided a pool of variation upon which selection pressure has been imposed to isolate unique forms of clones. This variation, known as somaclonal variation, has become important for plant improvement, but the genetic basis of this variation remains largely unknown. In this chapter we will explore different types of variation and then discuss relevant investigations that have been used to elucidate the genetics of the variation.

2. Tissue culture as a source of variation

The amount of variation arising *in vitro* depends on the clone, age of the culture, use of mutagenic agents, and selection pressure applied to single cell clones for stress conditions such as salt level, microorganism or their byproducts, and specific metabolites. It is even possible to select plants for cold tolerance *in vitro* [29]. The exploitation of such natural and induced variation seems especially applicable to long established woody plant cultivars, such as 'Bartlett' pear and 'Delicious' apple (introduced in 1770 and 1893, respectively), which could be expected to have accumulated large numbers of mutant cells that may have stabilized into cell mixtures (chimeras) of various complexity [229].

2.1. *Natural variation and 'sports' as sources of variation*

The use of asexual methods for clonal improvement of woody plants is not new. In fact, many trees and treelike monocots, which are grown extensively

M. R. Ahuja (Ed.), Micropropagation of Woody Plants, 121–152.

for their fruits and oils, are propagated vegetatively [95]. The improvement of these horticultural and plantation crops, such as oil palm, coconut, rubber, peach, apple, walnut, relies heavily on cloning of unique genotypes that have arisen through bud mutations called 'sports' [229]. For instance, over 150 bud sports of 'Delicious' apple have been named as unique cultivars [75]. The genetic basis of certain sports have been studied and some traits are transmitted through the sexual cycle. For instance, the 'thornless' condition of 'Thornless Evergreen' and 'Thornless Loganberry' blackberry obtained *in vitro* [97,160,161] is due to dominant genes which have named '*Ste*' and '*Sfl*', respectively [96,208]. The topic has been reviewed by McPheeters *et al.* [161a].

Some mutations have proven to be so unique that they are of little value to the geneticist for inheritance studies. For instance, it is reported that the Seedless Navel orange arose as a bud sport on a seeded [246] orange tree in Brazil. The seedless character has proven to be extremely important economically, but its sterility limit its importance as a female parent. Recently hybrids of Navel and mandarin orange were produced by somatic hybridization [134]. Perhaps this hybrid plant can be used to elucidate the genetics of parthenocarpy.

There are many ways that variability can be obtained *in vitro* [85]. We discuss several types and sources of variation and then to briefly review literature related to the topic. Finally, we present examples to illustrate how the topic either has been used for genetic analysis or how such studies could begin.

2.2. *Types of tissue culture variation*

There are many types of variation that one can encounter *in vitro* [125a]. These include changes in growth habit of callus, habituation [162,163], variability in ability to regenerate, biochemical requirements and sensitivity (see Rains *et al.* [189] for a review of salt tolerance), disease resistance [209], changes in chromosome constitution [39,146,175] and changes manifested in whole plants. The subject has been reviewed by several authors [26,131,141,143, 153,163,227,230]. Entire books on the subject have been published [53,100,253].

2.3. *Sources of variation*

Cellular variation can result from either genetic mutation, epigenetic change, or a combination of both. Meins [163] suggests that genetic mutations involve random alterations in genetic constitution such as point mutations, deletions, duplications, and rearrangements of the genetic material. Epigenetic changes primarily involve time or tissue-specific 'selective gene expression rather than sorting out of genetic determinants' [163].

Meins [163] suggests that genetic mutations can be differentiated from epigenetic changes through several simple steps: (1) epigenetic changes are directed, i.e. it occurs regularly in response to specific inducers. Under these conditions, the rate of epigenetic change is high, greater than 10^{-3} per cell generation. In contrast, well-characterized gametic mutations in plants occur at rates of less than 10^{-5}; (2) the variant phenotype may be stable but it is potentially reversible and the reversion occurs at high rates; (3) The range of phenotypes generated are limited by the genetic potential of the cell; and (4) epigenetic changes are not transmitted meiotically. For instance, Dix and Street [48] reported a callus line of tobacco that was resistant to chilling. When the callus was subcultured, some clones retained the chilling resistance through two subclones; most showed no enhanced resistance. This reversion to the wild type suggests that the chilling effect was epigenetic, not genetic [163].

When variant cells remain totipotent, mutations can be distinguished from epigenetic changes by regenerating plants from variant clones and assaying tissues from the *R* generation and its selfed offspring. For a good test of heritability, it may be necessary to continue the selfing into at least the *R*2 generation [163].

Stable epigenetic changes can be used to obtain unique characteristics such as variegation, modified growth habit, lack of thorns, etc. Since this type of variation is not sexually based, it is useful only if it can be maintained by asexual propagation systems such as cuttings, layerage, or tissue culture. It cannot be passed sexually to another cultivar. Epigenetic changes are not permanent because they involve a network of gene(s) that are 'turned on' at some specific stage of development and turned 'off' later, however a plant may remain in the activated state indefinitely. The process of meiosis resets the gene network to a 'ground state'. On the other hand, changes with a genetic basis, can be maintained asexually or they can be exploited via the sexual system to improve many cultivars through breeding systems.

Where does variation come from? Gould [85] has reviewed this topic. In some cases the variation may be preexisting. For instance, when a complex explant is placed *in vitro*, it is unlikely that all cells will proliferate equally. For this reason, selection can begin very quickly *in vitro*. This is especially noticeable in new cultures where cultures often change in color, friability, and growth rate within the first few subcultures.

With time one cell line can be preferentially selected over another, resulting in the loss of the original genotype. It is also possible that a presumed pure culture can consist of more than one cell type. For instance, cytokinin habituated cells of tobacco can produce sufficient growth factors to support the proliferation of non-habituated cells in the same tissue [241].

The source of variability *in vitro* also can be traced to changes which are associated with passage through the tissue culture environment [66,67,68,168, 214]. Morrison *et al.* [168] suggest that 'much of this variation may be related to substances in the culture media that act as mutagens or maintain a degree of cell division for the plant cell's repair mechanisms cannot keep pace'. For instance, Evans and Bravo [64] report that ornamental tobacco plants regenerated from media with 6-benzylaminopurine exhibited more variation than cells grown without this cytokinin.

In other cases, the variation can be induced by mutagenic agents or by the tissue culture process itself [19,64]. These possibilities are discussed later in this chapter.

Although there have been numerous reports of 'mutation' in culture, few have been verified by breeding experiments (for a sample of the literature on the subject see Refs. 4 and 100).

To study the genetics of micropropagated plants, most reserachers follow the recommendation of Cheleff [26] who suggested that plants derived from cell and tissue culture be called the '*R*' generation. Progeny obtained by self-fertilization of the regenerated plants can then be designated as the '*R*1' or '*R*2' generation etc. for each successive sexual generation. This designation will be used in this chapter.

2.4. *Accessing variation via* in vitro *techniques*

To access natural or induced variation, it is necessary that a regeneration system be developed whereby whole plants can arise adventitiously from single cells or small groups of cells. Furthermore, the use of tissue culture or genetic engineering systems for genetically improving woody plant species requires a reliable regeneration system. Unfortunately, many woody plant species are recalcitrant.

Although adventitious organogenesis has been achieved in several woody species [7,114,195,231], most regenerants have been from juvenile tissues of seedling origin. Direct improvement of woody plant cultivars requires the control of regeneration from mature tissues. Promising results have been obtained from apple [61,70,118,122,256], *Prunus* [50,117] and pear [31,173a]. The subject of somatic embryogenesis in woody plants has been reviewed by Tulecke [249]. The introduction of thidiazuron (a cytokinin-like substance) has made many previously recalcitrant woody crops yield to the tissue culturist [31,70,129]. As the ability to regenerate becomes more common among the woody plants, the importance of tissue culture biotechnologies for woody plant improvement will increase.

The sources of variation which we have chosen to discuss in further detail below are (a) somaclonal variation, (b) genetic manipulation via genetic engineering, (c) embryo culture or rescue, (d) *in vitro* pollination, (e) haploidy, (f) chimeral segregation, and (g) protoplast fusion.

3. Somaclonal variation

Somaclonal variation is defined as genetic variation observed among progeny of plants regenerated from somatic cells cultured *in vitro*. Somaclonal variation is now known to be widespread among tissue culture-derived plants [25,26, 141,153,161,163,173c,174,227,230], but its genetic basis largely remains unknown in both woody and herbaceous species. The genetic basis of the phenomenon is seldom discussed, particularly in early reports [168]. Although the cause of somaclonal variation is often not known, this variation has proved useful in classical breeding programs [67,68,96,97,98].

Somaclonal variation has been observed among woody plant regenerants. The subject has been reviewed by Ahuja [3a]. Lester and Berbee [150] observed variation in height, number of branches, leaf traits, and chromosome number among callus-derived plants of *Populus nigra* and *P.* × *euramericana*. Callus cultures and regenerated shoots of *Citrus* and grape [228a] have shown tolerance to sodium chloride [15,234]. In a field test, micropropagated loblolly pine plants have shown a higher mortality, less shoot growth and more curved stems then seedlings [143a]. Micropropagated aspen (*Populus tremula*) cultures showed differences in root morphology [2a].

Many traits have been indentified from somaclonal variation studies [168], but few of these have been analyzed genetically. The lack of genetic analyses is not due to lack of interest by the researcher. Genetic analyses can be slow and time consuming, particularly for woody tree species. Although the genetic basis of somaclonal variation has not been subjected to thorough analysis for woody plants, its nature has been explored in herbaceous crops.

Herbaceous species have been used to elucidate the genetic nature of somaclonal variation. Some of this research will be discussed as a model for those interested in woody plants. The best work to date has been done with tomato (*Lycopersicon esculentum*), an herbaceous perennial that is self pollinated, true breeding and is usually seed propagated (see refs. [62,66,69,226] for more detail).

Evans and Sharp [66] began their studies by establishing cultures of tomato from leaves of germinated seedlings. Among the regenerants were detected chromosomal variants, single cell changes, and cytoplasmic genetic variants [168].

Some of the somaclones proved to be mutants of the parental cultivar. In one experiment 13 single gene mutations were observed among 230 regenerated plants. The traits ranged from changes in fruit color to jointless pedicels [66]. Genetic analyses were performed by either evaluating selfed *R*2's that had been selected in the *R*1 cycle or by crossing with known mutants. For instance, in the case of fruit color several progeny had yellow instead of red fruit. The yellow fruited variant was crossed to known yellow-fruited mutants to identify the gene's location in the tomato genome. In this manner, it was demonstrated that the somaclonal yellow mutation was on chromosome 3. Similar strategy was used to identify orange fruit color on Chromosome 10 and jointless pedicels and resistance to *Fusarium oxysporum* reside at opposite ends of Chromosome 11 [68,168]. Morrison *et al.* [168] also reported that some of their variability resulted from mitotic crossing over. They demonstrated this fact by using a tomato line heterozygous at four marked loci on chromosome 6. Among 61 regenerants, 19 exhibited recombination for one or more of the markers. Lee and Phillips [146] have published an entire review on the chromosomal basis of somaclonal variation.

Another source of variation may have been changes in organelle DNA. For instance, Kemble and Shepard [128] detected changes in mitochondrial DNA but not chloroplast DNA among potato plants regenerated from protoplasts. The importance of such variation is discussed elsewhere in this chapter.

An additional form of variation among the tomato regenerants involved changes in a regulatory gene which caused virescence in the leaves of tangerine-colored fruit. This character proved to be a single recessive gene which probably regulated the amount of chlorophyll in young leaves but not older leaves [168]. One somaclone had resistance to *Fusarium* race 2, due to a single dominant gene mutation [63.].

In some cases, genetic analysis may be impossible. Shepard and his colleagues, for instance, gained recognition for their work with protoclones of the herbaceous perennial 'Russet Burbank' potato (*Solanum tuberosum*) and its extreme variability. Variant somaclones included such diverse characters as stem length, number of blooms, weight of #1 tubers, tuber specific gravity, number of tubers, etc. [215,221,222)]. The variability was field tested and some types were stable [193]. These clones were perfect candidates for genetic analysis, but, unfortunately, the 'Russet Burbank' cultivar has sterility problems. Although classic genetic analyses were not possible, Shepard's group did examine the chromosomal status of their clones and concluded that much of the variability could be attributed to chromosomal rearrangemenets [128]. An interesting conclusion is that the somaclonal variation system is suited well for improvement of plants which are fully or partially sterile but maintained as a clone.

Some types of sterility can be analyzed sexually. For instance, some types of seedless fruits actually set seed, but the embryo aborts prior to maturity. However, it lives long enough to produce sufficient hormones to cause fruit expansion. Embryo rescue systems have been used to obtain sexually-derived seedlings of grape [83] and *Prunus* [192]. Modified versions of this technique could be used to rescue embryos of sterile somaclones. Bajaj and Gill [9] have carried the system one step further by returning embryo rescued *Gossypium* hybrid seedlings to tissue culture to induce somaclonal variation in the hybrid.

Somaclonal variation among forest trees has been briefly reviewed by Ho and Zsuffa [110]. Lester and Berbee [150] reported variation in height, branching and leaf traits of a *Populus* hybrid; variability was associated with unstable chromosome counts. Cheng and Smeltzer [30] found loblolly pine regenerants with changes in needle morphology and number.

3.1. *Selection and screening of disease resistant plants*

Woody plant cultivars are constantly attacked by various insects and fungal and bacterial diseases. Control requires pesticide applications throughout the growing season. The use of tissue culture and other techniques of biotechnology may be useful to either isolate genotypes with reduced susceptibility to disease or transfer disease resistance genes without altering other characters such as fruit quality. The strategies of selecting and evaluating for disease resistance *in vitro* have been discussed by Miller and Maxwell [165] and Larkin and Scowcroft [142] who used their sugar cane cultures to obtain somaclonal variation for *Helminthosporium* resistance. Shepard [221] reported potato protoclones with increased resistance to early and late blight. Evans [63] reports a tomato somaclone with a dominant gene that gives it resistance to *Fusarium* race 2. The subject has been reviewed by Sacristan [209].

Larkin and Scowcroft [142] reported on the isolation of *Saccharum* somaclones with increased resistance to *Helminthosporium sacchari*. Sexual analysis of the trait was impossible due to very high polyploidy and sterility problems. Therefore, the authors chose to assay the stability of the character through several vegetative generations (of 85 somaclones, 73% were stable in their reaction to toxin; 8% reverted to some degree; and 19% segregated). Six toxin-tolerant lines were carried through a second generation of tissue culture and maintained as callus for 3 to 6 months prior to regeneration. Three of the progeny segregated; three were stable. Among 60 somaclones derived from the second cycle of tissue culture, 40% had tolerance similar to the parent clone, 22% were more tolerant and 38% were more susceptible. This suggests that through repeated cycles of tissue culture, it may be possible to continue to select desirable characters while retaining previously selected characteristics.

Somaclonal variation has been used to obtain a putative *Septoria* leafspot-canker tolerant hybrid poplar. These are being greenhouse and field tested to assess the true situation [60]. Barlass *et al.* [12b] have tried to screen for resistance to Downy Mildew in grapes by establishing dual cultures of grape and pathogen. Joung *et al.* [123] has screened apple cultures for resistance to cedar-apple rust *in vitro*.

3.2. *Selection and screening of herbicide- and salt-resistance*

3.2.1. *Herbicide resistance*

The control of weeds is a worldwide problem that has proved ideal for *in vitro* manipulations. As long as a researcher has access to a regeneration system, it is possible to grow cells or organs in contact with a specific herbicide (or toxin, metabolite, salt, etc.) and select cell lines that have improved tolerance or, perhaps, resistance to the compound. The possibility has stirred much interest. The status of such research has been reviewed [27,37,113,257].

The best known example of herbicide resistance is that reported by [219] who were able to select a line of tomato with resistance to glyphosphate *in vitro*. The nature of the tolerance was later shown to due to amplified gene copies of 5-enolpyruvylshikimate-3-phosphate (EPSP) synthase, the enzyme that is inhibited by glyphosphate. Transgenic plants containing the EPSP synthase gene produced more of the enzyme and are relatively resistant to glyphosate. In another case, a bacterial gene which confers resistance to glyphosphate was isolated from *Salmonella* [36]. Glyphosate tolerane has been transferred to many species [257] such as tomato and potato, via the Ti plasmid. Glyphosate-tolerant *Populus* plants have been reported by Fillatti *et al.* [72]. The herbicide tolerance factor can be expressed in the whole plant, but not always at agriculturally significant levels [37,257].

3.2.2. *Salt tolerance*

Salt-tolerant cell lines of many herbaceous species have been selected. The subject has been reviewed by Rains *et al.* [189]. Salt tolerant *Citrus* [15,135,234] and grape [228a] cell lines have been selected *in vitro*. Labrun *et al.* [144] showed that cotyledon and hypocotyl tissues of grape embryos were more tolerant of NaCl than root tissue. Salt tolerant plantlets were derived from *Citrus* callus selected on relatively high [15] and low [234] levels of NaCl. The plantlets selected on high salt were so abnormal (without internodes) than they could not be cloned for further assessment. Plantlets selected on lower NaCl levels were more normal in appearance and expressed some salt tolerance.

4. Genetic manipulation via genetic engineering

Virtually all plant genetic engineering involves tissue culture or micropropagation at some stage, hence a brief discussion of genetic engineering is relevant to a general discussion of genetics and micropropagation. The subject has recently been reviewed for woody plants by Ahuja [4]. The relative difficulty of manipulating woody species by classical genetic systems, coupled with their ability to be vegetatively propagated on a commercial scale, makes the application of genetic engineering to woody plants attractive. In the following section, we first address the status of genetic engineering in herbaceous plants; we then summarize progress in woody angiosperms and gymnosperms.

4.1. *Candidate genes*

Genetic engineering is currently limited to characteristics that are primarily controlled by the expression of single genes (see [84] and [78] for reviews). However, most agronimically important traits are controlled by multiple genes, and the role of specific genes in basic growth processes is largely unknown.

As a consequence, a common strategy for genetic engineering in plants is to introduce genes that confer novel functions. For example, resistance to the herbicide bromoxynil is conferred by a bacterial gene (*bxn*, isolated from *Klebsiella ozaenae*) encoding a nitrilase that detoxifies bromoxynil by converting it to 3,5-dibromo-4-hydroxybenzoic acid [236]. A similar strategy has been used to engineer resistance to the herbicide phosphophinothricin [42,244]. Genes encoding the insecticidal toxin from the bacterium *Bacillus thuringsiensis* have been introduced into several plants where they confer resistance to insect herbivory [13,74,251]. Finally, plants containing genes that encode viral coat protetins from tobacco mosaic virus appear resistant to viral infection [14,45,198].

While genes that impart a novel function or product often originate from prokaryotes, they may also come from plants. Genes encoding proteinase inhibitors have been isolated from potato [210,211] and cowpea [107]. The cowpea gene confers resistance to herbivorous insect pests in transgenic plants [107]. Resistance to glyphosate in plants may result from introduction of a gene from *Salmonella* [36], or from a plant gene. A gene encoding 5-enol-pyruvylshikimate-3-phosphate (EPSP) synthase (the enzyme inhibited by glyphosate) has been isolated from tomato (discussed earlier). Transgenic plants containing the EPSP synthase gene produced more of the enzyme and are relatively resistant to glyphosate [219]. In this instance, herbicide resistance is mediated by modifying an endogenous function, i.e. increasing the abundance of a specific protein.

It is more difficult to reduce or repress expression of endogenous genes, but methods are now becoming available. One approach involves introduction of genes encoding antisense RNA, RNA transcripts that are complementary to those of an endogenous gene. Antisense RNA is believed to interfere with processing of normal RNA such that normal gene expression is inhibited [55]. In plants, antisense RNA appears effective in reducing the expression of introduced [44] as well as endogenous [206,220] genes. More recently, RNA enzymer (ribozymes) with endoribonuclease activity have been engineered to have high specificity for target sequences [103]. Antisense RNA and ribozymes both offer the possibility of introducing genes causing dominant loss-of-function phenotypes.

A final technology offers unique opportunities to identify and isolate genes with important biological functions [213]. Gene-tagging involves the insertion of a foreign segment of DNA into a gene which in turn disrupts its normal expression (see Shepherd [224] for review). If the inserted DNA segment has previously been cloned, and if the phenotype is noticeably altered by the disruption, the altered gene can be indentified using the inserted segment as a tag. Once cloned, the altered gene can be used to isolate a copy of a normal gene. Because of their ability to move in the genome, transposable elements are ideal for gene tagging. While transposable elements have been cloned from only a few plants, some transposable elements (e.g. *Ac* from maize [10,11] retain their ability to move after transfer into plants belonging to a different species from which they were isolated. Therefore, gene tagging using transposable elements may be broadly applicable in plants, even in species for which transposable elements have yet to be identified and cloned.

In terms of promising candidates for genes to manipulate, woody plants suffer the same difficulties of their herbaceous counterparts, but the problem is exacerbated by their typically long generation intervals. Relatively few single-gene traits have been characterized by traditional genetic analyses in woody plants, and even fewer have potential for improvement of commercially important traits [145,216], but see Thompson [245], for an interesting candidate). Application of genetic engineering to woody plants will endoubtedly benefit from experience with herbaceous plants. However, for the many growth processes that are unique to woody plants (e.g. maturation, leaf senescence, nutrient translocation, perennial habit, wood formation, etc.), research on herbaceous species is not sufficient. Efforts must be made to direct research towards the unique characteristics of woody plants.

The best example to-date of the application of genetic engineering to a woody species is the introduction of herbicide resistance into *Populus* [72], although other systems are currently being explored, e.g. walnut [157] and apple [116].

4.2. *Regulating transferred genes*

Once a specific gene has been identified and transferred, optimal utility can be realized only if its expression is properly regulated. Among the many factors regulating gene expression, cis-acting DNA sequences such as those associated with enhancers, promotors, or introns are all being investigated [24,71,126,154, 177,178,254]. Similarly, proteins that interact with DNA directly are being characterized [81,87,88,119,120]

While mechanism regulating gene expression are incompletely understood, empirical studies have yielded useful generalizations. Genes that are highly regulated (either developmentally or in response to environmental stimuli) normally are also highly regulated following their transfer to other plants of the same species or closely related species [184]. It is not clear how similar target and donor species must be for a transferred gene to function properly. However, some regulatory processes appear more sensitive to differences in genetic relatedness than others [57,81,112,127,151,255]. Furthermore, most regulatory functions are intrinsic to the specific sequences (i.e. regulatory sequences) that regulate expression of the protein or RNA (i.e. structural) coding sequences with which they are associated. Hence, regulatory sequences from a light-inducible gene such as the small subunit of RUBP-carboxylase can be used to control expression of structural genes for which a similar pattern of expression is desired [35,247]. These generalizations are encouraging for future applications because it will be possible to engineer DNA molecules that contain coding sequences for a desired attribute expressed in a desired fashion.

4.3. *Gene transfer methods*

Use of an effective vehicle to move a desired gene into plants is a crucial and often limiting step in genetic engineering. To date most transgenic plants have resulted from use of *Agrobacterium tumefaciens* [80,105]. A related species, *A. rhizogenes*, has also been used). These organisms harbor a tumor-inducing (Ti) plasmid, and as part of their pathogenic life cycle, transfer a portion (transfer-DNA or T-DNA) to the chromosomes of their plant host. Isolated T-DNA can be modified *in vitro* to include genes of interest, and reintroduced into bacterial cells. As these cells subsequently infect a plant, host cells receive the modified T-DNA that includes the foreign gene. Problems related to the use of *Agrobacterium* result from host range limitations, as well as in regenerating plants from transformed cells.

Several approaches have been devised to circumvent the biological limitations to *Agrobacterium*. Perhaps the most popular has been to introduce DNA directly into protoplasts (see Shillito and Saul [225] for review of general methods). While assays of transient gene expression can be done for any plant for which viable protoplasts can be isolated, transfer of foreign genes into plants is limited to species for which plants can be regenerated from protoplasts. Because of its broad success in animals, microinjection into plant cells appeared promising. Unfortunately, there are only a few reports of successful gene transfers [38,199,200,201], and these were done using protoplasts. Recently, bombardment of intact cells with microprojectiles has stirred considerable interest [20,32,132,133,158]. The method requires relatively little preparation of the recipient specimen, it appears amenable to a variety of organs and species, and can be used to transfer genes into different organelles. In addition to transformation of recalcitrant species such as soybean [32,158], microprojectiles have now been used to transfer genes into Douglas fir [82] and loblolly pine (A. Weisberg, North Carolina State University, pers. comm.).

Transferring genes and ensuring their regulation are intimately intertwined, so results and progress in these areas for woody plants are discussed together. Experience using *Agrobacterium* in woody plants is limited, but growing. In angiosperms, transformation has been reported for *Populus* [73,176,187], English walnut [40,157], olive [239], peach [99], citrus [166], apple [8,116], grape [12a,33,138], and willow [252]. Transformed plants have also been regenerated [73,116,157,239]. *Agrobacterium*-mediated transfer of genes into other woody angiosperms is being actively pursued.

The host range of *Agrobacterium* had been reported to exclude most gymnosperms [43]. A survey of additional *Agrobacterium* strains has identified several capable of infecting recalcitrant conifers such as pine [217,238], and gene transfer has been verified (Loopstra *et al.* in prep.). Agrobacterium-mediated transformation of spruce has been reported [56,111], and studies involving a number of other conifers are under way.

These studies demonstrate that routine procedures for herbaceous plants can also be applied (often with somewhat greater difficulty) to woody plants.

Similarly, DNA constructs (i.e. a regulatory sequence coupled with an appropriate structural sequence) that work well in herbaceous plants often work well after transfer to cells of woody plants, including conifers [16,82, 92,159,218]. These results are encouraging, but caution is warranted because they are limited to gene constructs under the control of relatively strong promoters that are constitutively expressed in herbaceous plants. Highly regulated genes have not yet been studied.

5. Embryo culture and rescue

Embryo culture techniques have direct value to the plant breeder who is trying to make crosses among plants with fertilization barriers of various sorts. It is also important as a support system to facilitate genetic studies with families, genera, species, and cultivars which are normally not sexually compatible [203]. Embryo culture and rescue systems can also aid in germination of recalcitrant seeds [91]. Barriers such as ploidy differences, chromosome alterations, and genetic incompatibility could lead to embryo abortion. The situation is often seen in hybridization studies of interspecific or intergeneric type [6,110]. Embryo culture can also be used directly for biochemical studies such as that reported for cocoa butter accumulation in *Theobroma cacoa* [136]. Dunwell [51] has summarized the literature concerning the use of embryo culture to rescue interspecific hybrids.

Embryo culture can be accomplished by either *in ovulo* (where the embryo and its supporting tissues are grown *in vitro* or embryo rescue systems. About 50 interspecific and intergeneric hybrids have been obtained through *in ovulo* embryo culture systems [110]. Most of the earliest work was done with herbaceous species; there are recent reports of success with woody perennials. *In ovulo* techniques have had the most value in the breeding of seedless grapes. For instance Emershad and Ramming [58] reported the technique which was further developed by Spiegel-Roy *et al.* [235], Gray *et al.* [86] and later by Emershad *et al.* [59]. In this manner genotypes which could not have been used as female parents, due to seedlessness, can be used for genetic studies.

The method has also been used to obtain interspecific hybrids between incompatible species as *Carica papaya* and *C. cauliflora* [260]. Ramming [192] has used the technique to culture small embryos from early maturing *Prunus* genotypes which normally abort prior to fruit ripening. Kouider *et al.* [137] and Savka *et al.* [212] have used the system to rescue embryos of *Populus deltoides*.

Limited success with pine proembryos [139,242] and Norway spruce [140] has been reported. Proembryo degradation is thought to be due to genic and cytoplasmic incompatibility. Special techniques such as 'nurse cultures' have been adopted to overcome the difficulties encountered in culturing proembryos [242].

Nucellar embryony is a phenomenon commonly observed in species such as *Citrus* and *Mangifera* [194]. Nucellar embryos of these species can develop into virus-free juvenile progeny. The nucellar system has been expanded to grape [170] and monoembryonic *Citrus* [194].

6. *In vitro* pollination and fertilization

In vitro pollination and fertilization are most useful when the two parents are incompatible due to problems associated with reproductive organs or flower abscission prior to seed maturity [110]. The first successful use of this system was demonstrated for poppy (*Papaver*) [125]. Although the *in vitro* culture and pollination of floral parts may have value, we are not aware of any reports where these systems have been successfully for woody plants. The system has been used succesfully for a number of herbaceous species including maize [106] and cucumber [169].

6.1. *Haploids in tissue culture*

Haploid plants are of importance for plant breeding for several reasons: (1) homozygous plants can be obtained in one generation, even from self-sterile plants; (2) large numbers of haploid progeny can be obtained. Because dominance cannot mask expression of desirable alleles, the time required to select desirable genotypes can be reduced [167]; (3) haploid plants are better plants for inducing and detecting mutations than diploid plants [47] (the name given to this type of variation is gametoclonal [69]); and (4) for those plants that develop from single pollen grains, the problem of chimera induction via mutagenesis is reduced. (The lack of chimera formation has also been reported with protoplasts [18].)

Substantial progress has been made in the production of haploids during the last decade. The procedures have been used widely to produce disease-resistant and high yielding varieties [172,258]. Although haploid plants have been reported in many angiosperm species, there have been only a few reports on tree crops [108].

Haploids are usually obtained by anther culture. However, plants derived from anthers can be either maternal or pollen-derived. Distinguishing the two types is possible by counting chromosomes, but haploid cells frequently double spontaneously. Without markers the doubled haploids and the maternal-derived shoots are difficult to distinguish. A more reliable method to obtain haploid lines is by pollen culture. The status of this field has been reviewed [17,18,104,152].

Production of haploid plants of woody species appears to be much more difficult than for herbaceous plants [188]. Radojevic and Kovar [188] noted only 25 species of woody plants had been investigated, with successful results from only two of these. Typical results are those reported for grape (*Vitis sp.*)

where only three of 26 clones obtained from anthers survived [89]; none was haploid. Chang-jie and Peifen [28] reported the obtention of haploid grape plants with many aneuploid cells, but no geneticc studies were reported on these plants. Rajasekaran and Mullins [190,191] produced plants from grape anthers but all of the plants were diploid.

Production of haploid plants of several *Populus* species and hybrids have been reported by Hyun *et al.* [115] and Uddin *et al.* [250]. These workers described procedures for plant production from anthers cultures of *Populus deltoides*. They noted differences in leaf morphology and growth of established plants from the same clone. Chromosome counts of regenerates showed more variability than reported by Ho and Raj [109].

6.2. *Gymnosperm haploids*

To date only a few gymnosperm species have yielded plantlets through androgenesis and gynogenesis [18]. One would expect that it will be easier to obtain vigorous haploid plantlets from megagametophyte cultures of tree species such as *Pinus resinosa* than those of more heterozygous species [77a,227a]. However, progress in this area has been slow due to the poor regenerative capacity of cells and tissues of gymnosperms [173]. Perhaps this is to be expected given the extent of allelic heterozygosity and large genetic loads in conifers [78a,233a].

6.3. *Genetic studies with haploid-derived plantlets*

The first homozygous clone of apple was obtained *in vitro* by Milewska-Pawliczuk and Kunicki [164] from microspores. Later Lespinasse *et al.* [148,149] and Zhang *et al.* [261] obtained haploid plants from anthers of apple using a unique genetic based anthocyanin marker system. Duron and Lespinasse [52] have reported that one haploid plant was grown in a phytotron until it was old enough to flower (2 years). The pollen has been used for breeding.

6.4. *Problems of working with haploids*

The problems associated with haploid culture have been discussed by Bonga *et al.* [17,18]. The major limitations to haploid breeding are (1) lack of techniques to obtain large numbers of verifiable haploids since tissue culture is known to drive cell populations rapidly to polyploidy and aneuploidy [171,248]; (2) most trees are highly heterozygous and contain large numbers of lethal and semi-lethal recessive genes. Therefore homozygous plants could be of low vigor and may not grow to reach sexual maturity; (3) the importance of haploids for tree crop improvement has yet to be demonstrated; and (4) diploidization is difficult in some species.

7. Chimeras in tissue culture

Many woody plants are known to be chimeral in nature. A chimera consists of cell sectors or tissues which differ in genetic constitution but have developed from a meristem containing layers or sectors of mutated tissue [102]. Dermen [46] has described many chimeras. (See Whitham *et al.* [259] for an interesting discussion of evolutionary value of chimera formation and maintenance in woody plants.)

The arrangement of the genetically different tissue within the plant meristem affects chimeral stability. The most stable chimera is the periclinal chimera, however, changes sometimes occur by cell displacement and replacement [237]. It also has been noted that the rate of cell division in the apical meristem can be altered by environmental factors [182], this can affect the chimeral status [12]. Pillai [180] and Reeve [197] reported that stability in the tunica layer of woody plant apices can be a seasonal occurrence.

The most common cause of chimeral dissociation is adventitious shoot production [156]. Adventitious shoots can be produced either *in vivo* and *in vitro*. Adventitious shoots usually arise from a single cell or a few cells that could be derived from specific tissue [21,22]. If adventitious shoots can be traced to a single cell, then any system that yields adventitious shoots will result in chimeral segregation.

Dermen [46], Stewart and Dermen [237], and Dayton [40a,41] have used a technique known as disbudding (removal of all preformed buds on shoots) to stimulate formation of adventitious shoots and separate the components of the cytochimeral apple trees. Other methods to separate chimeras include the use of irradiation [121].

Chimeras, when separated into their component parts, have been used to produce new strains of standard fruit cultivars [96,97,98,160]. However, the lack of a system to separate fruit tree chimeras into pure types has limited their use for clonal improvement. The development of procedures whereby shoots could be formed adventitiously from single cells could speed the development of new types of standard cultivars.

Tissue culture techniques, may offer more control over phenotype selection and propagation than disbudding or other methods. Various explants can be stimulated to produce shoots adventitiously rather than by a pre-formed bud system, thus a chimeral plant, subjected to this system, will segragate into its component genotypes with a chance of new chimeral rearrangements

Dommergues and Gillot [49] were able to separate the histogenic layers of an unstable *Dianthus* chimera to form a new type that was stable enough to be more commercially tolerable. Opatry and Landa [175] utilized a chlorophyll chimera of tobacco to regenerate phenotypically pure plants from explants taken from variegated leaves. Kameya [124] was able to separate the green and white portions of variegated leaves. Skirvin and Janick [232a,233] produced nonchimeral *Pelargonium* plants through *in vitro* culture of stem tissue. One of these tissue culture-derived plants was introduced as a named cultivar, 'Velvet Rose' [232b]. Johnson [121] was able to separate a periclinal carnation chimera by using a macerated shoot tip culture technique. Bush *et al.* [23] found *Chrysanthemum morifolium* 'Indianapolis' (a periclinal chimera) was unstable in both shoot tip and callus cultures. Another periclinal chimera of 'Indianapolis' chrysanthemum cv 'CF # 2 In. Bronze' was not stable in shoot tip culture with both multiple stem system (shoot proliferation) or callus system [54].

8. Genetic studies

McPheeters and Skirvin [160,161] have used tissue culture to produce over 1000 plants from shoot tip of 'Thornless Evergreenn' (TE) blackberry. TE is a periclinal chimera in which the histogenic layers of the apical meristem differ in genotype. The outer layer (LI), which produces the epidermis has mutated to a thornless phenotype, while the internal layers (LII and LIII) retain the wild thorny genotype. Parental TE produces thorny rootsuckers; shoots derived from the LI of the TE were nonchimeral and produced pure thornless rootsuckers. About half of the tissue culture-derived propagules showed a normal vining growth habit, the remainder were dwarfed due to the shortened internodes. Adventitious shoots (subepidermal origin) were produced from isolated root segments of the regenerants: full sized plants produced thorny suckers while dwarf plants produced thornless suckers. This indicated that the dwarf plants had originated from the thornless maternal epidermis. In addition both dwarf plants and their pure thornless root suckers possessed a unique bent petiole morphology that could be used as a marker for the pure thornless condition. Hall *et al.* [97] have determined that the bent petiole character is linked to a dominant thornless gene (*Ste*) that is transmitted sexually.

A similar dominant thornless gene has been identified from 'Thornless Loganberry' [96] and named *Sf1* [208]. The pure thornless type was released to the public as 'Lincoln Logan' [98].

The field performance and phenotypic stability of thornless blackberries propagated by tissue culture and tip layer or stem cuttings were compared by Swartz *et al.* [240] they found that tissue cultured plants were phenotypically like tip layered plants and they performed similarly in the field. However, some differences in leaf size and shape were observed.

The genetic stability of micropropagated 'Loganberry' plants was followed for three years by Rosati *et al.* [207] by selfing and evaluating of the *R*1 progeny. 'Thornless Logan' is a periclinal chimera in which the LI involves the thornless character and both LII and LIII are thorny. The LII carries the thorny character as homozygous dominant. They found that the micropropagation did not affect the stability of the character in LI with 1.5% of the plants with a thorny sector were found in 1982, 1.6% in 1983 and 4.6% in 1984. The stability of the thorny character in LII was assessed by selfing one flower per plant. They found that all except 6 progenies had thorny seedlings. Two progenies of the 6 had thornless seedlings. For LIII all the suckers observed in 358 micropropagated plants were completely thorny.

Adventitious shoot propagation of 'Meunier' grape vine (a periclinal chimera) resulted in three different plant types, one group of plants resembling the 'Meunier' another resembling the 'Pinot Noir' and the other group bearing tomentose leaves with hairless sectors. This micropropagation system allowed the separation of the chimeral 'Meunier' into its genotypes [228].

Decourtye [43a] and Chevreau *et al.* [31] have reported that 'Louise Bonne Variegated' and 'Comice Variegated' pear cultivars have chlorophyll deficiences that are under cytoplasmic control. Decourtye [43a] found that the 'Max Red Bartlett' cultivar has a single dominant gene coding for anthocyanin, the same gene found in the cultivar 'Royal Red Hardy'.

Abu-Qaoud *et al.* [1] were able to separate two chimeral pears into their component genotypes using adventitious regeneration system. With this system 'Louise Bonne Panachee' (a periclinal variegated pear) segregated into pure green and albino plants. 'Red Hardy', a chimeral pear with anthocyanins found in LII only, segregated into pure green and red shoots. The red color was assessed by sugar differential media and a total anthocyanins measurement.

9. Mutation induction

The amount of variation observed *in vitro* may be limited by the amount of variation that preexisted in the parental plant and the mutation rate of the cells *in vitro*. The lack of variability in some cultivars may be overcome by inducing somatic mutations in small plant parts such as leaves or cells with chemicals or irradiation. Mutagens can increase the frequency of variation in culture. The subject has been reviewed briefly by Meins [163] and Skirvin [230] and extensively by Flick [77] and Bourgin [19]. Mutations that occur can be analyzed in the manner discussed earlier.

9.1. *Mutagen-induced chimeras and their separation into pure types*

The phenomenon of chimerism is one of the major obstacles to the use of spontaneous or induced mutations in fruit tree improvement. Chimeras are formed when a somatic mutation occurs within the stratified structure of the apical meristem. Such a mutated apex can give rise to mature tissues of independent origin and different phenotypes. Vegetatively propagated chimeras frequently revert to the non-mutated phenotype. Methods of selecting and screening fruit crop chimeras have been reviewed by Pratt [185] and discussed earlier.

10. Somatic hybridization

For several years sexual hybridization has been used to improve cultivated crops. However, both intraspecific and interspecific incompatibility have limited the value of the sexual hybridization [66]. The use of cell culture technology has provided a wide potential of using several techniques for crop improvement. For less than 20 years protoplasts have been isolated from many plant genera. The utilization of protoplast fusion in somatic hybridization has been also reported. This technique (somatic cell or protoplast fusion [181], provides a potential for crop improvement. It can be used to overcome incompatibilities that limit successful hybridization, to induce cytoplasmic variability [79,179], to transfer organelles and for genetic transformation [2,65]. For a review of protoplasts and crop improvement see Puite *et al.* [186].

Somatic cell hybridization includes many steps: isolation, fusion, cloning and regeneration of the hybrid [155,183]. The genetic variation, as well as induced variation by a pre-irradiation for cybrid production, may alter the hybridization products. A powerful selection method is necessary to isolate somatic hybrids. Many methods have been described, they vary from general observation to DNA analysis. For instance, Guri and Sink [93] were able to produce a true somatic hybrid between *Lycopersicon esculentum* and *Solanum nigrum*. The hybridity was assessed by both a nuclear genome and a mitochondrial DNA analysis. Fitter *et al.* [76] used monoclonal antibodies which combined with plasma membrane antigens to identify heterokaryons of protoplast fusion.

Very recently Ochatt *et al.* [173b] published a most exciting report of a somatic hybridization of sexually incompatible wild pear (*Pyrus communis* var. *pyraster* L.) and colt cherry (*Prunus avium* × *pseudocerasus*). The hybrid cells regenerated to yield whole plants that were confirmed to be hybrids by morphological features, chromosome complement, and isozyme analysis. The hybrid also exhibits *in vitro* graft compatibility with the both the pear and cherry parents, suggesting its eventual use as a universal rootstock.

Most of the work with protoplasts isolation and fusion has been done with herbeceous plants, mainly from Solanaceae family (for reviews see [2,101,179]. Two examples of herbaceous plants will be discussed. In *Solanum* species, two atrazine-resistant somatic hybrid plants were produced following protoplast fusion between *Solanum melongena* (eggplant) and an atrazine-resistant biotype of *Solanum nigrum* [94]. In another example, a somatic hybrid plant was recovered from a protoplast fusion of *Brassica oleracea* (a cytoplasmic male sterile) and *B. campestris* (atrazine resistant). The genetic analysis showed that the hybrid carried the *B. campestris* chloroplast, while the mitochondria were a combination of both species.

Protoplasts have been isolated from woody plants [3,132], but the ability for protoplast fusion and hybridization is still limited. For instance, Redenbaugh *et al.* [196] were able to isolate protoplasts from cotyledons of different *Ulmus* species but had little success with protoplast fusion except in one species.

In spite of the problems whole plants have been recovered from isolated protoplasts of some woody plants. A new somatic hybrid citrus plant reecently was produced by protoplast fusion between navel orange (*C. sinensis*) and satsuma mandarian (*C. unshiu*) [134]. The hybrid was amphiploid. Similarly, Grosser *et al.* [90] were able to produce and allotetraploid somatic plant from the fusion of 'Key' lime (*Citrus aurantifolia*) with 'Valencia' sweet orange (*Citrus sinensis*) protoplasts. Revilla *et al.* [202] described a simple strategy for the isolation of mesophyll protoplasts from woody species within Betulaceae, Juglandaceae, Rosaceae and Rutaceae families, however, the recovery of plants from the isolated protoplasts was not mentioned. In another example, protoplasts from two *Vitis* spp were succesfully isolated and moved to a culture media. Only callus growth was obtained [147].

In conclusion, somatic hybridization provides a potential for plant improvement, but there are many basic problems to be solved before it will be used regularly.

11. Summary and conclusions

In conclusion, the long generation cycle of tree crops limits the geneticist's ability to make small improvements. However, when a unique plant is obtained, it can be maintained indefinitely by asexual techniques such as grafting, cuttings, layerage, or micropropagation.

The use of tissue culture biotechnologies will facilitate the isolation and identification of improved genotypes for use by the industry and plant geneticists. Initial goals of somaclonal variation programs will include seaches for changes in growth habit, stem and branching habit, flowering and fruiting characteristics. A few years ago, some overly optimistic tissue culturists claimed that the variation available through tissue culture could eliminate the need for traditional plant breeding via crosses and selection pressure. Today, most scientists agree that the variation associated with *in vitro* systems is useful by itself, but it will have even more importance as it is combined with traditional plant breading [204].

References

1. Abu-Qaoud H, Skirvin RM, and Chevreau E (1990) *In vitro* separation of chimeral pears into their component genotypes. Euphytica 48:189–196.
2. Ahuja MR (1982) Isolation, culture and fusion of protoplasts: Problems and prospects. Silvae Gent 31:66–77.

2a. Ahuja MR (1983) Somatic cell differentiation and rapid clonal propagation of aspen. Silvae Genet 32:131–135.

3. Ahuja MR (1984) Protoplast research in woody plants. Silvae Genet 33:32–37.

3a. Ahuja MR (1987) Somaclonal variation. In: Bonja JM and Durzan DJ (Eds.) Cell and Tissue Culture in Forestry. Vol. 1. General Pinciples and Biotechnology. Martinus Nijhoff Publishers, Dordrecht, the Netherlands. pp. 272–285.

4. Ahuja MR (Ed.) (1988a) Somatic Cell Genetics of Woody Plants. Kluwer Acad. Publ. Dordrecht, the Netherlands. 225 pp.

5. Ahuja MR (1988b) Gene transfer in forest trees. In: Hanover JW and Keathley DE (Eds.) Genetic Manipulation of Woody Plants, Plenum Press, New York. pp. 25–41.

5a. Ahuja MR (1988c) Molecular genetics of transgenic plants. In: Hallgren JE (Ed.) Proc. Frans Kempe Symposium 'Molecular Genetics of Forest Trees'. Swedish Univ. of Agric. Sci., Umea. pp. 127–145.

6. Altman DA (1989) Exogenous hormone applications at pollination for *in vitro* and *in vivo* production of cotton interspecific hybrids. Pl Cell Repts 7:257–261.

7. Ammirato PV (1986) Gymnosperm tissue culture: a rare success. Biotechnology 4:607.

8. Atkinson RG and Gardner RC (1989) Gene transfer to New Zealand horticultural crops. J Cell Biochem Suppl. 13D:M100, p. 252, UCLA Symposia on Molecular and Cellular Biology, Abstr.

9. Bajaj YPS and Gill MS (1985) *In vitro* induction of genetic variability in cotton (*Gossypium* spp.). Theor Appl Genet 70:363–368.

10. Baker B, Coupland G, Federoffd N, Starlinger P, and Schell J (1987) Phenotypic assay for excision of the maize controlling element Ac in tobacco. EMBO J 6(6):1547–1554.

11. Baker B, Schell J, Lorz H, and Federoff N (1986) Transposition of the maize controlling element 'Activator' in tobacco. Proc Natl Acad Sci USA 83:4844–4848.

12. Balkema GH (1972) Diplontic drift in chimeric plants. Radiation Botany 12:51–58.

12a. Baribalult TJ, Skene KGM, and Scott NS (1989) Genetic transformation of grapevine cells. Plant Cell Repts 8:137–140.

12b. Barlass M, Miller RM, and Antcliff AJ (1986) Development of methods for screening grapevines for resistance to infection. I. Dual culture *in vitro*. Amer J Enol Vitic 37:61–66.

13. Barton KA, Whiteley HR, and Yang N (1987) *Bacillus thuringiensis* delta-entodoxin expressed in transgenic *Nicotiana tabacum* provides resistance to lepidopteran insects. Plant Physiol 85:1103–1109.

14. Beachy R, Register JC, Stark DA, Nejidat A, Clark WG (1989) Coat protein protection in transgenic plants: more facts and a better understanding of mechanism(s). J Cell Biochem Suppl 13D:M035, p. 245, UCLA Symposia on Molecular and Cellular Biology, Abstr.

15. Ben-Hayyim G and Goffer Y (1989) Plantlet regeneration from NaCl-selected salt-tolerant callus cultures of Shamouti orange (*Citrus sinensis* L. Osbeck). Plt Cell Repts 7:680–683.

16. Bekkaoi F, Pilon M, Laine E, Raju DSS, Crosby WL, and Dunstan DI (1988) Transient gene expression in electroporated *Picea glauca* protoplasts. Pl. Cell Repts 7:481–484.

17. Bonga JM (1987) Tree tissue culture applications. In: Maramorosch K (Ed.) Advances in Cell Culture. Academic Press, Inc., Orlando, Florida. pp. 209–239.

18. Bonga JM, von Aderkas P, and James D (1988) Potential application of haploid cultures of tree species. In: Hanover JW and Keathley DE (Eds.) Genetic Manipulation of Woody Plants, Plenum Press, New York, pp. 57–77.

19. Bourgin JP (1986) Isolation and characterization of mutant cell lines and plants: auxotrophs and other conditional lethal mutants. In Vasil IK (Ed.) Cell Culture and Somatic Cell Genetics of Plants. Vol. 3. Plant Regeneration and genetic Variability. Academic Press, Inc., Orlando. pp. 475–498.

20. Boynton JE, Gillham NW, Harris EH, Hosler JPl, Johnson AM, Jones AR, Randolph-Anderson RL, Robertson D, Klein TM, Shark KB, and Sanford JC (1988) Chloroplast transformation in *Chlamydomonas* with high velocity microprojectiles. Science 240:1534–1538.

21. Broertjes C, Haccius B and Weidlich S (1968) Adventitious bud formation on isolated leaves and its significance for mutation breeding. Euphytica 17:321–344.
22. Broertjes C and Keen A (1980) Adventitious shoots: do they develop from one cell?. Euphytica 29:73–87.
23. Bush SR, Earle ED, and Langhans RW (1976) Plantlets from petal segments, petal epidermis, and shoot tips of periclinal chimera, *Chrysanthemum morifolium* 'Indianapolis'. Amer J Bot 63:729–737.
24. Callis J, Fromm M, and Walbot V (1987) Introns increase gene expression in culturd maize cells. Genes & Development 1:1183–1200.
25. Cassells AC (1985) Genetic, epigenetic and non-genetic variation in tissue culture derived plants. In: Schafer-Menuhr A (Ed.) *In vitro* techniques propagation and Long Term Storage. Martinus Nijhoff/Dr. W. Junk Publ., Dordrecht, the Netherlands. pp. 111–119.
26. Chaleff RS (1981) Genetics of Higher Plants. Applications of Cell Culture. Cambridge Univ. Press, London.
27. Chaleff RS (1986) Isolation and characterization of mutant cell lines and plants: herbicide-resistant mutants. In: Vasil IK (Ed.) Cell Culture and Somatic Cell Genetics of Plants. Vol. 3. Plant Regeneration and Genetic Variability. Academic Press, Inc., Orlando. pp. 499–512.
28. Chang-jie Z and Pei-fen L (1981) Induction of pollen plants of grape (*Vitis vinifera* L.) Acta Bot Sinica 23(1):79–81.
29. Chen THH and Gusta LV (1986) Isolation and characterization of mutant cell lines and plants: cold tolerance. In: Vasil IK (Ed.) Cell Culture and Somatic Cell Genetics of Plants. Vol. 3. Plant Regeneration and Genetic Variability. Academic Press, Inc., Orlando. pp. 527–535.
30. Cheng, W and Smeltzer RH (1981) Comparison of tissue culture propagules and seedlings of *Pinus taeda* grown in a greenhouse. For Prod Res Intern Paper Co Tech Notes 58. 11 pp (as reported by Ho and Zsuffa (1985)).
31. Chevreau E, Skirvin RM, Abu-Qaoud H, Korban SS, and Sullivan JG (1989). Adventitious shoot regeneration from leaf tissue of three pear (*Pyrus* sp.) cultivars *in vitro* Pl. Cell Repts 7:688–691.
32. Christou P, McCabe D, and Swan W (1988) Stable transformation of soybean callus by DNA-coated gold particles. Plant Physiol 87:671–674.
33. Colby SM, Stamp JA, and Meredith CP (1989) *Agrobacterium*-mediated transformation of *Vitis vinifera* leaves and regeneration of transformants expressing beta-glucuronidase. J Cell Biochem Suppl 13D:M110, p. 255, UCLA Symposia on Molecular and Cellular Biology, Abstr.
34. Coleman GD and Ernst SG (1989) Targeting cells of *Populus deltoides* explants for transformation by *Agrobacterium tumefaciens*. J Cell Biochem Suppl 13D:M111, p. 255, UCLA Symposia on Molecular and Cellular Biology, Abst.
35. Colot V, Robert LS, Kavanagh TA, Bevan MW, and Thompson RD (1987) Localization of sequences in wheat endosperm protein genes which confer tissue-specific expression in tobacco. EMBO J 6(12):3559–3564.
36. Comai L, Facciotti D, Hiatt WR, Thompson G, Rose RE, and Stalker DM (1985) Expression in plants of a mutant aroa gene from *Salmonella typimuruym* confers tolerance to glyphosphate. Nature 317:741–744.
37. Crocomo OJ and Ochoa-Alego N (1983) Herbicide tolerance in regenerated plants. In: Evans DA, Sharp WR, Ammirato PV, and Yamada Y (Eds.) Plant Cell Culture-Techniques for Propagation and Breeding. Macmillan Publ. Co., New York. pp. 770–781.
38. Crossway A, Oakes JV, Irvine JM, Ward B, Knauf VC, and Shewmaker CK (1986) Integration of foreign DNA following microinjection of tobacco mesophyll protoplasts. Mol Gen Genet 202:179–185.
39. D'Amato F (1978) Chromosome number variation in cultured cells and regenerated plants. In: Thorpe TA (Ed.) Frontiers of plant tissue culture. Intern Assoc Pl Tissue Cult Univ Calgary, Calgary, Canada. pp. 287–295.

40. Dandekar AM, Martin LA, and McGranahan G (1988) Genetic transformation and foreign gene expression in walnut tissue. J Amer Soc Hort Sci 113(6):945–949.
40a. Dayton D (1969) Genetic heterogeneity in the histogenic layers of apple. J Amer Soc Hort Sci 94:592–595.
41. Dayton D (1970) New apple stains developed by forcing shoots on disbudded trees. Illinois Res 112:10.
42. De Block M, Botterman J, Vandewiele M, Dockx J, Thoen C, Gossele V, Rao Movva N, Thompson C, Ban Montagu M, and Leemans J (1987) Engineering herbicide resistance in plants by expression of a detoxifying enzyme. EMBO J 6(9):2513–2518.
43. DeCleene M and DeLey J (1976) Host range of crown gall. Bot Rev 42:389–466.
43a. Decourtye I (1973) Heredite de quelques caracteres marqueurs et possibilities offertes par la mutagenese. Angers (France) 4–8 Sep. 1972:93–110.
44. Delauney AJ, Tabaeizadeh Z, and Verma DPS (1988) A stable bifunctional antisense transcript inhibiting gene expression in transgenic plants. Proc Natl Acad Sci USA 85:4300–4304.
45. Deom CM, Olivaer MJ, and Beachy RN (1987) The 30-kilodalton gene product of tobacco mosaic virus potentiates virus movement. Science 237:389–394.
46. Dermen H (1960) Nature of plant sports. Amer Hort Mag 39:123–173.
47. Devreux M and de Nettancourt D (1974) Screening mutations in haploid plants. In: Kasha KJ (Ed.) Haploids in Higher Plants. University of Guelph, Guelph, Ontario, Canada. pp. 309–322.
48. Dix PJ and Street HE (1976) Selection of plant cell lines with enhanced chilling resistance. Ann Bot 40:903–910.
49. Dommergues P and Gillot J (1973) Obtention de clones genetiquement homogenes dans toutes leurs couches ontogeniques a partir d'une chimere d'ceillet americain. Ann Amelior Plantes 23:83–93.
50. Druart P (1980) Plantlet regeneration from root callus of different *Prunus* species. Sci Hort 12:339–342.
51. Dunwell JM (1986) Pollen, ovule and embryo culture as tools in plant breeding. In: Withers LA and Alderson PG (Eds.) Plant Tissue Culture, Its Agricultural Applications. Butterworths, London. pp. 375–404.
52. Duron M and Lespinasse Y (1985) *In vitro* propagation of new apple genotypes. In: Schafer-Menuhr A (Ed.) *In vitro* Techniques. Martinus Nijhoff/Dr. W. Junk Publishers, Dordrecht, The Netherlands. pp. 35–39.
53. Earle E and Demarly Y (1982) Variability in Plants Regenerated from Tissue Culture. Praeger Press, New York. 392 pp.
54. Earle ED, Langhans RW (1974) Propagation of Chrysanthemum in vitro. I. Multiple plantlets from shoot tips and the establishment of tissue culture. J Amer Soc Hort Sci 99:128–132.
55. Ecker JR and Davis RW (1986) Inhibition of gene expression in plant cells by expression of antisense RNA. Proc Nat Acad Sci 83:5372–5376.
56. Ellis D, Roberts D, Sutton B, Lazaroff W, Webb D, and Flinn D (1989) Transformation of white spruce and other conifers by *Agrobacterium tumefaciens.* Pl. Cell Repts 8:16–20.
57. Ellis JG, Llewellyn DJ, Dennis ES, and Peacock WJ (1987) Maize Adh-1 promoter sequences control anaerobic regulation: Addition of upstream promoter elements from constitutive genes is necessary for expression in tobacco. EMBO J 6:11–16.
58. Emershad RL and Ramming DW (1984) In-ovulo embryo culture of *Vitis vinifera* L. cv. 'Thompson Seedless'. Amer J Bot 41:1117–1127.
59. Emershad RL, Ramming DW, and Serpe MD (1989) *In ovulo* embryo developments and plant formation from stenospermic genotypes of *Vitis vinifera*. Amer J Bot 76:397–402.
60. Ettinger TL, Read PE, Hackett WP, Ostrav ME, and Skilling DD (1986) Development of resistance in *Populus* to *Septoria musiva* utilizing somaclonal variation. In: Caron F, Corriveau AG, and Boyle IJB (Eds.) Proc 12th Meeting of the Canadian Tree Improvement Assoc, Part 2, 1985. Canadian Tree Improvement Association, Canadian Forest Service, Ottawa, Canada. pp. 83–90.

61. Evaldsson I (1985) Induction, growth and differentiation of callus from stem segments of *in vitro* cultured apple shoots (*Malus domestica* Borkh.). Swedish J Agric Res 15:119–122.
62. Evans DA (1986) Case histories of genetic variability *in vitro*: tomato. In: Vasil IK (Ed.) Cell Culture and Somatic Cell Genetics of Plants. Vol. 3. Plant Regeneration and Genetic Variability. Academic Press, Inc., Orlando. pp. 419–432.
63. Evans DA (1989) Somaclonal variation – genetic variation and breeding applications. Trends in Genetics 5(2):46–50.
64. Evans DA and Bravo JE (1986) Phenotypic and genotypic stability of tissue cultured plants. In: Zimmerman RH, Griesbach RJ, Hammerschlag FA and Lawson RJ (Eds) Tissue Culture as a Plant Production System for Horticultural Crops. Martinus Nijhoff Publishers, Dordrecht, The Netherlands. pp. 73–94.
65. Evans DA and Flick CE, (1985) Protoplast fusion: Agricultural application of somatic hybrid plants. In: Kosuge T, Meredith CP, and Hollander A (Eds.) Genetic Engineering of Plants. An Agricultural Perspective, Plenum Press, New York, pp. 271.
66. Evans DA and Sharp RW (1983) Single gene mutations in tomato plants regenerated from tissue culture. Science 221:949–951.
67. Evans DA and Sharp WR (1986a) Somaclonal and gametoclonal variation. In: Evans DA, Sharp WR, and Ammirato PV (Eds.) Handbook of Plant Cell Culture. Vol. 4. Macmillan Publishing Co., New York. pp. 97–132.
68. Evans DA and Sharp WR (1986b) Applications of somaclonal variation. Bio/Technology 4:528–532.
69. Evans DA, Sharp WR, and Medina-Filho HP (1984) Somaclonal and gametoclonal variation. Amer J Bot 71:759–774.
70. Fasolo F, Zimmerman RH, and Fordham I (1989) Adventitious shoot formation on excised leaves of *in vitro* grown shoots on apple cultivars. Plant Cell Tissue Organ Cult 16:75–87.
71. Ferl RJ, Nick HS, and Laughner BH (1987) Architecture of a plant promoter: S1 nuclease hypersensitive features of maize Adh1. Plant Molec Biol 8:299–307.
72. Fillatti JJ, Hassig B, McCown B, Comai L, and Riemenschneider D (1988) Development of glyphosphate-tolerant *Populus* plants through expression of a mutant *aroA* gene from *Salmonella typhimurium*. In: Hannover JW and Keathley DE (Eds.) Genetic Manipulation of Woody Plants. Plenum Press, New York. pp. 243–249.
73. Fillatti JJ, Sellmer J, McCown B, Hassig B, and Comai L (1987) *Agrobacterium* mediated transformation and regeneration of *Populus*. Mol Gen Genet 206:192–199.
74. Fischoff DA, Bowdish KS, Perlak FJ, Marrone PG, McCormick SM, Niedermeyer JG, Dean DA, Kretzmer K, Mayer EJ, Rochester DE, Rogers SG, and Fraley RT (1987) Insect tolerant transgenic tomato plants. BioTechnology 5:807–813.
75. Fisher DV and Ketchie DO (1981) Survey of literature on red strains of 'Delicious'. Washington State Univ Coll Agr Res Ctr Bull 0898. pp. 1–17.
76. Fitter MS, Noman PM, Hahn MG, Wingate VPM, and Lamb CJ (1987) Identification of somatic hybrids in plant protoplast fusion with monoclonal antibodies to plasma-membrane antigens. Planta 170:49–54.
77. Flick CE (1983) Isolation of mutants from cell culture. In: Evans DA, Sharp WR, Ammirato PV, and Yamada Y (Eds.) Handbook of Plant Cell Culture, Volume 1. Techniques for Propagation and Breeding. Macmillan and Co., New York. pp. 393–441.
77a. Fowler DP and Morris RW (1977) Genetic diversity in red pine: evidence for low genic heterozygosity. Can J For Res 7:343–347.
78. Fraley RT, Frey NM, and Schell J (Eds.) (1988) Genetic improvements of agriculturally important crops: Progress and issues. Curr Comm in Molec Biol Cold Spring Harbor.
78a. Franklin EC (1972) Genetic load in loblolly pine. Amer Natur 106:262–265..
79. Galun E and Aviv D (1986) Organelle transfer. In: Methods in Enzymology. Academic Press, New York, 118:595–611.
80. Gasser CS and Fraley RT (1989) Genetically engineering plants for crop improvement. Science 244:1293–1299.

81. Giuliano G, Pichersky E, Malik V, Timko M, Scolnik P, and Cashmore A (1988) An evolutionary conserved protein binding sequence upstream of a plant light-regulated gene. Proc Natl Acad Sci 85:7089–7093.
82. Goldfarb B, Strauss SH, Bailey LM, Zaerr JB, and McCabe DE (1989) Transient expression of microprojectile-introduced DNA in Douglas-fir. J Cell Biochem Suppl 13D:M121, p. 259, UCLA Symposia on Molecular and Cellular Biology, Abst.
83. Goldy R, Emershad R, Ramming D, and Chaparro J (1988) Embryo culture as a means of entrogressing seedlessness from *Vitis vinifera* to *V. rotundifolia*. HortSci. 23:886–889.
84. Goodman RM, Hauptli H, Crossway A, and Knauf VC (1987) Gene transfer in crop improvement. Science 236:48–54.
85. Gould AR (1986) Factors controlling generation of variability *in vitro*. In: Vasil IK (Ed.) Cell Culture and Somatic Cell Genetics of Plants. Vol. 3. Plant Regeneration and Genetic Variability. Academic Press, Inc., Orlando. pp. 549–567.
86. Gray DJ, Fisher LC, and Mortensen JA (1987) Comparison of methodologies for *in ovulo* embryo rescue of seedless grapes. HortSci. 22:1334–1335.
87. Green PJ, Yong MH, Cuozzo M, Kano-Murakami Y, Silverstein P, and Chua NH (1988) Binding site requirements for pea nuclear protein factor GT-1 correlate with sequences require for light-dependent transcriptional activation of the rbcS-3A gene. EMBO J 7:4035–4044.
88. Green PJ, Kay SA, and Chua NH (1987) Sequence-specific interactions of a pea nuclear factor with light-responsive elements upstream of the rbcS-3A gene. EMBO J 6:2543–1449.
89. Gresshoff and Doy (1974) Derivation of haploid cell line from *Vitis vinifera* and the importance of the stage of meiotic development of anthers for haploid culture of this and other genera. Z Pflanzenphysiol 73:132–141.
90. Grosser JW, Moore GM, and Gmitter FG Jr. (1989) Interspecific somatic hybrid plants from the fusion of 'Key' lime (*Citrus aurantifolia*) with 'Valencia' sweet orange (*Citrus sinensis*) protoplasts. Sci Hortic 39:23–29.
91. Grout BWW (1986) Embryo culture and cryopreservation for the conservation of genetic resources of species with recalcitrant seed. In: Withers LA and Alderson (Eds.) Plant Tissue Culture, Its Agricultural Applications. Butterworths, London. pp. 303–309.
92. Gupta PK, Dandekar AM, and Durzan DJ (1988) Somatic proembryo formation and transient expression of a luciferase gene in Douglas fir and loblolly pine protoplasts. Plant Sci 58:85–92.
93. Guri A, Levi A, and Sink KC (1988) Morphological and molecular characterization of somatic hybrid plants between *Lycopersicon esculentum* and *Solanum nigrum*. Mol Gen Genet 212:191–198.
94. Guri A and Sink KC (1988) Organelle composition in somatic hybrid between an atrazine resistant biotype of *Solanum nigrum* and *Solanum melongena*. Plant Science 58:51–58.
95. Haissig BE, Nelson ND, and Kidd GH (1987) Trends in the use of tissue culture in forest improvement. Bio/Technology 5:52–57.
96. Hall HK, Cohen D, and Skirvin RM (1986a) The inheritance of thornlessness from tissue culture-derived 'Thornless Evergreen' blackberry. Euphytica 35:891–898.
97. Hall HK, Quazi MH, and Skirvin RM (1986b) Isolation of a pure thornless Loganberry by meristem tip culture. Euphytica 35:1039–1044.
98. Hall HK, Skirvin RM, and Braam WF (1986c) Germplasm release of 'Lincoln Logan', a tissue culture-derived genetic thornless 'Loganberry'. Fruit Var J 40:134–135.
99. Hammerschlag FA, Owens LD, and Smigocki AC (1988) *Agrobacterium*-mediated transformation of peach cells derived from mature plants that were propagated *in vitro*. J Amer Hort Sci 114:508–510.
100. Hanover JW and Keathley DE (Eds.) (1988) Genetic Manipulation of Woody Plants. Plenum Press, New York. 519 pp.
101. Harms CT (1985) Hybridization by somatic cell fusion. In: Fowke LC and Constabel F (Eds.) Plant Protoplasts. Chemical Rubber Co. Press. Baco Raton, Florida, 1985, pp. 169.

102. Hartmann HT and Kester DE (1983) Plant Propagation. Prentice-Hall, Inc., Englewood Cliffs, NJ. pp. 209.
103. Haseloff J and Gerlach WL (1988) Simple RNA enzymes with new and highly specific endoribonuclease activities. Nature 334:585–591.
104. Heberle-Bors E (1985) *In vitro* haploid formation from pollen: a critical review. Theor Appl Genet 71:361–374.
105. Herrrera-Estrella L and Simpson J (1988) Foreign gene expression in plants. In: Shaw CH (Ed.) Plant Molecular Biology: A Practical Approach. IRL Press, Oxford. pp. 131–160.
106. Higgins RK and Petolino JF (1988) *In vitro* pollination fertilization of maize: influence of explant factors on kernal development. Plant Cell Tissue Organ Cult 12:21–30.
107. Hilder VA, Gatehouse AMR, Sheerman SE, Barker RF, and Boulter D (1987) A novel mechanism of insect resistance engineered into tobacco. Nature 330:160–163.
108. Ho RH, Raj AY, and Zsuffa L (1983) Poplar plants through anther culture. In: Eckert RTG (Ed.) Proc 28th Northeast For Tree Improv Conf, Univ of New Hampshire, Durham, NH pp. 294–300.
109. Ho RH and Raj Y (1985) Haploid plant production through anther culture in poplars. For Ecol Man 13:133–142.
110. Ho RH and Zsuffa L (1986) Biotechnology in breeding for biomass production. In: Moo-Young M, Hasnain S, and Lamptey J (Eds.) Biotechnology and Renewable Energy. Elsevier App. Sci. Publ., London. pp. 23–33.
111. Hood EE, Ekberg I, Johannson T, and Clapham D (1989) T-DNA presence and opine production in *Agrobacterium tumefaciens* A281 induced tumors of Norway spruce. J Cell Biochem Suppl 13D:M125, p. 260, UCLA Symposia on Molecular and Cellular Biology, Abstr.
112. Howard EA, Walker JC, Dennis ES, and Peacock WJ (1987) Regulated expression of an alcohol dehydrogenase 1 chimeric gene introduced into maize protoplasts. Planta 170:535–540.
113. Hughes K (1983) Selection for herbicide resistance. In: Evans DA, Sharp WR, Ammirato PV, and Yamada Y (Eds.). Plant Cell Culture – Techniques for Propagation and Breeding. Macmillan and Co., New York. pp. 442–460.
114. Hutchinson JF and Zimmerman RH (1987) Tissue culture of temperate fruit and nut trees. Hort Review 9:273–349.
115. Hyun SK, Kin JH, Noh EW, and Park JI (1986) Induction of haploid plants of *Populus* species. In: Withers LA and Alderson PG (Eds.) Plant Tissue Culture and its Agricultural Applications. Butterworths, London. pp. 413–418.
116. James DJ, Passey AJ, Barbara DJ, and Bevan M (1989) Genetic transformation of apple (*Malus pumila* Mill) using a disarmed Ti binary vector. Plant Cell Rep 7:658–661.
117. James DJ, Mackenzie KAD, and Mahlotra SB (1987) The induction of hexaploidy in cherry rootstocks using *in vitro* regeneration techniques. Theor Appl Genet 73:589–594.
118. James DJ, Passey AJ, and Malhotral SB (1984) Organogenesis in callus derived from stem and leaf tissues of apple and cherry rootstocks. Plant Cell Tissue Organ Culture 3(4):333–341.
119. Jofuku KD, Okamuro JK, and Goldberg RB (1987) Interaction of an embryo DNA binding protein with a soybean lectin gene upstream region. Nature 328:734–737.
120. Jofuku KD and Goldberg RB (1988) Analysis of plant gene structure. In: Shaw CH (Ed.) Plant Molecular Biology: A Practical Approach. IRL Press, Oxford. pp. 37–66.
121. Johnson RT (1980) Gamma irradiation and *in vitro* induced separation of chimeral genotypes in carnation. HortSci 15:605–606.
122. Jones OP, Gayner JA, and Watkins R (1984) Plant regeneration from tissue cultures of the cherry rootstock Colt (*Prunus avium* × *P. pseudocerasus*) and the apple rootstocks M.25 (*Malus pumila*). J Hort Sci 59:463–467.
123. Joung H, Korban SS, and Skirvin RM (1987) Screening shoot cultures of *Malus* for cedar-apple rust infection by *in vitro* inoculation. Plan Dis 71:1119–1122.

124. Kameya T (1976) Culture of protoplasts from chimeral plant tissue of nature. Jap J Genet 50:417.
125. Kanta K, RangaSwamy NS, and Maheswari P (1962) Test-tube fertilization in a flowering plant. Nature 194:1214–1217.
125a. Karp A (1989) Can genetic instability be controlled in plant tissue cultures? Newsletter Intern Assn Plant Tiss Cult 58:2–11.
126. Kay R, Chan A, Daly M, and McPherson J (1987) Duplication of CaMV 35S promoter sequences creates a strong enhancer for plant genes. Science 236:1299–1302.
127. Keith B and Chua NH (1986) Monocot and dicot pre-mRNAs are processed with different efficiencies in transgenic tobacco. EMBO J 5(10):2419–2425.
128. Kemble RJ and Shepard JF (1984) Cytoplasmic DNA variation in a potato protoclonal population. Theor Appl Genet 69:211–216.
129. Kerns H and Meyer MM (1986) Tissue culture propagation of *Acer* × *freemanii* using thidiazuron to stimulate shoot tip proliferation. HortSci 21:1209–1210.
131. Kirby EG and David A (1988) Use of protoplasts and cell cultures for physiological and genetic studies of conifers. In: Hanover JW and Keathley DE (Eds.) Genetic Manipulation of Woody Plants. Plenum Press, New York. pp. 185–197.
132. Klein T, Harper E, Svab Z, Sanford J, Fromm M, and Maliga P (1988) Stable genetic transformation of intact *Nicotiana* cells by the particle bombardment process. Proc Nat Acad Sci 85:8502–8505.
133. Klein TM, Wolf ED, Wu R, and Sanford JC (1987) High-velocity microprojectiles for delivering nucleic acids into living cells. Nature 327:70–73.
134. Kobayashi S, Ohgawara T, Ohgawara E, Oiyama I, and Ishii S (1988) A somatic hybrid plant obtained by protoplast fusion between navel orange (*Citrus sinensis*) and Satsuma mandarin (*C. unshiu*). Plant Cell Organ Culture 14:63–69.
135. Kochba J, Gen-Hayyim G, Spiegel-Roy P, Saad S, and Neumann H (1982) Selection of stable salt-tolerant callus cell lines and embryos in *Citrus sinensis* and *C. aurantium*. Z Pflphysiol 106:111–118.
136. Kononowicz AK and Janick J (1984) *In vitro* development of zygotic embryos of *Theobroma cacao*. J Amer Soc Hort Sci 109:266–269.
137. Kouider M, Skirvin RM, Dawson JO, and Jokela JJ (1984) A method to culture immature embryos of *Populus deltoides in vitro*. Can J For Res 14:956–958.
138. Kovalenko PG and Galkin AP (1989) A rapid and simple method for transferring genes into grapevine cv. Cabernet Sauvignon. J Cell Biochem Suppl 13D:M133, p. 263, UCLA Symposia on Molecular and Cellular Biology, Abstr.
139. Kriebel HB (1975) Embryo development and hybridity barriers in white pines. Silvae Genet 21:39–44.
140. Krogstrup P (1986) Embryolike structures from cotyledons and ripe embryos of Norway spruce (*Picea abies*). Can J For Res 16:664–668.
141. Larkin PJ and Scowcroft WR (1981) Somaclonal variation – a novel source of variability from cell cultures. Theor Appl Genet 60:197–214.
142. Larkin PJ and Scowcroft WR (1983a) Somaclonal variation and eyespot toxin tolerance in sugarcane. Plant Cell Tissue Organ Culture 2:111–121.
143. Larkin PJ and Scowcroft WR (1983b) Somaclonal variation and crop improvement. In: Kosuge T, Meredith CP, and Hollaender A (Eds.) Genetic Engineering of Plants: An Agricultural Perspective. Plenum Press, New York.
143a. Leach GN (1979) Growth in soil of plantlets produced by tissue culture, loblolly pine. TAPPI 62:59–61.
144. Lebrun L, Rajasekaran K, and Mullins MG (1985) Selection *in vitro* for NaCl-tolerance in *Vitis rupestris* Scheele Ann Bot 56:733–739.
145. Ledig FT (1985) Genetic transformation in forest trees. For Chron 61(5):454–458.
146. Lee M and Phillips RL (1988) The chromosomal basis of somaclonal variation. Ann Rev Plant Physiol Plant Mol Biol 39:413–437.

147. Lee N and Wetzstein HY (1988) Protoplast isolation and callus production from leaves of tissue cultured *Vitis* spp. Plant Cell Report 7:531–534.
148. Lespinasse Y, Godicheau M, Noiton D, and Duron M (1982) L'hapoidie chez le pommier (*Malus pumila*, Mill). 2ieme Colloque sur les recherches fruitieres-Bordeaux, 1982. pp. 19–27.
149. Lespinasse Y, Godicheau M, and Duron M (1983) Potential value and method of producing haploids in the apple tree,. *Malus pumila* (Mill.). Acta Hortic 131:225–230.
150. Lester and Berbee (1977) Within-clone variation among black poplar trees derived from callus culture. For Sci 23:122–131.
151. Llewellyn DJ, Finnegan EJ, Ellis JG, Dennis ES, and Peacock WJ (1987) Structure and expression of an alcohol dehydrogenase 1 gene from *Pisum sativum* (cv. 'Greenfast'). J Mol Biol 195:115–123.
152. Maheshwari SC, Tyagi AK, Malhotra K, and Sopory SK (1980) Induction of haploidy from pollen grains in angiosperms - the current status. Theor Appl Genet 58:193–206.
153. Maliga P (1980) Isolation, characterization, and utilization of mutant cell lines in higher plants. Int Rev Cytol Suppl 11A:225–250.
154. Maniatis T, Goodbourn S, and Fischer JA (1987) Regulation of inducible and tissue-specific gene expression. Science 236:1237–1245.
155. Mantell SH, Matthews JA, and Mckee RA (1985) Principles of plant biotechnology. Blackwell Scientific Publication. 269 pp.
156. Marcotrigiano M (1990) Genetic mosaics and chimeras: implications in biotechnology. In: Bajaj YPS (Ed.) Biotechnology in Agriculture and Forestry. Vol. 11. Springer-Verlag, Berlin. pp. 85–111
157. McGranahan G, Leslie C, Uratsu S, Martin L, and Dandekar A (1988) *Agrobacterium*-mediated transformation of walnut somatic embryos and regeneration of transgenic plants. Bio/Technology 6:800–804.
158. McCabe D, Swain W, Martinell B, and Christou P (1988) Stable transformation of soybean (*Glycine max*) by particle acceleration. Bio/Technology 6:923–926.
159. McCown BH (1986) From gene manipulation to forest establishment: shoot cultures of woody plants can be a central tool. TAPPI J 68:116–119.
160. McPheeters KD and Skirvin RM (1983) Histogenic layer manipulation in chimeral 'Thornless Evergreen' trailing blackberry. Euphytica 32:351–360.
161. McPheeters KD and Skirvin RM (1989) Somaclonal variation among ex vitro 'Thornless Evergreen' trailing blackberries. Euphytica 42:155–162.
161a. McPheeters KD, Skirvin RM, and Hall HK (1989) Brambles (*Rubus* spp.). In: Bajaj YPS (Ed.) Biotechnology in Agriculture and Forestry 6. Crops II. Springer-Verlag, Berlin. pp. 104–123.
162. Meins F Jr (1982) Habituation of cultured plant cells. In: Kahl G and Schell JS (Eds.) Molecular Biology of Plant Tumors. Academic Press, Inc., Orlando. pp. 3–31.
163. Meins F Jr (1983) Heritable variation in plant cell culture. Ann Rev Plant Physiol 34:327–346.
164. Milewska-Pawliczuk E and Kunicki B (1977) Induction of androgenesis *in vitro* in *Malus domestica*. Acta Hortic 78:271–276.
165. Miller SA and Maxwell DP (1883) Evaluation of disease resistance. In: Evans DA, Sharp WR, Ammirato PV, and Yamada Y (Eds.) Handbook of Plant Cell Culture, Volume 1. Techniques for Propagation and Breeding. Macmillan and Co., New York. pp. 853–879.
166. Moore GA, DeWald MG, and Cline K (1989) *Agrobacterium*-mediated transformation of *Citrus*. J Cell Biochem Suppl 13D:M109, p. 255, UCLA Symposia on Molecular and Cellular Biology, Abstr.
167. Morrison RA and Evans DA (1988) Haploid plants from tissue culture: new plants varieties in a shortened time frame. Bio/Technology 6:684–689.
168. Morrison RA, Whitaker RJ, and Evans DA (1988) Somaclonal variation: its genetic basis and prospects for crop improvement. In: Conn EE (Ed.) Opportunities for Phytochemistry in Plant Biotechnology. Plenum Publ. Corp. New York. pp. 1–18.

169. Msikita W, Skirvin RM, Juvik JA, Splittstoesser WE, and Ali N (1990) Regeneration and flowering *in vitro* of 'Burpless Hybrid' cucumber cultured from excised seed. HortScience 25(4):474–477.
170. Mullins MG and Srinivasan C (1976) Somatic embryos and plantlets from an ancient clone of the grapevine (cv. Cabernet-Sauvignon) by apomixis *in vitro*. J Exp Bot 27:1022–1030.
171. Murashige T (1974) Plant propagation through tissue culture. Ann Rev Plant Physiol 25:135–166.
172. Nakamura A, Yamada T, Kadotani N, Itagaki R, and Oka M (1974) Studies on the haploid method of breeding in tobacco. SABRAO J 6:107–131.
173. Norstog K (1982) Experimental embryology of Gymnosperms. In: Johri BM (Ed.) Experimental Embryology of Vascular Plants. Springer-Verlag, Berlin. pp. 25–51.
173a. Ochatt SJ and Power JB (1988) Plant regeneration from mesophyll protoplasts of Williams Bon Chretien (syn. Bartlett) pear (*Pyrus communis* L.). Plant Cell Reports 7:587–589.
173b. Ochatt SJ, Patat-Ochatt EM, Rech EL, Davey MR, and Power JB (1989) Somatic hybridization of sexually incompatible top-fruit tree rootstocks, wild pear (*Pyrus communis* var. *pyraster* L.) and Colt cherry (*Prunus avium* × *pseudocerasus*). Theor Appl Genet 78:35–41.
173c. Orton TJ (1984) Somaclonal variation: theoretical and practical considerations. In: Gustafson JP (Ed.) Gene Manipulation in Plant Improvement. Plenum Press, New York. pp. 427–468.
174. Orton TJ (1986) Case histories of genetic variability *in vitro*: celery. In: Vasil IK (Ed.) Cell Culture and Somatic Cell Genetics of Plants. Vol. 3. Plant Regeneration and Genetic Variability. Academic Press, Inc., Orlando. pp. 345–366.
175. Opatry Z and Landa Z (1974) The regeneration of chlorophyll chimeras from leaf explants of *Nicotiana tabacum* L. Biol Plant 16:312–315.
176. Parsons TJ, Sinkar VP, Stettler RF, Nester EW, and Gordon MP (1986) Transformation of Poplar by *Agrobacterium tumefaciens*. Bio/Technology 4:533–536.
177. Paul AL, Vasil V, Vasil IK, and Ferl RJ (1987) Constitutive and anaerobically induced DNase-I-hypersensitive sites in the 5′ region of the maize Adh1 gene. Proc Nat Acad Sci USA 84:799–803.
178. Peacock WJ, Wolstenholme D, Walker J, Singh K, Llewellyn D, Ellis J, and Dennis ES (1987) Developmental and environmental regulation of the maize alcohol dehydrogenase 1 (Adh1) gene: Promoter-enhancer interactions. In: Plant Gene Systems and their Biology, Alan R. Liss (UCLA Symposium on Cellular and Molecular Biology). pp. 263–277.
179. Pelletier G, Primard C, Ferault M, Vedel F, Chetrit P, Renard M, and Delourme R (1988) Uses of protoplasts in plant breeding: Cytoplasmic aspects. Plant Cell Tissue Organ Culture. 12:173–180.
180. Pillai K (1963) Structural and seasonal activity of the shoot apex of some *Cupressus* species. New Phytol 62:335–340.
181. Poehlman JM (1987) Breeding field crops. AVI Press, Inc. Wesport, Connecticut. pp. 148.
182. Popham RA (1951) Principle types of vegetative shoot apex organization in vascular plants. Ohio J Sci 51:249–270.
183. Power JB and Chapman JV (1985) Isolation, culture and genetic manipulation of plant protoplasts. In: Dixon RA (Eds.). Plant Cell Culture: A Practical Approach. pp. 37–66.
184. Poulsen C, Fluhr R, Kauffman JM, Boutry M, and Chua NH (1986) Characterization of an rbcS gene from *Nicotiana plumbaginifolia* and expression of an rbcS-CAT chimeric gene in homologous and heterologous nuclear background. Mol Gen Genet 205:193–200.
185. Pratt C (1983) Somatic selection and chimeras. In: Janick J and Moore JN (Eds.) Methods in Fruit Breeding. Purdue Univ. Press, West Lafayette, IN, USA. pp. 172–185.
186. Puite KJ, Dons JJM, Huizing HJ, Kool AJ, Koornneeft M, and Krens FA (1987) Progress in Plant Protoplast Research. Kluwer, Dordreecht, the Netherlands.
187. Pythoud F, Sinkar VP, Nester EW, and Gordon MP (1987) Increased virulence of *Agrobacterium rhizogenes* conferred by the vir region of pTiBo542: Application of genetic engineering to poplar. BioTechnology 5:1323–1327.

188. Radojevic L and Kovoor A (1985) Induction of haploids. In: Bajaj YPS (Ed.) Biotechnology in Agriculture and Forestry. 1. Trees I. Springer-Verlag, Berlin. pp. 65–86.
189. Rains DW, Croughan SS, and Croughan TP (1986) Isolation and characterization of mutant cell lines and plants: salt tolerance. In: Vasil IK (Ed.) Cell Culture and Somatic Cell Genetics of Plants. Vol. 3. Plant Regeneration and Genetic Variability. Academic Press, Inc., Orlando. pp. 537–547.
190. Rajasekaran K and Mullins MG (1983a) Influence of genotype and sex-expression on formation of plantlets by cultured anthers of grapevines. Agronomie 3:233–238.
191. Rajasekaran K and Mullins MG (1983b) The origin of embryos and plantlets from cultured anthers of hybrid grapevines. J Enol Vitic 34:108–113.
192. Ramming DW (1985) *In ovulo* embryo culture of early-maturing *Prunus*. HortScience 20: 419–420.
193. Ramulu KS (1986) Case histories of genetic variability *in vitro*: potato. In: Vasil IK (Ed.) Cell Culture and Somatic Cell Genetics of Plants. Vol. 3. Plant Regeneration and Genetic Variability. Academic Press, Inc., Orlando. pp. 449–473.
194. Rangan TS, Murashige T, and Bitters WP (1968) *In vitro* initiation of nucellar embryos in monoembryonic *Citrus*. HortScience 3:226–227.
195. Rao AN and Lee SK (1986) An overview of the *in vitro* propagation of woody plants and plantation crops. In: Withers ED and Alderson PF (Eds.) Plant Tissue Culture and Its Agricultural Applications. Butterworths, London. pp. 123–138.
196. Redenbaugh K, Kasnosky DF, and Westfall RD (1981) Protoplast isolation and fusion in three *Ulmus* species. Can J Bot 59:1436–1443.
197. Reeve RM (1984) The 'tunica-corpus' concept and development of shoot apices in certain dicotyledons. Amer J Bot 35:65–75.
198. Register JC III and Beachy RN (1988) Resistance to TMV in transgenic plants results from interference with an early event infection. Virology 166:524–532.
199. Reich T, Iyer V, and Miki B (1986a) Efficient transformation of alfalfa protoplasts by the intranuclear microinjection of ti plasmids. BioTechnology 4:1001–1004.
200. Reich TJ, Iyer VN, Scobie B, and Mikki BL (1986b) A detailed procedure for the intranuclear microinjection of plant protoplasts. Can J Bot 64:1255–1258.
201. Reich TJ, Iyer VN, Haffner M, Holbrook LA, and Mikki BL (1986c) The use of fluorescent dyes in the microinjection of alfalfa protoplasts. Can J Bot 64:1259–1267.
202. Revilla MA, Ochatt SJ, Doughty S, and Power JB (1987) A general strategy for the isolation of mesophyll protoplasts from deciduous fruit and nut tree species. Plant Science 50:133–137.
203. Rhagavan V (1986) Variability through wide crosses and embryo rescue. In: Vasil IK (Ed.) Cell Culture and Somatic Cell Genetics of Plants. Vol. 3. Plant Regeneration and Genetic Variability. Academic Press, Inc., Orlando. pp. 613–635.
204. Riemenschneider DE, Haissig BE, and Bingham ET (1988) Integrating biotechnology into woody plant breeding programs. In:Hanover JW and Keathley DE (Eds.) Genetic Manipulation of Woody Plants. Plenum Press, New York. pp. 433–469.
205. Robertson D, Palmer JD, Earle ED, and Mutshchler NA (1987) Analysis of organelle genomes in a somatic hybrid derived from cytoplasmic male-sterile *Brassica oleracea* and atrazine-resistant B. *campestris*. Theor Appl Genet 74:303–309.
206. Rodermel S, Abbott M, and Bogorad L (1988) Nuclear-organelle interactions: nuclear antisense gene inhibits ribulose bisphosphate carboxylase enzyme levels in transformed tobacco plants. Cell 55:673–681.
207. Rosati P, Gaggioli D, and Giunchi L (1986) Genetic stability of micropropagated loganberry plants. J Hort Sci 61:33–41.
208. Rosati P. Hall HK, Jennings DL, and Gaggioli D (1988) A dominant gene for thornlessness obtained from the chimeral Thornless Loganberry. HortScience 23:899–902.
209. Sacristan MD (1986) Isolation and characterization of mutant cell lines and plants: disease resistance. In Vasil IK (Ed.) Cell Culture and Somatic Cell Genetics of Plants. Vol. 3. Plant Regeneration and Genetic Variability. Academic Press, Inc., Orlando. pp. 513–525.

210. Sanchez-Serrano J, Schmidt R, Schell J, and Willmitzer L (1986) Nucleotide sequence of proteinase inhibitor II encoding cDNA of potato (*Solanum tuberosum*) and its mode of expression. Mol Gen Genet 203:15–20.
211. Sanchez-Serrano JJ, Keil M, O'Connor A, Schell J, and Willmitzer L (1987) Wound-induced expression of a potato proteinase inhibitor II gene in transgenic tobacco plants. EMBO J 6:303–306.
212. Savka MA, Dawson JO, Jokela JJ, and Skirvin RM (1987) A liquid culture method for rescuing immature embryos of eastern cottonwood. Plant Cell Tissue Organ Culture 10: 221–226.
213. Schell JS (1987) Transgenic plants as tools to study the molecular organization of plant genes. Science 237:1176–1183.
214. Scowcroft WR (1985) Somaclonal variation: the myth of clonal uniformity. In: Hahn B and Dennis ES (Eds.) Genetic Flux in Plants. Springer Verlag, New York. pp. 112–156.
215. Secor GA and Shepard JF (1981) Variability of protoplast-derived clones of potato. Crop Sci 21:102–105.
216. Sederoff RR and Ledig FT (1984) Increasing forest productivity and value through biotechnology. Weyerhaeuser Science Symp Proceedings, pp. 253–276.
217. Sederoff R, Stomp AM, Chilton WS, and Moore LW (1986) Gene transfer into loblolly pine by *Agrobacterium tumefaciens*. Bio/Technology 4:647–649.
218. Seguin A and Lalond M (1988) Gene transfer by electroporation in Betulaceae protoplasts: *Alnus incana*. Plant Cell Rep 7:367–370.
219. Shah DM, Horsch RB, Klee HJ, Kishore GM, Winter JA, Tumer NE, Hironaka CM, Sanders PR, Gasser CS, Aykent S, Siegel NR, Rogers SG, and Fraley RT (1986) Engineering herbicide tolerance in transgenic plants. Science 233:478–481.
220. Sheehy R, Kramer M, and Hiatt W (1988) Reduction of polygalacturonase activity in tomato fruit by antisense RNA. Proc Nat Acad Sci 85:8805–8809.
221. Shepard JF (1981) Protoplasts as sources of disease resistance in plants. Annu Rev Phytopath 19:145–155.
222. Shepard JF, Bidney D, and Shanin E (1980) Potato protoplasts in crop improvement. Science 208:17–24.
223. Shepard JF (1982) Regeneration of potato leaf cell protoplasts. In: Earle E and Demarly Y (Eds.) Variability in Plants Regenerated from Tissue Culture. Praeger Press, New York. pp. 47–57.
224. Shepherd NS (1988) Transposable elements and gene-tagging. In: Shaw CH (Ed.) Plant Molecular Biology: A Practical Approach. IRL Press, Oxford. pp 187–220.
225. Shillito RD and Saul MW (1988) Protoplast isolation and transformation. In: Shaw CH (Ed.) Plant Molecular Biology: A Practical Approach. IRL Press, Oxford. pp. 161–186.
226. Sibi M (1982) Heritabale epigenic variations from *in vitro* tissue culture of *Lycopersicon esculentum* (var. Monalbo). In: Earle E. and Demarly Y (Eds.) Variability in Plants Regenerated from Tissue Culture. Praeger Press, New York. pp. 228–244.
227. Siegemund F (1981) Selektion von resistenzmutanten in pflanzlichen zellkulturen – eine übersicht. Biol Zentralbl 100:155–165.
227a. Simon JP, Bergeron Y, and Gagnon D (1986) Isozyme uniformity in populations of red pine (*Pinea resinosa*) in the Abitibi Region, Quebec. Can J For Res 16:1133–1135.
228. Skene KGM and Barlass M (1983) Studies on the fragmented shoot apex of grapevine. IV. Separation of phenotypes in a periclinal chimera *in vitro* J Exp Bot 34:1271–1280.
228a. Skene KGM and Barlass M (1988) Response to NaCl of grapevines regenerated from multiple shoot cultures exhibiting mild salt tolerance *in vitro*. Amer J Enol Vitic 39:125–128.
229. Skirvin RM (1977) Fruit improvement through single cell culture. Fruit Var J 31:82–85.
230. Skirvin RM (1978) Natural and induced variation in tissue culture. Euphytica 27:241–266.
231. Skirvin RM (1981) Fruit crops. In: Conger BV (Ed.) Cloning Agricultural Plants via *in vitro* Techniques. Chemical Rubber Company Press, Baco Raton, FL, pp. 51–139.

232. Skirvin RM, Chu MC, and Gomez E (1981) *In vitro* propagation of thornless trailing blackberries. HortScience 16:310–312.

232a. Skirvin RM and Janick J (1976) Tissue culture-induced variation in scented *Pelargonium* spp. J Amer Soc Hort Sci 101:281–290.

232b. Skirvin RM and Janick J (1976) 'Velvet Rose' *Pelargonium*, a scented geranium. HortScience 11:61–62.

233. Skirvin RM and Janick J (1977) Separation of phenotypes in a periclinal chimera. J Coll Sci Teaching 7:33–35.

233a. Sorenson EC (1969) Embryonic genetic load in coastal Douglas fir, *Pseudotsuga menziesii* var. *menzisii.* Amer Natur 103:389–398.

234. Spiegel-Roy P and Ben-Hayyim G (1985) Selection and breeding for salinity resistance *in vitro.* Plant and Soil 89:243–252.

235. Spiegel-Roy P, Sahar N, Baron J, and Lavi U (1985) *In vitro* culture and plant formation from grape cultivars with abortive ovules and seeds. J Amer Soc Hort Sci 110:109–112.

235a. Sriskandarajah S, Skirvin RM, Abu-Qaoud H, and Korban SS (1989) Factors involved in shoot elongation and growth of adventitious and axillary shoots of three apple scion cultivars *in vitro.* J Hort Sci 65(2):113–121.

236. Stalker D, McBride K, Malyj L (1988) Herbicide resistance in transgenic plants expressing a bacterial detoxification gene. Science 242:419–423.

237. Stewart R and Dermen H (1970) Determination of number and mitotic activity of shoot apical initial cells by the analysis of mericlinal chimeras. Amer J Bot 57:816–826.

238. Stomp A-M, Loopstra C, Sederoff R, Chilton S, Fillatti J-A, Dupper G, Tedeschi P, and Kinlaw C (1988) Development of a DNA transfer system for pines. In: Hanover JW and Keathley DE (Eds.) Genetic Manipulation of Woody Plants, Plenum Press, New York. pp. 231–241.

239. Strobel GA, Nachmias A and Hess WM (1988) Improvements in the growth and yield of olive trees by transformation with the Ri plasmid of *Agrobacterium rhizogenes.* Can J Bot 66(12):2581–2585.

240. Swartz H, Galetta G, and Zimmerman R (1983) Field performance and phenotypic stability of tissue culture-propagated thornless blackberries. J Amer Soc Hort Sci 108:285–290.

241. Tandeau de Marsac N and Jouanneau JP (1972) Variation de l'exigence en cytokinine de lignees clonales de cellules de tabac. Physiol Veg 10:369–380.

242. Thomas MJ (1972) Compartement des embryons de trois especes de pins, isoles au moment de leur elivage et cultives *in vitro*, en presence de cultures-neurrices. CR Acad Sci Paris 274:2655–2658.

243. Thomas B (1981) Bibliography of mutant isolation from plant cell cultures. Plant Mol Biol Newsl 2:77–89.

244. Thompson CJ, Rao Movva N, Tizard R, Crameri R, Davies JE, Lauwereys M, and Botterman J (1987) Characterization of the herbicide-resistance gene bar from *Streptomyces hygroscopicus.* EMBO J 6(9):2519–2523.

245. Thompson MM, Smith DC, and Burgess JE (1985) Nondormant mutants in a temperate tree species, *Corylus avellana* L. Theor Appl Genet 70:687–692.

246. Tilney-Bassett RAE (1986) Plant chimeras. Edward Arnold Publ., Ltd. London. 199 pp.

247. Timko MP, Kausch AP, Castresana C, Fassler J, Herrera Estrella L, Van den Broeck G, Van Montagu M, Schell J, and Cashmore AR (1985) Light regulation of plant gene expression by an upstream enhancer-like element. Nature 318:579–582.

248. Torrey J (1967) Morphogenesis in relation to chromosome constitution in long term plant tissue cultures. Physiol Plant 20:265–275.

249. Tulecke W (1987) Somatic embryogenesis in woody perennials. In: Bonga JM and Durzan DJ (Eds.) Cell and Tissue Culture in Forestry. Vol. 2. Specific Principles and Methods – Growth and Developments. Martinus Nijhoff, Publishers. Dordrecht, the Netherlands. pp. 66–91.

250. Uddin MR, Meyer MM, and Jokela JJ (1988) Plantlet production from anthers of eastern cottonwood (*Populus deltoides*). Can J For Res 18:937–941.

251. Vaeck M, Reynaerts A, Hofte H, Jansens S, DeBeuckeleer M, Dean C, Zabeau M, Van Montague M, and Leemans J (1987) Transgenic plants protected from insect attack. Nature 328:33–37.

252. Vahala T, Engstrom P, and Eriksson T (1989) Effects of T-DNA cytokinin and auxin genes on hormone concentrations, growth and differentiation of willow (*Salix* sp.) callus. J Cell Biochem Suppl 13D:M255, p. 291, UCLA Symposia on Molecular and Cellular Biology, Abstr.

253. Vasil IK (Ed.) (1986) Cell culture and somatic cell genetics of plants. Vol. 3. Plant Regeneration and Genetic Variability. Academic Press, Inc., Orlando. 657 pp.

254. Vayda ME and Freeling M (1986) Insertion of the Mul transposable element into the first intron of maize Adh1 interferes with transcript elongation but does not disrupt chromatin structure. Plant Mol Biol 6:441–454.

255. Walker JC, Howard EA, Dennis EA, and Peacock WJ (1987) DNA sequences required for anaerobic expression of the maize alcohol dehydrogenase 1 gene. Proc Nat Acad Sci, USA 84:6624–6628.

256. Welander M (1988) Plant regeneration from leaf and stem segments of shoots raised *in vitro* from mature apple trees. J Plant Physiol 132:738–744.

257. Weller SC, Masiunas JB and Gressel J (1987) Biotechnologies of obtaining herbicide in potato. In: Bajaj YPS (Ed.) Biotechnology in Agriculture and Forestry. Vol. 3: Potato. Springer-Verlag, Berlin. pp. 281–297.

258. Wenzel G (1980) Anther culture and its role in plant breeding. In: Rao PS, Heble MR, and Chadha MS (Eds.) Proceedings of a National Symposium on Plant Tissue Culture, Genetic Manipulation and Somatic Hybridization of Plant Cells. Bhaba Atomic Research Centre, Bombay, India.

259. Whitham TG, Williams AG, and Robinson AM (1984) The variation principle: individual plants as temporal and spatial mosaics of resistance to rapidly evolving pests. In: Price PW, Slobodchikoff CN and Gand WS (Eds.) A New Ecology: Novel Approaches to Interactive Systems. John Wiley & Sons, New York. pp. 15–51.

260. Yang JS (1986) Interspecific hybridization and immature embryo culture of *Carica* sp. Culture of *Carica* sp. Hort Abstr No. 8303 58(11):1019.

261. Zhang YX, Boccon-Gibod J, and Lespinasse Y (1987) Obtaining apple embryos (*Malus* × *domestica* Borkh.) from anther culture. Hort Abstr No. 8468 58(12):1037.

9. Acclimatization strategies for micropropagated plants

DANIELLE J. DONNELLY and LAURENCE TISDALL
Plant Science Department, Macdonald Campus of McGill University, 21, 111 Lakeshore Road, Ste. Anne de Bellevue, Quebec, Canada H9X 3V9

1. Introduction

The successful *ex vitro* acclimatization of micropropagated plants determines the quality of the end product and, in commercial production, the economic viability of the enterprise [6]. When shoots or plantlets are transplanted from culture to greenhouse conditions they may desiccate or wilt rapidly and can die as a result of the change in environment, unless substantial precautions are taken to accomodate them. In commercial micropropagation this step is often the limiting factor [53] and at best, is challenging, labour intensive and costly [6,7,10]. Methods which work for *ex vitro* establishment of one species are not necessarily satisfactory to ensure the survival of another [47].

The following discussion was not intended to be an exhaustive survey of the now extensive literature pertaining to acclimatization of micropropagated plants. The reader is directed to several excellent reviews [6,7,10,39,75]. We feel that overcoming *ex vitro* acclimatization problems is contingent on an improved general understanding of how the environment affects the anatomy and physiology of all plants subjected to environmental change. It is necessary to begin with a better understanding of the unique effects of the *in vitro* and the *ex vitro* environments on plant phenotype. It is information on this subject that is summarized herein. In this way we hope to provide some new insights into modern acclimatization strategies, applied both *in vitro* and *ex vitro*.

2. The culture-induced phenotype

Tissue cultured shoots and plantlets share certain characteristic features that are inconsistent with development under greenhouse or field conditions. The culture-induced phenotype (CIP) [16] reflects epigenetic variation [18]; acclimatization to environmental conditions which exist within the closed culture containers. *In vitro* environments are characterized by: a saturated atmosphere; relatively low light intensity (photosynthetic photon flux), averaging 12–70 μmol m^{-2}s^{-1}; relatively high and constant temperature (20–28 °C); low rates of gas exchange between the containers and the external atmosphere and high concentrations of carbohydrate and exogenous growth regulators in the

M. R. Ahuja (Ed.), Micropropagation of Woody Plants, 153–166.

medium. Although we are aware of some aspects of the CIP it can hardly be described as well defined. Our knowledge is mostly limited to temperate species and almost exclusively to angiosperms. While in some cases the environmental determinants are known, a direct relationship between certain aspects of the CIP and some component(s) of the culture environment remain obscure.

In vitro shoots and plantlets are invariably diminutive; much smaller than their greenhouse-grown counterparts. Blackberry leaves (*Rubus* sp.) in culture were only 1–2% of the area of greenhouse-grown control plant leaves [16]. In miniature red raspberry (*Rubus idaeus* L.) plantlets the proportion of foliar cell and tissue widths to total leaf width were the same in culture as for the large greenhouse-grown control plants [18]. Mature leaves of red raspberry plantlets always had palisade : epidermal cell ratios of 1 : 1 or 2 : 1, typical of very young control leaves prior to epidermal cell expansion and palisade cell division [21]. Microcultured Asian white birch (*Betula platyphylla* var *szechuanica* (Schneid) Rehd.) was also shown to be small, more from decreased cell division than reduced cell size [58]. The relatively high cytokinin concentration, especially in Stage II [52] media and the low water potential of most media tend to inhibit apical dominance and affect stature. Media that are more dilute or lacking in cytokinins (many Stage III media) promote the development of larger organ size (Donnelly, pers. obs.). *In vitro* shoots and plantlets have increased percentage water content and reduced dry matter accumulation per unit area compared to greenhouse-grown plants [2,19]. This is reflected in fragile organs with reduced mechanical support tissue and thin cell walls. Reduced mechanical support tissue formation occurred in all organs of red raspberry [21]. This could have been nutritionally based but could also have been influenced by the tranquil *in vitro* environment which inhibited cell wall deposition and sclerenchyma and collenchyma formation [21]. In some species vascular connections were fewer, thinner and poorly structured, as in the petioles of Asian white birch [58], the stems of carnation (*Dianthus caryophyllus* L.) [45] or the root-shoot interface of adventitious cauliflower (*Brassica oleraceae* var *botrytis*) plantlets [30].

The relatively low light levels and saturated internal atmosphere promote leaves *in vitro* that anatomically resemble both shade leaves [4,44,50] and hydrophytic plant leaves [31]. They often have reduced or absent epicuticular or cuticular wax, which can lack the characteristic crystalline structure, or differ in chemical composition from that of control plants [4,5,23,25,28,30,57, 59,60,63,64,73]. *In vitro* leaves had a thinner or somewhat collapsed epidermal layer [4,21,73,74] with a clearly defined [13,58] or an absent or limited [4,23, 30,31,73] palisade layer, sometimes with obconically-shaped palisade cells [13,18] and a loosely organized spongy mesophyll with an increased percentage air space [4,13,18,23,73]. Palisade development is related to light levels [22] and is reduced *in vitro* as the light levels are relatively low [4]. Increasing the light intensity to sweetgum (*Liquidambar styraciflua*) cultures increased leaf thickness, promoted palisade differentiation and decreased the percentage air space in the mesophyll [44]. In red raspberry, leaves in culture were simple, rather than compound. This may have resulted from incubation at relatively high, constant temperatures (27 °C) [21]. They also had fewer trichomes and an altered distribution of glandular and thick-walled unicellular hairs compared to greenhouse-grown plants. Increased light intensity in culture promoted filiform but not other trichome formation [18].

Stomatal frequency and density was higher [44,74] or lower [4] *in vitro* depending on the species, and the stomatal index was not greatly affected [5] or was lower [13] compared to control plants. Stomata on leaves in culture were more circular in shape [50,51], larger [5,50,51,74] and had larger substomatal cavities [21] than stomata on control leaves. Size of the substomatal cavities has been correlated to the amount of water stress; largest when the relative humidity is highest [48]. Stomatal aperture is usually larger *in vitro* than on control leaves [2] with guard cells raised above the epidermal layer [16,44,73,74] but can differ *in vitro* depending on the stage of culture or where the leaf is situated on the shoot. Stomatal aperture gradually decreased in chrysanthemum leaves towards the less mature leaves of the shoot apex [72]. *In vitro* stomata have slow response times or impaired function [2,50,63,64,72]; they do not close in response to stimuli such as darkness, abscisic acid application, solutions with high osmolarity (mannitol or sucrose) or when exposed to high levels of carbon dioxide [3,72,74,77]. In chrysanthemum, further opening was possible in CO_2-free air and higher light intensity or through cytokinin exposure and was followed by closure to their original aperture [72]. Interestingly, guard cell protoplasm was seen to react appropriately when leaves from culture were placed in solutions of various osmolarities or containing abscisic acid. So, impaired stomatal function may result from mechanical rather than physiological causes; reduced or altered distribution of cellulose microfibrils in the guard cell walls affecting cell wall elasticity [72,77]. Guard cell walls of *in vitro Prunus cerasus* L. were thinner [50,51] and lacked invaginations of the anticlinal epidermal cell walls next to the inter-guard cell wall ends, present in acclimatized or greenhouse-grown plants [51].

The hydathodes of rosaceous species *in vitro* were simpler and had fewer water pores with larger apertures and reduced epithem, the tissue that recovers solutes from the tracheary elements, compared to greenhouse-grown plants. This may result from the high relative humidity or the low water potential of the medium *in vitro* compared to greenhouse conditions [14,15,17].

In vitro plants rely principally on sugar as a carbon source [6] and CO_2 uptake capability is low [19,20,33,34,38]. Microcultured birch photosynthesized at one third of control plant levels [58] and red raspberry at about one quarter of control plant levels [19] at saturating light intensities. *In vitro* shoots and plantlets are mixotrophic in their mode of nutrition; they apparently alternate between carbohydrate use and CO_2 fixation. Carbohydrate use is stimulated by the high concentration of sugar and the presence of growth regulators in the medium and the relatively low light intensity during incubation. Carbon dioxide fixation is stimulated for a short time each day; the CO_2 concentration in the containers is rapidly depleted within about two hours of the start of the photoperiod to at or below the compensation point for the rest of the day [26,38]. Mixotrophy contributes to the recycling of cellular respiration and photosynthetic products and affects photosynthetic carbon metabolism.

Pigment synthesis and ribulosebisphosphate carboxylase (RubPcase) activity may be impaired in culture; photosynthetic pigment content was low-normal in cultured red raspberry [19] and cauliflower, which also had low RubPcase activity [33,35]. Some *in vitro* shoots and plantlets had starch in their chloroplasts [13,51,54,72], while others had little or no starch, as in *Leucaena leucocephala* (Lam) De Wit. [13] and sweetgum [43,73]. As sucrose concentration of the medium was augmented starch concentration in the chloroplasts increased [54]. Little starch was exported from the chloroplasts during the dark period and tended to accumulate. *In vitro* leaves exhibited flattened, disorganized chloroplasts, in some cases with swollen thylakoids [43,54,73]. Accumulation of starch and disorganization of thylakoid structure was attributed to the relatively low light levels in culture [54] or altered light spectrum resulting from glass containers [43]. However, cauliflower cultures had control levels of photosynthetic electron transport, indicating normal thylakoid structure and function [33].

Propagates are apparently acclimatized to *in vitro* conditions as their growth is extremely prolific. Furthermore, the *in vitro* environment affects a wide range of species in a similar way, morphologically and physiologically (Donnelly, pers. obs.). However, a change in a single climatic parameter or medium component may affect one or more of the CIP characteristics, which in turn affects *in vitro*, and subsequent *ex vitro* performance.

3. *Ex vitro* transplantation

Difficulties in successfully transplanting tissue cultured shoots and plantlets to soil are well documented [6,25,66]. They appear to be a direct result of the culture-induced phenotype which reflects adaptation to *in vitro* conditions but is inappropriate when shoots or plantlets are transferred to the greenhouse or field where the relative humidity tends to be less than 100%, the ambient light levels are much higher than in culture, there are fluctuating temperatures, the substrate has a much higher water potential and it is necessary to convert rapidly from a mixotrophic to a fully autotrophic mode of nutrition to survive.

Ex vitro plantlets have extreme evapotranspiration rates and may guttate copiously, demonstrating impaired ability to regulate water loss. Excessive evapotranspiration is affected by reduced or nonexistant stomatal control [2,3,5, 51,61,65,71,72,77], and large cuticular water losses [2,50] possibly due to poor epicuticular and cuticular wax formation [61] or reduced trichome numbers [18,60]. The major mechanism of water loss may depend on the species in question, as some species have large quantities of epicuticular wax *in vitro* but still have water regulation problems *ex vitro* [60,61,65]. No correlations have been established between *ex vitro* survival and the physical or morphological characteristics of foliar wax [60,65]. The *ex vitro* guttation rate may be affected by the large increase in substrate water potential at transplantation and tends to increase under conditions that depress the transpiration rate [15]. *Ex vitro* root function is uncertain at the time of transplantation, especially in adventitious propagation systems, and may contribute to water deficit in transplants [1,6,30].

To promote *ex vitro* survival and physiological competence; especially to guard against water stress and encourage autotrophy, a transitional environment is usually supplied for an acclimatization interval, ranging in duration from one to several weeks [3,4,6,23,34]. In this transitional environment the relative humidity is kept in the range of 70–100% via tenting, misting or fogging and the light level should not be too much greater than it was in culture. Red raspberry survival was optimal when the light intensity did not initially exceed a three-fold increase over that found in culture [21]. Growth was not limited by light since CO_2 uptake was not different in transplants grown at light intensities two- to three-fold higher than in culture [19]. Gradually, as the plantlets acclimatize, the relative humidity can be decreased and the light levels can be increased towards ambient.

4. Transplant phenotypes *ex vitro*

4.1. *The persistent leaves*

Leaves that developed in culture were retained after transplantation for a week to several months prior to senescing [18,19,21,32,34]. Persistence depended on the plant species and the degree of environmental stress *ex vitro*. These persistent leaves increased in size slightly, mainly due to cell elongation [23,31], and accumulated dry matter under certain conditions [19]. In some cases wax was deposited on the leaf surface after transplantation [23,30,50, 64,71,75]. Stomatal function (open-closure mechanism) was either improved [51] or was not established in persistent leaves [70]. In most cases stomatal function has been equated with closure [2–4,18,50,51,61,65,74]. However, stomatal closure may only indicate the collapse of the guard cell membranes in response to exposure to low levels of relative humidity [5] and need not indicate the stomatal capacity to reopen.

The role of the persistent leaves remains a controversial and important issue. Photosynthetic capacity appears to vary with plant species in culture and may determine the *ex vitro* contribution of persistent leaves. Cultured plants are divisible into photosynthetically non-competent and competent species [29]. For example, in the non-competent species group, cultured cauliflower [32] and strawberry [34] were net respirers both *in vitro* and after transplantation. In these species, leaves that developed in culture deteriorated rapidly after transplantation. These leaves contributed only those nutrients which could be resorbed by the transplant. Such leaves have been referred to as storage organs or pseudo-cotyledonary tissues [34,69,71]. Non-competence in strawberry has been attributed to irreversibly reduced levels of RubPcase activity in leaves developed in the presence of sucrose. Strawberry plantlets defoliated in the absence of sucrose in the medium were competent [29,35]. Dieffenbachia (*Dieffenbachia picta*) [33] as well as potato (*Solanum tuberosum*) and chrysanthemum *(Chrysanthemum morifolium*) [29] were photosynthetically competent *in vitro*. They achieved a positive carbon balance in culture and continued to contribute photosynthetically after transplantation. Leaves of competent species did not deteriorate rapidly after transplantation [29,33,69]. Persistent leaves of Asian white birch [58] and red raspberry [19] seem to fall into the competent group. Red raspberry plantlets photosynthesized at a low level after transplantation. However, persistent leaves shifted to become both net respirers and sinks for photoassimilates formed in the new leaves by one month *ex vitro* [19]. Retention time of these persistent leaves ranged up to three months *ex vitro* [18].

4.2. *The new leaves*

The phenotype of new leaves formed *ex vitro* varies with the species, the culture and transplant environments and the age of the transplant. New leaves of cauliflower (a non-competent species) that formed the second week after transplantation apparently exhibited greenhouse control levels of CO_2 uptake [30]. However, new leaves of red raspberry (a competent species) were transitional in the sense that weekly flushes of new leaves became progressively larger, eventually with control-type anatomy, functional stomata and improved CO_2 uptake capability [18–21]. Measured five weeks after transplantation leaves formed during the first week had activity levels much higher than in culture, but resembled the culture leaf phenotype while those formed the fifth week were operating at about half the control CO_2 uptake rates and anatomically resembled greenhouse-grown control plant leaves [20]. Transitional leaves have also been observed in other plant species, both competent and non-competent [13,23,50,73,74].

The number of transitional leaves produced by a transplant may depend on the number of immature leaf buds formed in culture. The degree of transition of these leaves and how closely they resemble those of control plants is probably a reflection of the stage of development of leaf primordia when the plantlet was transferred from culture and the conflicting stresses imposed on leaf development by both the culture environment and the new ambient environment [18,19,21]. It is likely too, that the retention of any culture-type organs on the transplant influences the physiological status of the rest of the plant [21].

5. Acclimatization strategies for micropropagated plants

5.1. *Acclimatization* ex vitro

Traditionally the acclimatization environment *ex vitro* is adjusted to accomodate transplants from culture; gradually weaning them towards ambient relative humidities and light levels. As previously mentioned, transplants must undergo a period of acclimatization, more specifically, a period of transitional development in which both anatomical characteristics and physiological performance escape the influence of the *in vitro* culture conditions [19,21]. While Stage III plantlets are generally easier to transfer to soil than Stage II shoots, whenever possible shoots are preferred due to economic considerations [6].

There are inherent limitations in the efficiency of conventional transplantation protocols. For non-competent species the transplanting risks are much greater than for competent species. Although, even the competent species can be slow to adjust to lower relative humidity and take time to become photosynthetically efficient.

Much has been written on the optimization of *ex vitro* transplant environments for tissue cultured shoots and plantlets [6,7,10,75]. Novel approaches to *ex vitro* acclimatization include CO_2 enrichment without [42] or with supplementary lighting [11]. These reduced the *ex vitro* acclimatization interval in controlled humidity chambers or greenhouses but failed to eliminate the requirement for habituation to low humidity [11]. Among the most sophisticated *ex vitro* acclimatization procedures utilizes the 'acclimatization unit' [36,40], an apparatus which emerged from the engineering science of climate-controlled greenhouses. The microcomputer controlled acclimatization unit can determine the relative humidity, temperature, light intensity, CO_2 concentration, air flow rate and even the temperature of the nutrient solution and has the potential to control almost every other feature of the environment. All facets of the environment can be made to change by increments over time; ranging from simulated *in vitro* conditions at transplantation to that of the greenhouse or the open field weeks later. In the beginning changes are made in small increments which are later increased. Special emphasis is placed on minimizing water stress in the early stages *ex vitro*. It is not surprising that in such a unit both transplant survival and growth rates are significantly increased. For now, this enviable research tool is beyond the reach of most scientists.

Antitranspirants have not proven useful in promoting *ex vitro* survival or performance; phytotoxicity and interference with photosynthesis were both cited as possible reasons [62]. Other leaf surface-covering agents such as glycerol, paraffin wax and grease promoted *ex vitro* survival of several herbaceous species but have not been evaluated over the long term or examined on woody species [55].

5.2. *Hardening-off* in vitro

Murashige [52] was first to promote hardening of plants during Stage III. This facilitates but does not, of course, preclude acclimatization *ex vitro*. Murashige recommended the reduction of medium nutrients, the use of auxins for rooting and increased light levels. Three major strategies have emerged that focus on substantially changing the *in vitro* environment, especially in the later stages of micropropagation, in order to modify the CIP towards improved storage capability, photosynthetic competence or water relations and thus facilitate transplantation.

The first strategy assumes that larger persistent leaves, packed with greater amounts of storage compounds, would contribute more after transplantation. Increasing the concentration of sugar in the medium might maximize the nutrient function of persistent leaves [11,34]. To some extent this strategy has been discounted as apt to heighten evapotranspiration losses in transplants [11,34]. However, it seems to hold promise for some plants [46,54].

The second strategy assumes that autotrophic cultures will have persistent leaves that live longer and would be more photosynthetically productive *ex vitro* [34]. The objective is to modify the CIP towards autotrophy in culture. To do this, the oxygen concentration can be reduced in the culture environment, which depresses the photorespiration rate [56]. Alternatively, the sugar is reduced or completely eliminated from the medium [33–35,38] while the photosynthetic photon flux [12,38,41] and the carbon dioxide concentration [12,41] are increased [8]. Increasing the light intensity alone cannot raise the net photosynthetic rate for cultures at their CO_2 compensation point. Such a ‘photoautotrophic tissue culture system’ (PTCS) has the added advantage that microbial contamination is less of a problem when sugar is eliminated from the medium [27]. In this system a gas permeable, clear plastic film is used as a vessel closure [38]. This plastic film improves gas exchange to the cultures; CO_2 enrichment or O_2 reduction; increases the light penetration to the container contents and decreases the relative humidity of the vessels. Strawberry shoots rooted in the PTCS unit had dry weights 1.7-fold greater and photosynthetic rates 4-fold higher than plantlets in the control, Stage III treatment [27]. No special *ex vitro* care was required for some plants [36,37] although water stress was still a problem for others [27]. In adopting the PTCS, *in vitro* culture has been exchanged for a hydroponic system. This has the advantage that contamination by bacteria and fungi is less of a problem *in vitro*, extensive climate control, larger vessels and robotization are possible and the *ex vitro* acclimatization stage is less problematic [39]. However, one must accept the concurrent loss of many of the advantages of micropropagation, such as axillary growth and miniaturization [8] and assume the problems of hydroponic systems, such as algae control and the requirements for chronic nutrient solution adjustments [27].

The third strategy assumes that plants developed under lower relative humidity will have fewer transpiration and translocation problems *ex vitro* and persistent leaves that look more like control plant leaves. The objective is to modify the CIP away from the characteristic hydrophytic-type anatomy and promote epicuticular and cuticular wax development, stomatal function and possibly overcome other deficiencies. Lowering the relative humidity *in vitro* has been done experimentally with varying results using desiccant in or around the culture container, by coating the medium with oily materials, or both [57,64,76], by opening culture containers into low relative humidity atmospheres [2], adjusting culture closures to reduce the relative humidity [57], using special closures that facilitate water loss [24] or by cooling container bottoms [67,68]. Generally the relative humidity could not be lowed to less than 80–85% without culture injury [57,76]. A relative humidity of 85% decreased the multiplication rate of carnation but increased the number of glaucus leaves, the pigment and protein content, decreased the percentage water content and improved *ex vitro* survival [76]. At 80% relative humidity, growth rates of cauliflower and chrysanthemum were similar to those of controls grown under 100% relative humidity but *ex vitro* transplantation was greatly facilitated by functional stomata and greater epicuticular wax deposition in plantlets rooted at the lower relative humidity [57]. Increasing the sugar or agar concentrations or adding osmotic agents such as polyethylene glycol to the medium will also lower the relative humidity and in some cases served the same purpose as desiccants [45,49,57,76]. Cooling container bottoms created a temperature gradient of about 5 °C, condensing water vapor on the surface of the medium and lowering the relative humidity at plant height inside the containers [67,68].

By decreasing the relative humidity in culture containers, both transpiration and translocation systems are presumably improved in cultured plants [7,8] with associated improvement in mineral ion uptake through the transpiration stream and other benefits [9]. The obvious disadvantage of more extreme relative humidity reduction was to decrease the multiplication rate [57,75,76], posing an obvious dilemma [76]. Ziv [75] recommended that relative humidity reduction should be considered *in vitro*, even at the expense of reduced propagation rates. As the propagation rate of cauliflower and chrysanthemum was not apparently compromised at 80%, relative humidity reduction was followed by the elimination of sucrose which successfully promoted autotrophy in Stage III cultures of both these plants [57]. This resulted in comparable photosynthetic rates for plantlets rooted at 80% relative humidity *in vitro* and seedling plants, and underlined the viability of this approach.

6. Conclusions and prospects

It is premature to advocate any one of the several, not mutually exclusive, acclimatization strategies presented above. Conventional *ex vitro* acclimatization works for many but not all micropropagated plants. When such protocols are refined, they succeed in shortening the establishment interval and promoting survival, but do not usually eliminate the necessity for low humidity habituation. The costs of facilities dedicated to *ex vitro* acclimatization can also be a limiting factor [29]. In order to deemphasize *ex vitro* acclimatization and reduce associated costs it may be necessary to precede this stage by *in vitro* treatments. *In vitro* hardening-off procedures may be appropriate for some plant species, but are apparently not advantageous for all. Modification to the CIP to promote improved photosynthetic competence and water regulation in shoots or plantlets at the later stages of micropropagation is deceptively simple. As more documentation results from the implementation of these varied acclimatization strategies, alone or in combination, it will be possible to make choices more accurately. These would probably be made on a plant by plant basis with attention to economic considerations.

Tissue culturists are in an ideal position to evaluate the phenotypic plasticity of plants and to sort out the environmental determinants of plant phenotype, since they have access to large numbers of genetically identical plants grown under climate-controlled conditions. In this way we can learn more about the adaptational responses of all plants to specific environmental cues and climatic changes, and can also meet an urgent industry objective, to successfully accomodate or manipulate the culture-induced phenotype in order to promote survival and performance *ex vitro*.

7. Summary

Ex vitro acclimatization of micropropagated plants can be difficult, costly and may limit commercial micropropagation. Solving *ex vitro* acclimatization problems is contingent on an improved understanding of the unique effects of the *in vitro* and *ex vitro* environments on plant phenotype.

Micropropagated plants possess a unique culture-induced phenotype (CIP), reflecting acclimatization to *in vitro* conditions. Plantlets are tiny with increased percentage water content and have fragile organs with reduced mechanical support tissue, thin cell walls and, in some species, fewer vascular connections.

The CIP is inappropriate when plants are transferred from culture to greenhouse or field conditions. *Ex vitro* transplants have extreme evapotranspiration rates due to reduced stomatal control and large cuticular water loss. Therefore, conventional *ex vitro* acclimatization involves a gradual weaning of transplants from culture conditions towards ambient relative humidity and light levels. This is more difficult for unrooted shoots and photosynthetically non-competent species than for plantlets or competent species.

Three major strategies for *in vitro* hardening have focused on changing the culture environment to modify the CIP towards improved storage capability, photosynthetic competence or water relations. Increasing the sugar concentration of the medium to maximize the storage capability of persistent leaves holds promise for some species. The promotion of autotrophy via CO_2 enrichment or O_2 reduction, reduced levels of sugar and increased light intensity is promising for some plants, although water stress remains a problem at transplantation. Lowering the relative humidity to 80–85% via several methods did little to compromise the multiplication rate for certain species and facilitated transplantation. Elimination of sugar from the medium, in conjunction with reduced relative humidity, promoted autotrophy for both a competent and a non-competent species. *In vitro* hardening-off procedures are relatively new and have not yet been widely evaluated.

References

1. Birchem R, Sommer HE, and Brown CL (1981) Scanning electron microscopy of shoot and root development in sweetgum callus tissue. Forest Sci 27:206–212.
2. Brainerd KE and Fuchigami LH (1981) Acclimatization of aseptically cultured apple plants to low relative humidity. J Amer Soc Hort Sci 106:515–518.
3. Brainerd KE and Fuchigami LH (1982) Stomatal functioning of *in vitro* and greenhouse apple leaves in darkness, mannitol, ABA and CO_2. J Expt Bot 33:388–392.
4. Brainerd KE, Fuchigami LH, Kwaitkowski S, and Clark CS (1981) Leaf anatomy and water stress of aseptically cultured 'Pixy' plum grown under different environments. HortScience 16:173–175.
5. Conner LN and Conner AJ (1984) Comparative water loss from leaves of *Solanum laciniatum* plants cultured *in vitro* and *in vivo*. Plant Science Lett 36:241–246.
6. Conner AJ and Thomas MB (1982) Re-establishing plantlets from tissue culture: a review. Proc Inter Plant Prop Soc 31:342–357.
7. Debergh PC (1986) Micropropagation of herbaceous plants. In: Alderson PG and Dullforce WM (Eds.) Micropropagation in Horticulture. Practice and Commercial Problems. pp. 27–36.
8. Debergh PC (1988) Control of *in vitro* plant propagation. International symposium on plant biotechnology. Piracicaba, Brazil Oct 25–28, 1988.
9. Debergh P (1988) Improving mass propagation of *in vitro* plantlets. In: Horticulture in High Technology Era. Symp. Tokyo, Japan. May 10–11, pp. 47–57.
10. Debergh P (1988) Micropropagation of woody species – state of the art on *in vitro* aspects. Acta Hort 227:287–295.
11. Desjardins Y, Gosselin A, and Yelle S (1987) Acclimatization of *ex vitro* strawberry plantlets in CO_2-enriched environments and supplementary lighting. J Amer Soc Hort Sci 112:846–851.
12. Desjardins Y, Laforge F, Lussier C, and Gosselin A (1988) Effect of CO_2 enrichment and high photosynthetic photon flux on the development of autotrophy and growth of tissue-cultured strawberry, raspberry and asparagus plants. Acta Hort 230:45–53.
13. Dhawan V and Bhojwani SS (1987) Hardening *in vitro* and morpho-physiological changes in the leaves during acclimatization of micropropagated plants of *Leucaena leucocephala* (Lam.) De Wit. Plant Sci 53:65–72.
14. Donnelly DJ and Skelton FE (1987) Hydathode structure of micropropagated plantlets and greenhouse-grown 'Totem' strawberry plants. J Amer Soc Hort Sci 112:755–759.
15. Donnelly DJ and Skelton FE (1989) Comparison of hydathode structure in micropropagated plantlets and greenhouse-grown 'Queen Elizabeth' rose plants. J Amer Soc Hort Sci 114:841–846.
16. Donnelly DJ, Skelton FE, and Daubeny HA (1986) External leaf features of tissue-cultured 'Silvan' blackberry. HortScience 21:306–308.
17. Donnelly DJ, Skelton FE, and Nelles JE (1987) Hydathode anatomy and adaxial water loss in micropropagated 'Silvan' blackberry. J Amer Soc Hort Sci 112:566–569.
18. Donnelly DJ and Vidaver WE (1984) Leaf anatomy of red raspberry transferred from culture to soil. J Amer Soc Hort Sci 109:172–176.

19. Donnelly DJ and Vidaver WE (1984) Pigment content and gas exchange of red raspberry *in vitro* and *ex vitro*. J Amer Soc Hort Sci 109:177–181.
20. Donnelly DJ, Vidaver WE, and Colbow K (1984) Fixation of $^{14}CO_2$ in tissue cultured red raspberry prior to and after transfer to soil. Plant Cell Tissue Organ Culture 3:313–317.
21. Donnelly DJ, Vidaver WE, and Lee KY (1985) The anatomy of tissue cultured red raspberry prior to and after transfer to soil. Plant Cell Tissue Organ Culture 4:43–50.
22. Esau K (1977) Anatomy of Seed Plants. 2nd ed. Wiley, New York.
23. Fabbri A, Sutter E, and Dunston SK (1986) Anatomical changes in persistent leaves of tissue-cultured strawberry plants after removal from culture. Scientia Hort 28:331–337.
24. Fari M, Andrasfalvy A, and Nemeth J (1987) Thin PVC foil covering (TPFC), an efficient method for culture and preacclimatization of *in vitro* plant cultures. Acta Hort 212:371–374.
25. Fuchigami LH, Cheng TY, and Soeldner A (1981) Abaxial transpiration and water loss in aseptically cultured plum. J Amer Soc Hort Sci 106:519–522.
26. Fujiwara K, Kozai T, and Watanabe I (1987) Fundamental studies on environments in plant tissue culture vessels. (3) Measurements of carbon dioxide gas concentration in stoppered vessels containing tissue cultured plantlets and estimates of net photosynthetic rates of the plantlets. J Agr Met 43:21–30 [Japanese with English summary].
27. Fujiwara K, Kozai T, and Watanabe I (1988) Development of a photoautotrophic tissue culture system for shoots and/or plantlets at rooting and acclimatization stages. Acta Hort 230:153–166.
28. Grout BWW (1975) Wax development on leaf surfaces of *Brassica oleraceae* var. Currawong regenerated from meristem culture. Plant Sci Lett 5:401–405.
29. Grout BWW (1988) Photosynthesis of regenerated plantlets *in vitro*, and the stresses of transplanting. Acta Hort 230:129–135.
30. Grout BWW and Aston MJ (1977) Transplanting of cauliflower plants regenerated from meristem culture. I. Water loss and water transfer related to changes in leaf wax and to xylem regeneration. Hort Res 17:1–7.
31. Grout BWW and Aston MJ (1978) Modified leaf anatomy of cauliflower plantlets regenerated from meristem culture. Ann Bot 42:993–995.
32. Grout BWW and Aston MJ (1978) Transplanting of cauliflower plants regenerated from meristem culture. II. Carbon dioxide fixation and the development of photosynthetic ability. Hort Res 17:65–71.
33. Grout BWW and Donkin ME (1987) Photosynthetic activity of cauliflower meristem cultures *in vitro* and at transplanting into soil. Acta Hort 212:323–327.
34. Grout BWW and Millam S (1985) Photosynthetic development of micropropagated strawberry plantlets following transplanting. Ann Bot 55:129–131.
35. Grout BWW and Price F (1987) The establishment of photosynthetic independence in strawberry cultures prior to transplanting. In: Ducate G, Jacobs M, and Simeon A (Eds.) Proc Symp Plant Micropropagation in Horticultural Industries. Arlon, Belgium. pp. 55–60.
36. Hayashi M and Kozai T (1987) Development of a facility for accelerating the acclimatization of tissue cultured plantlets and the performance of test cultivations. In: Ducate G, Jacobs M, and Simeon A (Eds.) Proc Symp Plant Micropropagation in Horticultural Industries. Arlon, Belgium. pp. 123–134.
37. Hayashi M, Nakayama M, and Kozai T (1988) An application of the acclimatization unit for growth of carnation explants, and for rooting and acclimatization of the plantlets. Acta Hort 230:189–194.
38. Kozai T (1988) Autotrophic micropropagation for mass propagation of *in-vitro* plantlets. Plant biotechnology seminar. Jakarta. Dec 12–13, 1988.
39. Kozai T (1988) High technology in protected cultivation – from environment control engineering point of view. In: Horticulture in High Technology Era. Special Lecture. Tokyo. pp. 1–43.

40. Kozai T, Hayashi M, Hirosawa Y, Kodama T, and Watanabe I (1987) Environmental control for acclimatization of *in vitro* cultured plantlets. (1) Development of the acclimatization unit for accelerating the plantlet growth and the test cultivation. J Agr Met 42:349–358. (Japanese with English summary.)
41. Kozai T, Oki H, and Fujiwara K (1987) Effects of CO_2-enrichment and sucrose concentration under high photosynthetic photon fluxes on growth of tissue-cultured *Cymbidium* plantlets during the preparation stage. In: Ducate G, Jacobs M, and Simeon A (Eds.) Proc Symp Plant Micropropagation in Horticultural Industries. Arlon, Belgium. pp. 135–141.
42. Lakso AN, Reisch BI, Mortensen J, and Roberts MH (1986) Carbon dioxide enrichment for stimulation of growth of *in vitro*-propagated grapevines after transfer from culture. J Amer Soc Hort Sci 111:634–638.
43. Lee N, Wetzstein HY, and Sommer HE (1985) Effects of quantum flux density on photosynthesis and chloroplast ultrastructure in tissue-cultured plantlets and seedlings of *Liquidambar styraciflua* L. towards improved acclimatization and field survival. Plant Physiol 78:637–641.
44. Lee N, Wetzstein HY, and Sommer HE (1988) Quantum flux density effects on the anatomy and surface morphology of *in vitro-* and *in vivo*-developed sweetgum leaves. J Amer Soc Hort Sci 113:167–171.
45. Leshem B (1983) Growth of carnation meristems *in vitro*: anatomical structure of abnormal plantlets and the effect of agar concentration in the medium on their formation. Ann Bot 52:413–415.
46. Maene L and Debergh P (1985) Liquid medium additions to established tissue cultures to improve elongation and rooting *in vitro*. Plant Cell Tissue Organ Culture 5:23–33.
47. Maene LJ and Debergh PC (1986) Optimization of plant micropropagation. Med Fac Landbouw Ghent 51:1479–1488.
48. Manning CE, Miller DG, and Teare ID (1977) Effect of moisture stress on leaf anatomy and water use efficiency of peas. J Amer Soc Hort Sci 102:756–760.
49. Marin JA and Gella R (1987) Acclimatization of the micropropagated cherry rootstock 'Masto de Montanana' (*Prunus cerasus* L.) Acta Hort 212:603–609.
50. Marin JA and Gella R (1988) Is desiccation the cause of the poor survival rate in the acclimatization of micropropagated *Prunus cerasus* L. Acta Hort 230:105–112.
51. Marin JA, Gella R, and Herrero M (1988) Stomatal structure and functioning as a response to environmental changes in acclimatized micropropagated *Prunus cerasus* L. Ann Bot 62:663–670.
52. Murashige T (1974) Plant propagation through tissue culture. Ann Rev Plant Physiol 25: 135–166.
53. Poole RT and Conover CA (1983) Establishment and growth of *in vitro*-cultured *Dieffenbachia*. HortScience 18:185–187.
54. Queralt MC (1989) Histological and ecophysiological study of the changes occurring during the acclimatization of *in vitro* cultured roses. Ph D Thesis, State Univ Ghent, Belgium. 98pp.
55. Selvapandiyan A, Subramani J, Bhatt PN, and Mehta AR (1988) A simple method for direct transplantation of cultured plants to the field. Plant Sci 56:81–83.
56. Shimada N, Tanaka F, and Kozai T (1988) Effects of low O_2 concentration on net photosynthesis of C_3 plantlets *in vitro*. Acta Hort 230:171–187.
57. Short KC, Warburton J, and Roberts AV (1987) *In vitro* hardening of cultured cauliflower and chrysanthemum plantlets to humidity. Acta Hort 212:329–334.
58. Smith MAL, Palta JP, and McCown BH (1986) Comparative anatomy and physiology of microcultured, seedling, and greenhouse-grown Asian white birch. J Amer Soc Hort Sci 111:437–442.
59. Sutter E (1984) Chemical composition of epicuticular wax in cabbage plants grown *in vitro*. Can J Bot 62:74–77.
60. Sutter EG (1985) Morphological, physical and chemical characteristics of epicuticular wax on ornamental plants regenerated *in vitro*. Ann Bot 55:321–329.

61. Sutter E (1988) Stomatal and cuticular water loss from apple, cherry, and sweetgum plants after removal from *in vitro* culture. J Amer Soc Hort Sci 113:234–238.
62. Sutter EG and Hutzell M (1984) Use of humidity tents and antitranspirants in the acclimatization of tissue-cultured plants to the greenhouse. Scientia Hort 23:303–312.
63. Sutter E and Langhans RW (1979) Epicuticular wax formation on carnation plantlets regenerated from shoot tip culture. J Amer Soc Hort Sci 104:493–496.
64. Sutter E and Langhans RW (1982) Formation of epicuticular wax and its effect on water loss in cabbage plants regenerated from shoot-tip culture. Can J Bot 60:2896–2902.
65. Sutter EG, Novello V, and Shackel K (1988) Physiological and anatomical aspects of water stress of cultured plants. Acta Hort 230:113–119.
66. Timmis R and Richie GA (1984) Progress in Douglas fir tissue culture. Int Symp Recent Advances in Forest Biotech. Traverse City, MT 1984. pp. 37–46.
67. Vanderschaeghe AM and Debergh PC (1987) Technical aspects of the control of the relative humidity in tissue culture containers. In: Ducate G, Jacobs M, and Simeon A (Eds.) Proc Symp Plant Micropropagation in Horticultural Industries. Arlon, Belgium. pp. 68–76.
68. Vanderschaeghe AM and Debergh PC (1988) Automation of tissue culture manipulations in the final stages. Acta Hort 227:399–401.
69. Wardle K, Dalsou V, Simpkins I, and Short KC (1983) Redistribution of rubidium in plants of *Chrysanthemum morifolium* Ram cv. Snowdown derived from tissue cultures and transferred to soil. Ann Bot 51:261–264.
70. Wardle K, Dobbs EB, and Short KC (1983) *In vitro* acclimatization of aseptically cultured plantlets to humidity. J Amer Soc Hort Sci 108:386–389.
71. Wardle K, Quinlan A, and Simpkins I (1979) Abscisic acid and the regulation of water loss in plantlets of *Brassica oleraceae* L. var. *botrytis* regenerated through apical meristem culture. Ann Bot 43:745–752.
72. Wardle K and Short KC (1983) Stomatal response of *in vitro* cultured plantlets. I. Responses in epidermal strips of *Chrysanthemum* to environmental factors and growth regulators. Biochem Physiol Pflanzen 178:619–624.
73. Wetzstein HY and Sommer HE (1982) Leaf anatomy of tissue cultured *Liquidambar styrraciflua* (Hamamelidaceae) during acclimatization. Amer J Bot 69:1579–1586.
74. Wetzstein HY and Sommer HE (1983) Scanning electron microscopy of *in vitro* cultured *Liquidambar styraciflua* plantlets during acclimatization. J Amer Soc Hort Sci 108:475–480.
75. Ziv M (1986) *In vitro* hardening and acclimatization of tissue culture plants. In: Withers LA and Alderson PG (Eds.) Plant Tissue Culture and its Agricultural Application. Butterworth, London, 1986, pp. 187–196.
76. Ziv M, Meir G, and Halevy AH (1983) Factors influencing the production of hardened glaucous carnation plantlets *in vitro*. Plant Cell Tissue Organ Culture 2:55–65.
77. Ziv M, Schwartz A, and Fleminger D (1987) Malfunctioning stomata in vitreous leaves of carnation (*Dianthus caryophyllus*) plants propagated *in vitro*; implications for hardening. Plant Sci 52:127–134.

SECTION TWO

Micropropagation

10. Propagation of apple *in vitro*

O. P. JONES

Physiology Department, Horticultural Research International, East Malling, West Malling, Kent ME19 6BJ, U.K.

1. Introduction

The cultivated apple (*Malus* × *domestica* Bork) [35] is probably the most widely-grown fruit crop in the world, being cultured throughout the northern and southern temperate zones.

Commercial apple trees are produced by budding or grafting a scion onto a rootstock. This method of tree raising is used because scion cultivars rarely come true from seed and are difficult or impossible to propagate by conventional shoot cuttings or layering. Rootstocks have been selected for their ease of propagation and useful effects in the orchard such as control of tree vigour and cropping and also resistance to soil-borne diseases.

Methods of plant tissue culture *in vitro* that provide rapid propagation of herbaceous plants have been regarded as difficult to apply to adult trees of proven long-term performance [65]. However, apple has been researched extensively in respect to such propagation with the result that *in vitro* methods have now been applied successfully to many rootstock and scion cultivars.

A number of review articles provide comprehensive overviews of progress with apple propagation *in vitro* [4,19,26,54,68] including application to genetic manipulation [21]. The present article does not attempt to duplicate the detail of these reviews but will concern general principles relating to methods and applications of propagation *in vitro* of apple with emphasis on significant developments during the last few years not covered in previous reviews. Such developments relate to simplified methods for rooting shoots, production of apparently rejuvenated clones, regeneration of adventitious shoots or embryos from tissue explants, callus or protoplasts, and *Agrobacterium*-mediated genetic transformation.

2. Propagation of apple *in vitro*

2.1. *General methods*

Shoot culture with the proliferation of axillary shoots is the generally accepted method of propagation *in vitro* to maintain the genetic integrity of a clone.

M. R. Ahuja (Ed.), Micropropagation of Woody Plants, 169–186.

Such methods have been applied extensively to apple during the last decade; about 70 apple rootstock and scion cultivars have now been reported as propagated successfully by shoot culture methods [19].

More recently, methods for regenerating adventitious shoots or embryos from explants, callus or protoplasts have met with increasing success with apple and may give potential for even more rapid propagation than by shoot culture. However, adventitious regeneration often gives rise to genetic variants or somaclones. These are undesirable for propagation but may be exploited for crop improvement if they have desirable agronomic characters [52]. Furthermore, adventitious regeneration is an essential prerequisite for application of gene transfer techniques via tissue culture *in vitro*.

2.2. *Juvenility*

This is of major significance in relation to vegetative propagation of trees by conventional or *in vitro* methods. The stage of growth of the seedling following germination is described as the juvenile phase. This is a phase of very active growth when the tree does not initiate flowers and vegetative propagation is usually achieved readily. This phase continues for a number of years that varies according to species, but eventually the tree attains the capacity to produce flowers, fruits and seeds. The tree now enters the adult phase, which is characterised by a decreased capacity for vegetative propagation, a feature that usually becomes progressively more marked as the tree ages further. However, the transition from the juvenile phase appears to be reversible to some extent and it is common for some tissues of the adult tree to have some of the physiological characteristics of seedlings such as high growth rates and high capacity for adventitious shoot production. Such tissues, described as 'rejuvenated', may occur naturally, as with shoots (suckers) arising directly from the roots of adult trees, or may be induced by various treatments such as pruning, vegetative propagation, grafting onto seedlings or spraying with growth regulators [10]. Knowledge of the physiological and biochemical basis of juvenility and rejuvenation is very incomplete [12].

2.2.1. *Propagation* in vitro *by shoot culture*

Methods for apple are similar to those for other plants. The propagation is initiated with explants which may be nodal buds or shoot tips of lengths which may range from about 0.3 mm to 1.0 cm. Explants are surface sterilised, usually by washing in solutions of sodium or calcium hypochlorite [14] and are then cultured in suitable vessels which contain a culture medium solidified with agar. This medium is based usually on that of Murashige and Skoog [41] and contains mineral salts, sucrose and growth regulators such as the cytokinin, 6-benzylaminopurine (BA), and possibly some auxin and gibberellin. Sorbitol, which is the major translocation form of carbon in *Malus* may be more effective than sucrose with some apple cultivars [47,64] whilst phloridzin (PZ), the most prominent phenolic compound of *Malus* and phloroglucinol (PG) which is a breakdown product of PZ, sometimes more than doubles shoot growth. However, this latter effect appears to depend upon cultivar and the physiological status of tissues [19,25,28] (see below).

Shoot cultures are maintained on culture medium in illuminated growth rooms where they grow rapidly and their axillary buds extend prematurely to produce new shoots. These are excised at intervals of about one month and are then transferred to fresh medium where further axillary shoot production occurs; shoot culture lines may be multiplied indefinitely by this method of sequential subculture. The production of plantlets is completed by excising shoots and then inducing them to produce adventitious roots. This has usually been achieved *in vitro* by transference of the excised shoots to another type of medium that contains the auxin indole-3-butyric acid (IBA) but has no cytokinin. The inclusion of PG in this medium sometimes results in much improved rooting but, as with shoot growth, the effect appears to depend upon cultivar and physiological status [28]. Exposure of shoots to periods of darkness can also improve rooting *in vitro* [67].

All manipulations during shoot culture and rooting *in vitro* are carried out under sterile conditions with the use of air-flow cabinets to eliminate micro-organisms which would quickly kill the cultures. Following the rooting stage, the small plantlets are transferred to pots of compost for growing-on in a glasshouse.

Recent research effort has been towards simplified rooting of shoots produced *in vitro*. A method for rooting shoots in darkness under non-sterile conditions in sugar solution with IBA was successful with several cultivars [69]. With the rootstock M.26, dipping shoots into IBA powder and rooting directly into potting compost was more successful than rooting *in vitro* and had the advantage of leading to improved establishment of plantlets in soil [53]. Similar results were obtained with the rootstock M.9. Furthermore, with M.9, etiolation of shoots by four weeks in darkness improved both *in vitro* and direct rooting but plantlet establishment was always poor following etiolation [60].

From the limited study thus far, direct rooting appears to be more effective than the *in vitro* method and now needs evaluation with more cultivars and on the large scale of commercial production.

2.2.2. *Vitrification*

Shoot cultures of many species, including apple, may become slow-growing with tightly-rolled translucent leaves, a process which has been described as 'vitrification' [42]. This condition has been alleviated in apple by modifying the culture medium with respect to the brand and concentration of agar [33], the type of carbon source [51], the concentration of NH_4^+ [38], or by the addition of PZ or PG [13]. However, vitrification appears to occur unpredictably in some laboratories and improved understanding of the phenomenon is required.

2.2.3. *Apparent rejuvenation as a result of culture* in vitro

An important feature of apple propagation *in vitro* by shoot culture is that success is achieved with subjects such as scion cultivars which are difficult or impossible to propagate by conventional methods. The basis of this *in vitro* advantage appears to be the progressive increase in propagation potential with repeated subculturing. The rate of this increase appears to be cultivar-dependent. Over 90% of shoots of cv Jonathan rooted by the ninth subculture whilst 31 subcultures were required for cv Delicious to reach 79% rooted shoots [55]. With the M.7 rootstock, fewer subcultures were required than with cv Greensleeves to reach 70% rooted shoots [32]. Shoots of rootstock Åkerö required eight subcultures before satisfactory rooting was achieved [62].

Effect of subculture has been studied extensively with the rootstock M.9 where there was a progressive increase in the rate of shoot proliferation with monthly over 21 months. Moreover, this enhanced shoot production was accompanied by a progressive increase in rooting ability so that a maximum of about 70% rooted shoots could be achieved after 10–11 months. An additional observation with M.9 was that shoot culture lines which originated from different shoot tip explants consistently exhibited different rates of shoot production and rooting ability irrespective of changes with subculture. This suggests that explants vary in their potential for propagation *in vitro* (Figure 1, [60]).

The basis of improved propagation with subculture remains obscure but there could be three possible explanations, viz: that culture *in vitro* leads to (a) genetic change, (b) elimination of viruses or other pathogens, or (c) tissue rejuvenation. Of these, rejuvenation seems to be the most likely since trees produced by shoot culture are generally true-to-type [19,26]. Elimination of viruses or other pathogens seems unlikely since the beneficial effects of subculture were apparent with culture lines initiated from, healthy, virus-free trees. Moreover, improvements with subculture occurred gradually whereas virus or pathogen elimination would be expected to occur immediately upon taking the initial shoot tip explants into culture [30].

3. Propagation *in vitro* through adventitious regeneration

Until recently almost all successful adventitious shoot or embryo regeneration with apple was with seedling material and this was of doubtful application to adult rootstock and scion clones. Reports of production of adventitious maternal-type embryos from cultures of apple nucellus have been regarded as equivocal since the embryos may have arisen from zygotic cells which were transferred inadvertently with nucellar tissue [21].

However, there has been substantial progress with the adult material in recent years; shoot or embryo regeneration has now been reported from tissues of about 20 rootstock and scion cultivars and in some cases this regeneration

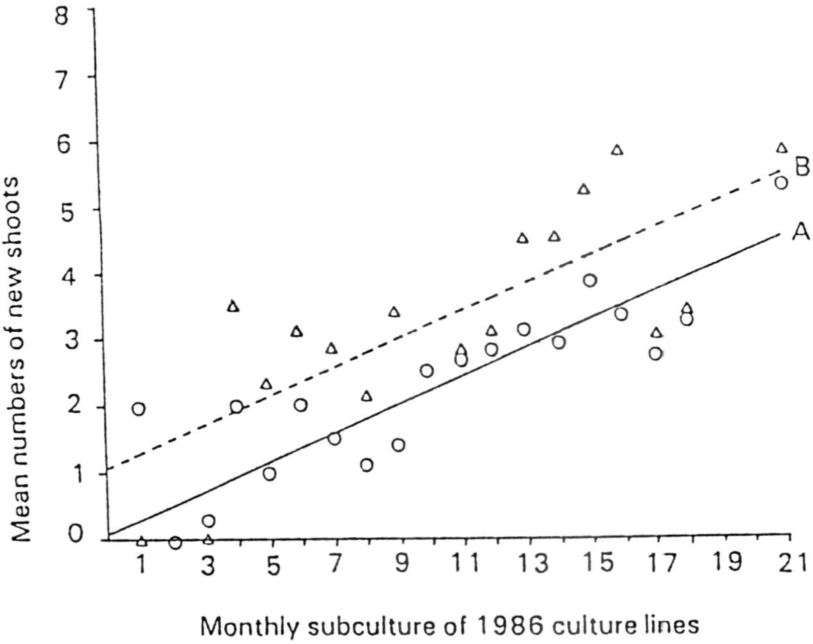

Fig. 1. Monthly shoot production from M.9 culture lines: mean numbers of new shoots from single shoot inocula.
○ -----Regression of numbers of new shoots produced by culture line A (Pr_A) on monthly subculture (sub): $Pr_A = 0.21$ sub $+ 0.11$.
Δ-----Regression of numbers of new shoots produced by culture line B (Pr_B) on monthly subculture (sub): $Pr_A = 0.21$ sub $+ 1.06$.
Culture lines A and B significantly different ($P < 0.001$).
(Webster and Jones, 1989 [60]. Reproduced with permission.)

was at high frequency [21]. Most of these reports were similar in that shoots were regenerated from leaf tissue with or without the intervention of a callus stage. Moreover, the culture media used were based usually on that of Murashige and Skoog with the addition of BA and the auxin naphthalene acetic acid (NAA). A notable exception to this was the formation of embryos on leaf tissue of five apple cultivars cultured on the N6 medium of Chu [5] with BA and NAA [63]. More recently, optimum regeneration of shoots was reported with several apple cultivars by culturing leaves from the distal portions of shoots *in vitro* for two weeks in the dark on N6 medium containing the cytokinin thidiazuron and NAA and then moving cultures to 16 h light daily [8].

However, in spite of the recent encouraging progress, there appears to be considerable variation in regeneration capacity between apple cultivars, and the pattern of response may well be under genetic control [21]. Improved understanding of the genetic and physiological factors that control regeneration capacity now appears to be essential for further progress.

3.1. *Anther and pollen culture*

There are only two reports, both from China, of successful anther culture with apple to produce haploid plants [9,66]. Haploid shoot cultures have also been produced from a single haploid embryo which was produced following pollination of blossom of cv Greensleeves with irradiated pollen. However, these shoots could not be rooted [21] and thus far, traditional breeding methods which do not involve culture *in vitro* have been the most successful in producing haploid apple trees [37].

3.2. *Protoplast culture*

Protoplast technology was considered to be still in its infancy with apple and other temperate fruit trees until very recently [19,21]. There were reports of isolation and culture of apple protoplasts to form callus but this could never be induced to regenerate shoots or embryos [6,17,22,36,44,50,57].

These earlier investigations were extended recently to achieve shoot regeneration from protoplast-derived callus of some apple rootstock and scion cultivars, followed in some cases by the production of autotrophic plants [45]. These successful studies related to protoplasts from leaves of shoot cultures *in vitro* of rootstocks M.9 and MM.106 and the scion cultivars Spartan and Bramley's Seedling. With the latter, protoplasts were also obtained from callus an cell suspensions.

Protoplasts from the four cultivars had different cultural requirements. Media based on that of Murashige and Skoog were best for the rootstocks whilst media based on that of Kao and Michayluk [34] were preferred for the scions. Murashige and Skoog media were also used for inducing shoot regeneration from the protoplast-derived callus of both rootstocks and also Spartan but again there were differences with cultivar. Thiamine, pyridoxine, nicotinic acid and casein hydrolysate were always needed in the medium for regeneration whilst the callus of both rootstocks also required PG, at 1 mM for M.9 and 0.1 mM for MM.106. There was no shoot regeneration with callus of Bramley's Seedling.

Shoots regenerated from the callus of M.9 and Spartan were multiplied *in vitro* and subsequently rooted followed by establishment of plantlets into soil; shoots of MM.106 were difficult to propagate *in vitro* and could not be rooted (Figure 2, [45]).

3.3. *Mode of action of PG*

This is of interest since PG has been implicated in shoot regeneration from protoplast-derived callus as well as in shoot growth, rooting and elimination of vitrification.

Fig. 2. Plant regeneration from apple protoplasts.
a. Isolated cell suspension protoplasts of Bramley's Seedling;
b. Dividing callus protoplast-derived cells of Bramley's Seedling after 15 days of culture;
c. Callus protoplast-derived cell colonies after 30 days of culture;
d. Shoots emerging from protoplast-callus of M.9 apple after 20 weeks from isolation;
e. Rooted plants (left-M.9; right-Spartan) at 28 weeks from isolation;
f. Sample of plants in soil (left-M.9; right-Spartan) after one month from *ex vitro* transfer.
Bars represent 40 μm (**a**), 125 μm (**b**), 175 μm (**c**), 1 cm (**d**; **e**), 5 cm (**f**).
(Patat-Ochatt *et al.*, 1989 [45]. Reproduced with permission.)

With some cultivars at least, the effect of PG appeared to depend upon duration, of subculture. With cv Northern Spy, PG promoted shoot growth during the first four months of subculture following establishment of a culture line but thereafter had no effect [18]. With the rootstock M.9, some newly established culture lines failed to survive in the absence of PG [27] and there were similar effects with some rootstocks of *Prunus* [2]. However, with M.9, PG had beneficial effects on shoot growth and rooting only during the first 10 and 15 months of subculture respectively [60]. Such dependence on duration of subculture led to the suggestion that effects of PG are related to the degree of tissue rejuvenation [26,27]. Direct support for this is provided by results with the plum rootstock Pixy (*Prunus insititia*) where enhanced shoot growth *in vitro* with the phenolic compound was apparent only with cultures from putatively non-rejuvenated hedges which had been produced from conventionally-propagated plants. There was no effect with shoot cultures from apparently rejuvenated hedges from plants propagated *in vitro* [15].

Studies with radiolabelled indole-3-acetic acid (2-^{14}C)-IAA) indicated that PG effects are not mediated through enhanced uptake of auxin by shoot cultures [20]. Moreover, PG stimulated rooting with the not-easily oxidised NAA and IBA and was also effective with supra-optimal levels of IBA which inhibited rooting in the absence of PG [29]. Therefore, an explanation other than the inhibition of auxin destruction should be sought for the results with PG.

PZ and PG had a remedial effect on vitrification of apple shoots *in vitro*, a condition which appeared to be due to deficient lignin biosynthesis [13]. It was proposed from study of activity of enzymes involved in lignin and flavonoid biosynthesis in vitrified shoots that PZ had a remedial effect because it was converted back into its normal precursor, p. coumaroyl-CoA, which could act as a precursor of lignin [46].

The mode-of-action of PZ and PG would appear to merit further study in view possible benefits to propagation *in vitro* and the possible role of these compounds in apple trees, particularly in relation to juvenility and maturation.

4. Applications of propagation *in vitro* with apple

Applications of propagation *in vitro* are most advanced with crops such as strawberry where the cycle from propagation to harvest and replanting is relatively short. With a tree crop like apple with a cycle of 20 years or more for the life of a commercial orchard, application of propagation *in vitro*, at least on a large commercial scale, must await long-term evaluation in the field of *in vitro*-propagated trees.

Aspects of such evaluation in relation to apple trees from shoot culture *in vitro* have been investigated, mainly in Europe and North America, during the last decade. This includes mass propagation of rootstocks, production of apparently rejuvenated clones to give improved conventional propagation from conventional shoot cuttings or stoolbeds, self-rooted scions and also rootstocks/scion trees produced by grafting small plants from propagation *in vitro*. By contrast, almost all propagation *in vitro* with apple relating to adventitious regeneration remains at the research laboratory stage of development and potential applications are only beginning to be explored.

Results of evaluation of apple material produced by shoot culture *in vitro* are now described.

4.1. *Mass propagation of rootstocks*

Mass propagation *in vitro* of all tree crops on a commercial scale is not as advanced as with herbaceous plants. Nevertheless, commercial laboratories which specialise in temperate fruit and other tree species have been established in Italy [39,71], Canada [7] and the USA [11]. Thus far, the most success has been with peach × almond hybrid rootstocks where the production of about 10 million plants annually by shoot culture *in vitro* has been an important contribution to the European fruit industry [70].

Propagation *in vitro* has been applied much less to apple rootstocks, mainly because the economic cost has been too high when set against their present low value. Moreover, confidence in the use of *in vitro*-propagated apple rootstocks may be undermined by reports that they sometimes have more burr-knots (nodules of callus on the bark) and suckers than conventionally-produced counterparts and this may lead to increased susceptibility to pathogens and more expensive tree management, respectively [56]. Such problems with propagation *in vitro* appear to be unique to apple amongst temperate fruit trees.

There now appears to be a need to apply to apple results of recent research which would lead to reduction in economic cost. The replacement of rooting *in vitro* by direct methods, possibly in conjunction with fogging systems [48], may now be feasible and should give major savings. Other cost reductions could come from the use of nutrient replenishment through adding fresh liquid culture medium to 'spent' agar medium and continually harvesting shoots from a culture vessel, thus avoiding constant subculturing. Automation or robotics used in combination with such cost-saving techniques could transform the economics of propagation *in vitro* [1].

The basis of increased burr-knot and sucker production also requires investigation. Both these problems may result from excessive cytokinin supply during the *in vitro* stage of shoot multiplication. Alternatively, enhanced sucker production may be the result of too shallow planting in soil of small trees from propagation *in vitro*.

4.2. *Improved conventional propagation from apparently rejuvenated clones*

A major advantage of propagation *in vitro* with fruit trees appears to be the production of apparently rejuvenated clones which remain exceptionally easy to propagate in the long-term by conventional methods.

Such apparent rejuvenation has been investigated most extensively with the plum rootstock Pixy. Field hedges of this rootstock established from plants propagated *in vitro* continued for at least nine years to produce winter hardwood cuttings with the juvenile character of much more rapid rooting than cuttings from conventionally-propagated counterparts. Physiological change seemed to be the most likely explanation of the improved propagation since the leaf and stem characteristics of the *in vitro*-propagated sources were typical of the cultivar involved. Rejuvenation *in vitro* as manifest by improved shoot growth and rooting with subculture would seem to be the likely basis of the juvenile characteristics of the plants in the field [16].

Much improved rooting has also been observed with both winter and summer cuttings from field hedges of the apple rootstock M.9 derived from plants propagated *in vitro*. M.9 is normally notoriously difficult to propagate from hedge cuttings. Thus far, this improved propagation has been sustained for three years and remains under investigation [61].

Conventional propagation from stoolbeds produced from micropropagated plants has also been studied with apple. Stoolbed shoots are pruned severely and etiolated at their bases by burying in soil. Such shoots generally root more readily than cuttings from hedges but stoolbeds have the disadvantage of producing fewer shoots and being more expensive to maintain than hedges. Such disadvantages may be obviated, at least in part, through the use of plants from propagation *in vitro*. Evaluation over five years indicated that stoolbeds produced from such plants of the rootstock M.27 produced between 1.5 and 5.0 times as many shoots as the conventional counterparts and these shoots were better rooted (Figure 3), [43]).

The finding that propagation *in vitro* leads to improved conventional propagation suggests that one of the most important future outlets for *in vitro*-propagated material will be the provision of field-grown stock plants from apparently rejuvenated clones. Rootstocks derived indirectly in this way from propagation *in vitro* may not suffer from the disadvantages of excessive suckering and burr-knot production.

It is also conceivable that apple rootstocks of proven commercial potential which presently is not being realised because of propagation difficulties, may become more amenable following culture *in vitro*. Improved conventional propagation following propagation *in vitro* has been demonstrated recently with a *Pyrus communis* rootstock which is normally almost impossible to propagate [31].

4.3. *Self-rooted scions*

The success of propagation *in vitro* with apple scion cultivars that are difficult or impossible to root by conventional cuttings makes feasible the production

Fig. 3. Harvest of rooted stoolbed shoots of M.27 (left-ex *in vitro*; right-standard). (Navatel *et al.*, 1988 [43]. Reproduced with permission.)

of self-rooted trees on a large scale. Such trees could be competitive economically with the traditional grafted rootstock/scion trees which take up to three years to produce and demand expensive nursery facilities and skills.

Results thus far from various studies with self-rooted apple trees are generally in agreement in that the trees are true-to-type and vigour and cropping efficiency are cultivar-specific, with most commercial cultivars being too vigorous for modern high-density orchards (see Ref. 68 for an overview of results from USA and Europe). Excessive sucker production has also been a problem with some varieties such as Golden Delicious (A.D. Webster, unpublished results). Nevertheless, some cultivars such as Malling Greensleeves and Gala which are naturally precocious in flowering appear to be satisfactory when self-rooted, at least under conditions in England ([58], A. D. Webster, personal communication).

Dwarf and spur-type apple cultivars may be an attractive proposition when self-rooted by micropropagation. Dwarf columnar types derived from McIntosh Wijick which generally lack lateral branches have been micropropagated successfully and have retained the compact habit after nine years in the field, although they had the defect of producing more lateral shoots at the bases of the trunks than conventionally-propagated counterparts on rootstocks (K. R. Tobutt, personal communication).

It would appear that further studies with self-rooted trees should concentrate on dwarf or spur types. However, it seems inlikely that self-rooted trees of this type will come into widespread use until more is known about the influence of micropropagation on long-term performance, particularly in relation to fruiting an fruit storage.

4.4. *Rootstock/scion trees from grafted micropropagules*

Apple trees have been produced by grafting small potted rootstock and scion plants, approximately 50 cm in height, and produced by propagation *in vitro*. Composite trees large enough for field planting may be produced about six months after grafting and an added advantage is that they can be produced at any time of the year.

The field performance of such grafted trees of cv Greensleeves on various rootstocks was compared with that of conventionally-propagated counterparts over five years in the field. The relative effects of rootstock on tree vigour followed the expected pattern with the trees propagated *in vitro* and there was no indication of any mutation as expressed by changed characteristics of shoots and fruits. However, the micropropagated trees displayed juvenile-like characters of very vigorous growth and delay in cropping by one year. The vigorous growth was sustained for at least five years. There were no consistent differences in burr-knot production but some of the *in vitro*-propagated trees produced more suckers than the conventional counterparts [30]. Studies of the cambial activity and wood structure indicated that differences in annual ring width between *in vitro* and the conventional counterparts were most marked in the third year after planting, but even after five years the cambium of the micropropagated trees produced about twice as much wood as that of the conventional counterparts (Figure 4, [40]). The increased vigour of *in vitro*-propagated trees could be of value in the nursery and the increased wood production could be particularly advantageous if applied to forest trees. The delay with regard to cropping and also increased sucker production could be disadvantageous for fruit growing.

Other studies have been carried out with only the scion or rootstock component of the grafted trees originating from *in vitro*. Juvenile-like characters of enhanced growth and delayed cropping were also a feature of two-year-old apple trees produced by grafting *in vitro*-propagated scions of cvs Cox's Orange Pippin or Bramley's Seedling onto conventionally-propagated MM.106 rootstocks [59]. In this case, the more vigorous growth represented an increase of about 30% over the conventional counterparts which is much less than the more than two-fold increase with the 'Greensleeves' trees described above.

There have also been studies of trees with rootstocks of M.26 from propagation *in vitro* grafted with conventially-propagated scions. Here, rootstocks performance with respect to tree vigour and cropping appeared to be similar to conventional counterparts. However, the *in vitro*-propagated trees produced more suckers and burr-knot production was more rapid, but after three years the number of burr-knots was the same as with the conventional counterparts ([43], Navatel, personal communication).

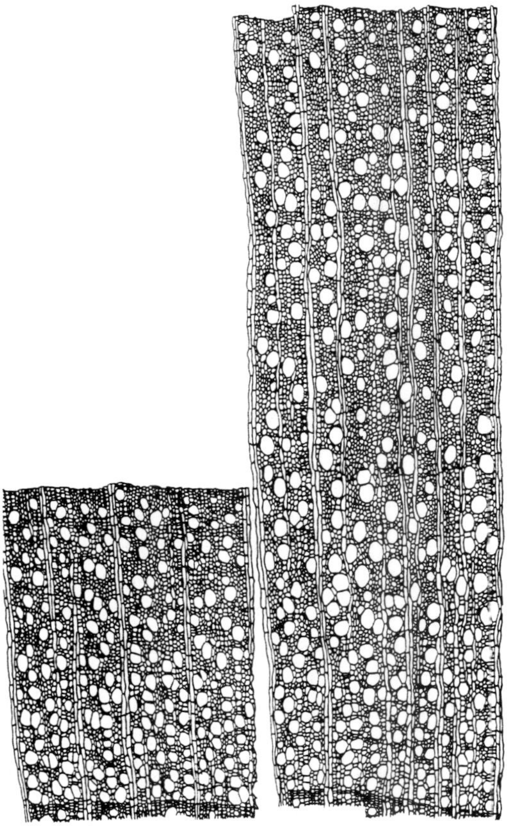

Fig. 4. Drawings of a portion of the outermost (fifth) annual ring of Malling Greensleeves/ M.27, from thin sections taken 10 cm above the graft union (left-conventionally propagated; right-micropropagated). (MacKenzie, 1989 [40] Reproduced with permission.)

The results of investigations described above differ in the degree of vigour exhibited by the grafted trees. Such variation may merely reflect cultivar differences or that both rootstocks and scion, or only one of these components, originated from propagation *in vitro*. Alternatively, there could have been different degrees of rejuvenation arising from different periods of subculture when producing the rootstocks and scions *in vitro*.

5. Conclusions and prospects for the future

Research and progress with propagation *in vitro* has probably been more extensive with apple than with any other type of tree and the results, apart from being of potential application to the cropping of apple trees, may serve as a model for other tree species.

Thus far, shoot culture methods are outstanding in their application to the widest range of apple rootstocks and scion cultivars, although commercial application has not been extensive for the economic and technical reasons described already. The most important application in the near future may well come from conventional propagation of apparently rejuvenated clones from propagation *in vitro*. There is now a need to evaluate the field performance of such apparently rejuvenated rootstocks when grafted with scions. Improved understanding of the physiological and biochemical basis of juvenility and maturation also merits some priority as a research topic which could lead to further exploitation of rejuvenation effects with tree crops.

Shoot culture methods are relatively crude with the need for division of the shoots, rooting and establishment in soil. Embryogenesis from callus or cell suspensions may offer more attractive possibilities in the long-term, possibly in conjunction with encapsulation of embryos to produce artifical seed [49]. The recent achievement of embryogenesis from apple leaf tissue [63] may be a first step towards the realisation of such methods with apple. The finding that embryos regenerated from cherry tissue produced true-to-type trees in the field gives further encouragement that large-sclae propagation by somatic embryogenesis will eventually become feasible with fruit trees [23].

Application of propagation *in vitro* to the production of new apple varieties could be of great value since breeding of trees by normal sexual hybridisation is very labour-intensive and takes many years. It is conceivable that these limitations will be overcome to some extent by the production of somaclones or polyploids from callus or protoplasts that are genetically variable or have been subjected to mutagenic agents [21]. In this context, some somaclones of apple, which were generated from leaf tissue, have been transferred to soil where they are showing increased resistance to the fireblight pathogen (*Erwinia amylovora*) (Anne Donovan, personal communication).

In the longer-term, application of gene transfer techniques could be of great value for the breeding of apple trees.

Potential applications of protoplast fusion with apple have already been described and include fusions between apple and other fruit tree types to transfer disease resistance or the control of size and cropping presently provided by grafting onto apple rootstocks [21].

With respect to *Agrobacterium*-mediated gene transfer the way now appears to be open to rapid progress with apple in view of the recent success with genetic transformation by a Ti-binary vector p BIN6 [3]. This vector was used in the transformation of leaf tissue of cv Greensleeeves followed by the regeneration of shoots to produce transgenic plants of normal phenotype. The presence of the gene for nopaline synthase and neomycin phosphotransferase (conferring resistance to the antibiotic kanamycin), inserted into the genome by the vector was confirmed by Southern blot analysis, the detection of nopaline synthase activity and rooting of shoots in the presence of kanamycin. The nopaline synthase gene continued to be expressed in glasshouse-grown plants for several months after removal from *in vitro* conditions. This initial success should now give the basis for the transfer into apple of

Fig. 5. In vitro pBIN6 transformed apple plantlet cv Greensleeves (left) compared to parent line (right). Bar = 1 cm. Plants approximately four weeks on rooting media. (James *et al.*, 1989 [24]. Reproduced with permission.)

genes of agronomic values, such as those for resistance to pests and disease (Figure 5, [24]).

6. Summary

General principles relating to methods and applications of propagation *in vitro* with apple are described. The emphasis is on recent developments in the last few years. Such developments relate to simplified methods for inducing shoots to form adventitious roots, production of apparently rejuvenated clones to give improved conventional propagation from conventional shoot cuttings or stoolbeds, regeneration of adventitious shoots or embryos from explants, callus or protoplasts and *Agrobacterium*-mediated genetic transformation.

References

1. Aitken-Christie and Jones C (1987) Towards automation: Radiata pine shoot hedges *in vitro*. Plant Cell Tiss Org Cult 8:185–196.
2. Baleriola-Lucas C and Mullins MG (1984) Micropropagation of two French prune cultivars (*Prunus domestica* L.). Agronomie 4:473–477.
3. Bevan MW (1984) Binary *Agrobacterium* vectors for plant cell transformation. Nucl Acids Res 12:8711–8721.
4. Boxus P and Druart P (1985) Mass propagation of fruit trees. In: Schaefer-Menuhr (Ed.) *In vitro* Techniques. Propagation and Long-Term Storage. Martinus Nijhoff/W. Junk, Dordrecht. pp. 29–33.
5. Chu C-C (1978) The N6 medium and its application to anther culture of cereal crops. Proc Sym Plant Tiss Cult. Science Press, Peking pp. 43–50.
6. Doughty S and Power JB (1988) Callus formation from leaf mesophyll protoplasts of *Malus* × *domestica* Borkh. cv Greensleeves, Plant Cell Rep 7:200–202.
7. Dunstan DI (1981) Transplantation and post-transplantation of micropropagated fruit-tree rootstocks. Comb Proc Int Plant Prop Soc 31:39–45.
8. Fasolo F, Zimmerman RH, and Fordham I (1989) Adventitious shoot formation on excised leaf *in vitro* grown shoots of apple cultivars. Plant Cell Tiss and Org Cult 16:75–87.
9. Fei, Kai-Wei and Xue, Guang-Rong (1981) Induction of haploid plantlets by anther culture *in vitro* in apple cv. Delicious. Scientia Agric Sinica 4:44.
10. Franclet A (1979) Rejeunissement des arbres adultes en vue de leur propagation végétative. Ann de recherches sylvicoles AFOCEL. Etudes et recherches, no. 12. Micropropagation d'arbres forestières, pp. 3–18.
11. Frecon J (1980) New tissue culture nursery opens in Oregon. Compact Fruit Tree 13:145.
12. Hackett WPO (1985) Juvenility, maturation and rejuvenation in woody plants. Hort Rev 7:109–155.
13. Hegedus P and Phan CT (1987) Activities of 5 enzymes of the phenolic metabolism on rooted acclimated vitreous plants in relation with phenolic treatments. Acta Hort 212(1):211–216.
14. Hennerty MJ, Upton Me, Furlong PA, Harris DP, Eaton RA ad James DJ (1988) Microbial contamination of *in vitro* cultures of apple rootstocks M.26 and M.9. Symposium on bacterial and bacteria-like contaminents of plant tissue cultures. Acta Hort No. 225:129–137.
15. Howard BH, Jones OP, and Vasek J (1989) Growth characteristics of apparently rejuvenated plum shoots. J Hort Sci 64:157–162.
16. Howard BH, Jones OP, and Vasek J (1989) Long-term improvement in the rooting of plum cuttings following apparent rejuvenation. J Hort Sci 64:147–156.
17. Hurtwitz CD and Agrios GN (1984) Isolation and culture of protoplasts from apple callus and cell suspension cultures. J Amer Sco Hort Sci 109:348–350.
18. Hutchinson JF (1984) Factors affecting shoot proliferation and root initiation in organ cultures of the apple 'Northern Spy'. Scientia Hort 22:347–358.
19. Hutchinson JF and Zimmerman RH (1987) Tissue culture of temperate fruit and nut trees. Hort Rev 9:273–349.
20. James DJ (1983) Adventitious root formation in apple rootstocks I. Factors affecting the length of the auxin sensitive phase in M.9. Physiol Plant 57:143–159.
21. James DJ (1987) Cell and tissue culture technology for the genetic manipulation of temperate fruit trees. Biotechnology and Genetic Engineering Reviews, 5:33–79. Intercept Limited, PO Box 402, Wimborne, Dorset, BH22, 9TZ, UK.
22. James DJ, Passey AJ, MacKenzie KAD, and Menhenick EC (1986). Protoplast culture and the use of apomixis *in vitro* for the regeneration of fruit plants. In Withers LA and Alderson PG (Eds.) Plant tissue culture and its agricultural applications. Butterworths, London, pp. 349–359.
23. James DJ, MacKenzie KAD, and Malhotra SB (1987) The induction of hexaploidy in cherry rootstocks using *in vitro* regeneration techniques. Theor and Appl Genetics, 73:589–594.

24. James DJ, Passey AJ, Barbara DJ, and Bevan M (1989) Genetic transformation of apple using a disarmed Ti-binary vector. Plant Cell Rep 7:658–661.
25. Jones OP (1976) Effect of phloridzin and phloroglucinol on apple shoots. Nature 262: 392–393 (Erratum 262:724).
26. Jones OP (1983) *In vitro* propagation of tree crops. In: Mantell AH and Smith H (Eds.) Plant Biotechnology. Soc Exp Biol Seminar Ser 18. Cambridge University Press, UK. pp. 139–159.
27. Jones OP (1985) The role of growth regulators in the propagation *in vitro* (micropropagation) or temporate fruit trees. British Plant Growth Regulator Group. Monograph 13 Growth regulators in horticulture. Menhenett R (Ed.) pp. 113–124.
28. Jones OP (1986) Endogenous growth regulators and rootstock/scion interactions in apple and cherry trees. Acta Hort No. 179. Fifth Intern Sym on growth regulators in fruit production Vol 1. Italy 1985. pp. 177–184.
29. Jones OP and Hatfield SGS (1976) Root initiation in apple shoots cultured *in vitro* with auxins and phenolic compounds. J Hort Sci 51:495–499.
30. Jones OP and Hadlow WCC (1989) Juvenile-like character of apple trees produced by grafting scions and rootstocks produced by micropropagation. J Hort Sci 64:395–401.
31. Jones OP and Webster CA (1989) Improved rooting from conventional cuttings taken from micropropagated plants of *Pyrus communis* rootstocks. J Hort Sci 64:429–434.
32. Jones OP, Marks TR, and Waller BJ (1982) Propagation *in vitro*. Rep E Malling Res Stn for 1981, p. 159.
33. Jones OP, Zimmerman RH, Fordham IM, and Hopgood ME (1985) Propagation *in vitro* of some dwarf apple trees. J Hort Sci 60:141–144.
34. Kao KN and Michayluk M (1975) Nutritional requirements for growth of *Vicia hajastana* cells and protoplasts at very low population density in liquid media. Planta 126:105–110.
35. Korban SS and Skirvin KM (1984) Nomenclature of the cultivated apple. HortSci 19: 177–184.
36. Kouider M, Hauptman R, Widholm JM, Skirvin RM, and Korban SS (1984) Callus formation from *Malus* × *domestica* cv Jonathan protoplasts. Plant Cell Rep 3:142–145.
37. Lespinasse Y, Godicheau M, and Duron M (1983) Potential value and method of producing haploids in the apple tree *Malus pumila* (Mill). Acta Hort 131:223–230.
38. Letouzé R and Daguin F (1987) Control of vitrification and hypolignification process in *Salix Babylonica* cultured *in vitro*. Acta hort 212(1):185–192.
39. Loreti F and Morini S (1982) Mass propagation of fruit trees in Italy by tissue culture: present status and perspectives. Comb Proc Int Plant Prop Soc 32:283–291.
40. MacKenzie KAD (1989) Cambial activity and wood structure of micropropagated *v* conventionally propagated trees. Ann Rep AFRC Inst Hort Research for 1988:pp.32–33.
41. Murashige T and Skoog F (1962) A revised medium for rapid growth and bioassays with tobacco tissue cultures. Physiol Plant 15:473–497.
42. Navatel JC (1980) L'utilisation des cultures *in vitro* pour la multiplication de quelques espèces legumières et fruitières. CR Acad Agric France 66:681–691.
43. Navatel JC, Nio M, Vaysse P, and Edin M (1988). Pommier. Comportement en marcottières et aux vergers des porte-greffes micropropagés de type *Malus*. Infos-Ctifl No 44, pp. 25–28.
44. Niizeki M, Hidano Y, and Saito K (1983) Callus formation from isolated protoplasts of apple *Malus pumila* Mill. Japan J Breed 33:369–374.
45. Patat-Ochatt EM, Ochatt SJ, and Power JB (1989) Plant regeneration from protoplasts of apple rootstocks and scion varieties (*Malus* × *domestica* Borkh). J Plant Physiol 133:460–465.
46. Phan CT and Hegedus P (1986) Possible metabolic basis for the developmental anomaly observed in *in vitro* culture, called 'vitreous plants'. Plant Cell Tiss and Org Cult 6:83–94.
47. Pua A-C and Chong C (1984) Requirement for sorbitol (D-glucitol) as carbon source for *in vitro* propagation of *Malus robusta* No. 5. Can J Bot 62:1545–1549.
48. Read PE and Fellman CD (1985) Accelerating acclimation of *in vitro* propagated woody ornamentals. Acta Hort 166:15–20.

49. Redenbaugh K, Paasch BD, Nichol JW, Kossler ME, Viss PR, and Walker K (1986) Somatic seeds: Encapsulation of asexual plant embryos. Biotechnology, 4:797–801.
50. Revilla MA, Ochatt SJ, Doughty S, and Power JB (1987) A general strategy for the isolation of mesophyll protoplasts from deciduous fruit and nut tree species. Plant Sci 50:133–137.
51. Rugini E, Tarini P, and Rossodivita ME (1987) Control of shoot 'vitrification' in olive grown *in vitro*. Acta Hort 212(1):177–183.
52. Scowcroft WR and Larkin PJ (1982) Somaclonal variation: a new option for plant improvement. In: Vasil IK, Scowcroft WR and Frey KJ (Eds.) Plant improvement and somatic cell genetics. Academic Press, New York, London. pp. 159–178.
53. Simmonds J (1983) Direct rooting of micropropagated M.26 apple rootstocks. Scientia Hort 21:233–241.
54. Skirvin RM (1981) Fruit crops. In: Conger BV (Ed.) Cloning agricultural plants via *in vitro* techniques. CRC Press, Boca Raton, FL. pp. 51–139.
55. Sriskandarajah S, Mullins MG, and Nair Y (1982) Induction of adventitious rooting *in vitro* in difficult-to-propagate cultivars of apple. Plant Sci Lett 24:1–9.
56. Villeneuve F, Ferré G, and Blanc R (1986) Les broussins, development et moyens de lutte. Arboriculture Fruitière 33:44–48.
57. Wallin A and Welander M (1985) Improved yield of apple leaf protoplasts from *in vitro* culture shoots using very young leaves and adding L-methionine to the shoot medium. Plant Cell Tiss and Org Cult 5:69–72.
58. Webster AD, Oehl VH, Jackson JE, and Jones OP (1985) The orchard establishment, growth and precocity of four micropropagated apple scion cultivars. J Hort Sci 60:169–180.
59. Webster AD, Sparks TR, and Blecher A (1986) The influence of micropropagation and chemical mutagens on the growth and precocity of Cox's Orange Pippin and Bramley's Seedling apple. Acta Hort 180:25–34.
60. Webster CA and Jones OP (1989). Micropropagation of the apple rootstock M.9: effect of sustained subculture on apparent rejuvenation *in vitro*. J Hort Sci 64:421–428.
61. Webster CA and Jones OP (1989) Effect of sustained subculture on apparent rejuvenation of the apple rootstock M.9 *in vitro* and *in vivo*. Ann. Sci. For., 46 suppl., 187–189. Forest Tree Physiology, Dreyer E *et al.* (Eds.) Elsevier/INRA.
62. Welander M (1985) *In vitro* root and shoot formation in the apple cultivar Åkerö. Ann Bot 55:249–261.
63. Welander M (1988) Plant regeneration from leaf and stem segments of shoots raised *in vitro* from mature apple trees. J Pl Physiol 132:738–744.
64. Welander M, Welander NT, and Brackman A (1989) Regulation of *in vitro* shoot multiplication in *Syringia*, *Alnus* and *Malus* by different carbon sources. J Hort Sci 64:361–366.
65. Wilkins CP, Cabrera P, and Dodd JH (1985) Tissue culture propagation of trees. Outlook on Agriculture, 14:2–13.
66. Wu J (1981) Obtaining haploid plantlets of crab apple from anther culture *in vitro*. Acta Hort Sinica 8:36 (abstr).
67. Zimmerman RH (1984) Rooting apple cultivars *in vitro*: interactions amongst light, temperature, phloroglucinol and auxin. Plant Cell Tiss Org Cult 3:301–311.
68. Zimmerman RH (1986) Propagation of fruit, nut and vegetable crops-overview. In: Zimmerman RH *et al.* (Eds.) Tissue culture as a plant production system for horticultural crops. Martinus Nijhoff, Dordrecht. pp. 183–200.
69. Zimmerman RH and Fordham I (1985) Simplified method for rooting apple cultivars *in vitro*. J Amer Soc Hort Sci 110:34–38.
70. Zuccherelli G (1979) Moltiplicazione *in vitro* die portainnesti clonali del pesco. Frutticoltura 41(2):15–20.
71. Zuccherelli G, Venturi V, and Damiano C (1978) Rapid propagation on a vast scale of Damasco 1869 rootstock by *in vitro* culture. Ann Inst Sper Frutt, Roma 9. pp. 21–23.

11. Regeneration and germplasm preservation in aspen-*Populus*

M. R. AHUJA
Institute of Forest Genetics, Federal Research Centre for Forestry and Forest Products, Sieker Landstrasse 2, 2070 Grosshansdorf, Germany

1. Introduction

The genus *Populus* (poplars and aspens) belong to the family Salicaceae, which is characterized by the dioecious nature of its trees, that is, male and female flowers are borne on separate trees. The somatic cells of poplars and aspens carry $2n$=38 chromosomes, although triploids ($2n$=57) and tetraploids ($2n$=76) have been reported in certain species of the genus *Populus*[35]. In a recent study, the chromosomes of *Populus tremula* have been characterized by employing C-banding technology [33]. The karyotype consists of rather small 36 chromosomes, and a relatively large and distinct chromosome pair, which is almost completely heterochromatic and may possibly be involved in sex determination [33]. However, detailed cytogenetic studies would be necessary in the genus *Populus* for a better understanding of the karyotype and sex determination.

Poplars and aspens occur interspersed in the forests of temperate and cold regions of northern hemisphere. They have a wide range of adaptability and can grow on marginal soils. There are five sections in the genus *Populus* [26,28], of which Section Leuce (consisting of aspens and white poplar) and Aigeiros (represented by 'true poplars') contribute major commercially important tree species. Both poplars and aspens possess suckering habit (sprouting from stump or roots), which makes them valuable in soil conservation plantations. However, aspens have not been as widely used in forest stands as compared to poplars. This may be partly due to the fact that clonal propagation of aspen is very difficult by woody cuttings.

Poplar and aspen wood is used in timber, pulp for paper, light packing material, and the matchstick and plywood industries. Because of their initial rapid growth in the first few years, poplars and aspens offer opportunities for total utilization of biomass for conversion into energy [20,32]. Because of their suckering habit, a new natural plantation can develop following removal of all wood that is above ground; a trait that would make poplars and aspens a source of so-called 'energy plantations' [40]. However, it would be important to know whether the second and third cycle root-suckers are as vigorous as the first cycle crop.

M. R. Ahuja (Ed.), Micropropagation of Woody Plants, 187–194.

Although hybridization is frequent between trees of different types and complementary sexes, cultivated poplars are not generally propagated by seed. They are conventionally propagated by vegetative means. This is in contrast to other tree species commonly used in forestry, which have been traditionally reproduced from seed. Poplars can be vegetatively propagated by woody cuttings. Aspens, on the other hand, are very difficult to propagate from woody cuttings, because the root primordia are absent in them. However, aspens can be propagated by root-suckers and green-cuttings.

Traditional tree breeding and improvement practices are long-term commitments, and their implementations require large areas of land for testing selected genotypes. Once improved genotypes have been identified, the limiting factors for mass scale utility of the selected material are: (1) the restricted availability of the improved genotypes as planting material; (2) the available nursery/field space; and (3) the number of plants that can be propagated in a tree species in a growing season by employing traditional methods of vegetative propagation. To overcome these problems, tissue culture technology offers opportunities not only for the mass propagation of selected genotypes throughout the year, but also for the genetic improvement of tree species. In this regard tissue culture can serve as a valuable adjunct to the time-honored practices of tree genetics, breeding and improvement.

In this article I shall review the scope of *in vitro* technology in regeneration and clonal propagation from juvenile and mature tissues, clonal fidelity of the regenerants, and germplasm preservation strategies in aspen. Micropropagation in aspen has been relatively recently reviewed by Ahuja [6,8,18] and therefore will only be briefly discussed in this paper. The non-aspen poplars and their hybrids are covered in two separate chapters by Ernst and Chun, respectively, in this volume.

2. Clonal propagation

Clonal propagation may be accomplished by macro- or micropropagation. Macropropagation involves traditional methods of vegetative propagation, while micropropagation is carried out with the help of tissue culture technology. Since it is very difficult to root woody cuttings in aspen, and propagation by grafting and root-suckers is rather labor intensive, macropropagation has not played a significant role in clonal forestry of aspen.

2.1. In vitro *plant regeneration*

Plant regeneration has been achieved in juvenile and mature aspen trees by tissue culture. In fact, the first plantlets in the tree species were regenerated in 1968 from the callus cultures of triploid aspen, *Populus tremuloides* [36,39]. Subsequently, however, plantlet regeneration has been achieved in a number of *Populus* species (see Ernst, this volume) and their hybrids (see Chun, this volume), as well as in a large number of forest tree species of angiosperms and gymnosperms [15,19,27]. Even though *in vitro* requirements for regeneration have been investigated in many tree species, the genus *Populus* has been perhaps the most extensively cultured *in vitro*. It has been employed in several studies as a model system for understanding the problems of regeneration of plantlets from organs, tissues, cells and protoplasts (see Ernst, Chun, Russel, this volume), germplasm preservation (see Chun, this volume, and refs [10,14,16], and genetic transformation (see Ostry and Michler, this volume). Because of problems in the macropropagation of aspen, *in vitro* technology would be particularly valuable for micropropagation and clonal forestry from selected mature trees.

Callus cultures used in earlier studies [22,37,38] have been largely replaced by organ cultures [2,4,6,8,24] for micropropagation. This may primarily be due to the fact that long-term callus cultures are generally more prone to genetic variation [31] than are organ cultures [6,8]. In particular, bud meristem cultures are relatively more stable and involve minimal genetic risk for micropropagation. By employing Aspen Culture Medium (ACM) [2], bud explants from a large number of European aspen (*Populus tremula*) and hybrid aspen (*P. tremula* × *P. tremuloides*) clones, plantlet regeneration was accomplished [2,4,6,8,18]. In addition to micropropagation from juvenile plants [21,23], mature selected genotypes were also successfully propagated by tissue culture technology [2,6,8]. A relatively simple two-step micropropagation method [4] involving induction and proliferation of microshoots on one medium and rooting microshoots in a potting mixture, was employed for regeneration of a large number of plants (more than 10 000) in selected genotypes of aspen and hybrid aspen. By employing a two-step micropropagation method, it is possible to reduce costs and minimize exposure of tissues to artificial *in vitro* conditions. There was almost direct differentiation of microshoots on the bud explants, with minimal callus formation. Chromosome analysis of randomly selected tissue-culture derived aspen and hybrid aspen plants revealed a normal diploid genetic complement [4].

Protoplasts have been isolated and cultured in several *Populus* species [1,3,5]. Regeneration of plants from protoplast cultures of forest tree species is somewhat difficult. Nevertheless, progress has been made in recent years towards the regeneration of plants from protoplast cultures of *Populus* and some other tree species (see Russell, this volume). Protoplasts derived from leaves/needles of mature trees are as recalcitrant as bud explants from old trees. Plant regeneration has been achieved from protoplasts derived from *in vitro* shoot cultures of *Populus* [34] or other juvenile tissues. Protoplasts from mature trees do not readily divide, and may usually undergo budding [3]. Protoplast systems, when functional, offer prospects for recovery of somaclonal variation, somatic hybridization and gene transfer studies (see Ostry and Michler, this volume). However, there are a number of problems and questions that need to be resolved regarding the growth and differentiation of forest tree species [17].

The regenerative potential of tissues depends, among other things, on the origin of the tissue, the physiological state of the tissue, the age of the donor tree, and the genotype. In addition, the sex of a tree, if male or female flowers are borne on different trees, may also play a role in the regenerative potential. We have examined some of these aspects using aspen as a model system. It seems that tissue differentiation and regeneration are rather complex processes, and probably depend on an interaction between several different factors. Nevertheless, some tentative conclusions can be drawn from the *in vitro* experimental data. As compared to other tissues, bud explants are generally more responsive in terms of induction and differentiation of microshoots. However, bud explants from juvenile plants are more responsive than are bud explants from mature trees. Apart from juvenility and maturation, the genotype of a donor plant also determines the regeneration response. In the aspen system, the genotype of the explant seems to play a greater role in the *in vitro* regeneration response, as compared to the maturation state ([2] and unpublished data). In other words, it is the genotype of an aspen clone that predominantly determines the regenerative potential under a given set of *in vitro* conditions.

In another study involving male and female aspen and hybrid aspen clones (16 – 30 years of age), there was no significant difference in the regenerative potential of tissues, whether the bud explants were taken from male or female trees (Ahuja, unpublished). Interestingly, there was a 1 : 1 segregation for shoot regeneration vs. no differentiation in the female as well as male aspen and hybrid aspen clones (Table 1). Taken together, these observations would suggest that the ability to regenerate microshoots *in vitro* on bud explants is predominantly determined by the genotype, and to a lesser extent by the sex and/or age of an aspen clone.

Another study explored the regenerative potential of hypocotyl, cotyledon, leaf segment and shoot tip from seedlings from different aspen and hybrid aspen clones. When cultured on the same medium with low concentrations of an auxin (NAA) and a cytokinin (BAP), shoot tips and hypocotyls showed a relatively greater ability to differentiate microshoots than cotyledons and leaf segments [30]. The genotype of a clone also played an important role in regeneration from juvenile explants.

Table 1. Segregation of shoot regeneration (+) vs. no shoot regeneration (–) on bud explants from female and male aspen and hybrid aspen clones.

Sex	Number of (+) clones	Number of (–) clones	χ^2	*P*
Female	8	9	0.06	0.8
Male	9	13	0.72	0.3–0.5
Total (Female + male)	17	22	0.64	0.3–0.5

3. Germplasm preservation

Aspen has been empoloyed as a model for investigating the preservation potential of pollen, seed, microshoots and dormant buds under controlled conditions. Pollen from aspen has been stored for a number of years at low temperature (–18 °C) under vacuum [29] without losing its viability. Seeds of forest tree species, including aspen, have been stored at low temperatures [0 ° to –20 °C]. At these temperatures, the seed can be stored for a limited period of time. In most cases, the seed begins to lose viability after certain periods, depending upon the storage temperature, forest tree species, genotype and the size of the seed. Cryopreservation (storage in liquid nitrogen at –196 °C), on the other hand, offers prospects for long-term storage of germplasm. Seeds of hybrid aspen (*Populus tremula* × *P. tremuloides*) could be stored in liquid nitrogen for one to six days without loss of viability [7,14].

3.1. *Dormant buds*

Dormant buds offer prospects for a long-term storage of germplasm if the dormant state can be prolonged without loss of viability and differentiation potential. We explored the storage potential of dormant buds at subzero temperatures that included –0 °C, –5 °C, –8 °C, –18 °C, –80 °C, and –196 °C (liquid nitrogen). Dormant buds of several aspen and hybrid aspen clones could be stored at subzero temperatures for up to two years. However, certain clones exhibited a better storage capacity, as monitored by microshoot differentiation on bud explants, than others at low temperatures [10,14]. After two years, dormant buds from all 14 aspen and hybrid aspen clones stored at subzero temperatures (up to –80 °C) lost their viability. Of the five temperatures tested, –80 °C seemed to be optimum for the short-term storage of dormant buds of certain aspen clones.

In another study dormant buds from 13 aspen and hybrid aspen clones were stored in liquid nitrogen (–196 °C) for 24 hours, by quick-freezing or by slow-freezing methods [16]. Dormant buds from all 13 clones were possibly killed by the quick-freezing and thawing procedures, as tested by shoot induction on the bud explants. On the other hand, on slow-freezing to –40 °C followed by storage in liquid nitrogen, the dormant buds remained viable. Shoots differentiated on bud explants of all clones retrieved from liquid nitrogen, although the degree of microshoot differentiation depended on the genotype [16]. The microshoots were rooted and plantlets were regenerated in each clone. Although dormant buds were cryopreserved for one day, theoretically it should be possible to preserve them at –196 °C (liquid nitrogen) on a long-term basis.

For preservation of forest tree germplasm, both *in situ* (preservation at site) and *ex situ* (preservation away from the original site) strategies are important. Here I have emphasized only the storage of seed and dormant buds at low temperatures under controlled conditions. Storage of seed is important for the preservation of genetic variability. Conservation of dormant buds or vegetative parts, on the other hand, would be valuable for the preservation of selected genotypes in forest tree species. Both approaches should go hand-in-hand for broad-spectrum conservation of forest tree germplasm under controlled conditions.

4. Outlook

Aspens are short-rotation, fast-growing forest trees, that are important for biomass production. Aspens can grow on marginal sites, and probably on wastelands. Since woody cuttings from aspen are very difficult to root, large-scale clonal propagation of selected trees could not be accomplished by traditional methods of vegetative propagation, either by root-suckers or green-cuttings. It has been demonstrated that clonal propagation of both juvenile and mature aspen trees can be achieved by employing tissue culture technology. This offers prospects for large-scale micropropagation of selected mature aspen clones for their utilization in clonal forestry programs. Since aspen and poplar tissues are morphogenetically labile, that is, can be induced to grow and differentiate *in vitro*, they have been employed as a model system for understanding problems in protoplast research [34], gene transfer [11,12,13] (also see Ostry and Michler, this volume), somaclonal variation [9]; and Michler (this volume), and germplasm preservation [7,10,14,16,25]. In addition, it can also serve as a model system for analysing molecular mechanisms in growth and differentiation, and juvenility, maturation and rejuvenation in woody plants.

5. Summary

The current status of *in vitro* regeneration in aspens has been reviewed. Aspens are generally difficult to propagate from woody cuttings. Vegetative propagation by root-suckers, green-currings, and grafting is possible but not practical for large-scale clonal propagation of mature selected genotypes. Tissue culture technology has been employed for micropropagation of juvenile as well as mature aspen clones. In recent years bud explant have been routinely employed for clonal propagation of aspen. Other tissues have also been investigated for their morphogenetic potential. Aspen was employed as a model system for examining regeneration potential of juvenile and mature trees, female and male trees, as well as for preservation of seed and dormant buds at low temperatures, including liquid nitrogen (–196 °C). Although regeneration problems of recalcitrant genotypes have not been resolved, advances in tissue culture technology will hopefully enable us to not only to overcome barriers to 'genetic-recalcitrance' and 'maturation-recalcitrance', but also provide approaches to rapid and economically feasible mass cloning of woody plants.

References

1. Ahuja MR (1982) Isolation, culture and fusion of protoplasts: problems and prospects. Silvae Genet 31:66–77.
2. Ahuja MR (1983) Somatic cell genetics and rapid clonal propagation of aspen. Silvae Genet 32:131–135.
3. Ahuja MR (1983) Isolation and culture of mega and normal protoplasts in aspen. Silvae Genet 32:225–227.
4. Ahuja MR (1984) A commercially feasible micropropagation method for aspen. Silvae Genet 33:174–176.
5. Ahuja MR (1984) Protoplast research in woody plants. Silvae Genet 33:32–37.
6. Ahuja MR (1986) Aspen. In: Evans DA, Sharp WR and Ammirato PJ (Eds.) Handbook of Plant Cell Culture. Macmillan Publishing Company, New York, pp. 626–651.
7. Ahuja MR (1986) Storage of forest tree germplasm in liquid nitrogen (−196° C). Silvae Genet 35:249–251.
8. Ahuja MR (1987) *In vitro* propagation of poplar and aspen. In: Bonga JM and Durzan DJ (Eds.) Cell and Tissue Culture in Forestry. Vol. 3. Martinus Nijhoff Publishers, Dordrecht, pp. 207–223.
9. Ahuja MR (1987) Somaclonal variation. In Bonga JM and Durzan DJ (Eds.) Cell and Tissue Culture in Forestry. Vol. 1 Martinus Nijhoff Publishers, Dordrecht. pp. 272–285.
10. Ahuja MR (1988) Differential growth response of aspen clones stored at sub-zero temperatures. In: Ahuja MR (Ed.) Somatic Cell Genetics of Woody Plants. Kluwer Academic Publishers, Dordrecht, pp. 173–180.
11. Ahuja MR (1988) Gene transfer in forest trees. In: Hanover JW and Keathley DE (Eds). Genetic Manipulation of Woody Plants. Plenum Press, New York, pp. 25–41.
12. Ahuja MR (1988) Gene transfer in woody plants: perspectives and limitations. In: Ahuja MR (Ed.) Somatic Cell Genetics of Woody Plants. Kluwer Academic Publishers, Dordrecht, pp. 83–101.
13. Ahuja MR (1988) Molecular genetics of transgenic plants. In: Hällgren JE (Ed.) Frans Kempe Symposium on Molecular Genetics of Forest Trees. Swedish Uni. Agric. Sci. Umeå, pp. 127–145.
14. Ahuja MR (1989) Storage of forest tree germplasm at sub-zero temperatures. In: Dhawan V (Ed.) Application of Biotechnology in Forestry and Horticulture. Plenum Press, New York, pp. 215–228.
15. Ahuja MR (1991) Biotechnology in forest trees. Plant Res Dev 23:106–120.
16. Ahuja MR (1991) Woody plant biotechnology: perspectives and limitations. In: Ahuja MR (Ed.) Woody Plant Biotechnology. Plenum Press, New York, pp. 1–10.
17. Ahuja MR (1991) Application of biotechnology to preservation of forest tree germplasm. In: Ahuja MR (Ed.) Woody Plant Biotechnology, Plenum Press, New York, pp. 307–313.
18. Ahuja MR, Krusche D, and Melchior GH (1988) Determination of plantlet regeneration capacity of selected aspen clones *in vitro*. In: Ahuja MR (Ed.) Somatic Cell Genetics of Woody Plants. Kluwer Academic Publishers, Dordrecht, pp. 127–135.
19. Ahuja MR and Muhs HJ (1985) *In vitro* techniques in clonal propagation of forest trees. In: Schäfer-Menuhr A (Ed.) *In Vitro* Techniques-Propagation and Long-term Storage. Martinus Nijhoff Publishers, Dordrecht, pp. 41–49.
20. Ahuja MR and Muhs HJ (1986) Biomassproduktion in forstlichen Plantagen für die Rohstoff- und Energiegewinnung. BMFT Forschungsbericht T 86-073, Bonn, pp. 1–38.

21. Barocka KH, Buas M, Lontke E, and Sievert F (1985). Tissue culture as a tool for *in vitro* mass propagation of aspen. Z Pflanzenzücht 94:340–343.
22. Chalupa V (1977) The use of regenerants from tissue culture of forest trees in tree breeding. In: Novák FJ (Ed.) Use of Tissue Culture in Plant Breeding. Czech Acad Sci Prague, pp. 183–193.
23. Chalupa V (1979) *In vitro* propagation of some broad-leaved forest trees. Commun. Inst. For. Czech. 11:159–170.
24. Christie CB (1978) Rapid propagatioon of aspen and silver poplar using tissue culture techniques. Int Plant Prop Soc 28:255–260.
25. Chun YW and Hall RB (1991) Influence of subculturing period and different culture media on cold storage maintenance of *Populus alba* × *P. grandidentata* plantlets. In: Ahuja MR (Ed.) Woody Plant Biotechnology. Plenum Press, New York, pp. 315–321.
26. Dickmann DI and Stuart KW (1983) The Culture of Poplars in Eastern North America. Michigan State University, East Lansing, pp. 168.
27. Dunstan DI and Thorpe TA (1986) Regeneration in forest trees. In: Vasil IK (Ed.) Cell Culture and Somatic Cell Genetics of Plants. Vol. 3. Academic Press, New York. pp. 223–241.
28. FAO (1979) Poplars and Willows in Wood Production and Land Use. Food and Agriculture Organization of the United Nations, Rome. pp. 328.
29. Hermann S (1976) Verfahren zur Konservierung und Erhaltung der Befruchtungsfähigkeit von Waldbaumpollen über mehrer Jahre. Silvae Genet 25:223–229.
30. Krein B und Ahuja MR (1991) Hybridaspen für den Anbau in Schnellwuchsplantagen als alternative Nutzung für Landwirtschaftliche Flächen: Entwicklung einer Massenvermerhrungsmethode mit Zellkulturtechniken für Hybridaspenklone. In: produktion- und Verwendungsalternativen für die Land- und Forstwirtschaft Nachwachsende Rohstoffe. Landwirtschaftsverlag, Münster-Hiltrup. pp. 130–138.
31. Lester DT and Berbee JG (1977) Within-clone variation among black poplar trees derivbed from callus cultures. For Sci 23:122–131.
32. Muhs HJ und von Wühlisch G (1991) Züchtung schnellwachsender Baumarten und deren Produktionspotential. II. Expertengespräch 'Lignocellulosen'. Mitt der BFH, Hamburg, N 166, pp. 3–12.
33. Ohri D and Ahuja MR (1991) Giemsa C-banding in *Fagus sylvatica* L., *Betula pendula* Roth. and *Populus tremula* L. Silvae Genet 40:72–75.
34. Russell JA and McCown BH (1988) Recovery of plants from leaf protoplasts of hybrid-poplar and aspen clones. Plant Cell Rep 7:59–62.
35. Seitz FW (1951) Chromosomenzahlenverhältnisse bei Holzplfanzen. Z Forstgenetik und Forstpflanzenzüchtung 1:22–32.
36. Winton LL (1968) Plantlet formation in aspen tissue culture. Science 160:1234–1235.
37. Winton LL (1970) Shoot and tree production from aspen tissue cultures. Amer J Bot 57:904–909.
38. Winton LL (1971) Tissue culture propagation of European aspen. For Sci 17:348–350.
39. Wolter KE (1968) Root and shoot initiation in aspen callus cultures. Science 219:509–510.
40. Zsuffa L, Sennerby-Forsse L, Weisgerber H and Hall RB (1993) Strategies for clonal forestry with poplars, aspens and willows. In: Ahuja MR and Libby WJ (Eds.) Clonal Forestry, vol 2, Conservation and Application. Springer Verlag, Heidelberg, New York (in press.)

12. *In vitro* culture of pure species non-aspen poplars[1]

STEPHEN G. ERNST

Department of Forestry, Fisheries and Wildlife, University of Nebraska, Lincoln, NE 68583-0814, U.S.A.

1. Introduction

Most species in this group, the pure species non-aspen (i.e., non-Leuce Section) poplars of the genus *Populus*, root very readily as stem cuttings [13]. Therefore, the development of *in vitro* micropropagation protocols will not be economically competitive for the commercial increase of desirable genotypes, except for the production of virus-free ramets, those few difficult to root species, or for recalcitrant genotypes within easily rooting species. However, *in vitro* manipulation can be an effective tool for better understanding the differentitation of adventitiously derived organs, the production of haploid plants, protoplasts, and somatic embryos for allied research, and for gene transfer systems. Therefore, the poriton of this chapter dealing with pure species non-aspen poplars of the genus *Populus* will focus on the development of these research tools and associated basic knowledge.

2. *In vitro* organogenesis

In this section, a chronological review of investigations of *in vitro* callus production and shoot and root organogenesis will be presented. Callus production is included here because the dedifferentiation of explant tissue is often a prerequisite for organogenesis. This coverage does not include studies of androgenesis and somatic embryogenesis, which will be discussed in later sections. Unless otherwise noted, mature (versus embryo or seedling) explant material and solid media (versus liquid) was used in the experiments surveyed for this review, and only one genotype per species was tested.

2.1. *Callus production response*

Gautheret [15] produced the first *in vitro* callus cultures of *Populus*, including *P. nigra*, using cambial tissue as explant material (as reported in Ref. 1). Pelet

[1] Journal Series No. 8974, Agricultural Research Division, University of Nebraska.

M. R. Ahuja (Ed.), Micropropagation of Woody Plants, 195–207.

et al. [28] compared *in vitro* callus growth from petiolar insect galls and normal stem tissues of *P. deltoides* (and woody species in three other genera) to determine the *in vitro* nutritional, pH and temperature requirements of both explants. Each type of tissue was grown on two different media: Hildebrandt's [16] C-medium, which includes Hildebrandt's Tobacco Medium plus 150 ml/L coconut milk supplemented with 0.1 mg/L napthaleneacetic acid (NAA), and Hildebrandt's [16] D-medium, which includes Hildebrandt's Tobacco Medium supplemented with 6 mg/L 2,4-dichlorophenoxyacetic acid (2,4-D). The *P. deltoides* normal stem and petiolar gall callus cultures grew best (eas measured by fresh weight gain over time) at a temperature of 28–32° C and an initial pH (before autoclaving) that was weakly acidic (i.e., just below pH 7.0).

Diaz-Colon *et al.* [12] utilized 5 mm thick transverse stem section explants from young shoots of *P. deltoides* to test the phytotoxicity of picloram, 2,4,5-trichlorophenoxyacetic acid (2,4,5-T), and dicamba in *in vitro* culture. The growth medium was a modified Murashige and Skoog [27] (MS) medium supplemented with 150 ml/L coconut milk, 0.1 mg/L 2,4-D, and 1.0 mg/L indole-3-acetic acid (IAA). Based on callus growth, the optimal initial pH for *P. deltoides* stem sections was between pH 5 and 6. Callus growth was reduced 80 to 95% by the addition of the herbicides at concentrations of 10^{-3} to 10^{-6}M to the culture medium, and the greatest uptake of herbicides occurred during the first 24 hours.

2.2. *Shoot and root organogenesis*

Bawa and Stettler [5] were the first to report successful organogenesis for a species in the pure species non-aspen poplars. The objective of their study was to induce newly formed female catkin primordia to grow and differentiate *in vitro*, but they also observed adventitious shoot production. Flower buds were removed from two mature female *P. trichocarpa* trees at about the same time the current year's catkins were ready to dehisce and shed seed. The bud scales were removed (essential for growth *in vitro*) and the female catkin primordia explants were grown on a slightly modified MS medium supplmented with 1.0 mg/L NAA, 0.2 mg/L kinetin, and either 0.5 or 1.0 mg/L IAA. Flower primordia in early stages of development (pistil primordia present only as adaxial mounds) had a 63 to 80% survival rate, while none of the primordia at later stages of development (pistil primordia mounds very evident) survived after 20 days in culture. The *in vitro* grown flower primordia increased in size and some produced additional bracts and pistils; a number of stamen primordia also differentiated. While both media permitted equal growth and survival of the primordia, none of the primordia attained full maturity *in vitro*. Four of the original 60 explants also produced leaf primordia, but after 7 to 14 days they dedifferentiated into callus. When primordia were grown on Wolter's medium [43] supplemented with 0.15 mg/L 6-benzyladenine (BA), most produced shoots in the axils of bracts rather than pistils, and the shoots grew and were easily rooted. Callus grown on Wolter's medium supllemented with BA also produced shoots.

Venverloo [40] produced adventitious shoots from internodal stem explants of *P. nigra* 'Italica' by first inducing callus growth. The callus induction medium consisted of a modification of MS medium supplemented with various concentrations of 2,4-D or NAA, and the shoot induction medium utilized the same modified MS medium supplemented with various concentrations of the cytokinins BA, 2-isopentyladenine (2iP), kinetin and zeatin. Callus production was optimal on 0.5 mg/L 2,4-D; root production optimal at low NAA or 2,4-D concentrations (0.01 to 0.05 mg/L); and shoot production best when explants were transferred from medium containing 0.1 mg/L 2,4-D to medium containing a cytokinin. Zeatin was the most active of the cytokinins tested (i.e., required lowest concentration), and kinetin the least active.

Chalupa [6] reported adventitious shoot regeneration from callus of *P. nigra* var. *typica*, in addition to similar successes with *P × euramericana*, *P. tremula*, and *P. × canescens*. However, details regarding explant source, optimal media, etc., were given only for *P. × euramericana*. Lester and Berbee [22] also successfully regenerated adventitious shoots from callus. Shoot tips of *P. nigra* 'Italica' and *P. × euramericana* were induced to produce callus on media (undefined) containing 0.5 to 1.0 mg/L 2,4-D. The callus was then subcultured on media containing various combinations of BA and either NAA or IAA for the production of adventitious shoots. The shoots were rooted and eventually established in the field.

Whitehead and Giles [42] utilized preformed axillary buds of *P. nigra* 'Italica' and *P. yunnanensis* (and *P × euramericana* 'Flevo'), which were elongated, sectioned and subcultured *in vitro*. Preformed shoots elongated from the proliferating axillary buds, and adventitious shoots from the cut and uncut explant surfaces. The growth medium was a slightly modified version of MS, supplemented with 0.2 mg/L BA for bud and shoot proliferation, 0.1 mg/L BA and 0.02 mg/L NAA for proliferation and elongation, and 0.01 mg/L BA plus 0.01 mg/L NAA for rooting. Greater numbers of adventitious buds and preformed and adventitious shoots were produced from *P. yunnanensis* relative to *P. nigra* 'Italica'. The leaves of the shoots were juvenile in shape, but changed to a mature shape when the shoots were rooted and transplanted to a pumice:peat soil mix. Using this proliferation system, and starting with a single axillary bud, it was estimated that over one million plantlets per year could be generated from any of the clones tested.

Saito [34] found that cambial tissue explants of *P. deltoides* (and *P. × euramericana*) could be used for callus formation and adventitious shoot regeneration. The callus and shoot induction media were relatively complex. The cambial tissue was grown on callus induction medium for 10–12 weeks, and then the callus was subdivided and grown on callus induction medium for an additional 10–12 weeks before being transferred to shoot induction medium. Approximately 100% of the subcultured explants produced shoots, with 1–20 shoots per explant.

Raj *et al.* [31] experimented with several explant types in developing a regeneration system for *P. nigra* (and also for *P. alba*, *P. grandidentata*, and *P. tremula*). Preformed axillary buds of *P. nigra* were elongated *in vitro*, and stem, leaf, and petiole explants were used for callus initiation and aventitious shoot regeneration. Bud break and shoot elongation was best on MS medium supplemented with 0.4 mg/L BA, with 70% of the buds elongating. The shoots derived from the preformed buds were then subdivided to proliferate axillary shoots and to initiate adventitious shoots on the same BA-supplemented medium. Approximately 60% of the stem and leaf explants and 8% of the petiole explants formed callus on MS medium supplemented with 0.1 mg/L BA and 0.05 mg/L NAA. Adventitious shoots were initiated from this callus (culture medium not given) and the shoots rooted on MS medium with one-half macronutrients, 0.002 mg/L NAA, and 2% sucrose before being transplanted to soil medium.

Douglas [14] utilized stem internodes from the non-aspen poplars *P. deltoides, P. nigra* var. *betulifolia, P. balsamifera, P. wilsonii,* and *P. trichocarpa*, as well as the Leuce species *P. alba* and the *P. trichocarpa* × *P. balsamifera* hybrid TT32, to determine how endogenous growth regulators and physiological conditions affect organogenesis *in vitro.* Internodal stem sections were excised from the fourth through seventh internodes of rapidly growing shoots from rooted cuttings of each species of hybrid and placed on a modified MS medium without added growth regulators. After ten days, adventitious buds were observed on the explants. An anatomical study showed that the buds arose from a wedge of parenchymatous tissue that formed after five days in culture, and that this organogenic tissue was derived from phloem and cambial cells in the cultured explant. There were marked species differences in the formation of adventitious buds and shoots; the percent of explants producing adventitious buds ranged from 100% for *P. deltoides* and *Populus* hybrid TT32 to 5% for *P. trichocarpa*, and the number of buds plus shoots per explant ranged from 8.4 for *P. deltoides* to 0.1 for *P. trichocarpa.* These results suggest that the stem internodes contain sufficient endogenous cytokinins at the time of excision for adventitious shoot production on a hormone-free medium, and that the response is species and genotype dependent. In addition, a concurrent experiment determined that the fourth through seventh internodes produced the most buds and shoots *in vitro.* This suggests that endogenous auxin, abscisic acid, and cytokinin may interact in the explant tissue, resulting in a gradient of potential organogenic response from the shoot tip downward.

Rutledge and Douglas [33] utilized preformed meristem explants (meristem tips, shoot tips, and nodal segments) to establish micropropagation protocols for four non-aspen species of *Populus* (*P. wilsonii, P. balsamifera, P. trichocarpa*, and *P. nigra* var. *butulifolia*) and eight *Populus* hybrid clones. For the meristem tip cultures, three different media and several different growth regulator combinations and concentrations were tested. However, the optimal medium was very species specific, and only 4% of all explants cultured (pure species and hybrids) resulted in viable shoot-producing cultures after 12 weeks. Callus was induced from the *P. nigra* meristem tip cultures, and adventitious shoots were regenerated from the callus. Shoot tip explants produced elongated shoots in 4–6 weeks, compared to the 10–12 weeks required for the meristem tip cultures. The optimal medium and growth regulator content was also very species specific, but 32% of all explants cultured resulted in shoot-producing cultures. For the nodal segment explants *P. nigra* and *P. wilsonii* were superior to *P. trichocarpa* and *P. balsamifera* in the production of viable shoot cultures. Overall, the shoot tip explants resulted in the highest frequency of shoot-producing cultures, in part due to difficulties in sterilizing the nodal segments and the lower viability of the smaller meristem tips.

Prakash and Thielges [29] described an adventitious shoot regeneration system which used leaf disks of *P. deltoides*. The leaf disks were initiated on a callus induction medium (MS medium supplemented with 1μM 2,4-D and 0.1 μM BA and grown in the dark), transferred to a second callus induction medium (MS medium supplemented with 1 μM NAA and 0.1 μM kinetin and grown in a 16-hour photoperiod), transferred to a shoot induction medium (WPM or MS supplemented with 1μM BA and grown in a 12-hour photoperiod), and finally transferred to a rooting medium. No regeneration efficiencies were reported in the abstract.

Riemenschneider *et al.* [32] utilized 3 cm long shoot tips and nodal sections of six clones of *P. deltoides* to determine how adventitious *in vitro* rooting is affected by genotype. The root induction medium was one-quarter strength MS supplemented with 1.0 mg/L NAA, and the explants were cultured in the dark for seven days followed by transfer to light. Root induction for each clone ranged from 8 to 90%, indicating that this response should be considered when selecting suitable genotypes for large-scale micropropagation *in vitro*.

Coleman and Ernst [11] developed a shoot regeneration system using internodal stem sections from rapidly growing shoots of greenhouse grown rooted cuttings of *P. deltoides*. WNA medium, a modification of Woody Plant Medium (WPM) [23] was supplemented with one of three cytokinins (BA, 2iP, or zeatin), each at five different concentrations, plus a phytohormone-free control. Among the three genotypes tested, the zeatin supplemented medium produced the best results, with 25 to 57% of the explants (across genotypes) regenerating adventitious shoots, and 1.4 to 2.3 shoots per explant. Explants rapidly became necrotic on medium supplemented with BA, indicating BA may be phytotoxic to *P. deltoides*. This observation may explain the relative recalcitrance of this species to *in vitro* culture. The optimal zeatin supplemented medium (0.5 mg/L) was then used to test the *in vitro* shoot regeneration response of 16 *P. deltoides* genotypes. A wide range of response was observed among the 16 genotypes, with 0 to 100% of the explants regenerating adventitious shoots and 0 to 4.3 adventitious shoots per explant. Continuous axillary shoot cultures initially derived from adventitious shoots for three genotypes have been maintained for over four years (Coleman and Ernst, unpublished data), and tests are currently underway to determine if they are stabilized according to the criteria of McCown and McCown [26].

2.3. *The role of competence in organogenesis*

Christianson and Warnick [9] have experimentally demonstrated that the time at which an explant is transferred from a culture medium which enhances its competence to an inducing medium (e.g., shoot or root inducing medium) can be as important as media composition in determining the efficiency of organogenesis. They suggested that three physiological phases are involved in organogenesis: (1) the acquisition of competence, whereby the cells become capable of induction in the presence of an inducer; (2) induction or determination of a cell or cells to differentiate and produce the organ of choice; and (3) growth and expression of that differentiation. For optimal organogenesis, the temporal requirements for each of these three phases must be empirically determined for a given species, and usually for different genotypes within a species.

Coleman and Ernst (in press) have demonstrated that manipulation of the competence satate of *P. deltoides* internodal stem explants can enhance shoot regeneration efficiencies. Internodal explants were transferred from WNA medium supplemented with 0.5 mg/L 2,4-D (callus induction medium, CIM, used to enhance the competence of the internodal stem explants by promoting initial dedifferentiation) to WNA medium supplemented with 0.5 mg/L zeatin (shoot induction medium, SIM) at intervals of 0 (placed immediately on SIM), 1, 2, 4, 6, 8, 10, 20, and 30 days. The number of shoots were counted after 60 days in culture and the results from two genotypes are shown in Figure 1. The number of days required on CIM before transfer to SIM to enhance the competence state of the explant and increase regeneration efficiency is genotype specific. For example, internodal stem explants of genotype 10, a highly regenerable genotype, were already highly competent but the competence state manipulation resulted in a two-fold increase in the number of adventitious shoots per explant (Figure 1B) and no real increase in the percent of explants that regenerated shoots (Figure 1A). Conversely, for genotype 301, which is recalcitrant to *in vitro* culture, the competence state manipulation resulted in a large increase in both the percent of explants regenerating shoots (Figure 1A) and the number of adventitious shoots per explant (Figure 1B).

3. Genotypic stability during *in vitro* organogenesis

The only well documented study of genotypic stability during *in vitro* organogenesis (vs. androgenesis, which will be discussed later) in a pure species non-aspen poplar is that of Lester and Berbee [22]. They utilized shoot tips

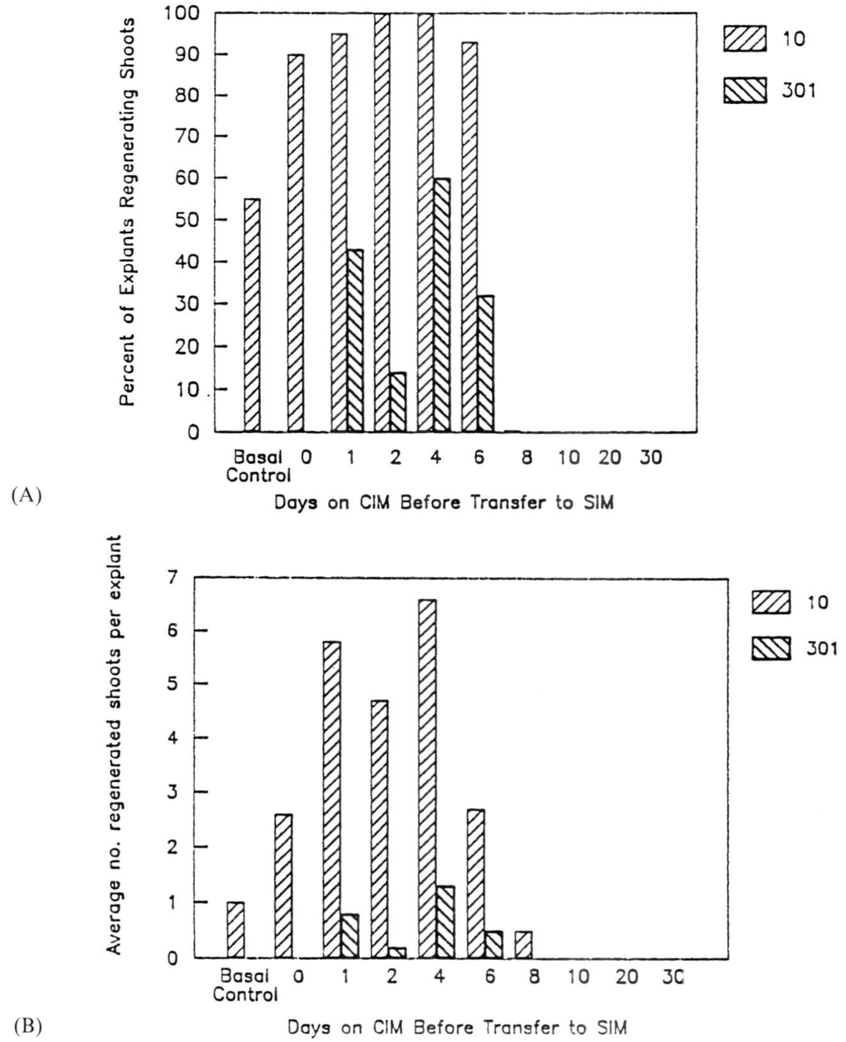

Figure 1. The effect of competence state manipulation on shoot regeneration from internodal stem explants of two genotypes of *Populus deltoides* (Coleman and Ernst, in press). A: The percentage of explants regenerating shoots. B: The average number of shoots per explant.

from rooted cuttings of one clone of *P. nigra* 'Italica' (and five clones of *P* × *euramericana*) as explants, and induced callus on a medium supplemented with 2,4-D. This callus was subcultured on a medium supplemented with BA and either NAA or IAA to induce adventitious shoots, and then the shoots were rooted and eventually established in the field. After one growing season, the plants were evaluated for variability in several growth characters (height, number of branches, and leaf characters), but only height was measured for the *P. nigra* clone. In addition, chromosome counts were made from greenhouse grown cuttings of the *in vitro* derived materials. For the *P. nigra* clone, the height of the *in vitro* derived ramets ranged from 56–144 cm after one year in the field, which the authors suggested was unsuitable for clonal planting stock. Ramets derived from the same callus were also quite variable in height growth. In addition, three of the six *P. nigra* ramets for which chromosome counts were done deviated from the normal diploid number of 38. It is now well accepted that prolonged exposure to 2,4-D *in vitro* and prolonged callus growth can cause genetic instability in plant cell cultures [1], and this may explain the variation in the chromosome number and height of the *P. nigra in vitro*-derived ramets. The variation in chromosome number suggests that at least some of the height differences among ramets could be due to genotypic instability. Many of the more recently developed regeneration protocols minimize the amount of time explants are on a 2,4-D supplemented medium and the associated callus growth stage (e.g. Refs. 2,11).

4. Androgenesis

Several excellent reviews on androgenesis for the production of haploid plants *in vitro*, including woody plants and specifically *Populus* species and hybrids, have been published [7,8,10,30]. Therefore, rather than repeat these reviews, a summary of the successful reports of pollen-derived callus and/or plants from species in the pure species non-aspen poplar group will be presented. In addition, evidence of genotypic instability in the *in vitro* androgenic systems will be addressed.

Table 1 lists those species in the pure species non-aspen poplar group for which haploid callus and/or plants have been produced from pollen *in vitro*. The developmental stage of the *Populus* anther explants was very important for successful induction of haploid callus. Uninucleate pollen grains, relative

Table 1. Pure specices non-aspen poplars for which pollen-derived callus and/or plants have been produced *in vitro*.

Species	Reference
P. deltoides	17, 18, 35, 39
P. euphratica	45
P. maximowiczii	35
P. nigra	17, 18, 41
P. pekinensis[a]	24 (from 10)
P. pseudo-simonii[a]	24 (from 10)
P. simonii	24 (from 10)
P. ussuriensis[a]	3, 4

[a] It could not be determined whether these represent true species, nor even to what Section they belong, based on the species list of Dickmann and Stuart (1983). Therefore, they may not belong to the pure species non-aspen poplar group.

to tetrad stage pollen mother cells or binucleate stage pollen grains, resulted in superior callus growth on a percent explant basis [17,39]. No pretreatment (e.g., cold) of *Populus* anthers has been reported. Different species, and different genotypes within a species, vary in their ability to form haploid callus from anthers [17,39]. *In vitro* andregenesis protocols in *Populus* typically involve an initial callus induction step (sometimes initially in the dark), followed by a shoot induction step, and then a root induction step [8,17,39]. No direct or indirect embryoid formation from anther explants has been reported for *Populus*. In one report [18], the callus originating from the anther cavities (i.e., pollen) turned green upon exposure to light, while callus derived from the inner wall of the anther, tapetal cells, or filaments generally did not. Based on the wide range of callus induction frequencies of *P. deltoides* genotypes on different media formulations (0–87% of the explants forming callus; [39], and the relatively low frequencies of callus induction in most reports on androgenesis in *Populus* (0–5%), much work remains to be done to optimize the callus and shoot induction media for the production of haploid plants in *Populus*.

Genotypic instability, as assessed by chromosome counts, is very common in *Populus* plants produced by *in vitro* androgenesis. Plants with diploid, haploid, polyploid, aneuploid, and less than haploid chromosome numbers have been reported [7,18,39]. Few of the plants remained haploid due to spontaneous diploidization, and chromosome numbers varied within a single plant [7]. Morphological variation of androgenically derived *Populus* plants has been reported to be both minimal [17] and extremely high [7]. More work is needed to increase the efficiency and genetic stability of haploid regenration systems in *Populus*, and to determine the utility of androgenesis in genetics and physiology research and applied breeding in *Populus*.

5. Somatic embryogenesis

There has been only one report of somatic embryogenesis in pure species non-aspen poplars. Layton and Ostry [20] reported that C. H. Micheler (USDA forest Service, Rhinelander, WI) has developed somatic embryogenesis protocols for *P. trichocarpa*, *P. deltoides*, and *Populus* hybrid clones. A paper describing these procedures for the *Populus* hybrids has been submitted (Michler, personal communication). No *Populus* species, regardless of Section, were mentioned in Tulecke's [38] review of somatic embryogenesis in woody plants.

6. Protoplast culture

There has been only one report of successful regeneration from protoplasts of a species in the pure species non-aspen *Populus* group. Lee *et al.* [21]

regenerated plantlets from protoplasts derived from callus (initiated from petiole tissue) of *P. nigra*. A yield of 1.4×10^6 protoplasts per gram of callus tissue was realized. The protoplasts were cultured on WPM supplemented with sugars for heterotrophic growth and osmoticum requirements. After four weeks in culture in the dark, optimal cell division and colony formation occurred with a plating density of 3.7×10^4 protoplasts/mL. Shoots were regenerated from the microcalli using WPM medium supplemented with 1.0 μM NAA and 0.1 μM BA, elongated on WPM supllemented with 0.2 mg/L BA, and roots induced on WPM supplemented with 0.2 mg/L indole-3-butyric acid (IBA). In the review by McCown and Russell [25], there was no mention of successful protoplast culture for a species in the pure species non-aspen *Populus* group, although regenerable protoplast cultures have been produced for several Leuce species and *Populus* hybrid clones.

7. Immature embryo (seed) rescue

Immature embryo, or immature seed, rescue has been reported for two species in the pure species non-aspen *Populus* group; for *P. deltoides* [19,37] and *P. heterophylla* [36]. The immature seed rescue technique was developed to enhance the recovery of control-pollinated seed or plants after forced female flowers on cut branches were pollinated in the greenhouse, because the flowers do not always live long enough for the seed to mature. By improving the protocol of Koudier *et al.* [19], Savka *et al.* [37] achieved germination rates in excess of 80% when culturing immature *P. deltoides* embryos collected 25 days after pollination. The best germination rates were achieved using a modified liquid MS medium with no growth regulators. Germination rates were much lower (28%) when the medium was agar-solidified. Using a similar protocol, germination rates in excess of 80% were achieved using immature seed collected 14 days after pollination for two full-sib crosses in *P. deltoides*. However, no germination was observed when immature seed from one of these two crosses was collected 19 days after pollination and placed on a water-saturated germination blotter (Ernst, unpublished data). Savka and Dawson [36] also reported 64% germination for immature *P. heterophylla* embryos collected approximately 40 days after pollination.

8. Field testing of *in vitro*-derived pure species non-aspen poplar plants

There is only one report of multiple-year (i.e., greater than one growing season) field testing of species in the pure species non-aspen *Populus* group, and that was with androgenically derived plantlets. As reported by Chen [7], Wu *et al.* [44] found that pollen-derived trees of *P. simonii* began to flower after seven to eight years in the field. Although the trees were quite variable in morphology and growth rate, it was possible to select several promising genotypes. Obviously, much more field testing of *in vitro*-derived *Populus* trees needs to be done in the future.

9. Summary

While preformed meristems have been utilized successfully by several investigators as initial explants to produce viable shoots by elongation *in vitro*, adventitious organ formation was common to all studies surveyed for this review. Cut surfaces and appropriate media provided the necessary physical and physiological conditions for adventitious organ formation. A wide variety of explant types were represented (meristem and shoot tips, nodal and internodal sections, axillary buds, cambial tissue, leaf and petiole segments, and flower primordia), and all resulted in the successful production of shoots *in vitro*. However, the number of shoots or roots generated depended on the species, genotype, and culture medium. The importance of genotype on regeneration response is very evident, with a range of responses found among different genotypes within a species, and some species more responsive to *in vitro* culture than others. Competence state manipulation can be used to increase the efficiency of regeneration of adventitious shoots from both easily cultured and recalcitrant genotypes. Increased long-term field testing of *in vitro* derived plants is needed. Also, while there has been some success in obtaining pollen-derived plants in *Populus*, much more work is needed to increase the efficiency of the androgenic haploid systems.

References

1. Ahuja MR (1987) *In vitro* propagation of poplar and aspen. In: Bonga JM and Durzan DJ (Eds.) Cell and Tissue Culture in Forestry, Vol. 3, Martinus Nijhoff Publ., Dordrecht. pp. 207–223.
2. Ahuja MR (1984) A commercially feasible micropropagation method for aspen. Silvae Genet 33:174–176.
3. Anonymous, Breeding Lab., Institute of Forestry, Acad. of Forestry, Heilunkiang (1976) Haploid plants of poplar induced from anther culture *in vitro*. Acta Genet Sinica 3(2):145–149. (Engl. summary.)
4. Anonymous, Tree Improv. Lab., Heilunkiang Institute of Forestry (1975) Induction of haploid pollen plants from anther culture *in vitro*. Sci Sinica 18(6):769–777.
5. Bawa KS and Stettler RF (1972). Organ culture with black cottonwood: morphogenic response of female catkin primordia. Can J Bot 50:1627–1631.
6. Chalupa V (1974) Control of root and shoot formation and production of trees from poplar callus. Biologia Plantarum 16(4):316–320.
7. Chen ZH (1986) Induction of androgenesis in woody plants. In: Hu H and Yang H (Eds.) Haploids of Higher Plants *In Vitro*, Springer-Verlag, New York. pp. 42–66.
8. Chen ZH (1987) Induction of androgenesis in hardwood trees. In: Bonga JM and Durzan DJ (Eds.) Cell and Tissie Culture in Forestry, Vol. 2, Martinus Nijhoff Publ., Dordrecht pp. 247–268.
9. Christianson ML and Warnick DA (1988) Physiological genetics of organogenesis *in vitro*.

In: Hanover JW and Keathley DE (Eds.) Genetic Manipulation of Woody Plants, Plenum Press, New York,. pp. 101–115.

10. Chu C-C (1982) Haploids in plant improvement. In: Vasil IK, Scowcroft WR and Frey KJ (Eds.) Plant Improvement and Somatic Cell Genetics, Academic Press, New York. pp. 129–158.
11. Coleman GG and Ernst SG (1989) *In vitro* shoot regeneration of *Populus deltoides*: effect of cytokinin and genotype. Plant Cell Rep 8:459–462.
12. Diaz-Colon JD, Bovey RW, Davis FS, and Baur JR (1972) Comparative effects and concentration of picloram, 2,4,5-T and dicamba in tissue culture. Physiol Plant 27:60–64.
13. Dickmann DI and Stuart KW (1983) The culture of poplars in eastern North America. University Publications, Michigan State Univ., East Lansing. 168 pp.
14. Douglas GC (1984) Formation of adventitious buds in stem internodes of *Populus* species cultured *in vitro* on basal medium: influence of endogenous properties of explants. J Plant Physiol 116:313–321.
15. Gautheret RJ (1934) Culture du tissu cambial. CR Acad Sci Paris 198:2195–2196.
16. Hildebrandt AC (1958) Stimulation or inhibition of virus infected and insect-gall tissues and single-cell clones. Proc Natl Acad Sci USA 44:354–363.
17. Ho RH and Ray Y (1985) Haploid plant production through anther culture in poplars. For Ecol and Mgmt 13:133–142.
18. Ho RH, Raij AY, and Zsuffa L (1983) Poplar plants through anther culture. Proc 28th Northeastern Forest Tree Improv Conf, pp. 294–300.
19. Kouider M, Skirvin RM, Saladin KP, Dawson JO, and Jokela JJ (1984) A method to culture immature embryos of *Populus deltoides in vitro*. Can J For Res 14:956–958.
20. Layton PA and Ostry ME (1988) Biotechnology of poplars: a US perspective. In: Proc Intl Energy Agency Joint Workshop on Willow Breeding and Biotechnology Development, Long Ashton Res. Station, Bristol, UK, Sept. 28–Oct. 1, 1988.
21. Lee JS, Lee SK, Jang SS, and Lee JJ (1987) Plantlets regenerated from callus protoplasts of *Populus nigra*. Inst For Genet Res Rep 23:143–148.
22. Lester DT and Berbee JG (1977) Within-clone variation among black poplar trees derived from callus culture. For Sci 23:122–131.
23. Lloyd GB and McCown BH (1980). Commercially feasible micropropagation of mountain-laurel, *Kalmia latifolia*, by use of shoot tip culture. Proc Intl Plant Prop Soc 30:421–427.
24. Lu ZH, Zhang XF, and Liu YX (1978) Preliminary observation on induction of poplar pollen plants and their genetic performance. Proc Symp Anther Culture (1977), Science Press, Beijng, pp. 195–197. (In Chinese.)
25. McCown BH and Russell JA (1987) Protoplast culture of hardwoods. In: Bonga JM and Durzan DJ (Eds.) Cell and Tissue Culture in Forestry, Vol. 2, Martinus Nijhoff Publ., Dordrecht. pp. 16–30.
26. McCown DD and McCown BH (1987) North American hardwoods. In: Bonga JM and Durzan DJ (Eds.) Cell and Tissue Culture in Forestry, Vol. 3, Martinus Nijhoff Publ., Dordrecht. pp. 247–260.
27. Murashige T and Skoog F (1962) A revised medium for rapid growth and bioassays with tobacco tissue cultures. Physiol Plant 15:473–497.
28. Pelet F, Hildebrandt AC, Riker AJ, and Skoog F (1960). Growth *in vitro* of tissues isolated from normal stems and insect galls. Am J Bot 47:186–195.
29. Prakash CS and Thielges BA (1988) Plantlets from leaf discs of *Populus deltoides*. Poster abstract. In: Hanover JW and Keathly De (Eds.) Genetic Manipulation of Woody Plants, Plenum Press, New York. p. 482.
30. Radojevic L and Kovoor A (1986) Induction of haploids. In: Bajaj YPS (Ed.) Biotechnology in Agriculture and Forestry, Vol. I, Springer-Verlag, Berlin. pp 65–86.
31. Raj AY, Ho RH, and Zsuffa L (1983) *In vitro* propagation of forest trees by tissue culture. Proc. 28th Northeastern For Tree Impr Conf. pp. 218–293.
32. Riemenschneider D, Haissig B, and Michler C (1988) Genetic effects of adventitious rooting

in vitro in *Populus deltoides*. In Vitro 24 (3, Part II): 52A (Oral Presentation Abstract 161).
33. Rutledge CB and Douglas GC (1988) Culture of meristem tips and micropropagation of 12 commercial clones of poplar *in vitro*. Physiol Plant 72:367–373.
34. Saito A (1980) Medium for shoot formation from somatic callus tissues in *Populus*. J Jap For Soc 62(7):270–272.
35. Sato T (1974) Callus induction and organ differentiation in anther culture of *Populus*. J Jap For Soc 56:55–62.
36. Savka MA and Dawson JO (1985) Culture of ovules containing immature embryos of *Populus heterophylla* L. *in vitro*. Forestry Res Rep No 85-1, Dept. Forestry, Agric Expt Stn Univ Illinois. 4 pp.
37. Savka MA, Dawson JO, Jokela JJ, and Skirvin RM (1987) A liquid culture method for rescuing immature embryos of eastern cottonwood. Plant Cell, Tissue Organ Culture 10:221–226.
38. Tulecke W (1987) Somatic embryogenesis in woody perennials. In: Bonga JM and Durzan DJ (Eds.) Cell and Tissue Culture in Forestry, Vol. 2, Martinus Nijhoff Publ., Dordrecht. pp. 61–91.
39. Uddin MR, Meyer MM Jr., and Jokela JJ (1988) Plantlet production from anthers of eastern cottonwood (*Populus deltoides*). Can J For Res 18:937–941.
40. Venverloo CJ (1973) The formation of adventitious organs. I. Cytokinin-induced formation of leaves and shoots in callus cultures of *Populus nigra* L. 'Italica'. Acta Bot Neerl 22(4):390–398.
41. Wang CC, Chu ZC, and CS (1975) The induction of *Populus* pollen plants. Acta Bot Sin 17:56–59. (English Summary).
42. Whitehead HCM and Giles KL (1977) Rapid propagation of poplars by tissue culture methods. New Zeal J For Sci 7:40–43.
43. Winton LW (1968) Plantlets for aspen tissue culture. Science 160:1234–1235.
44. Wu KX, Xu MZ, Ma HP, Li W, and Zhu WH (1981) Preliminary analysis on characteristics of pollen plants of *Populus simonii* Carr. 'Linye Keji' 2:5–6. (In Chinese.)
45. Zhu XY, Wang RL, and Liang, Y (1980) Induction of poplar pollen plantlets. Sci Silvae Sinica 16:1191–1197.

13. Clonal propagation in non-aspen poplar hybrids

YOUNG WOO CHUN

Department of Forestry, College of Forestry, Kookmin University, Seoul, 136–702 Republic of Korea

1. Introduction

1.1. *Background*

Species and hybrids of *Populus* are considered ideal species for pulpwood, lumber production, and woody biomass production because of their rapid growth, ease of establishment through stem or root cuttings, and relative ease of coppice regeneration [19]. Trees of this genus also represent wide genetic diversity, exhibit a wide range of site requirements, and respond well to cultural inputs [52]. Many new, high-quality, rapidly-growing poplar clones have been developed in various breeding programs from many parts of world [29]. Vegetatively propagated *Populus* species and hybrid clones have been planted extensively in many countries [63]. For the difficult-to-root poplar (section Leuce), the application of tissue culture techniques for micropropagation has become a primary alternative to more conventional propagation procedures.

1.2. *History of* Populus *hybrid utilization*

Species and hybrids of *Populus* (poplar) respond well to cultural inputs, propagate relatively well vegetatively, coppice from stump or root sprouts usually easily, and have demonstrated phenomenal growth rates in temperate climates. Many superior clones of poplar hybrids have been developed. *Populus deltoides, P. nigra*, and their hybrids have proven most useful for obtaining high yields on bottom land sites in the USA [28]. *P. euramericana* is a well known example of a natural poplar hybrid that represents various clones of *P. deltoides* and *P. nigra* parentage. These hybrids have been utilized over 300 years and have already established their value in many parts of world [19]. For example, *P. euramericana* I-476 and I-214 are being planted on a large scale (about 80% of whole poplar plantation area) in the Republic of Korea. It also seems likely that *P. trichocarpa, P. maximowiczii* and *P. balsamifera* will be useful in producing successful hybrids for a variety of purposes [28]. Already in Belgium and Canada, extensive selection and breeding programs have utilized *P. deltoides, P. nigra, P. trichocarpa*, and *P. maximowiczii* to develop many superior hybrid clones [29,54,63].

M. R. Ahuja (Ed.), Micropropagation of Woody Plants, 209–222.

2. Micropropagation

Among the forest tree genera, the genus *Populus* is one of the most widely-studied in the area of tissue culture [16]. The tissue of *Populus* exhibits a high degree of developmental plasticity, similar to that of tobacco in the herbaceous species. The development of *in vitro* cloning techniques provides an integral basis for biotechnological applications [16]. McCown [41] has stressed that the shoot cultures of woody plants could be a central tool for ranging from gene manipulation to forest establishment.

Since the earlier studies by Wolter [62], and Winton [58,59], several species and hybrids of *Populus* have been successfully propagated through *in vitro* techniques (Table 1). For more detailed information on tissue culture of *Populus* hybrids and species, the reader is referred to other reviews by Ahuja [1] and Chun *et al.* [16].

Table 1. Tissue culture research with *Populus* hybrids.

Populus hybrids	Culture[a]	Regen.[b]	Ref.
P. euramericana	C	S R	10
P. deltoides × *P. nigra*	C	S R W	7
P. simonii × *P. nigra*	A	S R	31
P. tristis × *P. balsamifera*	AM	S R W	55
P. deltoides × *P. nigra*	S	S R W	56
P. harbinensis × *P. pyramidalis*	A	S R W	47
P. maximowiczii × *P. deltoides*	A	S R	32
P. trichocarpa × *P. tacamahaca*	E	S R W	21
P. nigra × *P. trichocarpa*	SV	S R W	23
P. deltoides × *P. trichocarpa*	S, IE	S R W	44
P. nigra × *P. trichocarpa*	S, IE	S R W	44
P. deltoides × *P. maximowiczii*	S, IE	S R W	44
P. nigra × *P. trichocarpa*	SV	S R W	48
P. maximowiczii × *P. trichocarpa*	SV	S R W	45
P. nigra × *P. maximowiczii*	E, SE	S R W	49
	S, SV	S R W	18
	N	S R W	17
P. koreana × *P. nigra*	S, E	S R W	Chung and Chun[c]
P. euramericana	N	S R W	17

[a] Culture = type of tissue culture. C: callus culture; A: anther culture; AM: apical meristem culture; E: cultured explant culture; S: Shoot tip (axillary bud) culture; SV: induction and selection of somaclonal variation; IE: *in vitro* culture for international exchange; SE: somatic embryogenesis; N: nodule culture.

[b] Regen. = S: shoot regeneration; R: root regeneration; W: whole plant regeneration.

[c] Chung and Chun = unpublished.

2.1. *Axillary bud culture vs. adventitious bud culture*

Two routine methods have been developed for experimental micropropagation of *Populus* hybrids and species. These experimental methods are axillary bud production (also termed axillary shoot production) and adventitious bud initiation [24,30]. The distinctive features of axillary bud culture and adventitious bud culture are summarized in Table 2.

Several species of poplar have been rapidly produced by *in vitro* techniques [1,16]. In *Populus* hybrids, Whitehead and Giles [56] have shown that *P. deltoides* × *P. nigra* and other *Populus* species can be mass propagated by axillary bud culture. Thompson and Gordon [55] described a tissue culture system for apex culture propagation of hybrid poplar, *P. tristis* × *P. balsamifera*. Like other aspen or hybrid aspen, axillary branching in *Populus* hybrids could be enhanced on the Murashige and Skoog [46] (MS) medium containing benzyladenine (BA) 0.2–0.5 mg/l. It is well known that BA, serving as a cytokinin, produces marked effects, promoting the outgrowth of axillary branches of cultured buds [33]. In some poplar hybrids, an appropriate balance of auxin and cytokinin must be applied for enhancing axillary branching [11,56]. In the other cases, auxin must be omitted to generate axillary branching shoots from cultured poplar shoots [13,36,37]. Nevertheless, some poplar hybrid apparently require only higher BA concentrations for axillary shoot multiplication. *P. nigra* × *P. maximowiczii* and *P. koreana* × *P. nigra* do not need auxin to stimulate proliferation of axillary shoots (Figure 1).

Shoots may be regenerated directly on explants from various *Populus* hybrids (Figure 2). Adventitious shoots can be induced from *in vitro* cultured leaf, internode, and petiole explants that normally never reproduce vegetatively *in vivo*. Adventitious regeneration *in vitro* may generate a much higher rate of shoot production than is possible by proliferation of axillary shoots. Plant regeneration from stem internodes has been accomplished in *P. trichocarpa* × *P. tacamahaca* [21]. Park and Son [49] demonstrated that an average of 178 shoots were directly produced from the punctured leaf of *P. nigra* × *P. maximowiczii* after 6 weeks in culture. Many adventitious shoots formed at the basal part of axillary shoot tips of *P. nigra* × *P. maximowiczii* and *P. koreana* × *P. nigra* when cultured on MS medium with 0.2 mg/l BA. Generally, morphogenetic responses of *in vitro* cultured explants depend upon the species, the explant source, and the combinaton of exogenously applied plant growth regulators. To develop more generalized organogenesis techniques for explant culture of poplar, further suggestions for additional research include: (1) application of physical treatments (e.g. puncturing the leaf surface as exemplified by Park and Son) and various plant growth regulator substitutions (e.g. 2,4-D vs. NAA) in explant culture medium need to be tested to determine their effect on direct organogenesis and somatic embryogenesis from various *Populus* hybrids; and (2) the optimum culture conditions for explant source material need to be defined in order to optimize organogenesis or somatic embryogenesis.

Table 2. Distinctive features of axillary bud culture and adventitious bud culture.

Features	Axillary shoot culture	Adventitious shoot culture
Plant sources	Axillary bud	Cultured explants such as leaf, internode, petiole, root or original organs
Meristem mode	Existing meristem	Adventitious meristem
Multiplication rate	Relatively low × 5–× 10 / 5–10 wks	Relatively high × 10–× 100 / 5–10 wks
Plant growth regulator supplements	Cytokinin only or occasionally combination of high cytokinin with low auxin concentration	Involves a combination of high cytokinin with low auxin concentration
Genetic stability on a clonal basis	Limited spontaneous mutation rate similar to that of conventional vegetative propagation	Spontaneous mutation may exist at a very low rate
Practical utilization	- Micropropagation - International exchange of germplasm - Germplasm preservation through cold storage	- Induction and selection of somaclonal variation - Integral part of genetic transformation - Micropropagation

Figure 1. Four-week-old culture showing axillary branching shoots of *P. nigra* × *P. maximowiczii* on shooting medium.

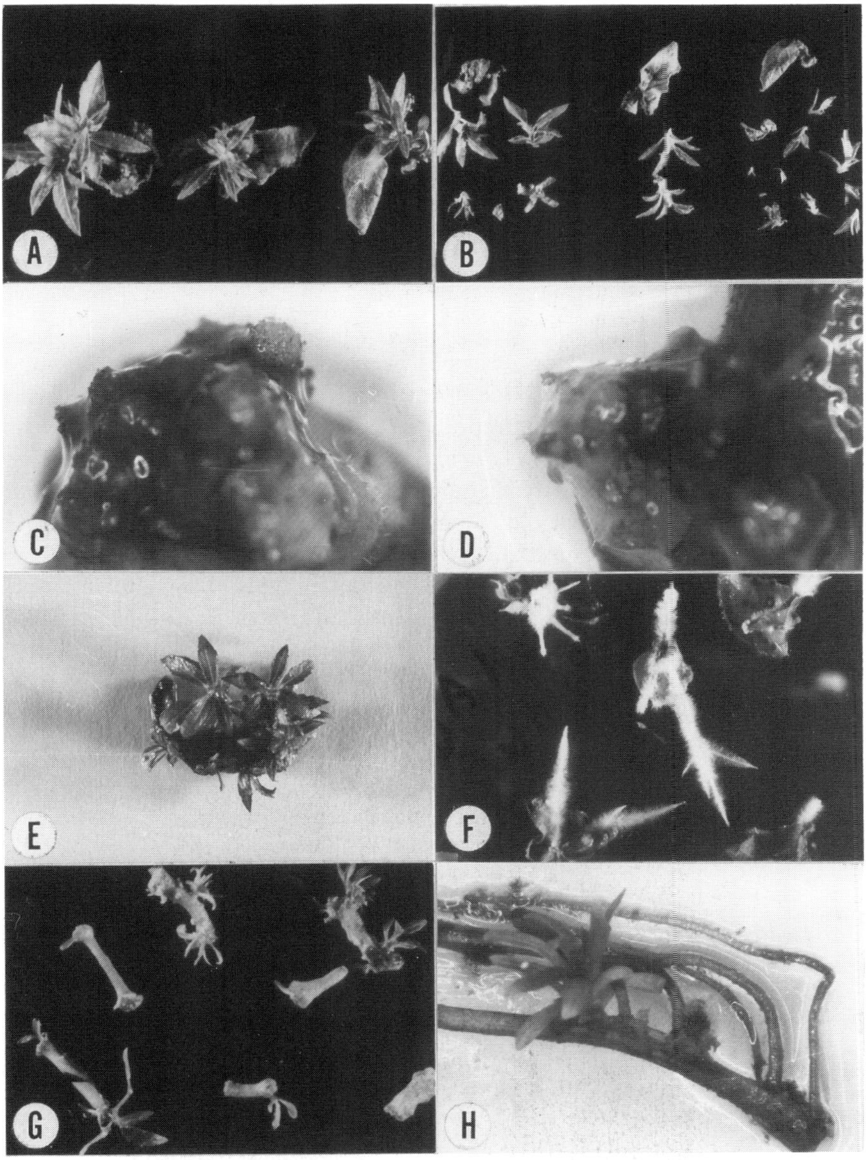

Figure 2. Six-week-old culture showing adventitious shoot regeneration on the entire leaf with petiole (left), terminal half (middle), and basal half with petiole (right) (A), and different number of adventitious shoots regenerated from three leaf explant sources (B). Two-week-old culture showing the adventitious meristem zone at the petiole (C), and proximal cut of leaf segment (D). Four-week-old culture showing the regeneration of adventitious shoots from the marginal meristem zones on both abaxial and adaxial leaf surface (E). Root formation at the end of the petiole or cut surface in adaxial side leaf culture. Adventitious shoot regeneration on the internode segments after six weeks of culture (G). After four weeks of culture, adventitious shoot regeneration on a root segment (H). All plates originated from the explant culture of *Populus alba* × *P. grandidentata.*

Theoretical rates of clonal multiplication are very large. Kim *et al.* [36] and Garton *et al.* [26] suggest that it is possible to produce over ten million plantlets per year from one axillary bud. It has been estimated that culture numbers can be increase by 5–10 times over a period to 4–8 weeks for many species of *Populus* [11,54]. Such rapid increases have particular application to woody species that are multiplied only slowly by such conventional propagation methods as greenhouse or nursery cuttings.

2.2. *Somatic embryogenesis*

Even though somatic embryogenesis has been demonstrated in several woody plants including sweetgum, horse chestnut, sandalwood, *Paulowina*, and loblolly pine, there is only one report in which somatic embryogenesis was induced from leaf explants of *Populus* spp. Park and Son [49] obtained embryoids on callus derived from punctured leaves of *P. nigra* × *P. maximowiczii*. The highest frequency of embryoids were induced to form when a plant growth regulator combination of 0.44 μM BA and 2.26 μM 2,4-D was used. Using nodule culture system, Chung and Chun [17] have shown embryoids on nodule derived from cell suspension cultures of *P. euramericana* and *P. nigra* × *P. maximowiczii* (Figure 3).

Figure 3. A culture of nodules showing various stages of embryoid development on *P. euramericana*.

3. Production cost and potential for automation

Commercial scale propagation of Leuce hybrids is performed through tissue culture in Germany [53]. Economic factors, however, currently restrict the large-scale utilization of tissue culture for mass propagation of poplar species because of the high production cost of *in vitro* cloning in comparison with that of conventional propagation by rooted cuttings [27]. The cost of tissue culture derived plantlet will vary with production techniques and a number of other production factors. Production cost will also vary from country to country, especially with respect to labor costs. Generally, the cost of a plantlet is higher than that of a seedlings in case of *Liquidambar styraciflua* [9], *Pinus taeda* [*43*]. Kolison [38] estimated that the production cost of a poplar plantlet produced by *in vitro* micropropagation methods was 64 cents, and costs of producing a poplar plant by unrooted hardwood cutting and rooted hardwood cutting methods were about 3 cents and 29 cents, respectively. For more detailed information on determining and minimizing production cost, the reader is referred to other reviews by Donnan [20] and Jones [34].

Various methods have been attempted to reduce the high production cost of *in vitro* cloning, including somatic embryogenesis [51], automatic shoot hedge production [3], nodule culture by organogenic cell culture [42] and minimization of transplanting frequency, including direct rooting of tissue cultured shoot cuttings in potting medium [1,13]. Even though somatic embryogenesis has been proposed as a labor saving method for *in vitro* cloning, more intensive studies are required to develop a generalized technique for somatic embryogenesis with *Populus* species. Ahuja [1] suggested that the production costs of *in vitro* micropropagation for poplar can be minimized by reducing rates of transplanting at various tissue culture stages. Anderson *et al.* [4] suggested eliminating the rooting stage of proliferated microshoots as a means of reducing production costs. Aitken-Christie and Jones [2] demonstrated that the cost of shoot production using the hedging method for *Pinus radiata* was approximately seven times cheaper than that of normal transfer methods. McCown *et al.* [42] reported that nodule culture originated from organogenic cell culture can be considered an automated micropropagation system.

Chun *et al.* [15] have found that for *in vitro* proliferation of the hybrid poplar, liquid medium produced more shoots with greater fresh weight than did agar solidified medium. Although the liquid medium was better than solid medium for *in vitro* proliferation, vitrification symptoms appeared in the cultures of liquid medium. To reduce the vitrification symptoms of *in vitro* proliferation in liquid medium, various cellulose rods (Sorbarod System, from Baumgartner Papiers Co., Switzerland) have great potential. Our results on *Populus nigra* × *P. maximowiczii* indicated that axillary bud and adventitious shoots that were cultured on cellulose rods in liquid medium produced as many multiple shoots as those cultured directly in liquid medium without vitrification. This simple technique offers potential for automatic shoot hedge production within single culture vessels. Through weekly or monthly liquid nutrient replenishment, shoots could be periodically harvested without the need to transfer the cultures [12].

4. Germplasm preservation

Two major methods available throygh *in vitro* technologies for germplasm preservation are: (1) slow growth by growth suspension; and (2) cryopreservation [60,61]. The slow growth method for germplasm preservation has several advantages compared with the cryopreservation method. The slow growth method is a very simple and economical approach. In addition, plantlets can be maintained free from pests and disease. This method does not require the highly sophisticated techniques necessary for cryopreservation steps involving freezing, storage, and thawing of cultures. Compared with clone banks in the greenhouse or field, this method requires relatively little space for a large number of plantlets. This method does not require frequent maintenance in contrast with various cultural practices of the greenhouse or field, and consequently saves both labor and space. Therefore, we could directly apply this method for germplasm preservation of *Populus* species.

4.1. *Non-frozen, low temperature storage*

To minimize time and expense for routine culture maintenance, preservation of plant material in a state of growth suspension has been attempted by various methods [57]. Such growth suspension methods in plant tissue culture can be divided into three categories: non-frozen cold storage, alteration of the basic culture media, and addition of growth retardants. Among these tissue culture storage methods, temperature reduction has been demonstrated to be very effective for slowing growth of *in vitro* cultured plantlets [60]. There are many species that can be conserved by temperature reduction, especially with non-frozen, low temperature storage methods. These species include: *Malus* spp. [39], *Musa* spp. [5], *Prunus* spp. [40], *Pinus radiata* [2], *Trifolium repens* [8], *Allium sativum* [22] and *Populus* spp. [12,14]. Chun [12] could store *in vitro* cultured hybrid poplar, *Populus alba* × *P. grandidentata*, at 4° C in darkness over 5 years and still recover suitable multiplication potential (Figure 4). Chun [12] found that a subculturing period preceding cold storage, plantlet condition, and culturing medium all had an important influence on survival at 4° C in darkness. Chun [12] also showed that, after 24 months storage, a 70% survival percentage was obtained with plantlets possessing 4–6 axillary branching shoots that were subcultured on shoot proliferation medium for one month preceding cold storage.

Because most clones of aspen or poplar are routinely grown in greenhouses or fields as a source of germplasm, the relatively simple and inexpensive cold

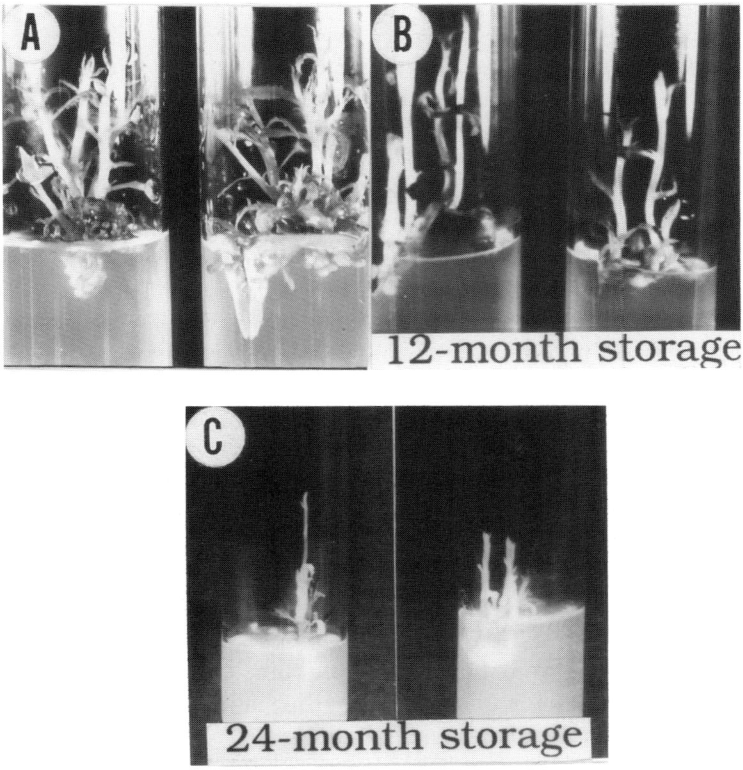

Figure 4. Shootlets with 4–6 axillary shoots on shoot proliferation medium for one month subculturing period preceding cold storage (A), and after 12 months (B), and 24 months (C) cold storage.

storage method could reduce the costly greenhouse or field maintenance of these poplar stocks. In addition, the cold storage technique could facilitate germplasm shipments, as well as alleviate the maintenance problems of established micropropagation systems. Continuous subculturing is the initial requirement for establishing a micropropagation system. Maintenance of established cultures can require substantial space, equipment, chemicals, and labor.

To succesfully conserve germplasm through cold storage of *in vitro* cultured plantlets, however, factors other than subculturing period preceding cold storage and plant growth regulator supplements need to be tested. Additional factors worthy of consideration include: storage temperature, nutrient composition of the culture medium, photoperiod, and light intensity.

5. Tools of international exchange

The exchange of disease and insect-free plant material across national boundaries is very important. Several catastrophic events have occurred in the history of forestry as a result of the transfer of diseased plant materials across national boundaries [6,25]. The plant quarantine aspects of the international exchange of plant tissue culture for scientific or commercial purposes are well discussed by Khan [35] and Parliman [50]. These authors demonstrated the role of tissue culture in providing safeguards that lower risks and facilitate international movement of plant material.

Clonal *Populus* material suitable for exchange is readily available from breeding programs for biomass and pulpwood production, and for *in vitro* micropropagation. *In vitro* micropropagation as a solution to disease-free requirements for international exchange of woody plant materials was attempted by McNabb *et al.* [44] at Iowa State University, U.S.A. Through the International Energy Agency program, 21 clones representing the best current European selections and some of their parents were introduced from the Rijksstation voor Populierenteelt, Geraardsbergen, Belgium under special arrangement as axenic tissue cultured plantlets [44]. The original importation of 26 poplar clones including *Populus deltoides* × *P. trichocarpa, P. nigra* × *P. trichocarpa*, *P. deltoides* × *P. maximowiczii* among other poplar hybrids was the first such exchange from Europe in over three decades. *Populus* species have become candidates for *in vitro* micropropagation of disease-free plant material because of their current and aniticipated role in woody biomass for energy and fiber production worldwide.

6. Other considerations

Species and hybrids of *Populus* are promising candidates as the target species for tissue culture biotechnology applications because recent work has demonstrated rapid progress in specific area such as micropropagation, protoplast culture, and induction of beneficial somaclonal variation [16]. This recent progress should stimulate more rapid incorporation of specific improvements with conventional tree breeding programs within a short period of time. There are still many basic problems, however, to overcome for utilization of biotechnology in *Populus* species. These problems include the limitation in: (1) somatic embryogenesis; (2) generalized regeneration technique for protoplast culture; (3) uncontrolled somaclonal variation leading to a failure of true-to-type propagation in tissue culture.

7. Summary

Clonal propagation and utilization, and the development of tissue culture for non-aspen poplar hybrids were reviewed. The micro-propagation system is probably not as cost-effective as propagating species and hybrid of poplars by unrooted hardwood cutting and rooted hardwood cutting methods. For the difficult-to-root poplar, however, the application of tissue culture techniques for clonal micropropagation has become a primary alternative to more conventional propagation procedures. The high production cost of poplars by the *in vitro* micropropagation method will be solved through the introduction of refined automated micropropagation systems in the near future.

The established clonal micropropagation systems of non-aspen hybrid poplar are very useful for (1) adventitious shoot culture through direct morphogeneisis from cultured explants. And eventually, we could apply this for genetic transformation through co-cultivation with genetically engineered *Agrobacterium* vector system. (2) Germplasm preservation through non-frozen, low temperature storage. And (3) international exchange of disease- and insect-free plant material. In addition, the established cultures of non-aspen poplar hybrids provide also a very useful source material for tissue culture biotechnology applications including induction and selection of somaclonal variation, and protoplast isolation and culture.

Acknowledgments

The author is grateful to Professor Richard B. Hall (Iowa State Univ.) for his initial encouragement. I also thank Dr. Ned B. Klopfenstein at the Department of Forestry, Iowa State University for his critical reading of the manuscript and Mrs. Kyung Ho Chung and Yun Seop Um at Kookmin University for technical assistance. Financial support for this research was provided by the Institute of Forest Science, Kookmin University.

References

1. Ahuja MR (1987) *In vitro* propagation of poplar and aspen. In: Bonga JM and Durzan DJ (Eds.) Cell and Tissue Culture in Forestry. Martinus Nijhoff Publishers, Boston. pp. 207–223.
2. Aitken-Christie J and Gleed JA (1984) Uses for micropropagation of juvenile radiata pine in New Zealand. Proc International Symp of Recent Advances in Forest Biotechnology, Traverse City, Michigan. pp. 47–57.
3. Aitken-Christie J and Jones J (1987) Towards automation: radiata pine hedges *in vitro*. Plant Cell Tissue Organ Culture 8:185–196.
4. Anderson WC, Meager GW, and Nelson AG (1977) Cost of propagating broccoli plants through tissue culture. HortScience 12:543–544.
5. Banerjee N and de Langhe E (1985) A tissue technique for rapid clonal propagation and storage under minimal growth conditions of *Musa* (banana and plantain). Plant Cell Reports 4:351–354.
6. Benedict WV (1981) History of white pine blister rust control. USDA For. Serv. FS-355. Government Printing Office, Washington, D.C. 47 pp.

7. Berbee FM, Berbee JG, and Hildebrandt AC (1972) Induction of callus and trees from stem tip culture of a hybrid poplar. *In Vitro* 7:269 (Abstr.)
8. Bhojwani SS (1981) A tissue culture method for propagation and low temperature storage of *Trifolium repens* genotypes. Physiol Plant 52:187–190.
9. Brown CL and Sommer HE (1982) Vegetative propagation of dicotyledonous trees. In: Bonga JM and Durzan DJ (Eds.) Tissue Culture in Forestry. Martinus Nijhoff/Junk Publ, The Hague. pp. 109–149.
10. Chalupa V (1974) Control of root and shoot formation and production of trees from poplar callus. Biol Plant 16:316–320.
11. Christie CB (1978) Rapid propagation of aspen and silver poplars using tissue culture techniques. Proc Int Plant Prop Soc 28:255–260.
12. Chun YW (1987) Biotechnology applications of *Populus* micropropagation. PhD Diss, Iowa State University, Ames, Iowa. 149 pp.
13. Chun YW (1989) Effects of benzyladenine on *in vitro* shoot proliferation of hybrid poplar using two different explants cultured on different media. Forest and Humanity 1:25–34. Kookmin Univ., Korea.
14. Chun YW and Hall RB (1986) Low temperature storage of *in vitro* cultured hybrid poplar, *Populus alba* × *P. grandidentata* plantlets. In: Sommers DA, Gengenbach BG, Biesboer DD, Hackett WP, and Green CE (Eds.) Proc Sixth International Congress of Plant Tissue and Cell Culture. Univ of Minn, Minneapolis, MN. p. 13.
15. Chun YW, Hall RB, and Stephens LC (1986) Influences of medium consistency and shoot density on *in vitro* proliferation of *Populus alba* × *P. grandidentata*. Plant Cell Tissue Organ Culture 5:179–185.
16. Chun YW, Klopfenstein NB, McNabb HS Jr., and Hall RB (1988) Biotechnological applications in Populus species. J Kor For Soc 77:467–483.
17. Chung KH and Chun YW (1991) Plant regeneration of hybrid poplar using nodule culture system. J Kor For Soc 80:1–8.
18. Chung KH and Chun YW (1990) Variation in aluminium tolerance among 5 species of *in vitro* cultured *Populus*. J Kor For Soc 79:26–32.
19. Dickmann DI and Stuart KW (1983) The Culture of Poplars in Eastern North America. Michigan State University, East Lansing. 168 pp.
20. Donnan A Jr (1986) Determining and minimizing production costs. In: Zimmerman RH *et al.* (Eds.) Tissue culture as a plant production system for horticultural crops. Martinus Nijhoff Publishers, Hague. pp. 167–174.
21. Douglas GC (1984) Formation of adventitious buds in stem internodes of *Populus* spp. cultured *in vitro* on basal medium: influence of endogeneous properties of explants. Plant Physiol 116:313–321.
22. El-Gizawy AM and Ford-Lloyd BV (1987) An *in vitro* method for the conservation and storage of garlic (*Allium sativum*) germplasm. Plant Cell Tissue Organ Culture 9:147–150.
23. Ettinger TL, Ostry ME, Hackett WP, Read PE, and Skilling DD (1986) Utilization of somaclonal variation in the development of *Septoria musiva*-resistant hybrid *Populus*. In: Sommers DA, Gengenbach BG, Biesboer DD, Hackett WP, and Green CE (Eds.) Proc 6th Intl Congress of Plant Cell Culture. Univ of Minn, Minneapolis, MN. p. 214.
24. Faltonson RR, Chun YW, Robison TL, and Hall RB (1984) The role of vegetative propagation in tree improvement. Proc of Forestry Comm Great Plains Agric Counc (Publ No 112) 36:202–213.
25. Forest Pest Control: an assessment of present and alternative technolgies (1975) Forest pest control. Vol. 4. National Academy of Science, Washington, D.C. 170 pp.
26. Garton S, Hosier MA, Read PE, and Farnham RS (1981) *In vitro* propagation of *Alnus glutinosa* Gaertn. HortScience 16:758–759.
27. Haissig BE, Nelson ND, and Kidd GH (1987) Trends in the use of tissue culture in forest improvement. Bio/Technology 5:52–59.
28. Hall RB (1988) Personal communication. Forestry Department, Iowa State University, Ames, Iowa.

29. Hall RB (1985) Breeding strategy for *Alnus, Populus*, and *Salix*. In: Mitchell CP, Nilsson PO, and Zsuffa L (Eds.) Proc Joint IEA Forestry Energy Program and FAO/Cooperative Network on Rural Forest Energy Conf and Workshops on Research in Forestry for Energy. Rungstedgaard, Denmark. Oct 1985. Swedish Univ of Ag Sci N 49/1986. pp. 75–95.
30. Hartmann HT and Kester DE (1983) Plant Propagation-Principles and Practices. Fourth edition. Prentice- Hall, Inc., Englewood Cliffs, New Jersey, 727 pp.
31. Heilungkiang Inst For Tree Improv Lab (1975) Induction of haploid poplar plants from anther culture *in vitro*. Sci Sinica 18:669–677.
32. Ho RH (1982) Haplois plant culture in poplar. In: Proc 7th North American Forest Biology Workshop, July 26–28, 1982. Univ of Kentucky, Lexington, Kentucky. pp. 210–213.
33. Hussey G (1986) Vegetative propagation of plants by tissue culture. In: Yeoman MM (Ed.) Plant Cell Culture Technology. Blackwell Scientific Publications, Boston. pp. 29–66.
34. Jones JB (1986) Determining markets and potential of Horticultural crops. In: Zimmerman RH *et al.* (Eds.) Tissue Culture as a Plant Production System for Horticultural Crops. Martinus Nijhoff Publishers, Hague pp. 175–182.
35. Kahn RP (1986) Plant quanratine and international shipment of tissue culture plants. In: Zimmerman RH *et al.* (Eds.) Tissue Culture as a Plant Production System for Horticultural Crops. Martinus Nijhoff Publishers, Hague. pp. 147–164.
36. Kim JH, Lee SK, and Chun YW (1981) Mass propagation of tree species through *in vitro* culture. I. Bud culture of *Populus alba* × *P. glandulosa* F1. Institute of Forest Genetic Research Report 17:57–63.
37. Kim JH, Shim SY, Noh EW, and Park JI (1982) Mass propagation of selected poplar clones through bud culture. Institution of Forest Genetic Research Report 18:80–85.
38. Kolison SH (1986) Costs associated with *in vitro* micropropagation in forestry. MS thesis, Iowa State Univ, Ames, Iowa. 61 pp.
39. Lundergan C, and Janick J (1979) Low temperature storage of *in vitro* apple shoots. HortScience 14:514.
40. Marino G, Rosati P, and Sagrati F (1985) Storage of *in vitro* cultures of *Prunus* root stocks. Plant Cell Tissue Organ Culture 5:73–78.
41. McCown BH (1985) From gene manipulation to forest establishment: shoot cultures of woody plants can be central tool. TAPPI J 68:116–119.
42. McCown BH, Zeldin EL, Pinkalla HA, and Dedoplh RR (1988) Nodule culture; a developmental pathway with high potential for regeneration, automated micropropagation, and plant metabolite production from woody plants. In: Hanover JW and Keathley DE (Eds.) Genetic Manipulation of Woody Plants. Plenum Press, New York. pp. 149–166.
43. Mckeand SE and Weir RJ (1984) Tissue culture and forest productivity. J For Soc 82: 212–218.
44. McNabb HS Jr., Serres RA, Cram MM, and Klopfenstein NB (1986) Biological constraints in the exchange of planting material for energy production. Proceedings of the FAO/CNRE/El 3.-Forestry Biomass for Energy, IEA/Forestry Energy Agreement, IUFRO sponsored conference and workshop entitled 'Research in Forestry for Energy', Rungstedgaard, Denmark; 28 Oct-1 Nov 1985. Dept. of Operational Efficiency, Swedish University of Agricultural Sciences, Uppsala, Sweden. pp. 96–97.
45. Michler CH and Bauer E (1987) Selection for herbicide tolerance in *Populus* to sulfometuron methyl *in vitro*. In: Hanover JW and Keathley DE (Eds.) Genetic Manipulation of Woody Plants. Plenum Press, New York. p. 479.
46. Murashige T and Skoog F (1962) A revised medium for rapid growth and bio-assays with tobacco tissue cultures. Physiol Plant 15:473–497.
47. North East For Acad, Gr Tree Breed (1977) Induction of haploid poplar plantlets from pollen. Acta Genet Sinica 4:49–54.
48. Ostry ME and Skilling DD (1988) Somatic varaiation in resistance of *Populus* to *Septoria musiva*. Plant Disease 72:724–727.

49. Park YG and Son SH (1988) *In vitro* organogenesis and somatic embryogenesis from punctured leaf of *Populus nigra* × *P. maximowiczii*. Plant Cell Tissue Organ Culture 15:95–105.
50. Parliman BJ (1986) Tissue culture techniques and plant introduction/quarantine procedures. In: Zimmerman RH *et al.* (Eds.) Tissue Culture as a Plant Production System for Horticultural Crops. Martinus Nijhoff Publishers, Hague. pp. 271–282.
51. Rao PS and Ozias-Akins P (1985) Plant regeneration through somatic embryogenesis in protoplast cultures of sandalwood (*Santalum album* L.). Protoplasma 124:80–86.
52. Schreiner EJ (1974) *Populus* L., poplar. Seeds of Woody Plants in the United States. US Forest Service Agriculture Handbook. No 450. pp. 645–655.
53. Schultzke R (1981) Tagung der Westeropaischen Papperlarbeitsgemeinschaft. Holzzucht 35 (3/4):33.
54. Teissier du Cros E (1984) Breeding strategies with poplars in Europe. For Ecol and Manage 8:23–39.
55. Thompson DG and Gordon JC (1977) Propagation of poplar by shoot apex culture and nutrient film technique. TAPPI Conference Papers; Forest Biology Wood Chemistry Conference 1977. pp. 77–82.
56. Whitehead MCM and Giles KL (1977) Rapid propagation of poplars by tissue culture methods. NZJ For Sci 7:40–43.
57. Wilkins CP and Dodds JH (1983) The application of tissue culture techniques to plant genetic conservation. Sci Prog Oxf 68:259–284.
58. Winton L (1968) Plantlet formation from aspen tissue culture. Science 160:1234–1235.
59. Winton L (1971) Tissue culture propagation of European aspen. For Sci 17:348–350.
60. Withers LA (1985) Long-term storage of *in vitro* cultures. In: Schafer-Menuhr A (Ed.) *In vitro* Techniques: Propagation and Long Term Storage. Martinus Nijhoff/Dr. W. Junk Publishers, Boston. pp. 137–148.
61. Withers LA (1986) Cryopreservation and genebank. In: Yeoman MM (Ed.) Plant Cell Culture Technology. Blackwell Scientific Publ. pp. 96–142.
62. Wolter KE (1968) Root and shoot initiation in aspen callus cultures. Science 219:509–510.
63. Zsuffa L (1985) Concepts and experiences in clonal plantations of hardwoods. Proc 19th Meeting of Clonal Forestry: The Canadian Tree Improvement Association, Can For Serv pp. 12–25.

14. Micropropagation of birch

MARGARETA WELANDER
Department of Horticultural Science, The Swedish University of Agricultural Sciences, S-230 53 Alnarp, Sweden

1. Introduction

1.1. *Distribution of* Betula *species*

The genus Betula probably has its origin in the NE of Asia and consists of about 60 species [20]. The birch species are widely distributed over the Northern Hemisphere, but they are limited to the temperate and arctic zones. According to Regel [32], two groups can be distinguished; *Eubetula and Betulaster*. The latter group consist of a few species mainly found in Japan and in the SW and the middle of China. The *Eubetula* group is divided in to three sub-groups, *Costatae Regel* (yellow birches), *Albae Regel* (white birches) which are tree forms, and *Nanae Regel* representing bush and dwarf forms. The yellow birches grow wild only in North America and in Asia, whereas the white birches are distributed throughout Europe, Asia and North America. Four birch species are native to Europe. Two of them – *Betula pendula* (Roth) and *Betula pubescens* (Ehrh) – are tree forms and the others – *Betula nana* (L) and *Betula humulis* (Schrank) – are bush forms. The birch species are well known for their peculiar bark, striking trunks and graceful branches, which make them valuable as garden trees.

1.2. *Industrial use*

Several features of birch wood make it very useful for veneer and plywood production. According to the prognosis of FAO the veneer production in Europe will increase by 3–4% per year. Most of the pulpwood produced from birch is used in the cellulose industry. However, high quality semi-chemical pulpwood can be produced because of the relative stability of its hemicellulose and the low lignin content. The fuel value of birch is high, thus constituting a great potential as a source of energy. Birch can also be used in the furniture industry, for the production of fiber boards and several chemical and biochemical products as rosin, cattle food, acetone and glycerol. Today it is well known that repeated tree-generations of coniferous reduce the production capacity of the land, whereas the decidious trees have a positive

M. R. Ahuja (Ed.), Micropropagation of Woody Plants, 223–246.

influence on the physical and chemical qualities as well as the pH of the soil. Therefore it is very probable that decidious trees like birch will be more frequently used both in mixed forests as well as in pure stands in the future.

1.3. *Conventional propagation*

Birch is generally propagated via seeds. The limitations of sexual reproduction are seed crop failure, low germination rate and genetically non-uniform offspring. Vegetative propagation has been practised in the selection of mother trees for breeding programs and the establishment of seed orchards with controlled pollination. Various methods of vegetative propagation have been used, such as budding, grafting and rooting of cuttings. However, many problems are associated with vegetative propagation.

Grafting techniques are expensive and are frequently associated with such difficulties as incompatibility and delayed graft failure. The rooting capacity of cuttings declines with the age of the mother tree, which is about 5 years for birch [1]. Micropropagation is therefore an interesting alternative.

2. *In vitro* approaches

2.1. *Genetic improvement by tissue culture techniques*

More intensive forest management practices, such as short rotation forestry, improved selective harvesting techniques and planned reforestration are necessary to increase the productivity of commercial forests [39]. A considerable potential for increasing productivity is by planting selected genotypes with improved growth rates, form, wood quality and insect and disease resistance [31]. Until now the plant characteristics in woody species have mainly been altered by utilizing natural selection or selective breeding. In the future genetic improvement will certainly be obtained by application of *in vitro*-based biotechnologies. However, in order to capture the unique genotype achieved by genetic engineering a reliable clonal propagation technique is necessary.

2.1.1. *Use of haploid plants*

There are several possible uses of haploid or double-haploid plants. Haploid plants obtained from microspores will provide oppertunities to develop homozygous lines in a short period of time by doubling haploid chromosome sets. Homozygous diploid lines could then be used for controlled hybridization and tested for heterosis and disease resistance. Because there is no dominance in haploids they can be used to reveal recessive genetic variation and study the inheritance of qualitative traits [2]. Haploids can also be used to transfer genes from polyploid to diploid species. In herbaceous species the production of relatively homozygous lines has been achieved through several generations of inbreeding. However this has not been practiced for tree improvement because of the length of time required to produce successive inbred generations [3,19]. Chinese researchers have been leaders in the field of the anther culture of trees and they have obtained haploid trees from several species. Huthinen [12] reported haploid callus from anther culture of birch (*Betula pendula* Roth). Plant regeneration from the callus was obtained but no evidence for haploid plants was given.

2.1.2. *Protoplasts*

Many of the techniques from protoplast research on agronomic crops could be applicable to forest trees. Protoplast fusion might be useful in hybridizing species which are sexually incompatible, to cross sexually sterile triploids and alleviate ploidy-level differences. Protoplasts may also be useful to induce new traits by genetic engineering. Protoplasts have been isolated from *in vitro* shoot cultures of *Betula platyphylla szechuanica* [37] and callus tissue from *Betula* × *'Boerner'* and *Betula papyrifera* [38] and from juvenile leaf tissue from *Betula pubescens* and *Betula pendula* [4]. The authors pointed out the importance of the developmental stage of the explant source, culture environment and inhibitory substances released during protoplast isolation to protoplast yield. However, according to my knowledge, no plants have been regenerated from birch protoplasts.

2.1.3. *Somatic embryogenesis*

A breakthrough in the tissue culture propagation of forest trees would be when trees can be regenerated from liquid suspension cultures and the development of artificial seeds. The potential use of artifical seeds is the production of a large number of indentical embryos at a low cost, competitive with the cost of true seeds. The artificial seed coating would allow incorporation of beneficial adjuvants such as growth promoting rhizobacteria, plant nutrients and pesticides [30]. Several tree species can be grown in cell suspensions but very few forest species have been regenerated via embryogenesis from suspension cultures [43].

2.1.4. *Genetic engineering*

Molecular biology may be even more applicable to long lived woody plants, such as forest trees, compared to agronomic crops because of difficulties in breeding to achieve genetic improvement. With the discovery of restriction enzymes and the new molecular cloning techniques, the molecular biologists have a powerful tool to introduce new beneficial characteristics without time consuming selection and back-cross programs.

There are essentially three methods for transferring isolated genes into plant cells. One is based on the capacity of protoplasts to incorporate exogenous DNA. Unfortunately very little success has been obtained with the regeneration of plants from protoplasts of woody species. The other method is based on the natural gene transfer capacity of *Agrobacterium tumefaciens* in which regenerable tissue can be used. Because of this *A. tumefaciens*-mediated transformation is very useful for producing transgenic plants for species within the host range. The third method involves delivery of DNA directly to the target cells by micropojectile bombardment. The advantage with this technique is that there are no host range limitations.

There are very few reports on transgenic plants from woody species [14,26,28] and several problems still have to be solved. First, a very good regeneration technique is required for the explants used. Secondly, the infectivity varies among different strains of *Agrobacterium* and could be specific to the species. A third problem is the efficacy of the selectable marker used for a particular plant species or genotype [10]. For example, regenerated shoots can grow in the presence of kanamycin without integration of the genes expressing neomycin phosphotransferase which confer resistance to kanamycin. Therefore a second selection for the ability to root in the presence of kanamycin might be necessary. In *Betula papyrifera* crown gall tumorigenesis, integration and expression of T-DNA encoded genes from wild-type strains of *Agrobacterium tumefaciens* have been demonstrated. However viable shoots produced from tumors were not transformed [23]. It has been shown for other species that transformed callus does not necessarily regenerate transformed shoots [18].

3. Micropropagation

3.1. *Summary of past and present work*

Table 1 shows that several *Betula* species have been micropropagated from both juvenile [21,24,29,41] and adult trees [5,13–15,33–35,40,42,44]. Several basal culture media and growth regulators have also been used during the different growth phases. Clonal propagation through axillary branching has been obtained from seedlings and young trees using shoot tips, apical and axillary buds as well as nodal segments as explant source [21,24,44]. Plantlet regeneration, through a callus phase, has been obtained from both juvenile and adult tissues using explants as roots and leaves from *in vitro* plants [41], internodes [11], nodes [29], peeled twigs [9], and anthers [12]. Adventitious bud formation directly on the explant has also been reported [34]. The initiation of cultures from adult material usually involves adventitious bud formation with or without callus. When buds are used to initiate the cultures the explants

Table 1. Summary of micropropagation procedures.

		Culture media. Basal media plus growth regulators mg/l					
Species	Age	Explants	Shoot-induction	Shoot-multiplication	Mode of regeneration	Rooting	Ref.
B.schmidtii	J	shoot tips ax.buds	WPM BAP (0.5) NAA (0.001) Myo (200)	WPM BAP (2)	ax.shoots	GD IBA (0.5)	[21]
B.costata	J	shoot tips ax.buds	WPM K (0.5)	BTM BAP (2)	ax.shoots	GD IBA (0.5)	[21]
B.davurica	J	shoot tips ax.buds	WPM BAP (2) NAA (0.01) Myo (100)	WPM BAP (2) NAA (0.01) Myo (100)	ax.shoots	GD IBA (0.5)	[21]
B.platyphylla, var.azechuanica	J	shoot tips nodal seg.	Mod.GD BAP (0.9)	Mod.GD BAP (0.9)	ax.shoots	In vivo	[24]
B.celtiberica	J	nodal seg.	MS BAP (0.6)	MS BAP (0.6) BAP (0.1)	ad+ax. shoots elongation	1/2MS IBA (0.2)	[29]
B.pendula	J	shoot buds	MS BAP (5) Ad (30) IAA (2)	MS BAP (5) Ad (30) IAA (2)	callus, plants	As shoot-multipli.	[41]
B.pendula	J	Roots	MS 2iP (5) Ad (30) NAA (2)	MS 2iP (5) Ad (30) NAA (0.5)	callus, plants	As shoot-multipli.	[41]
B.pendula	J	leaves	MS Z (2) Ad (30) IAA (2)	MS Z (5) Ad (30) IAA (2)	callus, plants	As shoot-multipli.	[41]

Table 1 (continued).

Species	Age	Culture media. Basal media plus growth regulators mg/l					
		Explants	Shoot-induction	Shoot-multiplication	Mode of regeneration	Rooting	Ref.
B.pendula early flow.	1	internodes	MS K (0.5) IAA (25)	MS K (0.5) IAA (25)	callus, ad.shoots	MS 2,4–D(0.1)	[11]
B.costata	3	peeled twigs	MS,WPM BAP (0.8–1.2)	MS,WPM BAP (1.2) IBA (0.02)	callus, adv.shoots	MS,WPM IBA(0.5) NAA(0.02)	[9]
B.pendula f.purpurea	?	leaves	Mod N6 K (0.5–1) 2,4–D(2–5) CH (1000)	Mod N6 Z,ZR (5–10) NAA,NOA (0.1–0.2)	callus, ad.shoots	Mod N6	[36]
B.platyphylla skatchev var. japonica Hara	A	leaf-petioles	Mod IS BAP (0.8)	Mod IS BAP (0.8)	ad.shoots	Mod MS NAA (0.02)	[34]
B.platyphylla skatchev var. japonica Hara	A	peeled twigs	Mod MS BAP (0.8–1.2)	Mod MS BAP (0.8–1.2)	callus, ad.shoots	Mod MS IBA (0.5) or NAA (0.02–0.2)	[35]
B.pendula	A	catkins	DWM K (5) IAA (2) CH (1000)	DBM K (6) Ad (40) IAA (2)	callus, plants	As shoot- multiplic.	[40]
B.pendula	23	apical- ax.buds	Mod N6 BAP (2) NAA (0.01) CH (1000)	Mod N6 BAP (2) NAA (0.01) CH (1000)	ad.shoot, ax.shoots	Mod N6 NAA (0.1)	[42]
B.pendula (verrucosa EHRH)	2–20	shoot tips, internodal segments	Mod MS BAP (0.6) IBA (0.05)	Mod MS BAP (0.2–0.3)	ad.shoots	1/2Mod MS IBA or NAA (0.1)	[5]

Table 1 (continued).

Species	Age	Culture media. Basal media plus growth regulators mg/l Explants	Shoot-induction	Shoot-multiplication	Mode of regeneration	Rooting	Ref.
B.grossa S et Z.	12	dormant buds	Mod MS BAP (0.4–1.2) GA_3 (0.5)	Mod IS NAA (0–0.2)	ax.shoots f.nod.seg	As shoot-multipli.	[13]
B.pendula var.carelica	20–50	apical-ax.buds	Mod N6,MS BAP (1)	MS BAP (10) NAA (0.2)	Callus, ad.shoots	1/2 MS	[33]
				1/2 MS BAP (0.5) IAA (0.5)	elongation		
B.pendula *B.costata* *B.jacquemontii* *B.albosinensis* *B.ermanii*	15–45	buds growing and dormant	MOD M6 BAP (0.5–1) NAA (0,0.001)	WPM BAP (0.5) NAA (0.001)	ad.,ax shoots	1/5 WPM IBA (0.1)	[15]
B.pendula	1–20	buds growing and dormant, shoot tips	MOD M6 BAP (1) NAA (0.001)	WPM BAP (0.5) NAA (0.001)	ad.,ax shoots	1/5 WPM IBA (0.1)	[44]
B.pendula	A	anthers	MS K (0.1–1) 2,4–D (1–10) CH (1000)	MS BAP (0.1–1) NAA (1–5)	callus, ad.shoots	–	[12]

Abbreviations: BTM, broad-leaved trees medium [21], DWM, defined White's medium [40] DBM, defined basal medium [40] Mod GD, modified Gresshof and Doy [24] Mod IS, Saito and Ide [34], MS, Murashige and Skoog, 1962 [27] Mod N6, macronutrients of Chu *et al.*, 1975 [7], and micronutrients and vitamins of MS., WPM, Lloyd and McCown, 1980 [22].

Ad, adenine (6-aminopurine); BAP, ^{6}N-benzylaminpurine; CH, casein hydrolysate; 2.4–D, 2.4-dichlorophenoxyacetic acid; IBA, indole-3-butyric acid; IAA, indole-3-acetic acid 2iP, 6-(3-methyl-2-butenyl-amino)purine; K, kinetin (6-furfurylaminopurine); Myo, myoinositol; NAA, α-naphthaleneacetic acid; **Z**, zeatin (6-(4-hydroxy-3-methyl-*trans*-2-enyl)aminopurine).

usually develop into a shoot from pre-formed organs. From the base of the shoot, or the leaf margins in contact with the medium, adventitious shoots are induced within 5–10 weeks and the original explant deteriorates and turns brown [15,44]. There is only one report on true axillary shoot formation from the initial explant of adult material [13]. In this investigation GA_3 was necessary for shoot development. The induction of axillary shoots usually requires low concentrations of cytokinins and auxins, whereas high concentrations are needed for induction of adventitious buds from callus (Table 1). When low concentrations of growth regulators are used the callus formation can be reduced or avoided, thus minimizing the risks of chromosomal deviations and aberrant plants. Most *Betula* species form adventitious buds quite easily and this is probably why adult material can be propagated without major difficulties in tissue culture. The time required to initiate the cultures varies with the type of explant [41], the age of the mother tree [24,34], the species [15] and the clones of the same species (Welander, unpublished results). Normally 4–10 weeks are required for bud induction. Although adventitious buds are formed on the initial explant, further shoot multiplication can always be obtained from axillary shoots by using shoot tips and nodal segments. The number of axillary shoots per shoot tip range from 3–8 [42,44]. However, in most of the reports, there is no distinction between axillary and adventitious shoots and a multiplication rate based only on axillary shoots can not be given. The rooting percentage for *in vitro* produced shoots is very high, ranging from 80–100%. Low salt concentration in the medium, as well as addition of low auxin concentrations (0.02–0.5mg/l), improves rooting [5,15,29,33,44]. Roots usually appear after 7–10 days on the rooting media. Shoots can be rooted both *in vivo* and *in vitro* but for commercial purposes *in vivo* rooting is recommended because exclusion of a rooting medium saves time and expense [25], (Kemira, personal communication). Acclimatization of *in vitro* plants is easy and if the quality of the transplants is maintained 90–100% survival is obtained. The micropropagation system for birch seem to be quite reliable concerning the clonal fidelity of different genotypes. Phenotypically uniform plants from organ culture as well as callus have been reported for *B. pendula* and *B. pendula f. purpurea* [3,6,36]. Three thousand birch plants from 200 clones have been produced and no aberrant plants have been observed so far (Welander unpublished results). In Finland thousands of birch plants have also been produced commercially using a similar technique and only one aberrant phenotype was observed [17].

3.2. *Factors affecting culture initiation*

3.2.1. *Climate conditions*

In *Betula pendula* the initial growth and further development of the explants from both juvenile and adult material were affected by several factors [40].

Table 2. Development of buds at different positions from LD- and SD-treated trees.

Stock plant treatment	Bud position	Number of excised buds	Days in culture 7 Number of green buds	14 Number of open buds	21 Number of buds developed to a shoot	21 Number of shoots with adventitious buds
LD	A	40	5	5	2	1
	B	40	15	11	7	0
	C	35	13	10	6	3
	D	38	35	34	32	5
SD	B	40	35	30	26	23
	C	40	33	26	20	17
	D	40	39	39	39	28

The values are based on two separate experiments. For explanation of A–D, see legend to Figure 1. Data are average values from duplicate experiments. From [44] with permission.

Figure 1. Bud positions on LD- and SD-treated trees. (A) Elongated terminal and lateral shoot tips. (B) Buds from middle branches. (C) Buds from upper stem (0–30 cm). (D) Buds from lower stem (0–30 cm). From [44] with permission.

In juvenile trees both stockplant treatment and bud position influenced culture initiation. Trees exposed to short day (SD) conditions, which induces dormancy in birch (Figure 1) resulted in the most suitable explants for *in vitro* growth (Table 2). In actively growing trees long day (LD) conditions there was a pronounced gradient in morphogenetic response in respect to *in vitro* growth due to bud position which was not observed in SD-treated trees (Figure 2). During LD-conditions buds from the lower part of the stem were superior to buds at other positions. Regarding adult material, best results were obtained with elongated shoot tips (Figure 3) from twigs forced to flush in a climate chamber. Dormant buds were less useful because they were more difficult to sterilize and at this stage vegetative and generative buds could not be distinguished. Prolonged cold storage (>4 Months) resulted in reduced bud break after flushing and increased infection rate.

3.2.2. *Phenolic content*

A serious problem encountered in tissue culture of woody species is the phenolic compounds exudated upon excision. Welander [44] showed that the initial phenolic exudation depended on age, bud position, type of explant and stock plant treatment. In adult material elongated shoot tips exudated less phenolics

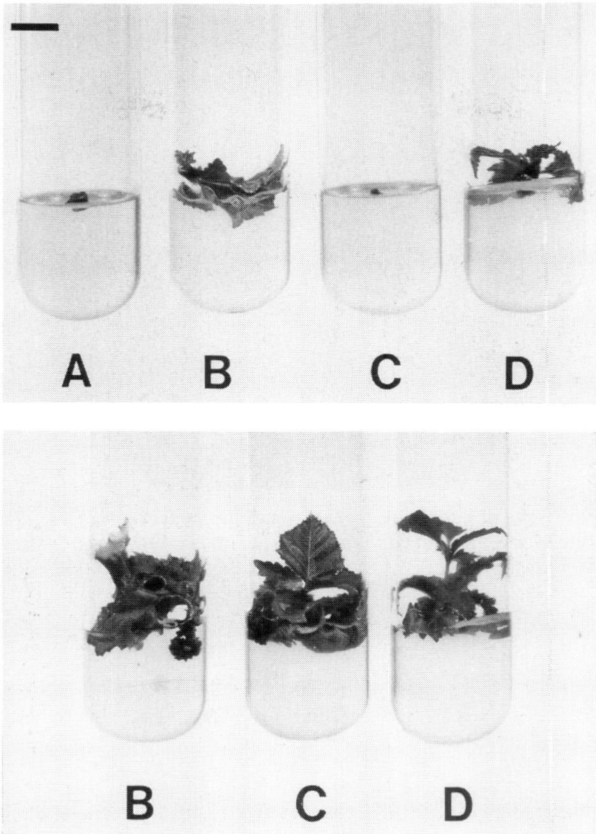

Figure 2. Development of buds *in vitro* that were excised from different positions on LD- and SD-treated trees. The cultures are five weeks old. See legend to Figure 1 for description of bud positions corresponding to A, B, C, and D. Scale bar = 10 mm. From [44] with permission.

compared to dormant buds. As shown in Figure 4 buds from LD-treated trees contained a much greater amount of phenolic compounds compared with buds from SD-treated trees. It was also shown that buds exhibiting good growth *in vitro* had a similar pattern of phenolic compounds, as shown by TLC (Table 3). Not all phenolics are inhibitory. Extracts of dormant buds from adult trees which were then tested on bud growth *in vitro* showed that the most inhibitory substances were present in the ethyl acetate phase whereas some promoting substances were present in the chloroform phase (Figure 5).

3.3. *Anatomical and metabolic status*

Anatomical studies revealed that buds from SD-treated trees contained a

Figure 3. Different types of bud development from adult material of *Betula pendula* (Roth) after flushing (1) Female catkin; (2) elongated shoot tip; (3) short shoot. From [44] with permission.

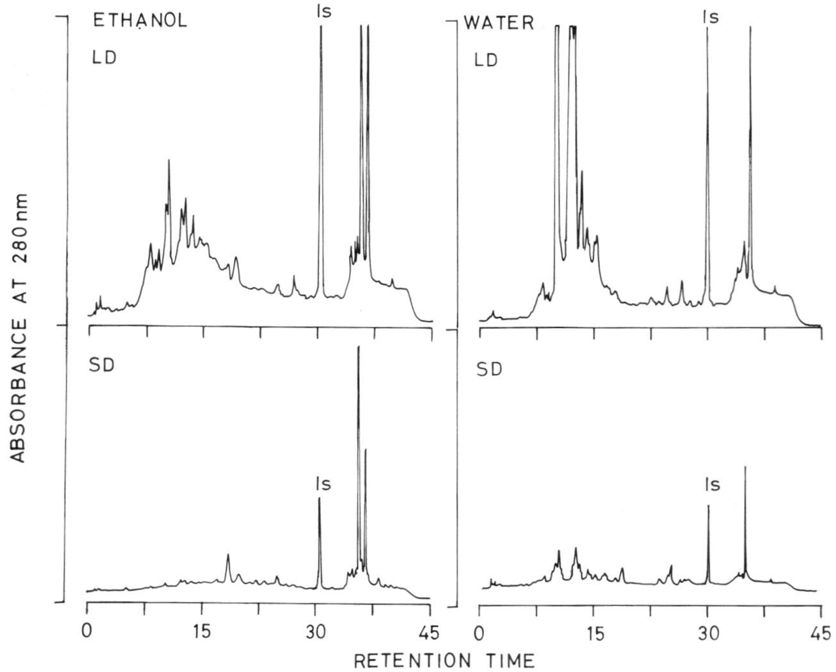

Figure 4. Separation by HPLC of phenolic compounds in ethanol and water extracts of B buds from LD- and SD- treated trees. Is, Internal standard (2,4-dichlorophenol). From [44] with permission.

Table 3. Thin-layer chromatography separation of ethanol extracts from buds at different positions from LD- and SD-treated trees.

Stock plant treatment	Bud position	Phenolic compounds			
		a	b	c	d
LD	B	XXX	t	X	XX
	C	XXX	t	X	XX
	D	XX	t	XXX	t
SD	B	X	XX	XXX	t
	C	X	X	XXX	t
	D	X	t	XXX	t

Note: Spots were detected by spraying with diazotized sulpanilic acid plus sodium carbamate, and after spraying with sulfuric acid. Thin-layer chromatography was performed on Merck's silica gel plates with fluorescent indicator and solvent system (chloroform:methanol:water). For explanation of B–D, see legend to Figure 1. Score: t (trace) to XXX (highly stained).

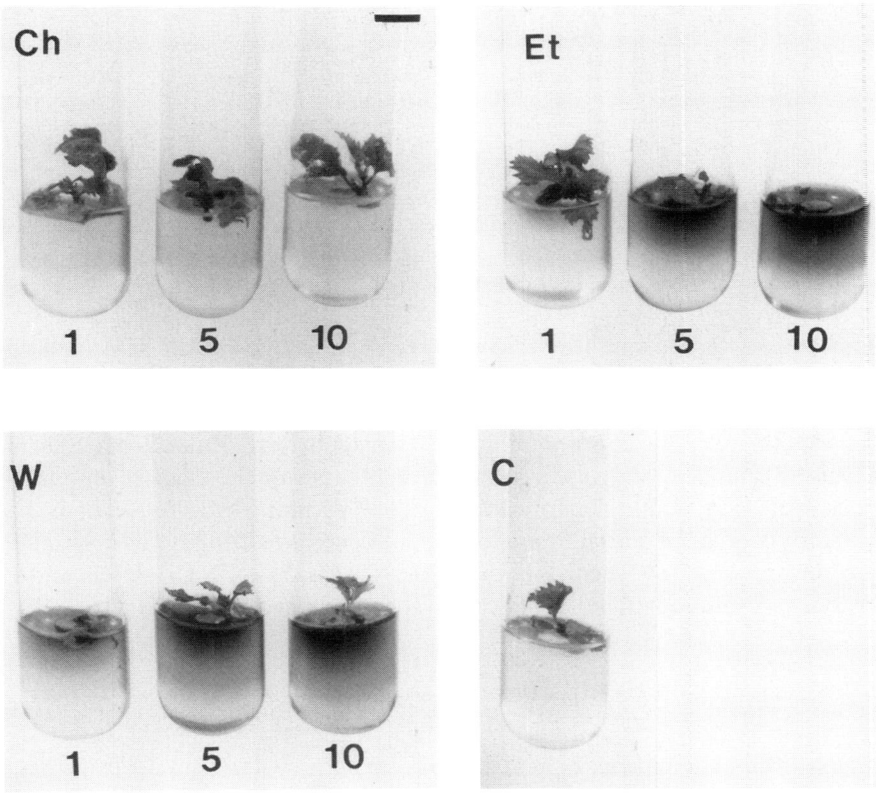

Figure 5. The effect of chloroform (CH), ethyl acetate (Et), and water (W) extracts on growth and development of buds *in vitro*. The extracts were obtained from buds of adult material in dormant stage. The buds tested consisted of buds from adult material after flushing and established one week on the initiation medium before being subjected to the different extracts. The amounts of extract added correspond to one, five, and ten buds. C, Control. From [42] with permission.

Table 4. Contents of glucose, sucrose, starch, and nitrogen in buds at different positions from LD- and SD-treated trees.

Stock plant treatment	Bud position	Glucose (mg/g FW)	Sucrose (mg/g FW)	Starch (mg/g FW)	Nitrogen (mg/g FW)
LD	A	4.9±0.6	1.5±1.7	NA[a]	NA
	B	9.5±1.7	1.5±1.1	0.0±0.0	15.9±2.1
	C	8.1±0.4	1.9±1.1	0.0±0.3	14.7±3.3
	D	8.7±2.2	3.1±1.8	13.2±2.4	14.8±0.6
SD	B	13.7±4.9	4.4±1.9	11.7±3.1	26.6±1.6
	C	10.2±2.8	3.1±2.0	14.5±2.7	27.4±1.8
	D	10.3±1.1	3.3±1.1	11.8±3.0	21.8±1.9

Note: For explanation of A–D, see legend to Figure 1. Data, given as mean ± S.D., are average values from duplicate experiments. FW, Fresh weight; DW, dry weight.
[a] Not analyzed.

Table 5. Contents of glucose, sucrose, starch, and nitrogen in buds excised on different occasions from adult material.

Type of bud development	Glucose (mg/g FW)	Sucrose (mg/g FW)	Starch (mg/g FW)	Nitrogen (mg/g DW)
Buds from dormant stage	10.7±0.8	2.94±1.6	5.80±0.7	12.9±2.2
Buds after flushing	6.2±0.3	0.03±0.04	0.00±0.0	17.5±1.4
Shoot tips after flushing	5.4±0.2	0.00±0.00	0.00±0.0	32.3±2.2

Note: For explanation of type of bud development, see legend to Figure 3. Data, given as mean ± S.D., are average values from duplicate experiments.

substantially larger number of leaves compared to buds at the same position on LD-treated trees [40]. This might be one reason for the more rapid development *in vitro* of buds from SD-treated trees. Biochemical analyses showed that the most pronounced differences between buds from SD and LD were found in starch and nitrogen content with the highest amount in buds from SD (Table 4). In adult material when the vegetative buds had burst no content of starch or sucrose could be detected in the buds. However the nitrogen content increased three-fold (Table 5). A high starch and/or nitrogen content might be an indication for rapid development *in vitro*.

3.4. *Sterilization*

Stem or branch segments 2–3 cm long can be surface sterilized in 70% ethanol for 1 min followed by 15 min in 7% calcium hypochlorite containing 0.1% Tween 20. Shoot tips do not require ethanol treatment and only 10 min of calcium hypochlorite. The explants were then rinsed three times in sterile distilled water. In order to avoid browning of the tissue and the culture medium, the segments were placed in a solution of 0.1 mM *L*-cystein and the buds excised under a dissecting microscope. The bud scales were removed and the buds placed in a fresh *L*-cystein solution for 24h or 3h in sterile distilled water. Dormant buds from adult material were much more difficult to sterilize and required an additional sterilization in 7% calcium hypochlorite for 3 min after the bud scales had been removed.

3.5. *Culture initiation medium*

In our experiments the most suitable medium for culture initiation was a modified N6 medium containing macroelements according to Chu *et al.* [7], microelements and vitamins based on Murashige and Skoog's medium [27], 10 mM Fe-EDTA, 2% sucrose, 4.4 μM 6-benzylaminopurine (BAP), 0.005μM α-naphthaleneacetic acid (NAA) and 0.6% Difco Bacto agar. The pH was adjusted to 5.5 and the medium autoclaved for 15 min at 120 °C. The buds were cultured in 25 mm × 150 mm test tubes with 10 ml medium and sealed with plastic caps. After three weeks buds were transferred to fresh medium and cultured for an additional three weeks. The cultures were maintained at a 16h photoperiod with 42 $\mu E\ m^{-2}\ s^{-1}$ from cool-white fluorescent lamps a 24/22°C. During culture initiation most of the buds from juvenile material develop into a shoot and form both axillary and adventitious shoots. However buds from adult material usually deteriorate at the end of the culture period and new shoot formation has to be based on adventitious buds formed at the base of the explant or leaves in contact with the medium (Figure 7). We have found large clonal differences between species regarding the time needed as well as the number of explants forming adventitious buds [15].

3.6. *Shoot multiplication*

In our experiments the best medium for shoot multiplication was the WPM medium [21] used with 2.2 μM BAP, 0.005 μM NAA, 2% sucrose and 0.6% Difco Bacto agar at pH 5.5. Shoots produced on WPM medium were thin and had small leaf lamina, wheras on N6 medium the shoots were more compact with short internodes and large leaf lamina (Figure 6A). The shoot multiplication took place in 200 ml glass jars, containing 40 ml of medium, sealed with plastic caps. The material was subcultured every fourth week. The culture conditions were as for culture initiation. Clonal propagation by axillary branching is favourable because of the genetic stability of the shoot apex under culture conditions resulting in offspring true to type. For adult material of birch we have found that it is not possible to base a rapid micropropagation technique initially on axillary shoots only. However when low concentrations of cytokinins and auxins are used, adventitious shoots can be induced directly on the explant (Figure 7) thus avoiding a callus phase and the risks for

Figure 6. The effect of two basal media consisting of mod. N6 and WPM on A: Shoot morphology and B: root morphology of *Betula pendula* (Roth). A: Few shoots with short internodes and large leaf lamina developed on mod N6 medium (a) and many thin shoots with small leaf lamina on WPM medium (b). B: Roots on mod N6 medium (a) were thick and short whereas roots developed on WPM medium (b) were long, thin and with several lateral roots. From [15] with permission.

chromosomal deviations and aberrant plants. Further multiplication can be based on shoot tips and nodal segments giving a satisfactory multiplication rate of 3–8 new shoots per shoot tip [42,44].

3.7. *Rooting*

Rooting of birch microcuttings is generally easy both *in vitro* and directly under non-sterile conditions. It is important that shoots used for rooting are in good growing condition. Shoots used for *in vitro* rooting should exceed 1cm and 2cm for *in vivo* rooting in order to abtain high survival rate during acclimatization. A rooting medium based on WPM medium with macronu-

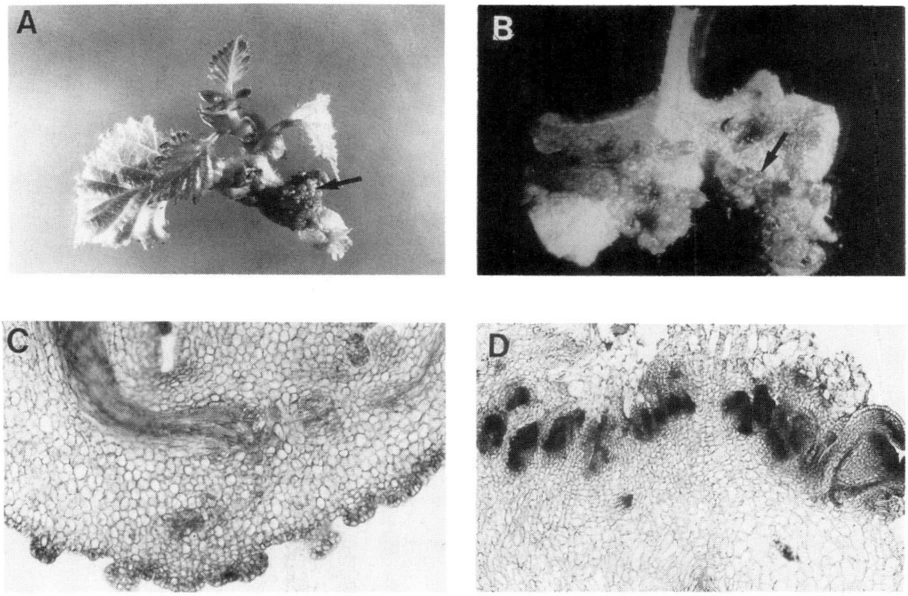

Figure 7. Formation of adventitious buds without intervening callus from A: Stem base of a shoot B: Leaf explant bud explant after 4 days in culture D: adventitious buds and meristems after 15 days in culture. From [44] with permission.

trients reduced to one fifth plus 0.1 mg l^{-1} indole-3-butyric acid (IBA) usually gives 95–100% rooting. When comparing the WPM medium with the N6 medium we observed big differences in root morphology (Figure 6B). On N6 medium the roots became short and thick whereas roots produced on WPM medium were long, thin and with several laterals.

3.8. *Acclimatization*

After about 10 days on the rooting medium plantlets were washed to remove the agar and transferred to a mixture of peat and perlite 1 : 1. The plantlets were covered with plastic beakers or clear plastic sheet to maintain a high relative humidity and placed in a climate chamber at 20 °C, 15 h photoperiod with a light intensity of 35 $\mu E\ m^{-2}\ s^{-1}$ from cool-white fluorescent tubes. After 1–2 weeks the humidity was gradually reduced to normal values. With this treatment about 100% survival was obtained.

3.9. *Field performance*

Field performance of micropropagated plants is an essential part of the propagation system. This information is extremely important for forest trees

Table 6. Shoot multiplication rate during four subcultures from buds at different positions from LD- and SD-treated trees.

Stock-plant treatment	Bud position	Subculture 1		2		3		4	
		Ax	Ad	Ax	Ad	Ax	Ad	Ax	Ad
		Number of shoots per shoot tip							
LD	A	1.0±0.0	–	1.0±0.0	Nodules	Dead	3.0±0.0	Dead	3.0±0.8
	B	1.0±0.0	–	1.0±0.0	Nodules	Dead	2.8±1.0	Dead	2.5±1.0
	C	1.4±0.5	–	2.5±0.5	Nodules	Dead	Dead	Dead	Dead
	D	2.0±0.9	Nodules	1.8±0.7	3.8±2.9	1.6±0.5	3.5±3.0	1.6±0.9	2.7±1.6
SD	B	1.4±0.9	Nodules	1.4±0.9	5.7±1.5	1.3±0.9	4.2±1.4	1.2±0.5	3.3±0.9
	C	1.0±0.0	Nodules	1.0±0.0	4.5±2.0	Dead	5.1±2.0	Dead	3.8±1.3
	D	2.2±0.2	Nodules	2.3±0.3	5.6±2.2	2.4±1.1	3.5±1.1	1.3±0.5	2.9±0.6

Note: Shoots from axillary (Ax) and adventitious (Ad) origin are kept separated. For explanation of A–D, see legend to Figure 1. Data, given as mean ± SD, are average values from duplicate experiments.

Table 7. Field experiments with micropropagated plants and seedlings of *Betula pendula* (Roth). Ad = plants with adventitious origin, Ax = plants with axillary origin, S = seedlings, – = not recorded.

Recording date	Origin of the plants								
	Ax	Ad	S	Ax	Ad	S	Ax	Ad	S
	Heigth cm ± SE			No of branches ± SE			Stem diameter cm ± SE		
870925	34.0±0.99	38.9±0.96	51.7±0.78	2.5±0.23	1.93±0.19	2.89±0.24	–	–	–
880509	46.7±2.09	53.9±1.7	96.7±2.1	2.5±0.49	11.10±0.61	19.50±0.43	–	–	–
890509	103.3±3.23	112.2±2.9	174.9±3.16	19.5±0.63	17.40±0.75	26.70±0.53	8.3±0.42	9.08±–0.40	18.3±0.48

because of the investment required for establishment of a plantation and the length of time needed before tree charateristics such as growth rate, form, wood quality, insect and disease resistance can be evaluated. Micropropagated trees should be phenotypically identical and genetically stable. Organ cultures are generally genetically stable whereas callus cultures are regarded as more unstable, with increasing risks of genetic abberations resulting in new genotypes. This is especially true for prolonged callus cultures. As pointed out earlier, adult birch cannot be propagated *in vitro* without rejuvenation either by mother plant treatment or induction of adventitious shoots. How this will affect trees grown in the field is not known. In order to study the genetic stability of micropropagated birch trees (*Betula pendula* Roth.) in comparison with seedlings a coordinated international research group consisting of 10 European countries within the COST-87 program of the EEC was initiated, which allows field trials at various places. Shoot cultures were initiated from a 60 year old tree and half-sib family seeds of the same origin were utilized. Field performance was undertaken with 100 plants originating from prolonged callus cultures, 100 plants from shoot tips and nodal segments and 100 seedlings. In the Swedish experiments the plants were planted in the field in May 1988. Planting distances were 2.5×2 m consisting of 10 divisions with 3×10 plants of each origin. Plants of adventitious and axillary origin were smaller than seedlings at planting date (Table 7) and these differences persisted the follwoing year. The number of surviving plants was also correlated with the initial height. The height of the plants at planting date is probably very important and it might be a minimum size of the plants in order to get a good start. In the English experiment all plants, irrespective of origin, had an initial height over 130 cm at planting date in the field and in these experiments there were no differences in height between plants of different origin. However, as in the Swedish experiments, the girth was somewhat larger for the seedlings. The most promising results were the uniformity amongst the micropropagated plants both from adventitious and axillary origin. In a separate experiment, where micropropagated plants from seedlings was compared with seed propagated plants, the micropropagated plants grew faster and were much more uniform than the seedlings (Figure 8). Studies on birch species hybrids have shown that most of the changes in growth rate occur within the first 10 years [16] indicating that this might be the time required before the production capacity of micropropagated plants can be evaluated.

4. Summary

Betula species have been micropropagated from both juvenile and adult trees. Propagation through axillary branching has been obtained from seedlings and young trees using shoot tips, apical and axillary buds and nodal segments as explant source. Culture initiation from adult material usually involves adventitious bud formation with or without callus. Buds and shoots tips can

Figure 8. Field experiments with micropropagated plants from a seedling (A) and seed propagated plants (B) of *Betula pendula* (Roth). The trees were planted 861008.

regenerate adventitious buds from developing leaves and shoot base in contact with the medium without any intervening callus formation. The original explant usually deteriorates and turns brown. The time required for initiating the cultures varies with the type of explant, the age and physiological condition of the mother tree, the species and the clone within the same species. Normally, 4–10 weeks are required for bud induction. Although adventitious buds are formed on the initial explant, further shoot multiplication can be obtained by axillary shoots, using shoots tips and nodal segments thus minimizing the risks of aberrant plants. Induction of axillary shoots usually requires low concentrations of cytokinins and auxins, whereas high concentrations are needed for adventitious bud formation from callus. The rooting percentage for *in vitro* produced shoots is very high, ranging from 80–100%. Low salt concentration and low auxin concentration in the medium improve rooting. Shoots can also be rooted *in vivo*, and this is recommended for commercial purposes. The survival during acclimatization of the plants is 90–110%, if the quality of the transplants is maintained. The micropropagation system for birch seems to be quite reliable as to the clonal fidelity of different genotypes. Phenotypically uniform plants from both axillary and adventitious shoots have been reported.

References

1. Alden T, Dormling I, Ehrenberg C, Kellerstam H, and Persson S (1977) Some methods for vegetative propagation. In: Vegetative Propagation of Forest Trees-Physiology and Practice. Lectures from a Symposium in Uppsala, Sweden Feb 16–17. pp. 137–147.
2. Bonga JM, von Aderkas P, and James D (1988) Potential application of haploid cultures of tree species. In: Hanover JW and Keathley DE (Eds.) Genetic manipulation of Woody Plants. Plenum Press New York, London. pp. 57–77.
3. Bonga JM (1982) Vegetative propagation in relation to juvenility, maturity and rejuvenation. In: Bonga JM and Durzan DJ (Eds.) Martinus Nijhoff/Dr. W. Junk Publishers, The Hague. pp. 387–412.
4. Butt AD (1985) A general method for the high-yield isolation of mesophyll protoplasts from deciduous tree species. Plant Science 42:55–59.
5. Chalupa V (1981) *In vitro* propagation of birch (*Betula verrucosa* EHRH). Biolodia Plantarum (PRAHA) 23:472–474.
6. Chalupa V (1987) European hardwoods. In: Bonga JM and Durzan DJ (Eds.) Cell and Tissue Culture in Forestry Vol. 3. Martinus Nijhoff Pub. Dordrecht/Boston/Lancaster. pp. 224–246.
7. Chu CC, Wang, CC, Sun CS, Hsii C, Yin KC, Chu CY, and Bi FY (1975) Establishment of an efficient medium for anther culture of rice through comparative experiments on the nitrogen sources. Sci Sin 18:659–668.
8. Compton ME and Preece JE (1986) Exudation and explant establishement. IAPTC Newsletter No. 50, 9–18.
9. Hong HS, Shim SY, Park HS, Kwon OW, and Lee SJ (1986) *In vitro* plantlet regeneration from adventitious buds induced on cuttings of peeled twigs of *Betula costata* traut. Res Rep Inst For Gen Korea 22:35–39.
10. Horsch R, Fraley R, Rogers S, Fry J, Klee H, Shah D, McCormick S, Niedermayer J, and Hoffman N (1987) Agrobacterium-mediated transformation of plants. In: Green CE,

Sommers DA, Hackett WP, Biesboer DD (Eds.) Plant Tissue and Cell Culture. Alan R. Liss, Inc., New York. pp. 317–329.

11. Huthinen VO and Yahyaoglu Z (1974) Das frühe Blühen von aus Kalluskulturen herangezogenen Pflänzchen bei der Birke (*Betula pendula* Roth.). Silv Genet 23:1–3.
12. Huthinen VO (1978) Callus and plantlet regeneration from anther culture of *Betula pendula* Roth. In: Fourth Int Cong Plant Tissue and Cell Culture, Calgary, Canada. Abstract No. 1740, p. 169.
13. Ide Y (1987) *In vitro* clonal propagation of mature japanese cherry birch. J Jpn For Soc 69:161–16.
14. James DJ, Passey AJ, Barbara DJ, and Bevan M (1989) Genetic transformation of apple (*Malus pumila* Mill.) using a disarmed Ti-binary vector. Plant Cell Rep 7:658–661.
15. Jansson E and Welander M (1990) Micropropagation of some adult *Betula* species. Report 55, Swedish Univ of Agric Sci Dep of Hort Sci Box 55 S-230 53 Alnarp, Sweden.
16. Johnsson H (1966) Avkommeprövning av björk. (Progeny testing of birch.) Föreningen Skogsträdsförädling, årsbok.
17. Jokinen K, Tormola T, and Virta U (1989) Clonal fidelity of mass propagated silver birch (*Betula pendula* Roth). IUFRO somatic cell genetics working party S2-04-07, October 15–19, Placerville, California.
18. Jordan MC and McHughen A (1988) Transformed callus does not necessarily regenerate transformed shoots. Plant Cell Rep 7:285–287.
19. Karnosky DF (1981) Potential for forest tree improvement via tissue culture. BioScience 31(2):114–119.
20. Krüssman G (1976) Handbuch der Laubgehölze. Verlag Paul Parey.
21. Lee BC, Kim JH, Park JI, and Lee SK (1986) Rapid micropropagation of *Betula* spp. through *in vitro* tissue culture. Res Rep Inst For Gen Korea 22:132–138.
22. Lloyd G and McCown B (1980) Commercially feasible micropropagation of mountain laurel, *Kalmia latifolia* by use of shoot tip culture. Int Plant Prop Combined Proc 30:421–426.
23. Mackay J, Séguin A, and Lalonde M (1988) Genetic transformation of 9 *in vitro* clones of *Alnus* and *Betula* by *Agrobacterium tumefaciens*. Plant Cell Rep 7:229–232.
24. McCown B and Amos R (1979) Initial trials with commercial micropropagation of birch selections. Int Plant Prop Soc 29:387–393.
25. McCown DD and McCown BH (1987) North american hardwoods. In: Bonga JM and Durzan DJ (Eds.) Cell and Tissue Culture in Forestry Vol. 3. Martinus Nijhoff Pub. Dordrecht/Boston/Lancaster. pp. 247–260.
26. McGranahan GH, Leslie CA, Uratsu SL, Martin LA, and Dandekar AM (1988) Agrobacterium-mediated transformation of walnut somatic embryos and regeneration of transgenic plants. Biotech 6:800–804.
27. Murashige T and Skoog F (1962) A revised medium for rapid growth and bioassays with tobacco tissue cultures. Physiol Plant 15:473–497.
28. Parsons TJ, Sinkar VP, Stettler RF, Nester EW, and Gordon MP (1986) Transformation of poplar by Agrobacterium tumefaciens. Biotech 4:533–536.
29. Perez C and Postigo P (1989) Micropropagation of *Betula celtiberica*. Ann Bot 64:67–69.
30. Redenbaugh K, Viss P, Slade D, and Fujii JA (1987) Scale-up. Artificial Seeds pp. 473–493.
31. Rediske J (1978) Vegetative propagation in forestry. Proc 26th North-Eastern Forest Tree Improvement Conf State College, PA. pp. 1–12.
32. Regel E (1861) Monographia Betulacearum huiusque congitarum. Mém Soc Natur Moscow 13.
33. Ryynänen L and Ryynänen M (1986) Propagation of adult curly-birch succeds with tissue culture. Silva Fennica 20:139–147.
34. Saito A and Ide Y (1985a) *In vitro* plantlet regeneration from adventitious buds induced by petiole culture in Japanese white birch. J Jap For Soc 67:373–375.
35. Saito A and Ide Y (1985b) *In vitro* plantlet regeneration from adventitious buds induced on cuttings of peeled twigs of japanese white birch. J Jap For Soc 67:282–284.

36. Simola LK (1985) Propagation of plantlets from leaf callus of *Betula pendula* f. purpurea. Scien Hortic 26:77–85.
37. Smith MAL and McCown BH (1983) A comparision of source tissue for protoplast isolation from three woody plant species. Plant Sci Lett 28:149–156.
38. Smith MAL (1984) Application of microculture and protoplast techniques to physiological analysis of woody crops. PhD Thesis, Univ of Wisconsin, Madison.
39. Spurr SH (1979) Silviculture. Sci Am 240:76–91.
40. Srivastava PS and Steinhauer A (1981) Regeneration of birch plants from catkin tissue cultures. Plant Sci Lett 22:379–386.
41. Srivastava PS, Steinhauer A, and Glock H (1985) Plantlet regeneration in leaf and root cultures of birch (*Betula pendula* Roth.). Plant Sci 42:209–214.
42. Särkilahti E (1988) Micropropagation of mature colchicine-polyploid and irradiation-mutant of *Betula pendula* Roth. Tree Physiol 4:173–179.
43. Tulecke W (1987) Somatic embryogenesis in woody perennials. In: Bonga JM and Durzan DJ (Eds.) Cell and Tissue Culture in Forestry, Vol. 2. Martinus Nijhoff Pub. Dordrecht/Boston/Lancaster. pp. 61–92.
44. Welander M (1988) Biochemical and anatomical studies of birch (*Betual pendula* Roth) buds exposed to different climatic conditions in relation to growth *in vitro*. In: Hanover JW and Keathley DE (Eds.) Genetic Manipulation of Woody Plants. Plenum Publishing Corporation. pp. 79–99.

15. Micropropagation of teak

ANTHONY F. MASCARENHAS*, SHUCHISHWETA V. KENDURKAR and SURYAKANT S. KHUSPE

Biochemical Sciences Division, National Chemical Laboratory, Pune 411 008, India

1. Distribution and biology

Teak (*Tectona grandis L.f.*) (family Verbenaceae) is one of the exotic, hardwood timber trees. Teak is indigenous to South East Asia (India, Burma, Thailand and Indonesia). Under favourable conditions the tree attains a large size with a tall, clean and cylindrical bole.

Teak has adapted well to moist, dry and semimoist sites as defined by Seth and Khan 1958 [84], but it flourishes well in a fairly moist, warm, tropical climate [5]. In the moist parts of the west coast, where it reaches larger dimensions than in the drier parts of the Indian peninsula, the absolute maximum shade temperature is 35–38° C and the absolute minimum shade temperature is 13–17° C. Normal annual rainfall for its best development varies from 1240 mm to 3750 mm [88]. It can withstand drought very well but it is frost sensitive [65]. Growth of teak depends largely on the depth, drainage, moisture and fertility of the soil [5]. The tree is a pronounced light demander, which affects its growth, branching and flowering [72].

Teak is an open pollinated species, self incompatibility being 96–100% [45,46]. The period between pollination and fertilization is very short but seed maturity takes a longer period (6–7 months) after the flowering period. Some observations indicate poor fruit formation, emptiness of teak fruits and failure of viable seed formation [9,21,35].

The tree is affected by numerous diseases and pests. More than 50 insect pests attack the living trees, the most devastating being the leaf defoliator, *Hyblea purea* and the leaf skeletonizer *Haplia machaerates* [5].

2. Economic importance

Teak wood has always enjoyed the reputation of being a high class timber. It combines qualities like strength, durability and resistance to termites [65]. It is used for bridge construction, marine construction, furniture making and other construction works. Various uses of teak wood have been described

Communication No: 4755. * To whom all correspondence should be addressed.

M. R. Ahuja (Ed.), Micropropagation of Woody Plants, 247–262.

[5,61]. The quality of teak woods differs with location and the woods fetch high prices for their colour (rose, golden yellow), texture, decorative grains, and ornamental figuring [9].

3. Conventional methods of propagation

Natural regeneration of teak is largely by seeds. Large scale propagation of teak for various afforestation programmes is mainly achieved by seeds. There is a considerable variation in seed numbers per fruit and in seed germination [9,19,24,35,48]. Germination studies carried out by the Indo-Danish project, Hyderabad, India on 36 sources of teak seeds confirmed earlier findings [41], that dormancy in teak seeds is of three main types, viz.:

(a) presence of water soluble germination inhibitor in the outer spongy coat of the teak fruit;
(b) a nutrient imbalance in the seeds;
(c) a physiological dormancy known as 'after ripening'.

Because of the hard coats several pretreatments have been used for weathering of teak fruits, the most common methods being alternate soaking in water and drying for a month before sowing, or soaking in Sachs nutrient solution for 24 h and drying for 24 h alternately for a month before sowing [9]. Mechanical extraction of seeds has also been found to improve germination [19]. Other techniques involve scorching of seeds in a light fire or hot sand or subjecting them to acid, alkali and bacterial action. Recently, use of the fungus belonging to the species *Scytalidium* has proved successful in the rapid decomposition of the epicarp from the fruits within 21 days. These extracted seeds exhibit a higher percentage of germination as compared to other treatments [22].

Seeds suffering from 'after ripening' do not respond to any type of pretreatments and do not germinate immediately after collection. These seeds require storage for 1–2 years after collection [9].

Teak plantations are raised either by directly using the nursery grown seedlings or by stumps. For easy transportation and storage, stumps are prepared from 1–2 year old nursery grown seedlings having good tap roots [53,76].

The importance of vegetative propagation for clonal seed orchards has long been realised and its feasibility has been proved for the large scale establishment of seed orchards [47,49]. Grafting and budding are the most widely practiced methods in India and other teak growing countries.

The 'Forkert budding' method on 1 year old root stocks [15,51], 'patch budding' on 2–3 year old field grown stocks [79], and stumps [59] have been applied with varying degrees of success.

'Cleft grafting' [79] and 'root grafting' [20] have also been practiced. In 'root grafting' a flowering shoot is grafted on to the root. This ensures flowering within 1 or 2 years after grafting. Rooting of cuttings [54] and rooting of buds cut from stock stumps raised in polypots [60] have also been obtained. These methods have been described in detail [64].

4. Role of tissue culture in teak

The need for selection and breeding in teak for improvement of its genetic pedigree has been recognised for many years [50]. Due to its wide geographic distribution, outbreeding habit and seed propagation teak shows a high degree of inter- and intra-population variation [9] such as:

1 ecological variation;
2 variation in bole form and timber characteristics;
3 morphological variation;
4 variation in floral biology;
5 variation in seed biology;
6 variation in disease and pest resistance (e.g. 'Teli' variety of teak has been found resistant to leaf skeletonizer [51]).

This variation can be exploited for building up a wide genetic base by improving and cloning the superior genotypes. Selection of 'Plus' teak trees is based on superior growth in length and diameter, clean bole height, crown height, crown diameter, cylindrical unfluted trunk and self pruning habit. Some other characteristics, like wood quality, resistance to diseases and pests such as leaf defoliators and leaf skeletonizers, and ability to produce regular crops of viable seeds are also considered [50].

Provenance tests of teak provide a relatively quick and cheap way of identifying the superior varieties which already exist. These require a good stock of planting material, including large quantities of seeds from trees of proven performance. Seed orchards have already been established in India using various traditional methods of vegetative propagation [49].

Tissue culture (micropropagation) can complement these conventional methods of vegetative propagation as the same plantation targets can be achieved by rapid replication of selected genotypes.

5. Micropropagation

Mass production of selected clones by *in vitro* techniques – 'micropropagation' – is of great importance in clonal forestry to overcome the constraints like scarce seed supplies, germination problems, long regeneration time etc. [56]. It has long been apparent that genetic gains can be captured by clonal propagation [30,87]. A growing number of tree species are being found which are amenable to micropropagation [6,7,10–13,29,43,52,89]. The different pathways of micropropagation have already been discussed [33,44,52,89]. Shoot culture has been a method of choice for micropropagation since, in shoots originating from preformed meristems, genetic integrity is very high, which is very important for maintaining clonal fidelity [66].

At the National Chemical Laboratory (NCL), Pune, India several mature 'elite' trees of teak [40], *Eucalyptus* species [38,39], and bamboo [70] have been successfully micropropagated. Tissue culture studies on teak were undertaken at NCL in collaboration with the Forest Development Corporation of Maharashtra, India. Methods for the micropropagation of teak were described for the first time, using plant material both from seedlings and mature trees [40]. Viable plants were produced from elite trees over 80 years old, growing in the forests of Allapally in central India. These plantlets were successfully grown in the field. The protocols for multiplication of seedlings and mature trees have already been described in detail [61].

The method has been extended to several other elites with further refinements, both in media composition and reduction in the number of stages. These modifications are described below. The latest results of statistically designed comparative field trials being conducted on tissue culture raised plants at NCL have also been included in this chapter.

5.1. *Culture media*

The following media were used.

MS: Modified Murashige and Skoog's medium [69] (edamin and IAA omitted and sucrose added at 2%) was the basal nutrient solution used.

MS-1: MS with Kinetin (0.15 mg/l), benzyl amino purine (0.15 mg/l), solidified with agar-agar (0.4%, Qualigens Fine Chemicals, India). MS-1 was used for initiation and proliferation of shoot cultures.

MS-2: Half strength MS liquid medium provided with a filter paper support, and a mixture of auxins (indole-3-acetic acid, indole-3-butyric acid, indole-3-propionic acid) each added aseptically at 2 mg/l concentration.

MS-3: Half strength MS liquid medium with a filter paper support. MS-2 and MS-3 were used for *in vitro* rooting of shoots.

5.2. *Collection of material*

Fresh green apical buds were collected from terminal branches in the sprouting season. The season is very critical, since buds can be obtained only once in a year. The season may vary with the location. In Allapally forests (Maharashtra, Central India) the best sprouting season was found to be from April to June. Grafts from about 75 elite trees from various locations in India are growing in the NCL teak genetic garden, from which vegetative buds were collected for raising shoot cultures.

5.3. *Surface sterilization and inoculation*

Buds (two leaves and a single apical bud) were surface sterilized as follows: (1) detergent treatment with Teepol for 2–3 minutes and washing with distilled water (2–3 times). (2) 70% Alcohol rinse, washing off with distilled water (3–4 times), (3) Sterilizing with 0.15% mercuric chloride for 10 minutes and washing with sterile distilled water under aseptic conditions.

Buds were then dissected to 3 to 5 mm size and inoculated on MS-1 medium. Initial blackening at the base of the explant was minimised by repeated shifting to MS-1 medium. The percentage of bud sprouting and the conditioning time required vary and have been found to be phenotype/genotype dependent.

5.4. *Proliferation and elongation of shoots*

Following culture and elongation of buds to a length of 4–6 cm, segments were excised to include the apical and nodal regions. Each segment formed 1 to 2 shoots by the proliferation of apical and axillary meristems in MS-1 medium. The number of shoots was increased by regular subcultures at 20 day intervals on MS-1 medium prior to the rooting stage.

5.5. *Rooting* in vitro

For *in vitro* rooting, healthy, erect, elongated shoots (4–6 cm) with 1–2 internodes were selected to avoid non-erect, crooked plantlets. These shoots were treated with an auxin mixture (IAA, IBA, IPA 2 mg/l each) on liquid MS-2 medium, for 48 h. Root elongation took place on MS-3 medium. Fully developed plantlets having healthy roots with laterals could be obtained within 30 days. Rooting ability and percentage increasesd with subculture. Shoots which had undergone 2 to 3 passages gave a higher rooting response.

5.6. *Rooting* ex vitro

To reduce the steps in the micropropagation process, it is essential to root shoot cultures *ex vitro*. Here, the commonly used rooting substrates are peat, perlite, vermiculite, pumice, sand, soil and rock wool. Commercial rooting compounds are either applied at the cut end of the shoot base or rooting substrates are saturated with these compounds [33]. In aspen and poplar, shoots could be directly rooted in peat perlite substrates [3], which helped in shortening a multistep procedure. In our laboratory, direct rooting has been achieved in different *Eucalyptus* species and *Salvadora persica* using commercial rooting compounds and different rooting substrates (unpublished).

In teak, initial attempts using a commercial hormone mixture in combination

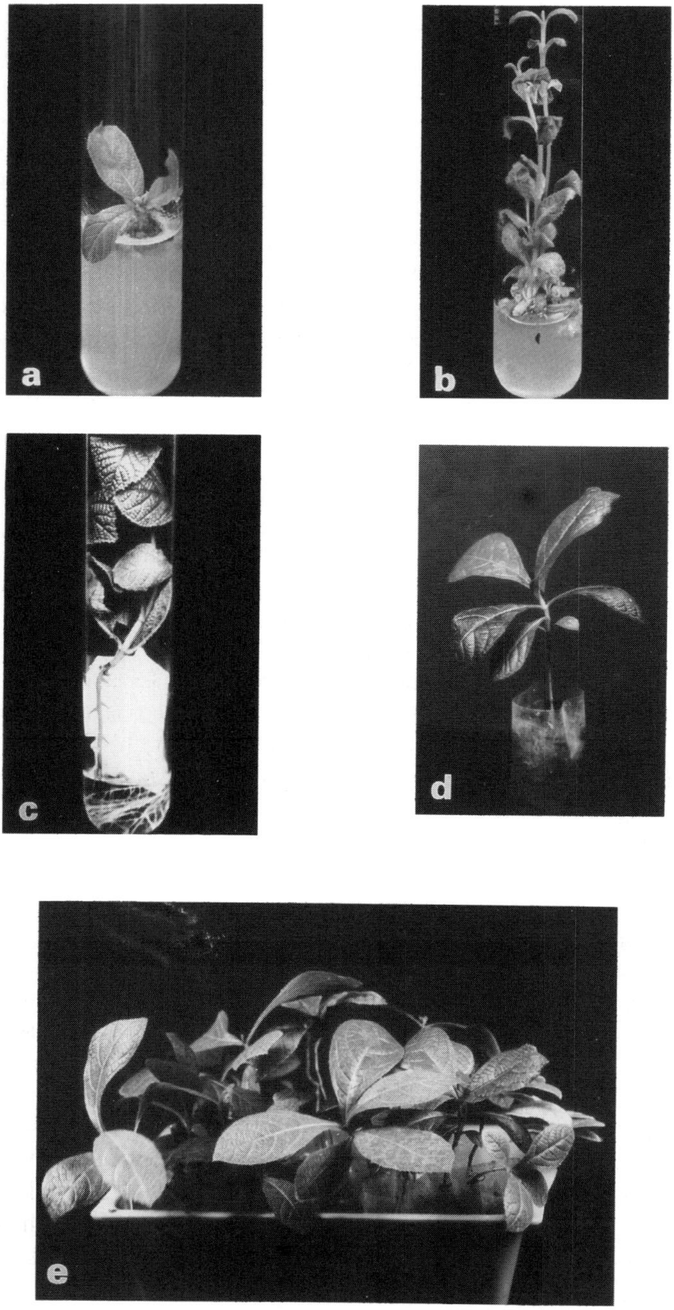

Figure 1. Stages in micropropagation of teak from apical buds. (a) Sprouted bud; (b) Multiple shoots; (c) Rooted plantlet; (d) Plantlet in polybag; (e) Plants in polybags.

with a fungicide and talc powder have been successful. Sixty percent of shoots formed roots. Further experiments are being conducted to optimize the conditions for rooting *ex vitro*.

5.7. *Acclimatization*

In vitro rooted plantlets were transplanted into polybags (5 × 10 cm) containing soil : sand : compost (3 : 3 : 1, V : V : V). These polybags were kept in plastic containers and covered with glass sheets to maintain high humidity (80–90%). Gradual acclimatization with decreasing humidity was found to be necessary to obtain good survival (70–80%). Hardened plants were transplanted to polybags (15 × 25 cm) and kept in a glass house before transfer to field (Figure 1).

5.8. *Field planting*

Plantlets which had attained a height of 40–60 cm were sown in 1980, in pits (30 cm × 30 cm × 30 cm) containing a mixture of good soil (river silt) and farmyard manure. The distance between pits in each row was 3 m and between rows was 3.5 m. Copper fungicide was used to avoid fungal attack. Irrigation of plants was initially done at 3 day intervals and later after a month in the winter and once a fortnight in summer till the plantlets were 7 years old. Since 1987 watering has been discontinued in winter, but two protected irrigations were given in middle of April and May. Initially urea was used for manuring the plants.

6. Field evaluation

A field trial in a randomised block design with seven replications was conducted in 1980 to study the growth performance of tissue culture raised plantlets of seedling (T2) and elite tree (T3) origin in comparison with seedlings raised from seeds (T1). In earlier observations [61,62], we reported significant increases in height and girth measurements in T2 and T3 as compared to T1. The height increase has more or less levelled out in the last two years, although the girth increments are significant in all treatments (Fig. 2). Flowering in tissue culture plants occured one year earlier in comparison with seed raised plants. This observation could be useful for raising seed orchards. The results are summarised in Table 1.

Figure 2. Comparative field trials of micropropagated teak plants along with controls from seedlings. This trial was started in June 1980.

Table 1. Comparative growth data. Plants per treatment: 2. Replicates: 7. Treatment No.: 3. Spacing: 3.5 m × 3 m. Statistical design: Random block design.

	Height increment (cm)					DBH increment (cm)			
	1982	1983	1984	1987	1989	1983	1984	1987	1989
T1	1.67	0.90	1.16	2.29	1.20	5.30	1.93	2.32	9.98
T2	1.98	1.05	1.39	1.98	1.34	5.80	2.01	2.25	10.97
T3	2.16	1.37	1.40	1.87	1.16	6.97	2.15	2.42	11.06

T1 = Control (Seed raised), T2 = Tissue Culture (Seedling), T3 = Tissue Culture (Mature trees).
DBH = Diameter at Breast Height.
Flowering dates – T1: 1984, 1985; T2 & T3: 1983.

7. Factors affecting micropropagation of teak

7.1. *Phenolics*

The problem of phenolics is encountered during establishment of buds and proliferation of shoots in initial subcultures which cause inhibition of growth. Blackening could be reduced by pretreatment with H_2O_2 (5.0%) or ascorbic acid (0.005%), the best response was obtained by treating the shoots with 0.75% polyvinyl pyrrollidone (insoluble, Mol. Wt. 36 000, crosslinked, Sigma Chemical Ltd.) and 2% sucrose for 30 min. Addition of antioxidants and charcoal to the culture medium was not effective. Various other techniqeus to reduce phenol exudation during explant establishment have been reviewed [17].

7.2. *Temperature*

The optimum temperature was found to be 28 ±2° C. At lower temperatures suppression of growth and at higher temperature browning of shoots was observed.

7.3. *Light*

A 16 h photoperiod (1000 lux) was found suitable. Higher multiplication rates and healthy vigorous growth of shoots could be obtained under a continuous light regime (1000 lux), although the percentage of rooting and survival of these shoots was very low.

7.4. *Culture vessels*

With a view to developing a large scale multiplication procedure, various types of culture vessels such as Borosil glass test tubes (25 × 150 mm), conical flasks (250 ml), autoclavable plastic bottles with screw caps (6 × 10 cm), jam bottles etc. have been used. Higher multiplication rates of shoots with broad leaves were obtained in conical flasks (250 ml) as compared to test tubes (Figure 3). In plastic bottles the cultures appeared vitrified, possibly due to higher humidity and condensation of water.

7.5. *Seasonal variation*

In most cultures, grown *in vitro*, the different stages leading to plantlet formation is a continuous process, carried out round the year. Teak cultures in winter

Figure 3. Multiple shoot formation and growth of teak cultures in flasks.

months (November to March) grown under controlled conditions of incubation (light, temperature, humidity, media) showed a considerable suppression of growth. Leaf fall, typical of teak trees growing in the forests during the winter months, was also observed.

The shoots regained the original vigour with a change of season. As this is a major problem, particularly in the scaling up operations and for year-round production of plants, various conditions are being tested to overcome this problem.

8. Somatic embryogenesis

Although shoot tip culture is the most commonly practiced method for micropropagation, somatic embryogenesis is preferred for the production of large numbers of plantlets in fewer steps with a concomitant reduction in labour, time and cost [86]. It is evident from the increasing success of somatic embryogenesis for forest tree species [4,36,52,63,67,89] that it will gradually replace the shoot tip culture method, being more amenable to automation [42].

Somatic embryos (bipolar structures) could be successfully encapsulated to form somatic seeds which are capable of germination [8,37]. Artificial seeds are easy to handle and store. The coating also has the potential to hold and deliver beneficial adjuvants such as plant nutrients, growth control agents and pesticides for precise placement [81].

At NCL, globular callus has been obtained in teak by using different explants such as apical meristem, *in vitro* leaves, stem, shoot tips and axillary buds. This callus turns green in light. Experiments are being carried out to obtain further development to somatic embryos. A previous report [55] indicates the development of embryo-like structures in callus cultures obtained from shoots and endosperm of teak, although plantlet regeneration has not been reported.

9. Future prospects

Micropropagation has a major supporting role in biotechnology [42]. As micropropagation becomes a reality for an increasing number of mature trees, the immediate attention will be to convert this technique to a technology suitable for commercial propagation. Several refinements in the technique may be necessary [25]. Before scaling up of the process, the clonal fidelity if the micropropagated trees should be tested by extensive field trials. Isozyme analysis [2,14] and DNA probe fingerprinting may also help in monitoring genetic stability [82]. Isozyme analysis has an important role in disease indexing, too [14]. For commercial exploitation the cloning method must be extremely rapid, capable of generating several million plants annually, and economically competitive with the seed method [68]. Since micropropagation is a labour intensive process, mechanisation/robotics may be helpful for reduction of cost [18,26,34,57].

Rejuvenation of mature forest trees has been an essential need for obtaining regeneration. As maturation is usually associated with declining regenerative capacity, various pretreatments [1,27,28,58,80], serial rooting or grafting on juvenile root stocks [32] have been tried. But there are various tree species which do not respond to these treatments. Development of rejuvenation techniques can be achieved by involving either demethylation of DNA to reactivate inactive genes or manipulation of the cytoplasm for its rejuvenation using protoplast fusion methodology for cytoplasmic hybrid (cybrids) production [16].

Use of *Agrobacterium rhizogenes* could have an impact in propagating difficult-to-root woody species [78]. In the near future one can visualise further improvements in elite clonal material by introducing new useful genetic traits, using either somaclonal or gametoclonal technologies [73,74,75] or genetic transformation via vectors such as *Agrobacterium tumefaciens*. Various methods are now available to insert and integrate foreign genes [31]. Such genetic engineering is aimed at stress resistance (salts, herbicides, alkali and acid), insect resistance (BT toxin gene), and for secondary products. Transformations have been achieved in poplar [77], Douglas fir [23], loblolly pine [o83], neem [71], etc. using *Agrobacterium tumefaciens.*

In our laboratory tumor induction has been obtained in teak and *Eucalyptus tereticornis* using *Agrobacterium tumefaciens* strains K12 × 167 (from Calgene, Inc. Davis, California). This strain consists of derivatives of wild type Ti plasmid pTiA6 with a chimeric foreign gene encoding resistance to antibiotic kanamycin. All tumors were grown on phytohormone-free medium containing carbenicillin [62]. These tumors showed the presence of octopine and APH (3′)II (amino glycoside phosphotransferase) activity (unpublished).

Increased frequency of regeneration from individual cells and protoplasts will facilitate direct DNA insertion, organelle insertion and somatic hybridization. Currently restriction fragment length polymorphism (RFLP) techniques are developing as a direct method for selecting desirable genes via their linkage to easily detectable RFLP markers. This will have a major impact on breeding [85].

Taken together, these potential developments offer prospects for introducing new genotypes and for rapid multiplication of forest trees to combat deforestation and wasteland formation.

10. Summary

Teak (*Tectona grandis* L) is a large, deciduous, tropical hardwood tree, well known for its sterling wood qualities. A high degree of variation in progeny and several constraints in its natural/artificial regeneration clearly demand a need for a method for the faster multiplication of selected clones. Micropropagation of teak has been achieved using apical buds from different elite trees. Plants raised *in vitro* are presently being evaluated in field trials. The technique is also being refined for application to large scale production of plantlets.

Acknowledgement

We acknowledge the help of Dr D. K. Kulkarni and Mr S. V. Shirke for photography and Mrs Shobha Bhondve for typing the manuscript.

References

1. Abo-El-Nil MM (1982) Method for asexual reproduction of coniferous trees, US Patent No. 4.353,184.
2. Adams WT (1983) Application of isozymes in tree breeding. In: Tanksley SD and Orton TJ (Eds.) Isozymes in Plant Genetics and Breeding, Part A, Elsevier Science Publishers B.V., Amsterdam.
3. Ahuja MR (1987) *In vitro* propagation of poplar and aspen. In: Bonga JM and Durzan DJ (Eds.) Cell and Tissue Culture in Forestry Vol 3, Case Histories: Gymnosperms, Angiosperms and Palms. Martinus Nijhoff, Dordrecht. pp. 207–223.
4. Ammirato PV (1989) Recent progress in somatic embryogenesis, IAPTC News Letter, 57:2–16.
5. Anonymous (1976) The wealth of India, Vol. X, Publ. Inf. Directorate, CSIR, New Delhi, India. pp. 136–150.
6. Bajaj YPS (Ed.) (1986) Biotechnology in Agriculture and Forestry, Vol. 1 Trees I, Springer Verlag, Berlin.
7. Bajaj YPS (Ed.) (1989) Biotechnology in Agriculture and Forestry, Vol. 5 Trees II, Springer Verlag, Berlin.

8. Bapat VA and Rao PS (1988) Sandalwood plantlets from 'synthetic seeds'. Plant Cell Reports 7:434.
9. Bedell PE (1989) Preliminary observations of variability of teak in India. Ind. Forester 115:72–81.
10. Bonga JM and Durzan DJ (Eds.) (1987) Cell and Tissue Culture in Forestry, Vol. 1: General Principles in Biotechnology, Martinus Nijhoff, Dordrecht.
11. Bonga JM and Durzan DJ (Eds.) (1987) Cell and Tissue Culture in Forestry, Vol 2. Specific Principles and Methods: Growth and developments, Martinus Nijhoff, Dordrecht.
12. Bonga JM and Durzan DJ (Eds.) (1987) Cell and Tissue Culture in Forestry, Vol 3. Case histories: Gymnosperms, Angiosperms and Palms, Martinus Nijhoff, Dordrecht.
13. Boulay M (1987) *In vitro* propagation of tree species. In: Green CE, Somers DA, Hackett WP, and Biesboer DP (Eds.) Plant Tissue and Cell Culture, Alan R. Liss Inc. New York. pp. 367–382.
14. Burdon JJ and Marshall (1983) The use of isozymes in plant disease research. In: Tanksley SD and Orton TH (Eds.) Isozymes in Plant Genetics and Breeding, Part A, Elsevier Science Publishers, B.V., Amsterdam. pp. 401–412.
15. Cameron AL (1966) Genetic improvement of teak in New Guinea. Austral Forestry 30:76–87.
16. Cocking EC (1987) Plant Cell Biology in the 21st Century. The Needs of Plant Cell and Tissue Culture, In: Green CE, Somers DA, Hackett WP, and Biesboer DD (Eds.) Plant Tissue and Cell Culture, Alan R. Liss Inc. New York. pp. 3–15.
17. Compton ME and Preece JE (1986) Exudation and explant establishment. IAPTC Newslett 50:9–37.
18. Constantine DR (1986) Micropropagation in the commercial environment. In: Withers LA and Alderson PG (Eds.) Plant Tissue Culture and its Agricultural Application, Butterworth, London. pp. 175–187.
19. Dabral SL (1976) Extraction of teak seeds from fruits, their storage and germination. Ind Forester 102:650–658.
20. Dabral SL (1977) Propagation of teak by root grafts. Ind Forester 103:225–230.
21. Dabral SL and Amin PW (1975) Poor fruit formation in teak in Chanda forests of Maharashtra. Ind Forester 101:616–619.
22. Dadwal VS and Jamaluddin (1988) Role of fungi in weathering of teak fruits. Ind Forester 114:328–330.
23. Dandekar AM, Gupta PK, Durzan DJ, and Knauf V (1987) Genetic transformation and foreign gene expression in micropropagated Douglas fir (*Pseudotsuqa menzieii*). Bio/Technology 5:587–590.
24. Davidson J and Fairlamb J (1976) Germination of teak seeds – preliminary evidence of a chemical germination inhibitor. In: Seed problems, Proc 2nd Int IUFRO Symp Physiology of Seed Germination, Japan.
25. Debergh PC (1987) Improving micropropagation. IAPTC New Lett 51:2–10.
26. Debry L (1986) Robots in plant tissue culture: an insight. IAPTC News Lett 49:2–22.
27. De la Goublaye de Nantois H (1980) Reneunissement chez Ie Douglas (*Pseudotsuqa menziessi*) en vue de la propagation vegetative etude sur la flagiotropic des parties aeriewnes at rainaires DEA. Universite de Paris, VI.
28. Doorebos J (1953) Rejuvenation of *Hedera helix* in graft combinations. Preb 115, Wagenignen, November 28.
29. Dunstan DI and Thorpe TA (1986) Regeneration in forest trees. In: Vasil IK (Ed.) Cell Culture and Somatic Cell Genetics of Plants, Academic Press Inc. London. pp. 223–237.
30. Durzan DJ (1988) Application of cell and tissue culture in tree improvement. In: Bock G and Marsh J (Eds.) Application of plant cell and tissue culture, A Wiley Interscience Publication, John Wiley and Sons Chichester, Sussex UK. pp. 36–49.
31. Fraley RT, Rogers SG, and Horsch RB (1986) Genetic transformation in higher plants, CRC Crit Rev Plant Sci 4:1–46.
32. Franclet A, Boulay M, Bekkaoui F, Fourest Y, Verschoore-Martouzet B and Walker N (1987)

Rejuvenation. In: Bonga JM and Durzan DJ (Eds.) Cell and Tissue Culture in Forestry. Vol. 1: General Principles in Biotechnology, Martinus Nijhoff, Dordrecht. pp. 237–247.

33. George EF and Sherrington PD (1984) Plant Propagation by Tissue Culture, Handbook and Directory of Commercial Laboratories. Eastern Press Reading, Berks, Great Britain.
34. Giles KL and Morgan WM (1987) Industrial scale plant micropropagation. TIBTECH 5:35–39.
35. Gupta BN and Kumar A (1976) Estimation of potential germinability of teak fruits from twenty three Indian sources by cutting test. Ind Forester 102:808–813.
36. Gupta PK (1988) Advances in biotechnology of conifers. Current Sci 57:629–639.
37. Gupta PK and Durzan DJ (1987) Biotechnology of somatic polyembryogenesis and plantlet regeneration in Loblolly pine. Bio/Technology 5:147–151.
38. Gupta PK, Mascarenhas AF, Jagannathan V (1981) Tissue culture of forest trees: Clonal multiplication of mature trees of *E. citriodora* Hook by tissue culture. Plant Sci Lett 20:195–201.
39. Gupta PK, Mehta U and Mascarenhas AF (1983) A tissue culture method for rapid multiplication of mature trees of *Eucalyptus torelliana* and *Eucalyptus camaldulensis*. Plant Cell Reports 2:296–299.
40. Gupta PK, Nadgir AL, Mascarenhas Af, and Jagannathan V (1980) Tissue culture of forest trees – Clonal multiplication of *Tectona grandis* (Teak) by tissue culture. Plant Sci Lett 17:259–268.
41. Gupta SN and Pattanath PG (1975) Factors affecting germination behaviour of teak seeds of eighteen Indian origins. Ind Forester 101:584–588.
42. Haissig BE, Nelson ND, and Kidd GH (1987) Trends in the use of tissue culture in forest improvement. Bio/Technology 5:53–59.
43. Hanover JW and Keathley DE (Eds.) (1988) Genetic Manipulation of Woody Plants, Basic Life Sciences Vol 44. Plenum Press, New York and London.
44. Hazra S and Mascarenhas AF (1987) Cloning of forest trees by *in vitro* techniques. In: Islam AS (Ed.) Proc. Reg. Workshop on Tissue Culture of Crop Plants, Dhaka, pp. 47–62.
45. Hedegart T (1973) Pollination of teak (*Tectona grandis* L.) 2. Silvae Genetica 22:124–128.
46. Hedegart T (1976) Breeding systems variation and genetic improvement of teak (*Tectona grandis*). In: Burley J and Styles BT (Eds.) Tropical trees: Variation, Breeding and Conservation, Academic Press, London. pp. 109–121.
47. Hedegart T, Lauridsen EB and Keiding H (1975) Teak. In: Faulkner R (Ed.) Seed Orchards. Forestry Commission Bulletin No. 54. Her Majesty's Stationary Office, London. pp. 139–142.
48. Joshi MD, Kelkar SP (1971) Germination of Seed in dry teak (*Tectona grandis* L.) (i) Preliminary studies in fruit development and seed dormancy. Ind Forester 97:210–212.
49. Kedharnath S (1983) Genetics and forest tree breeding. In: Jaiswal PL (Ed.) Genetical Research in India, ICAR, New Delhi, India. pp. 181–190.
50. Kedharnath S and Mathews JD (1962) Improvement of teak by selection and breeding. Ind Forester 88:277–284.
51. Keiding HG and Boonkird S (1960) Vegetative propagation of teak. Unasylva 14:193–194.
52. Kendurkar SV and Mascarenhas AF (1987) Micropropagation of forest trees through tissue culture. In: Reddy GM (Ed.), Proc Symp Plant Cell and Tissue Culture of Economically Important Plants, Hyderabad, India. pp. 181–190.
53. Kushalappa KA (1977) Teak plantations in Thailand. Ind Forester 103:325–328.
54. Lahiri AK (1974) Preliminary study on rooting of green wood cuttings of teak. Ind Forester 100:559–560.
55. Lakshmisita G and Chatopadhayay S (1986) Improvement of forest trees by tissue culture of economically important plants. In: Reddy GM (Ed.) Proc. of Natl. Symp. Hyderabad, India. pp. 195–198.
56. Leakey RRB (1987) Clonal forestry in the tropics: A review of developments, strategies and opportunities. Commonw For Rev 66:61–75.
57. Levin R, Victorgaba, Tal B, Hirsch S, and Denola D (1988) Automated Plant tissue culture

for mass propagation. Bio/Technology 6:1035–1040.
58. Magalewski RL and Hackett WP (1979) Cutting propagation of *Eucalyptus ficofolia* using cytokinin-induced basal trunk sprouts. Proc Int Plant Prop Soc 29:118–124.
59. Mahmood Husain and Somasundaram TR (1975) Scions on stumps – an innovation in the vegetative propagation of teak (*Tectona grandis* L.) Southern Forest Rangers College Magazine 51:1–8.
60. Mahmood Husain, Somasundaram TR, Subramanian KN (1976) A recent advance in teak culture. Ind Forester 102:531–532.
61. Mascarenhas AF, Kendurkar SV, Gupta PK, Khuspe SS, and Agrawal DC (1987) Teak. In: Bonga JM and Durzan DJ (Eds.) Cell and Tissue Culture in Forestry, Vol. 3, Case Histories: Gymosperms, Angiosperms and Palms. Martinus Nijhoff, Dordrecht, pp. 300–315.
62. Mascarenhas AF, Khuspe SS, Nadgauda RS, Gupta PK and Khan BM (1988) Potential of cell culture in plantation forestry programms. In: Hanover JW and Keathley DE (Eds.) Genetic Manipulation of Woody Plants, Plenum Press, New York and London. pp. 391–412.
63. Mascarenhas AF and Muralidharan EM (1989) Tissue Culture of Forest Trees in India. Current Sci 58:606–613.
64. Mascarenhas AF and Muralidharan EM (1993) Clonal forestry with tropical hardwoods. In: Ahuja MR and Libby WJ (Eds.) Clonal Forestry, Vol. 1, Genetics and Biotechnology, Springer Verlag (In press).
65. Maslekar AR (1981) Forester's companion (Indian Forestry Handbook) Jugal Kishore and Company, India.
66. McCown BH (1984) From gene manipulation to forest establishment: shoot cultures of woody plants can be a central tool. In: Proc TAPPI Research and Development Conference. TAPPI Press, Atlanta CA. pp. 21-26.
67. Muralidharan EM, Gupta PK and Mascarenhas AF (1989) Plantlet production through high frequency somatic embryogenesis in long term cultures of *Eucalyptus citriodora*, Plant Cell Reports 8:41–43.
68. Murashige T (1977) Plant cell and organ cultures as horticultural practices. Acta Hort 78: 17–30.
69. Murashige T and Skoog F (1962) A revised medium for rapid growth and bioassay with tobacco tissue culture. Physiol Plant 15:473–493.
70. Nadgir Al, Phadke CH, Gupta PK, Parasharami VA, Nair S and Mascarenhas AF (1984) Rapid multiplication of bamboo by tissue culture. Silvae Genetica 33:221–223.
71. Naina NS, Gupta PK and Mascarenhas AF (1989) Genetic transformation and regeneration of transgenic neem (*Azadirachta indica*) plants using *Agrobacterium tumefaciens*. Current Sci 58:184–187.
72. Nanda KK (1962) Some observation as growth, branching behaviour and flowering of teak (*Tectona grandis*) in relation to light. Ind Forester 88:207–218.
73. Nelson ND (1985) North Central Forest Experiment Station biotechnology programme – application to tree improvement. In: Proc. 18th Southern Forest Tree Improvement Conf Long Beach, MS University of Southern Mississippi, Gulf Park, MS, pp. 14–22.
74. Nelson ND, Haissig BE and Riemenschneider DE (1984) Applying the new somaclonal technology to forestry. In: Proc. TAPPI Research and Development Conference. TAPPI Press, Atlanta, CA, pp. 27–34.
75. Nelson ND and Haissig BE (1986) Herbicide stress: Use of biotechnology to confer herbicide resistance to selected woody plants. In: Hennessey TC, Dougherty PM, Kossuth SV and Johnson JD (Eds.), Stress Physiology and Forest Productivity, Martinus Nijhoff, Dordrecht. pp. 191–215.
76. Oka AF (1976) Teak nursery practices in Maharashtra, Research and Education Circle. Nagpur, India.
77. Parsons TJ, Sinkar VP, Stettler RF, Nester EW and Gordon MP (1986) Transformation of poplar by *Agrobacterium tumefaciens*. Bio/Technology 4:533–536.

78. Patena L, Sutter EG and Dandekar AM (1988) Root indcution by *Agrobacterium rhizogenes* in a difficult to root woody species. Acta Hort. 227:324–330.
79. Rawat MS and Kedharnath S (1968) Field grafting and budding in teak (*Tectona grandis* (L.) Ind Forester 94:260–261.
80. Read P (1985) Novel plant growth regulator delivery systems for *in vitro* culture of horticultural plants. In: Symposium on *In vitro* Problems Related to Mass Propagation of Horticultural Plants, Abstr 5, Gembloux, Belgium.
81. Redenbaugh K, Viss P, Slade D, and Fujii JA (1987) Scale up artificial seed. In: Green CE, Somers D, Hackett WF, and Biesboer DD (Eds.) Plant Tissue and Cell Culture, Alan R. Liss Inc. New York. pp. 473–493.
82. Riemenschneider DE, Haissig BE and Bingham ET (1987) Biotechnology in woody plant breeding programmes. In: Hanover JW and Keathley DE (Eds.) Genetic manipulation of Woody Plants. pp. 433–449.
83. Sederoff R, Stomp AM, Chilton WS and Moore LV (1986) Gene transfer into loblolly pine by *Agrobacterium tumefaciens*. Bio/Technolgoy 4:647–649.
84. Seth SK, Waheed Khan MA (1958) Regeneration of teak forests. Recommendations of teak study tour and Symposium, December 1957–January 1958 Forest Research Institute, Dehra Dun, India.
85. Tanksley SD, Young ND, Paterson AH, Bonierbale MN (1989) RFLP mapping in plant breeding: New tools for an old science. Bio/Technology 7:257–264.
86. Thorpe TA (1988) *In vitro* somatic embryogenesis. In: ISI Atlas of Sciences: Animal and Plant Sciences. pp. 81–88.
87. Timmis R, Abo El-Nil M and Stonecypher RW (1987) Potential genetic gains through tissue culture. In: Bonga JM and Durzan DJ (Eds.) Cell and Tissue Culture in Forestry, Vol. 1: General Principles and Biotechnology, Martinus Nijhoff, Dordrecht. pp. 198–215.
88. Troup RS (1921) Silviculture of Indian trees. Vol. II, The Clarendon Press, Oxford.
89. Von Arnold S and Wallin A (1988) Tuissue culture methods for clonal propagation of forest trees. IAPTIC New Letter 58:2–13.

16. Micropropagation of *Eucalyptus*

G. LAKSHMI SITA
Department of Microbiology and Cell Biology, Indian Institute of Science, Bangalore 560 012, India

1. Introduction

A study of forest resource management in India and elsewhere shows that forest resources always have been subjected to socio-economic compulsions from time to time. Forestry is geared to the production of matchwood, plywood and construction timber. The dependence of common man on forests for his daily needs, like fuelwood, has increased. During recent years forest resources underwent a change from conservation-oriented management to forestry for the people. Population growth and industrialization have enhanced the demand for wood products in India, accentuating the pressure on forest resources. Already demand is far in excess of supply and this gap is expected to widen considerably. Several strategies were adopted to increase the supply of wood to meet the industrial and domestic demands. One step taken was to introduce rapidly growing species in plantation forestry which, till then, was dominated by slow growing hardwoods. With this, plantation forestry took a new dimension in India [36]. Cultivation of rapidly growing species commenced on a large scale from the early sixties. The most important tree species, which has been raised extensively, is *Eucalyptus*, a native of the Australian continent. It was first introduced into India as a garden tree in 1790 on Nandi Hills in Karnataka State. Later, from 1856, it was raised in the Nilgiris in an appreciable area to meet the firewood requirements. Within a short period *Eucalyptus*, commonly called Eucalypt, became the largest hectarage, next to teak, aggregating half a million hectares of forest and farm lands in the country [62].

The potential benefits of establishing forests with selected vegetatively propagated clonal planting stock have long been recognized. In the past decade, the potential for application of *in vitro* techniques for the vegetative propagation of forest trees has attracted a great deal of attention. The importance of tissue culture in forestry has recently been emphasized by Bonga and Durzan [8] and Bajaj [2]. Three excellent chapters have been written on *Eucalyptus* alone giving much details [1,21,53]. The work is briefly described below.

The pioneering work of Fossard [21,23–25] and his group in Australia [11–19] and Franclet and Boulay [20] in France has indicated the possibility of *in vitro* micropropagation of some species of *Eucalyptus*. Work started in AFOCEL in 1973 with a view to mass producing clones of *Eucalyptus* selected

M. R. Ahuja (Ed.), Micropropagation of Woody Plants, 263–280.

for their cold resistance. This success encouraged the work with young plants produced from seeds grown *in vitro*, or juvenile sprouts from the stock plants. Plants of *E. dalrympleana* were successfully planted in the field in 1975. Later on 2-year old plants were selected for cold tolerance and micropropagated. Intensive research was carried out to successfully develop 25 000 plants per year. Work was concentrated on four species of Eucalyptus (*E. gunnii*, *E. dalrympleana*, *E. pauciflora* and *E. delegatensis*). Micropropagation of the methods followed by AFOCEL are given in detail by Durand-Cresswell *et al.* [21]. Methods to further simplify the costs are being worked out. This is probably the first work on commercialization of *Eucalyptus* species. Preliminary experiments on organ culture of *Eucalyptus* were performed in Australia [11]. Initially, cultures of apical tips from adult field grown *Eucalyptus* were tried and were found to be unsuccessful. However, apical explants from aseptic seedlings of *Eucalyptus* species (*E. bacroftii*, *E. grandis* and *E. deglupta*) produced roots [11]. Much progress has been made in the organ culture of *Eucalyptus*. Plantlets have been developed by the nodal culture of seedling material of the following *Eucalyptus* species: (*E. grandis*, *E. gunnii*, *E. dalrympleana*, *E. pauciflora* and *E. ficifolia*) [17]. Nodes were also cultured to induce multiple buds which can later be separated and rooted. Using this method on seedlings of *E. camaldulensis*, *E. curtissii*, *E. ficifolia*, *E. grandis*, *E. obtusifolia* and *E. globulus* subspecies *E. bicostata* and *E. regans* were multipled [31]. Very high multiplication rates for adult *E. ficifolia*, *E. camaludulensis* and *E. polybractea* have been reported using a combination of rejuvenation of adult tissues by coppicing and rooting of coppice shoot cuttings (Durand-Cresswell, Personal Communication). Initial work on *Eucalyptus* also concentrated on callus cultures for regeneration from seedlings. [35]. As many as 175 combinations have been tried in *E. bacroftii* [48]. Although nodular callus was obtained, no organogenesis was observed. Rather than repeating too much of the earlier work or reiterating the critical aspects covered so well in other reviews [21,30,53], I would like to concentrate, in this chapter, on the three major different approaches used for mass propagation, namely: micropropagation by enhanced axillary budding; organogenesis; and somatic embryogenesis. I would also like to give details of origin, distribution and utilization, for the benefit of people who are new to *Eucalyptus*. Since so much has already been written about the work at AFOCEL [9,21] as well as the work in Australia [53], emphasis will be on some species of currently of interest in India and efforts to mass propagate.

2. Distribution

Eucalyptus comprises more than 625 species. Quite a few of them have been raised successfully all over the world under a wide range of climatic and edaphic conditions. Eucalypts as exotics are known for their fast growth and high biomass productivity even on poor sites. The natural distribution of

Eucalyptus species is largely in Australia with a limited occurence of species of minor importance in New Guinea, Timore, Java and Philippines. The species has become naturalized in several countries, but in some countries, like Brazil, the plantations have to be carefully managed to minimize insect attack. Although Brazil has over 1 million hectares of plantations, naturalization seems unlikely. In India Eucalypts were introduced a couple of centuries ago [62,67], large scale planting dates back to 1856, where *Eucalyptus globulus* was taken up in the Nilgiris (Southern India). Simultaneously, *E. robusta* was also introduced there, though on a much smaller scale. Plantations of *E. citriodora* came into existence in the mid-thirties, particularly in parts of Karnataka and Kerala States. The 1960s heralded the introduction of *E. tereticornis* – the Mysore hybrid or the Mysore gum – on a massive scale throughout the country from Kanyakumari to Himalayas. This species was planted in all agro-climatic zones, in forest plantations as well as avenues. The other commercial species of importance are *E. torreliana* and *E. camaldulensis*, *E. grandis.*

3. Utilization

The major use of *Eucalyptus* in India is as pulpwood. Blue gum (*E. globulus*) and rose gum (*E. citriodora*) are used exclusively for rayon pulp, and Mysore gum *E territicornis* is used for paper pulp. The wood of some species is so hard and durable that it can be put to many uses in addition to timber and fuel wood. The lops and tops, including bark, are used as firewood. The leaves of blue gum yield eucalypt oil. The leaves of *E. citriodora* are used mainly for the manufacture of citriodora oil. All species give poles, pit props, electric transmission poles, and fence posts. Large trees of blue gum are used as timber in the Nilgiris (Southern India). Due to a shortage of timber, Mysore gum over 70 cm in girth is also used as timber. Another most important use of *Eucalyptus* wood is in making packing cases which are mainly used for industrial packing. The wood is also being utilized for making charcoal. In some parts of India, *Eucalyptus* wood is also being supplied to the copper industry and is used in copper smelters as a reducer to regulate the oxidation of copper in the molten stage [10].

4. Conventional method of propagation and genetic improvement studies

Many species of the genus *Eucalyptus* have been planted all over the world, mainly because of their rapid rate of growth, low susceptibility to disease and browsing and better utility as pulpwood. *Eucalyptus* is naturally regenerated by seed which usually results in highly variable progeny. To cope with the ever increasing demand for wood, for fuel and industrial purposes, intensive forestry practices are necessary to develop superior clones for mass propagation.

A large variation in woodyield has been found in many species. Cloning by grafting is often used in the genetic improvement of forest trees for establishing clone archives, clone banks and productive seed orchards. Grafted seed orchards have been established in some countries. However graft incompatibility and rejection have proved a major problem with eucalyptus. Using root stock grown from half-sib seed of the scien parent helps decrease graft failures. Other alternatives are establishing seedling orchards and using air-layers or rooted shoot cuttings for the purpose. Leafy cuttings taken from young seedlings or shoots from lignotubers of most eucalyptus can be easily rooted in sand. However the capacity to root from cuttings falls off very rapidly as the plant gets older [61]. Failure to root, shoot cuttings is attributed to a rooting inhibitor in adult tissue [60]. Coppicing, however restores physiological juvenility [25] and cuttings from rejuvenated coppice shoots easily root. This method has been used on a large scale in Brazil with *E. grandis*. Since *Eucalyptus* is refractory to vegetative propagation, seed continues to be the mainstay of propagation. But being a predominantly cross-pollinated species with different mechanisms of pollination in operation, seedlings raised from seeds abound in variation. This necessitates identification of a realiable method for use in plant breeding programmes for tree improvement and subsequent mass propagation. Techniques for vegetative propagation can be a very useful tool in tree improvement programmes to bring together all desirable clones for clonal testing and breeding. In afforestation and re-forestation programmes, clonal plantation of uniform, superior clones is a promising method to achieve high levels of genetic gain in the shortest possible time by capturing all additive and dominance gene effects [7]. *In vitro* propagation may provide better multiplication rates than conventional methods of air layering or rooting of cuttings, as with many horticultural crops. It has also been observed that callus cultures are a rich source of genetic resources. The callus tissues on prolonged culturing undergo endomitosis chromosome loss, translocations, polyploidy, aneuploidy gene amplification, mutations and other genetic changes leading to wide diversity. Successful plant improvement programmes can be developed from such variations. However there is no record of this type of work in trees [3]. There is an urgent need to increase the quality and quantity of fast growing timbers in future and *in vitro* multiplication of *Eucalyptus* can play a significant role.

5. *In vitro* approaches

Tissue culture studies on *Eucalyptus* were initiated nearly two decades ago. Already reviews have appeared on this subject at greath length. It is not intended to review every single paper that has been published. Rather, the achievements of the *in vitro* research will be highlighted, the emphasis being on the currently used species. Cloning by conventional or *in vitro* techniques is especially valuable for the propagation of heterozygous, sexually incompatible and sterile genotypes. This can be achieved by tissue culture techniques. Rapid asexual miltiplication can be achieved by micropropagation through: (1) enhanced axillary branching; (2) organogenesis; and (3) somatic embryogenesis.

5.1. *Micropropagation through enhanced axillary branching*

Shoot multiplication by enhanced axillary and terminal meristem from seedling as well as mature materials is the most commonly exploited and successful technique in *Eucalyptus* species. Shoot cultures are means of long term cultivation of higher plants in an organized condition. In contrast to callus cultures, they retain their regenerative capacity over long periods. Hence they are particularly appropriate as stocks for the clonal propagation of crops, ornamentals and trees with a high multiplication rate and sufficient genetic stability. Even though some trees can be propagated from nodal segments collected form mature trees, many others can be propagated only from tissues of seedling materials like shoot apical meristem, and axillary meristems. However, cloning of mature trees is generally preferred over cloning of seedling explants because it is often not possible to determine whether these embryos or seedlings have the genetic potential to develop the desired qualities later in their life cycle [6,7]. Standardization of techniques with seedling material, however, helps to some extent to determine the media conditions for use with adult materials. The principle of enhanced axillary branching is as follows. Each leaf has an axillary bud in its axils, which has the potential to develop into a shoot. In nature, these buds remain dormant for various periods, depending on the growth pattern and environmental conditions. However, by culturing nodal segments on media containing appropriate concentrations of cytokinins, it is possible to break this dormancy and subsequent development of multiple shoots from nodal segments of aseptic seedlings. A few species will be discussed in detail.

5.1.1. *E. grandis*

Barker *et al.* [4] reported the formation of single plants from nodal segments of seedling, coppice and young trees. No success was achieved with adult material. Nodal cultures obtained from forest trees invariably had to be discarded because of multiplication. Lakshmi Sita *et al.* [42–45] have reported the formation of multiple shoots from meristems of seedling as well as nodal segments obtained from five year old trees. Shoot proliferation was consistantly obtained with BAP (0.5–2.0 mg/l) and NAA (0.1–1.0 mg/l). After initial multiple shoot formation, very low cytokinin concentrations were adequate for further proliferation. In a later work, Lakshmi Sita and Sukanya Chattopadhya [46] have used several combinations of lower concentrations of kinetin, BAP and NAA. Among those, MS medium supplemented with 0.1 mg/l BAP and 0.01 mg/l NAA was found to be the best. The approximate number of shoots in a tube in four weeks time was more than 100 shoots. Outer shoots in the clumps tend to be longer than the others, which look like miniatures. These miniature shoots can either be routinely multiplied by dividing, or plantlets can be obtained by separating the shoot clumps and transferring them to media containing 1/2 MS + 1 or 2mg/l NAA or 1/2 MS + 1.0 IBA. Shoots which look like miniatures will show the normal habitat within 3–4 weeks after reducing the mineral strength. Hartney [31], and Hartney and Barker [32] have also reported shoot multiplication and rooting of shoots. Several laboratories are routinely producing plantlets but enhanced axillary by proliferation.

5.1.2. *E. citriodora*

One of the earliest reports [1] of shoot regeneration of plantlets from seedlings and lignotubers was in *E. citriodora*. But this paper does not give any details of regeneration of plantlets. Multiple shoot production followed by rooting was reported by Lakshmi Sita and Vaidyanathan [41] from shoot meristem of germinating 5-day-old seedlings. Shoot apices were cultured on media containing (0.5–2 mg/l) BAP. Shoots numbers up to 100 were obtained after two passages into higher concentration of BAP (2 mg/l). Grewal *et al.* [26] have also reported shoot multiplication from aspetic seedlings. Gupta *et al.* [27] for the first time have reported shoot multiplication from nodal segments of mature 20-year-old selected elite trees. Successful multiple shoot formation can also be initiated from terminal buds of 10–20-year-old trees mature plants as reported by Mascarenhas *et al.* [52].

Nodal segments from mature trees were cultured on solid MS salts with vitamins and glycine and supplemented with kinetin 0.2 mg/l and BAP 0.3 mg/l + calcium panthothenate 0.1 mg/l and biotin 0.1 mg/l. After 30–35 days shoot buds sprouted from nodal segments in 35–40% cultures. Subculturing at this stage to fresh medium increased the number of shoots within 10–15 days. Within 25 days shoot elongation was found to be 40–45 mm. Shoots were excised and rooted. For successful establishment, gradual simplification of the nutrient medium and increased light intensities were found to be necessary.

5.1.3. *E. camaldulensis*

Hartney [33], Hartney and Barker [32], and Diallo and Duhox [20] have reported micropropagation of *E. camaldulensis* from seedling segments. Gupta *et al.* [29] have reported the formation of multiple shoots from the nodal segments of 20-year-old trees. Fast growing elite trees were selected and nodal segments from lateral branches were surface sterilized and multiple shoots were initiated from axillary meristems on MS solid media supplemented with 0.2 mg/l kinetin and 0.5 mg/l BAP and calcium panthothenate 0.1 mg/l and 0.1 mg/l biotin. Cold treatment was found to be necessary for 72 hours at

15° C prior to transfer to a growth chamber. After 30–35 days shoot buds sprouted, and up to 6–10 multiple shoots were obtained. On subsequent subculturing, the individual excised shoots were placed on MS supplemented with 0.05 mg/l kinetin, 0.1 mg/l BAP and calcium pantothenate 0.1 mg/l for elongation on this medium. These were later rooted on unusually high concentration of IAA, IBA, IBA & NAA (10 mg/l each) in liquid media in the dark at 23° C for 72 hours. Tissue culture plants have established well in the ground.

5.1.4. *E. globulus*

Hartney [33] has reported plantlet production from seedling segments. Mascarenhas *et al.* [52] and Gupta and Mascarenhas [30] have successfully reported the establishment of plantlets from nodal segments excised from lateral branches of fast growing adult trees. Conditions similar to *E. camaldulensis* were used.

5.1.5. *E. torreliana*

Shoot multiplication from 10–20-year-old trees using terminal buds was successful [29]. Here the conditions used were the same as those used for *E. camaldulensis* and *E. globulus.* Rooting conditions were however slightly different. Shoots were pretreated with half-strength MS salts and vitamins and 2.0 mg/l NAA for 48 hours and then transferred to half-strength MS medium with activated charcoal 0.25% w/v on solid medium. Shoots have rooted within 10–15 days. For successful establishment, adaptation to the soil environment is usually required, involving gradual simplification of the media. Over 1000 plants of *E. torreliana* were established.

5.1.6. *E. territicornis*

In *E. territicornis* multiple shoots were successfully obtained from fresh sprouted lateral branches from elite trees. Primary shoots could be obtained on MS media containing 0.5–2 mg/l BAP in about 4–5 weeks [46]. These shoots on subculture to MS media supplemented with 0.2 mg/l BA and 0.1 mg/l kinetin developed profuse shoots filling up the flask. However these shoots were like miniatures, measuring no more than 10 mm and filling the whole flask (Figure 1). These, on subculture to half-strength MS mineral salts, 1% sucrose, vitamins and 2 mg/l NAA, produced normal shoots with profuse rooting measuring 7–10 cm in length (Figure 2). These plantlets or clumps of plantlets were transferred to MS half-strength liquid medium for hardening before planting. For proliferation of shoots we did not find it necessary to add calcium pantothenate and biotin. Multiple shoots could be successfully proliferated on other combinations of cytokinins like 0.1 mg/l BA + 0.01 mg/l NAA; 0.02 BA + 0.01 NAA; 0.5 BA + 0.1 Ki + 0.2 NAA. Figure 3 shows the different stages of development of multiple shoots. Mascarenhas *et al.* [52]

Figure 1. Multiple shoots of *E. territicornis.*

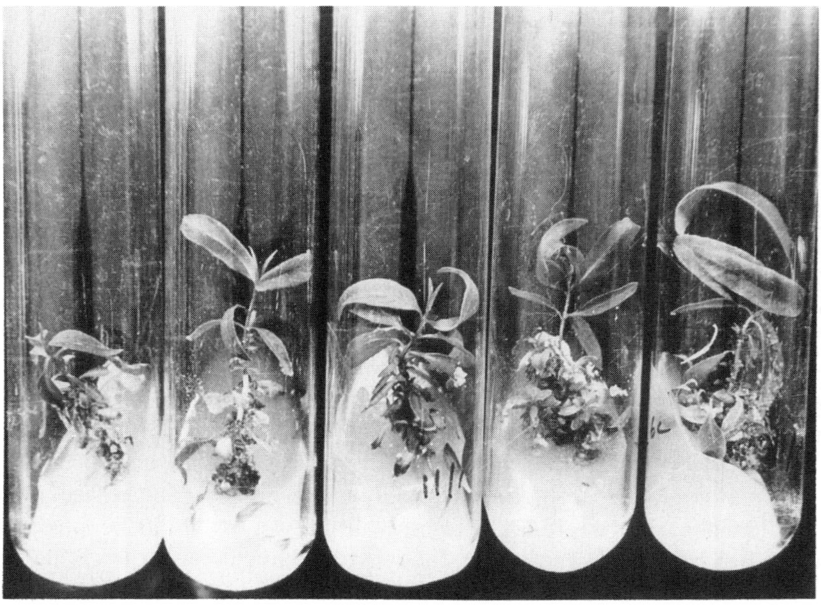

Figure 2. Elongated and rooted plants of *E. territicornis.*

Figure 3. Different stages of development in *E. territicornis.*

have reported successful shoot induce on media supplemented with low cytokinins (0.2 mg/l kinetin and 0.5 mg/l BAP).

6. Organogenesis

Several species have been cultured with a view to induce regeneration. Jacquiot [35] worked with five species of eucalyptus (*E. gunni*, *E. territicornis*, *E. cladocalyx*, *E. gamocephala* and *E. camaldulensis*) and found that *Eucalyptus* tissues could only be grown on media with added coconut milk and none of those cultured formed any organs. In a short communication [1], regeneration of plants from tissue cultures of *E. citriodora* was reported from tissues derived from lignotubers, but not from stem or root tissue. There was no follow-up of this work and reproducibility was not confirmed. A coconut milk medium was used and the concentrations of amino acids and vitamins used were not given. Fossard [14] has cultured five species of *Eucalyptus* (*E. bancroftii*, *E. grandis*, *E. laevopinae*, *E. nichollii* and *E. melliodora*) on media without coconut milk. Regeneration of roots was only obeserved in *E. grandis*. In a second paper Fossard *et al.* [16] have described the tissue cultures of stems and lignotubers, and the response of various auxin-cytokinin concentration. A cultural mixture of eucalyptus and tobacco was tried in the hope that a diffusable factor from the tobacco might induce the eucalypt to regenerate. This, however, did not occur, but in some treatments massive roots developed, unlike either those of tobacco or eucalyptus. Regeneration from such callus was not achieved. Kitahara and Caldas [37] have reported regeneration from hypocotyl callus in *E. alba*. Regeneration was obtained on White's medium supplemented with coconut milk, 15% v/v and IAA 1 mg/l. In *E. citriodora* regeneration was reported [40] from cotyledonary callus on MS medium supplemented with 1 mg/l zeatin and 0.2 mg/l IAA. Leaf and shoot callus obtained from one-year old or mature trees did not differentiate. However, callus showed nodular, embryo-like structures on media containing zeatin and IAA. Regeneration from callus cultures originated from mature tissues have been found to be difficult.

In isolated cases differentiation from compact calli of *E. grandis* was observed [45]. But these results could not be repeated. In *E. marginata* [5] Bennett and McComb have reported callus induction and regeneration from cotyledons, similar to Lakshmi Sita in *E. citriodora*. They were also successful in initiating callus from mature material by culturing stamen filaments. Conditions similar to regeneration form cotyledonary callus were used. The best hormone levels for shoot regeneration from filament callus were 0.087 mg/l IAA and 2.15 mg/l zeatin. They have also increased the shoot multiplication by using media suggested by Lakshmi Sita and Vaidyanathan [41], thereby increasing the number of shoots. This is the only report [5] of organogenesis from mature plant material in *Eucalyptus*. Regeneration form cotyledons and hypocotyledons was reported by Bennett and McComb (unpublished) in *E. diversicolor* on MS medium supplemented with 2.15 mg/l of kinetin and 1.75 mg/l of NAA or 2.15 mg/l zeatin and 0.10 mg/L μM of IAA. Mehra-Palta demonstrated regeneration from seedling [55] of *E. novaanglica*. Although regeneration of shoots has been reported for a number of species with juvenile material, callus from mature materials has not responded in the same way to the hormone treatments with adult material.

7. Somatic embryogenesis

Induction of somatic embryogenesis, either by direct induction on the explants or through callus cultures as in sandalwood [47] will probably be the most appropriate mode for faster propagation of *Eucalyptus*. Somatic embryogenesis, especially in cell suspensions, would enable their production in millions. Such embryos, which are regarded as seeds, could then be directly used for large-scale multiplication [49]. Somatic embryos can now be obtained for a number of tree species, including a large number of conifers [65]. It has been suggested [63,38] that perhaps such embryos could be encapsulated to form so-called artificial seeds, which can be stored until used. Indirect somatic embryogenesis refers to the development of embryos from callus or cell suspension or from cells or groups of cells of somatic embryos. The latter process is called repetitive somatic embryogenesis. Any cells which can develop into somatic embryos are said to possess embryogenic competence. Whether these cells are target cells or have this capacity is not yet known. For successful embryogenesis, selection of specific explant, conditioning of media, sequential transfers [47], and appropriate conditions are generally necessary. The primary events required for the induction of embryogenesis are still not clear. Researchers have been trying to induce somatic embryogenesis in eucalyptus from callus cultures almost for a decade. The earliest reports of embryogenesis are those of Lakshmi Sita [40,42] in *E. citriodora* and *E. grandis*. Ouyang *et al.* [58] and Ouyang *et al.* [59] in *E. leichow*. Recently, Muralidharan and Mascarenhas [56] and Muralidharan *et al.* [56a] have reported the induction of embryos and their subsequent development into plantlets. Details of all the three groups will be discussed and compared.

The first report in *E. citriodora* [40] showed the nodular calli resembling embryos. A later paper [42] reported the development of typical embryonal stages of globular and heart shaped stages. These structures failed to develop due to high phenolics in the medium. Embryo-like structures seem to develop from multicellular aggregates. Small groups of cells come together initially and are gradually enveloped by an outer wall forming globular, proembryonal structures. These structures further develop into heart- and torpedo-shaped structures. Embryogenesis was observed in media supplemented with MS + 1 mg/l GA; MS + 1 mg/l GA + 0.3 mg/l BA; MS + 1 mg/l GA + 0.3 mg/l IAA and MS + 2–5 mg/l BAP + 1.0 mg/l NAA. According to all evidence in the literature the induction of embryoids take place only in the presence of auxins. However, earlier in sandalwood and now in *Eucalyptus*, GA was found to induce embryogenesis. In addition, BAP supplemented with NAA also induced embryogenesis. Although cytokinins are often regarded as inhibitory to somatic embryogenesis, a few reports contradict the general opinion. In some cases even a promoting influence on somatic embryo formation by cytokinin has been reoorted [39]. In a recent paper [45], we have traced the developmental sequence of embryo-like structures. Cell suspensions from 2,4-D medium were transferred to MS supplemented with 1 mg/l each of BAP, kinetin, NAA and 2,4-D in Steward flasks. After two weeks fine granular structures were subcultured into MS supplemented with 3 mg/l NAA and 1 mg/l BAP, where they developed into typical embryonal structures 125–500 μ in length with green pigmentation and the flask was full of embryoids. Embryoid formation was also observed in the presence of high levels of kinetin, BAP, zeatin and GA. Embryoids were plated on several media suggested in the literature, including abscissic acid and gibberellic acid, but they could not be germinated.

In the absence of germination, we have tried to trace the formation of embryoids. Structures were found resembling zygotic embryos, two-celled, four-celled, eight-celled etc. These have developed from cell aggregates. These aggregates have highly meristematic regions. These cells seem to develop into embryos. All developmental stages can be seen in the same suspension. Ouyang *et al.* [58] reported the embryo derived plantlets. Suitable media were developed. With calluses obtained from young leaves, stem and buds, seed embryos could be established easily; only callus from seed embryos (lower embryonic axis) of sterile seedlings of *E. leickow* was found to be superior for embryo induction. The growth of seed/embryo callus was slow initially requiring one to three months. During multiplication of cell colonies numerous complete plantlets were formed, with a transplant success rate of 90%. (The author saw only the translation of the paper supplied by Jen McComb. The original paper was not seen.) Embryoids are usually red, or deep red in colour; they can be isolated by using a dissecting microscope. Early pro-embryonal stages along with fully developed plantlets were found in the same test tube. Transplantation from laboratory to land was found to be difficult. The results of this group are very interesting, but details of basal media etc. were not given. According to the paper the following results are given.

Basal media

B_5, H, and improved H media. pH 5.6–6.0.

1. *Callus inducing media*

(a) B_5 + 0.8 – 1.0 mg/l Kinetin + 2.0 mg/l 2,4,-D + Sucrose 40 gm/l.
(b) H + 1.0 mg/l Kinetin + 2.0 mg/l 2,4-D + Sucrose 50 gm/l.
(c) Improved H + 1.0 mg/l Kinetin + 2.0 mg/l 2,4.D + Sucrose 50 gm/l.

2. *Differentiation media*

(a) H + 0.1 mg/l BAP + 0.5 mg/l NAA + 50 gm/l sucrose.
(b) Improved H + 1.0 mg/l BAP + 0.5 mg/l NAA.
(c) Improved H + 0.5 mg/l IBA.

In a second paper [59] the same authors slightly modified the conditions and obtained thousands of plants. It would be most interesting to apply these experimental conditions to other eucalypts.

Muralidharan *et al.* [56a] have reported embryogenesis and subsequent plantlet development from *E. citriodora* with seed callus as the starting material. Direct regeneration of embryos was obtained from cotyledons on B_5 media. Embryogenic callus was maintained on B_5 media supplemented with casein hydrolysate 500 mg/l; glutamine 500 mg/l, NAA 5 mg/l and sucrose 30 mg/l in darkness at 27° C. Regeneration was obtained with reduced sucrose (20 mg/l) under continuous light with minerals and vitamins only. No growth hormones were used. Embryos could be multiplied by repetitive embryogenesis to form clusters of embryos. Embryo development was asynchronous and all stages could be found simultaneously in the same tube, similar to Ouyang *et al.* [59]. An average of 200 + 50 embryos were found per gram of embryogenic tissue. Embryogenesis potential remained for a long time, suggesting the potential for large scale multiplication.

While these results [56,58,59] are very interesting, embryos have originated from 2-day old germinating callus. The immature zygotic embryos and ovule nucellus are believed to be composed of pre-embryogenic determined cells (PEDCS) which require only disruption of the controls that integrate their behaviour as a group, permissive growth conditions, and mitotic stimulus to express their somatic embryogenesis [22,64]. In other words, embryogenesis from tissues without PEDCS is still a challenge. Work from our laboratory seems to be promising since it offers a method for the induction of embryos from tissue without PEDCS. In contrast, cells that are not programmed for embryogenesis, like shoot or leaf callus, require differentiation into callus or suspension culture and subsequently a redetermination event to induce the embryogenic state. The resulting induced embryogenic determined cells (IEDCS) are developmentally equivalent to PEDCS, but because of their passage through a destablished callus phase are more likely to give rise to somaclonal variants [63]. Immature embryos are known to be favourable material for the induction of stable totipotent cultures capable of regeneration by somatic embryogenesis [64,67]. The results obtained by Lakshmi Sita can probably form a basis for future research on the induction of embryogenesis from cell tissues originating from mature tissues. Even in embryogenesis the reported [45] media are some what similar to those used by Muralidharan *et al.* [56a] for *E. citriodora*. Sequential changes in our experiments resulted in the formation of embryoids. These, however, failed to germinate. Treatments with ABA and GA did not help. In view of its importance, development of reproducible methods of somatic embryogenesis will be the greatest challenge in *Eucalyptus*. Induction of direct somatic embryogenesis may restrict the level of variation in culture. It also provides an important tool for studies of embryogeny. The quality of embryoids [45] from callus of adult origin, needs to be improved to provide embryos with structural physiology and vigour comparable to zygotic embryos.

8. Anther culture/embryo culture/protoplast culture

Anther culture has not been successful by itself for obtaining haploids. Bennett and McComb, however, have induced callus by culturing stamen filaments and regeneration of shoots and plantlets was reported as described in the section on organogenesis. Embryo culture was done by Muralidharan *et al.* [56a] as a source material for callus induction and subsequent embryogenesis as reported in the section on embryogenesis. Embryo culture has not been done to rescue any interspecific or intergenic crosses. Work is in progress to culture protoplasts of *Eucalyptus* species at Rutgers University (unpublished). There is one report on the isolation of protoplasts from *E. gunni* and *E., pauciflora*. The precise method is not given, but good yield of protoplasts are reported from leaves of very young seedlings grown in the greenhouse or growth chambers [66]. Protoplast formation from field grown plants was poor. *In vitro* grown shoots may be a better source for protoplasts isolation.

9. Commercialization of *Eucalyptus*

The paper and pulp industry in India is interested in the clonal propagation of elite *Eucalyptus* species. Laboratories have been sent up by Tata tea and Co., Harihar Polyfribres, Unicorn Biotechnology, Bhadrachalam Paper Mills and South India Viscose based on the technology developed in this Country. *Eucalyptus*, in addition to its use in the paper and pulp industry, is also popular as a rapidly growing fuel wood. Hence there is a great interest, in spite of the great controversy over *Eucalyptus* in relation to the water table. It is suggested as one of the fast growing trees to be planted in non-agricultural lands.

10. Summary

In view of their astonishing growth characteristics, climatic adaptability and wide ranging usefulness, eucalypts are increasingly being regarded as amongst the most important trees available for human exploitation. Though eucalypts are considered to be Australian in origin, they have been introduced widely into many parts of the world, including India. In India, eucalypts were introduced around 1790. Since then, the genus has become an important commercial crop. Five most important species cultivated in India are *E. territicornis*, *E. grandis*, *E. citriodora, E. globulus*, *E. camaldulensis.* Of these, *E. globulus* (blue gum) is considered the rapid grower. In order to mass produce the superior clones successfully, tissue culture methods have been developed. Various techniques developed for the different species are described. Micropropagation by means of enhanced axillary branching is the most successfully exploited technique. Differentiation by organogenesis and somatic embryogenesis, although reported, has not yet reached the stage of commercialization. Extensive investigation is needed in all the five species which are commonly grown in order to exploit the potential of somaclonal variation, encapsulation of embryos and any other future programs, like gentic engineering. Existing technology available for eucalypts is adequate enough for commercialization. At least 4–5 private companies have already set up tissue culture labs. The National Bank for Agriculture and Rural Development (NABARD) is one of the major financial supporters of the extensive application of tissue culture for mass propagation of eucalypts.

The usefulness of eucalyptus as a fast growing species is recognized beyond doubt and eucalypts have come to stay in India as one of the main species for commercial planting. Intensive work is needed for selecting the species, matching each site and climate, and to evolve ideal techniques for planting, so that the full potential of the species can be utilized at each site. There is strong evidence to show that interspecific hybrids frequently will out-perform pure species in terms of desirable produce. The application of tissue culture for the multiplication of such hybrids will be most suitable. *Eucalyptus* is still underexploited in terms of basic research, such as protoplast, isolation and fusion. Any successful future genetic engineering programs will depend upon the demonstration of cell totipotency. In other words, research on regeneration from protoplast/cell cultures by somatic embryogenesis needs to be strengthened.

References

1. Aneja S and Atal CK (1969) Plantlet formation from tissue cultures from lignotubers of *Eucalyptus citriodora* Hook. Curr Sci 38:69–71.
2. Bajaj YPS (1986) Biotechnology in Agriculture and Forestry. Springer/Verlag Berlin, Heidelberg, New York, Tokyo.
3. Bajaj YPS (1986) In: Biotechnology of Tree Improvement for Rapid propagation and Biomass Energy Production. pp. 1–23.
4. Barker PK, Fossard De RA, and Bourne RA (1977) Progress toward clonal propagation of *Eucalyptus* species by tissue culture techniques. Int Plant Propagators Soc Comb Proc 27:546–556.
5. Bennett IJ and McComb JA (1982) Propagation of jarran (*Eucalyptus marginata*) by organ and tissue culture. Aust For Res 12:121–127.
6. Bonga JM (1977) Application of tissue culture in Forestry. In: Reinert J and Bajaj YPS (Eds.) Pant Cell Tissue and Organ Culture, Springer-Verlag, Berlin. pp. 93–108.
7. Bonga JM (1987) Clonal propagation of mature trees: problems and possible solutions. In: Bonga JM and Durzan D (Eds.) Cell and Tissue Culture in Forestry, Vol. 1. pp. 249–271.
8. Bonga JM and Durzan D (1987) Cell and Tissue Culture in Forestry, Specific principles and methods: growth and development Vol. 1, Vol. 2, Vol. 3.
9. Boulay M (1983) Micropropagation of frost resistant Eucalyptus. In: Proceedings of a workshop on Eucalyptus in California, June 14–16th, Sacramento, California. pp. 102–107.
10. Chowdry YP (1984) Utilisation of Eucalyptus wood in Punjab. In: Sharma JK, Nair TSC, Kedarnath S, and Kongas S (Eds.) Eucalyptus in India. pp. 60–67.
11. Cresswell RJ (1971) The veg propagation of eucalyptus by organ culture, University of New England, Honours thesis.
12. Cresswell RJ and Fossard De RA (1974) Organ culture of *Eucalyptus grandis*. Aust For 37:55–69.
13. Cresswell RJ and Nitsch C (1975) Organ culture of *Eucalyptus grandis* L. Planta 125: 87–90.
14. Fossard De RA (1974) Tissue culture of *Eucalyptus*. Aust For 37:43–54.
15. Fossard De RA and Bourne RA (1976) Vegetative propagation of *Eucalyptus ficifolia* F. Muell. by nodal culture *in vitro*. Int Plant Propagators Soc Comb Proc 26:373–378.
16. Fossard De RA, Nitsch C, Cresswell Ra, and Lee ECM (1974) Tissue and organ culture of eucalyptus, NZ J For Sci 4:267–278.
17. Fossard De RA, Barker PK, and Bourne RA (1977) The organ culture of nodes of four species of eucalyptus. Acta Hortic 78:297–302.
18. Fossard De RA, Bennett MT, Gorst JR, and Bourne RA (1978) Tissue culture propagation of *Eucalyptus ficifolia*. Int Plant Propagators Soc Comb Proc 28:427–435.
19. Fossard De RA (1978) Tissue culture propagation of *Eucalyptus ficifolia* F. Muell. In: Proceedings of a Symposium on Plant Tissue Culture. Science Press, Peking, China. pp. 425–438.

20. Diallo H and Duhoux E (1984) Organogenesis et multiplication *in vitro Eucalyptus camaldulensis*. J Plant Physiol 115:177–182.
21. Durand-Cresswell R, Boulay M, and Franclet A (1982) Vegetative propagation of *Eucalyptus*. In: Bonga JM and Durzan DJ (Eds.) Tissue Culture in Forestry. Martinus Nijhoff, The Hague, FAO, 1979 Eucalyptus for Planting. FAO. pp. 158–181.
22. Evans DA, Sharp WR, and Flick CE (1981) Growth and behaviour of cell cultures. Embryogenesis and organogenesis. In: Thorpe TA (Ed.) Plant Tissue Culture: Methods and applications in Agriculture, Academic Press. pp. 45–113.
23. Franclet A and Boulay M (1982) Micropropagation of frost resistant *Eucalyptus* clones. Aust J For Res 13:83–89.
24. Franclet A (1983) Rejuvenation-theory and practical experiences in clonal silviculture. Paper for Canadian Tree Improvement Association. 19th Biennial Meeting. Toronto.
25. Gorst JR (1981) Effect of indole-3-butyric acid and riboflavin on the morphogenesis of adventitious roots of *Eucalyptus ficifolia* F. Muell. grown *in vitro*. MSc Thesis, Univ. of New England, Armidale, Australia.
26. Grewal S, Ahuja A, and Atal CK (1980) Clonal multiplocation of medicinal plants by tissue culture. 5. *In vitro* proliferation of shoot apices of *Eucalyptus citriodora* Hook. Indian J Exp Biol 18:775–776.
27. Gupta PK, Mascarenhas AF, and Jagannathan V (1981) Tissue culture of forest trees – clonal propagation of mature trees of *Eucalyptus citriodora* Hook by tissue culture. Plant Sci Lett 20:195–201.
28. Gupta PK and Mascarenhas AF (1983) Essential oil production in relation to organogenesis in tissue culture of *Eucalyptus citriodora* Hook. In: Giles KL and Sen SK (Eds.) Plant Cell Culture in Crop Improvement, Plenum Press, New York. pp. 299–308.
29. Gupta PK, Mehta UJ, and Mascarenhas AF (1983) A tissue culture method for rapid propagation of *Eucalyptus camaldulensis* and *E, torelliana* from mature trees. Plant Cell Rep 2:296–299.
30. Gupta PK and Mascarenhas AF (1987) Eucalyptus. In: Bonga JM and Durzan DJ (Eds.) Cell and Tissue Culture in Forestry. Vol. 3. Martinus Nijhoff, Dordrecht/Boston/Lancaster, pp. 385–399.
31. Hartney VJ (1980) Vegetative propagation of the eucalyptus. Aust For Res 10:191–211.
32. Hartney VJ and Barker PK (1980) The vegetative propagation of eucalyptus by tissue culture. In: Fast Growing Trees IUFRO Congress, Sao Paulo, Brazil. pp. 1–7.
33. Hartney VJ (1981) Vegetative propagation of eucalyptus *in vitro*. In: Proc IUFRO Sect s2 01 5. Int Workshop 'In Vitro' Cultivation For Tree Species, Fontainebleau, France, AFOCEL. pp. 175–180.
34. Hartney VJ (1982) Tissue Culture of Eucalyptus. Int Plant Propagators Soc Comb Proc 32:98–109.
35. Jacquiot D (1964) Application de la technique de culture de tissus vegetaux a l'etude de quelque problemes de la physiology de l'arbre. Ann Sci For (Paris) 21:311–465.
36. Kedarnath S (1986) Preface. In: Eucalyptus in India Past Present and Future.
37. Kitahara EH and Caldas LS (1975) Shoot and root formation in hypocotyl callus cultures of *Eucalyptus*. For Sci 21:242–243.
38. Kitto SL and Janick J (1985) Production of synthetic seeds by encapsulation of asexual embryos of carrot. J Am Soc Hortic Sci 110:277–282.
39. Kohlenbach HW (1977) Basic aspects of differentiation and plant regeneration from Cell and Tissue Cultures. In: Barz W, Reinhard E, and Zenk M (Eds.) Plant Tissue and its Biotechnological Applications. Springer-Verlag. pp. 355–366.
40. Lakshmi Sita G (1979) Morphogenesis and Plant regeneration from cotyledonary cultures of Eucalyptus. Plant Sci Lett 14:63–65.
41. Lakshmi Sita G and Vaidyanathan CS (1979) Rapid multiplication of *Eucalyptus* by multiple shoot production. Curr Sci 48:350–352.
42. Lakshmi Sita G (1982) Tissue Culture of Eucalyptus. In: Rao AN (Ed.) Tissue Culture of Economically important plants. Singapore COSTED/ANBS. pp. 180–184.

43. Lakshmi Sita G and Shobharani B (1985) *In vitro* propagation of *Eucalyptus grandis* L. by tissue culture. Plant Cell Rep 4:63–65.
44. Lakshmi Sita G, Shobharani B, and Rao KS (1986) Propagation of *Eucalyptus grandis* by Tissue Culture. In: Eucalyptus in India: Past Present and Future. pp. 318–321.
45. Lakshmi Sita G (1986) Progress towards clonal propagation of *Eucalyptus grandis.* In: Withers LA and Alderson PG (Eds.) Plant Tissue Culture and its Agricultural applications. Butterworths. pp. 159–166.
46. Lakshmi Sita G and Sukanya C (1987) Improvement of forest trees by tissue culture. In: Reddy GM (Ed.) Plant Cell and Tissue Culture of economically important plants. Proc of the National Symposium. pp. 195–198.
47. Lakshmi Sita G (1986) Sandalwood. In: Bajaj YPS (Ed.) Biotechnology in Agriculture and Forestry, Springer-Verlag. pp.363–374.
48. Lee ECM and Fossard De (1974) The effect of various auxins and cytokinins on the *in vitro* culture of stem and lignotubers tissues of *Eucalyptus bancroftii* Maiden, New Phytol 73:707–717.
49. Lutz JD, Wong JR, Rowe J, Tricoli DM, and Lawrence RH Jr (1985) Somatic embryogenesis for mass cloning of crop plants. In: Henko RR, Hughes KW, Constantin MJ, and Hollaender A (Eds.) Tissue Culture in Forestry and Agriculture, Plenum Press. pp 105–116.
50. Maheswaran G and Williams EG (1984) Direct somatic embryoid formation on immature embryos of *Trifolium repens*, *T. pratense* and *Medicago Sativa*, and rapid clonal propagation of *T. repens.* Ann Bot. 54:201–211.
51. Mascarenhas AF, Gupta. PK, Kulkarni VM, Mehta U, Iyer RS, Khuspe SS, and Jagannathan V (1981) Propagation of trees by tissue culture. In Rao AN (Ed.) Proc COSTED Symp on Tissue Culture of Economically Important Plants. Singapore. pp. 175–179.
52. Mascarenhas AF, Hazara S, Potdar U, Kulkarni DK, and Gupta PK (1982) Rapid clonal multiplication of mature forest trees through tissue culture. In: Fujiwara A (Ed.) Plant Tissue Culture 1982. Proc 5th Int Cong Plant Tissue Cell Cult, Tokyo, Japan. pp. 719–720.
53. McComb JA and Bennett IJ (1986) In: Bajaj YPS (Ed.) Biotechnology in Agriculture and Forestry. pp. 340–362.
54. Meins F Jr. (1983) Heritable variation in Plant Cell Culture. Ann. Rev. Plant Physiol. 34:327–346.
55. Mehra-Palta A (1982) Clonal propagation of *Eucalyptus* by tissue culture. Plant Sci Lett 26:1–11.
56. Muralidharan EM and Mascarenhas AF (1987) Plant Cell reports 6:256–259.
56a. Muralidharan EM, Gupta PK, and Mascarenhas AF (1989) Plantlet production through high frequency somatic embryogenesis in long term cultures of *E. citriodora*. Plant Cell Reports 8:41–43.
57. Murashige T and Skoog F (1962) A revised medium for growth and bioassay with tobacco tissue culture. Physiol Plant. 15:473–497.
58. Ouyang Q, Li QQ, and Peng HZ (1980) Preliminary report on the development of embryoids from eucalyptus. Acta Phyto Physio Sin 6:429–432. (Chinese).
59. Ouyang Q, Peng HZ, and Li QQ (1981) Studies on the development of embryoids from *Eucalyptus* callus. Sci Silvae Sin 17:1–7. (Chinese).
60. Paton DM, Willings RR, Nicholls W, and Pryor LD (1970) Rooting of stem cuttings of *Eucalyptus* a rooting inhibitor in adult tissue. Aust J Bot 18:175–183.
61. Pryor LD and Willings RR (1963) The vegetative propagation of Eucalyptus – an account of progress. Aust For 27:52–62.
62. Sharma JK, Nair CTS, Kedarnath S, and Kondas S (1984) Eucalyptus in India Past, Present and Future, Kerala Forest Research Institute, Peechi, Kerala, India.
63. Sharp WR, Evans DA, and Sondahl MP (1982) Application of somatic embryogenesis to crop improvement. In: Fujiwara A (Ed.) Plant Tissue and Cell Culture. Maruzen, Tokyo. pp. 759–763.
64. Sharp WR, Sondahl MR, Caldas LD, and Maraffa SB (1980) The physiology of *in vitro* asexual embryogenesis. Hortic Rev 2:268–310.

65. Tulecke W (1987) Somatic embryogenesis in woody perennials. In: Cell and Tissue Culture in Forestry. Vol. 2. Specific principles and methods; growth and development. Martinus Nijhoff, Dordrecht, pp. 61–91.
66. Venkateswaran S and Gandhi V (1980) Protoplast isolation and culture from tree genera. In vitro 16:232 (abstract).
67. Williams WM and Maheswaran G (1986) Somatic embryogenesis: Factors influencing coordinated behaviour of cells as an embryogenic group. Ann Bot 57:443–462.
68. Yamaguchi T and Fukuzami T (1982) Volatile compounds from callus cells of woody plants. In: Fujiwara A (Ed.) Plant Tissue Culture 1982, Proc 5th Int Cong Plant Tissue Cell Cult, Tokyo Japan. pp. 287–288.

17. *In vitro* culture of *Liriodendron tulipifera*

SCOTT A. MERKLE, H. DAYTON WILDE and H.E. SOMMER
School of Forest Resources, University of Georgia, Athens, GA 30602, U.S.A.

1. Introduction

1.1. *Distribution and importance of* Liriodendron tulipifera

Liriodendron tulipifera L. (yellow-poplar, tulip-poplar, tuliptree), a native of eastern North American, is one of two extant species of *Liriodendron* (Magnoliaceae). The other, *L. chinense* (Hemsl.) Sarg. (Chinese tuliptree), is native to central mainland China. The natural range of yellow-poplar is throughout the eastern United States and into southern Ontario, Canada. Although it grows on a wide variety of sites, it is most abundant and reaches its largest size in the lower Ohio river valley and in the mountains of North Carolina, Tennessee, Kentucky and West Virginia [12]. Yellow-poplar is an early stage successional species on most sites, but its rapid growth rate and the large size of mature trees (up to 60 m), make it a dominant canopy tree. Yellow-poplar possesses many desirable traits which make it one of the most important hardwood species in the United States. Besides rapid growth, it is also known for its straight form, self-pruning ability and wood of exceptional working quality [49]. Large amounts of the wood are used for furniture (mostly hidden parts), plywood, corestock, millwork, siding and other light construction lumber. It is also used for pulping and for products such as chipboard and flakeboard [36]. Bees are a primary pollinator of yellow-poplar, producing a popular honey. Finally, with its symmetrical leaves and yellow, green and orange flowers, yellow-poplar is widely planted both in the United States and abroad as an ornamental species. A number of horticultural cultivars have been described for the species [39].

1.2. *Genetics and breeding*

Natural genetic variation within yellow-poplar species has been characterized for a number of phenotypic traits. The largest proportion of variation in wood specific gravity was found to be associated with individual trees, rather than among stands or physiographic areas [19,47,48]. Fiber length varied most highly within individual trees, versus among trees or stands [48]. Kellison [19] reported that yellow-poplar displayed clinal patterns of variation for a

M. R. Ahuja (Ed.), Micropropagation of Woody Plants, 281–302.

number of morphological characters among North Carolina populations. Using six polymorphic allozyme loci as markers, Brotschol *et al.* [4] reported that yellow-poplar populations from six locations in North Carolina were highly differentiated with respect to allele frequencies in both mature tree and seedling populations. However, provenance studies by Sluder [40] and Farmer *et al.* [10] indicated that height growth differences among seedlings from different geographic sources were not significant.

Although genetic improvement guidelines have been established for yellow-poplar, breeding programs have been limited and only a few states have production seed orchards for yellow-poplar. Clonal seed orchards can easily be established by grafting [49], and bee colonies are maintained in the orchards to obtain adequate pollination. Wilcox and Taft [49] developed alternative selection criteria for individual trees to be included in an improvement program for either lumber and veneer production or pulpwood production. For lumber and veneer production, highest priority was given to form and freedom from wood defects, while for pulpwood production, growth rate was given more weight. Heritability estimates for height growth of full-sib, selfed and open-pollinated progeny from yellow-poplar clones in two Tennessee seed orchards indicated that 15 to 28% genetic gain in 5-year height could be obtained by selecting the top 10% of full-sib families [11]. Rapid progress was also believed to be possible in changing wood specific gravity, since individual trees displayed high amounts of variation in this characteristic [49].

Another possible direction for yellow-poplar tree impovement lies in its potential for hybridization with Chinese tuliptree. Despite evidence from molecular marker analysis indicating that *L. tulipifera* and *L. chinense* have been separated for 10–16 million years [33], the two species are highly interfertile. There have been at least three programs of interspecific hybridization [7,32,38], all of which reported evidence of heterosis in the hybrid trees. Parks *et al.* [32] reported that interspecific hybrids displayed heterosis in the form of significantly greater biomass than intraspecific hybrids at 9 years of age, and produced a vigorous F_2. Parks *et al.* [32] also noted that in flowers produced by the interspecific hybrids, orange pigment was present in most of the outer petal surface, rather than being restriced to a narrow band toward the insertion of the petals, as is typical with *L. tulipifera*. Examination of the inheritance of isozyme markers in *L. tulipifera* × *L. chinense* hybrids and backcross progeny to *L. tulipifera* indentified 16 putative segregating loci [34]. One strong linkage and several weak linkages between these loci were demonstrated. Segregations followed Mendelian expectations and no evidence of abnormal genetic behavior was detected in the segregation of the isozyme alleles in the F_1 hybrids.

1.3. *Conventional propagation methods*

The major propagation method for yellow-poplar is via seeds. Yellow-poplar seeds are produced in samaras which are borne in terminal aggregates. Trees generally do not begin to produce seeds in significant quantities until they are at least 15 years old. Trees over 2 centuries old are known to still produce seeds [16]. Although very heavy seed crops are often produced, only approximately 10% of the samaras contain at least one filled seed. Since two seeds are contained in each samara, the true frequency of filled seeds is only 5%. Low filled seed percentages have been attributed to insufficient pollination, self-incompatibility and incompatibility between adjacent trees [18]. However, filled seed percentages can be raised substantially by artificial pollination, producing seeds that are up to 90% filled [5,45,46]. Radiography with soft X-rays has proven useful in seed testing [18]. The percentage of filled seeds indicated by this method has been shown to be highly correlated with the percentage of seeds germinating in greenhouse flats [45]. Yellow-poplar seeds have an internal dormancy which can be overcome by moist stratification at 2–3 °C for at least 60 days [3].

Yellow-poplar has a substantial capacity for natural vegetative regeneration due to its ability to sprout vigorously from stumps, forming clumps of trees. However, stands derived from sprouts are not as desirable as seedling-derived stands, as sprout-derived trees are more likely to develop heart rot than are those from seedlings [12]. Artificial vegetative propagation has been reported for yellow-poplar using rooted stem cuttings and grafting. Donovan [9] also listed the species as one that has been regenerated by root cuttings, but no details were given.

Although attempts to root cuttings from mature trees were initially disappointing, this barrier was overcome by using cuttings taken from stump sprouts, derived from juvenile dormant buds in the root collar [22]. Basal ends of the cuttings were dipped in 0.8% indolebutyric acid and the cuttings were maintained under intermittent mist, resulting in over 70% rooting. As an alternative method to cutting the tree to obtain stump sprouts, partial girdling [21] or severe crown pruning [23] were employed to remove inhibition of dormant juvenile buds without severely damaging the tree. The resulting epicormic branches could be rooted with a frequency of about 75%.

A number of grafting methods have been successfully applied to yellow-poplar, including chip and patch budding [13], whip grafts onto bare-rooted 1–0 seedlings and whip or side grafts onto 1–1 rootstocks [8,49]. Grafting has also been employed to produce rooted cuttings from trees up to 80 years old by first grafting scion material from the crowns of the trees to 1–0 rootstock and allowing the grafts to grow for 2–3 years. Then, cuttings could be taken from the new shoot growth and rooted with up to 16% success[1].

2. Callus culture and organogenesis

Micropropagation in the narrow sense (i.e. shoot or bud multiplication followed by rooting of shoots to obtain plantlets) has not been reported with yellow-poplar. In fact, prior to the somatic embryogenesis work discussed in the next section, there were only a few reports of attempts to culture yellow-poplar tissues *in vitro*. Unlignified internodes and petioles of yellow-poplar explanted onto Nitsch and Nitsch's [31] medium supplemented with 0.5 mg/l 2,4-D and 1 mg/l kinetin or Murashige and Skoog's [30] medium supplemented with 2 mg/l BAP produced abundant callus [44]. However, the callus failed to produce any differentiated structures and began to degenerate after 28 days, despite transfer to fresh medium. Furmanowa and Rzedowski [14] and Rzedowski [37] reported adventitious root production in cultures initiated from 1–3 year old stems and shoot tips on Nitsch and Nitsch's medium supplemented with 0.5 mg/l 2,4-D and 1 mg/l kinetin. Although the explanted shoot tips apparently produced leaves, it is unclear whether any complete plantlets were obtained.

3. Somatic embryogenesis

3.1. *Synopsis*

Somatic embryogenesis in yellow-poplar was reported by Merkle and Sommer [27]. Immature zygotic embryos explanted onto a modified Blaydes' [52] medium supplemented with 2,4-D and BA produced a fast-growing, pale-yellow nodular callus after 2 months in culture. Embryogenic callus transferred to basal medium produced somatic embryos within 1 month. Although conversion rates were low, some plantlets were produced when embryos with well-formed cotyledons were transferred to a Risser and White's [35] medium. More recently, Merkle *et al.* [29] reported an improved method for high-frequency production of yellow-poplar embryos from embryogenic suspension cultures. Proembryogenic masses (PEMs) fractionated on screens and plated on filter paper overlaying basal medium produced well-formed, mature embryos capable of up to 70% conversion.

3.2. *Initiation and maintenance of embryogenic cultures*

Embryogenic cultures are initiated from immature zygotic embryos. We have found that the optimum developmental stage for explanting zygotic embryos to initiate embryogenic cultures is the globular to early heart stage [43] (see section on embryogenic potential and markers). In the southeastern United States, this stage is most commonly found in seeds 8 weeks postpollination. Aggregates of tightly fused samaras are collected from trees and refrigerated.

Within a few days following collection, aggregates are dissected into individual samaras using a grafting knife. Then the samaras are dewinged and refrigerated until they are used, usually the same day. Dewinged samaras are disinfested using the following sequence: 70% ethanol (20sec), 10% Roccal (National Laboratories; 2 min), repeat ethanol step, repeat Roccal step, 100% Clorox (5.25% sodium hypochlorite; 5 min), water rinse (3 min), 0.01N HCl rinse (3 min), 3 more water rinses (3 min each). Samaras remain in the final water rinse until they are dissected to obtain the embryo and endosperm for culture.

Following disinfestation, samaras are dissected aseptically and embryos (approximately 0.5 mm long) and surrounding endosperm are placed on approximately 10 ml of semisolid induction medium in 60 mm plastic petri dishes. Yellow-poplar induction medium (referred to as 'conditioning medium' in Merkle and Sommer [27]) is a modified Blaydes' [52] medium, containing iron solution of Murashige and Skoog [30], Brown's minor salts [41], vitamins of Gresshoff and Doy [15], 1 g/l casein hydrolysate (enzymatic), 40 g/l sucrose, 2 mg/l 2,4-D, 0.25 mg/l BA, and solidified with 8 g/l Phytagar (Gibco). Six mg/l naphthaleneacetic acid (NAA) may be substituted for the 2 mg/l 2,4-D (Merkle, unpublished data). Explants are transferred to fresh induction medium monthly. Usually within 2 months of explanting, zygotic embryos begin to produce clusters of pale yellow PEMs, which in turn are made up of small (20μm diameter) densely cytoplasmic cells. Nonembryogenic callus is also produced by the explanted embryos, often from the same embryos that produce PEMs. This callus is composed of elongated vacuolate cells and shows no regeneration potential. In addition the endosperm explanted with the embryo rarely forms a colorless callus that grows slowly in comparison to the embryo derived calli.

Once produced, PEMs can be maintained by transferring monthly to fresh semisolid induction medium, or they can be used to initiate embryogenic suspension cultures by inoculating into liquid induction medium. Suspension cultures are maintained in Erlenmeyer flasks on a gryatory shaker at 90 rpm. Best proliferation of PEMs is maintained at 30°C. Embryogenic suspension cultures and cultures on semisolid medium can be maintained in either light or dark conditions, but some genotypes tend to accumulate darkening secondary metabolites when maintained in the light, and therefore appear to be healthier when kept in darkness. We have maintained some embryogenic yellow-poplar lines for as long as 5 years in suspension, although their capacity to actually produce somatic embryos has deteriorated with time.

3.3. *Somatic embryo production, maturation and conversion*

Somatic embryos can be produced from yellow-poplar embryogenic cultures simply by transferring PEMs from liquid or semisolid induction medium to basal medium (i.e. lacking growth regulators). PEMs transferred to semisolid basal medium begin to form somatic embryos usually within 4–6 weeks

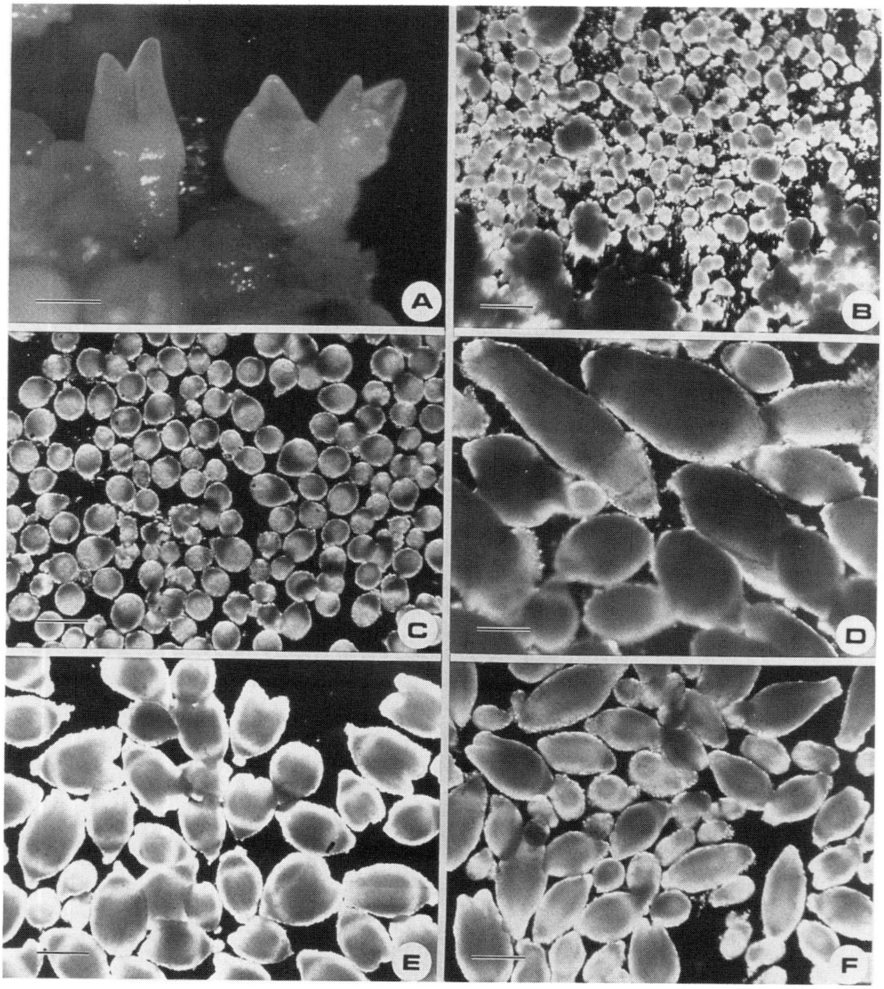

Figure 1. Effects of fractionation, ABA and sucrose on maturation of yellow-poplar somatic embryos. A. Yellow-poplar somatic embryos arising from proembryogenic masses (bar = 200 μm). B. Non-fractionated yellow-poplar proembryogenic masses (PEMs) in liquid basal medium one week following transfer from induction medium (bar = 500 μm). C. Globular stage somatic embryos derived from fractionated PEMs, 3 days following second sieving, in liquid basal medium (bar = 500 μm). D. Elongating somatic embryos derived from fractionated PEMs, 10 days following second sieving, in liquid basal medium (bar = 500 μm). E. Heart-stage somatic embryos derived from fractionated PEMs, 7 days following second sieving, in liquid basal medium with 5×10^{-7} M ABA. Note distinct cotyledons and proliferation of cells around periphery of embryos (bar = 500 μm). F. Torpedo stage somatic embryos derived from fractionated PEMs, 7 days following second sieving, in liquid basal medium with 10% sucrose (bar = 500 μm). (From Merkle *et al.* [29].

following transfer, and differentiation of new embryos continues in an asynchronous manner for up to 2 months (Figure 1A). Often fused clumps of embryos are produced. Most of the embryos forming on semisolid basal medium are malformed, often with fused cotyledons. Embryos with fused cotyledons continue to expand, followed by either radicle elongation (i.e. germination), callusing or production of secondary embryos, usually from the area of the radicle. Germination of malformed embryos usually is not followed by apical development, and thus these embryos do not form complete plantlets. Less than 1% of the total embryos appear to mature to a stage resembling that of zygotic embryos found in seeds. It is mainly these embryos that are capable of conversion to plantlets. PEMs transferred from liquid induction medium to liquid basal medium produce mainly clusters of fused embryos within a month following transfer, although some singularized embryos are formed as well (Figure 1A). As with semisolid medium, embryos allowed to differentiate from PEMs in liquid medium develop asynchronously, and do not appear to mature properly. Usually, distinct cotyledons do not develop before radicle elongation begins, resulting in malformed germinants.

Mature, well-formed embryos will eventually germinate if maintained on basal medium with CH. However, germination can be accelerated if mature embryos are transferred to basal medium lacking CH, which significantly lowers the osmotic potential of the medium [29]. Vigorous germination usually follows within 1 week of transfer. Germinants are then routinely transferred to a Risser and White's [35] plantlet development medium, with Murashige and Skoog's [30] iron solution, Brown's [41] minor salts, Gresshoff and Doy's [15] vitamins, 20 g/l sucrose and solidified with 8 g/l Phytagar. Embryos are placed either in 20 × 150 mm test tubes containing 20 ml of medium or Magenta boxes containing 100 ml of medium, with radicles or developing roots inserted into the medium. Plantlets are grown under cool white fluorescent light (16h/day) at 25 °C.

The problems with embryo maturation and plantlet production described above result in a relatively low efficiency of plantlet production, with less than 1% of the embryos forming plantlets. Therefore, a series of alternative methods for embryo and plantlet production from embryogenic suspension cultures was tested [29]. These experiments resulted in the protocol we currently employ to produce large numbers of mature, well-formed embryos capable of conversion to plantlets at high frequencies. Populations of roughly (60–70%) synchronous somatic embryos can be produced using two protocols, both of which employ fractionation of embryogenic suspensions on stainless steel sieves. In one method (Figure 2), 1 g of PEMs cultured in liquid induction medium is placed in liquid basal medium and sieved on a 140 μm screen. The fraction passing through this screen is then sieved on a 38 μm screen and the fraction remaining on the 38μm screen is backwashed from the screen into a 50 ml Erlenmeyer flask where PEMs are cultured in 20 ml of liquid basal medium. Cultured PEMs are maintained in the light at 30 °C on a gyratory shaker at 90 rpm. After 1 week, developing embryos derived from

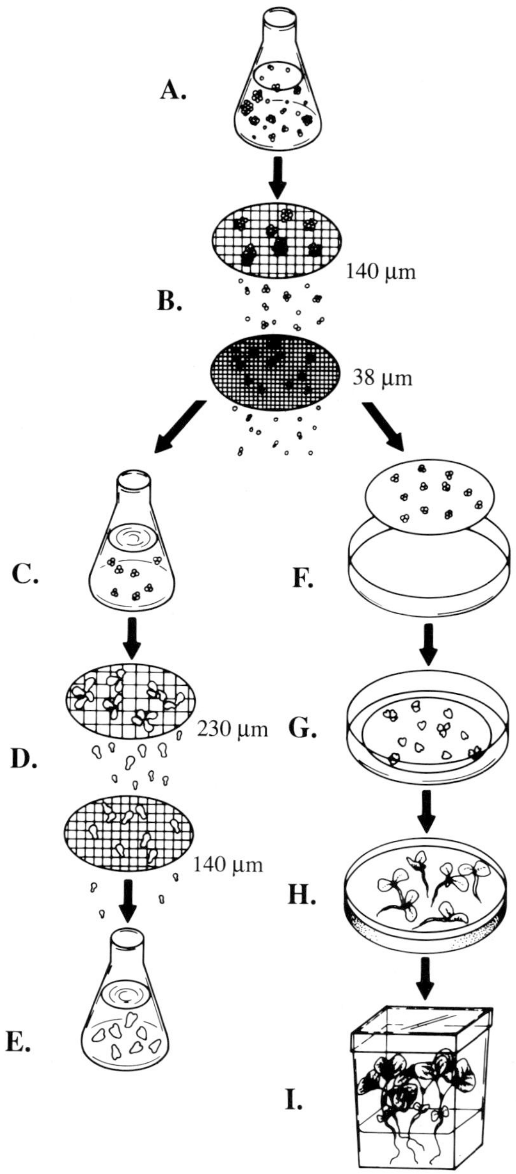

Figure 2. Fractionation/synchronization of embryogenic yellow-poplar suspension cultures. A. Embryogenic suspension cultures are grown in shaken flasks of liquid induction medium. B. PEMs are fractionated on stainless steel sieves, saving fraction that passes 140 μm but not 38 μm. C. Saved fraction is cultured for one week in liquid basal medium. D. Globular stage embryos are fractionated again to eliminate clusters and free cells, saving fraction that passes 230 μm but not 140 μm. E. Globular stage embryos are cultured an additional 7–10 days in basal medium to obtain synchronous heart-torpedo stage embryos. Alternatively, following first fractionation, F. PEMs are immediately plated in filter paper, which is placed on semisolid basal medium. G. PEMs are cultured on filter paper/basal medium for 12–14 days to obtain synchronous, mature embryos. H. Mature embryos are transferred to basal medium without filter paper to promote germination. I. Germinants are transferred to plantlet development medium, from which plantlets are ready for transfer to soil mix in 6–8 weeks. (From Merkle *et al.* [25]).

the PEMs are sieved on a 230 μm screen. The fraction passing this screen is sieved on a 140 μm screen and the fraction remaining on the 140 μm screen is cultured in fresh liquid basal medium. Fractionation of PEMs not only has the effect of synchronizing development of embryos, but also accelerates the differentiation of embryos from PEMs almost three-fold over the methods described above, yielding approximately 90% pure globular stage embryos within 10 days after the first sieving (Figure 1C). In addition, the double fractionation greatly reduces the number of clusters of embryos and free cells in the medium. However, embryos maintained in basal medium following the second fractionation still fail to mature properly. Few embryos proceed though the classic heart or torpedo stages of development, and instead begin elongation, in a similar manner to nonfractionated embryos (Figure 1D). Thus synchonization in itself does not appear to improve embryo maturation.

Development of embryos from synchronized PEMs in suspension culture can be modified using abscisic acid (ABA) or osmoticum treatments, both of which slow embryo expansion and promote maturation, at least in terms of the outward appearance of the embryos. Liquid basal medium was supplemented with ABA at levels from 10^{-7} M to 10^{-5} M to test its impact on the development of PEMs to embryos following fractionation. We found that at 10^{-5} M, ABA appears to block the development of yellow-poplar somatic embryos past the globular stage, and promote proliferation of cells around the perifery of the embryos. At 10^{-7} M, ABA has little effect on maturation. However, 10^{-6} M ABA appears to provide a balance between the two responses, allowing embryos to complete maturation while preventing premature expansion. Embryos cultured with this concentration of ABA will continue development to the heart or early torpedo stage within a few weeks following fractionation (Figure 1E). Similarly, 8% sorbitol or mannitol will slow or completely halt embryo development, although these osmotica do not appear to promote maturation at this or any level. Sucrose at 10% also slows embryo development, but unlike the other osmotica, allows continuing development and some degree of both maturation and elongation (Figure 1F).

Although yellow-poplar embryos that appear mature and well formed can be produced using a combination of synchronization and ABA or sucrose, the majority of these embryos fail to germinate when transferred to semisolid basal medium lacking CH. With few exceptions, those that do germinate fail to undergo shoot expansion, resulting in very few plantlets. It is possible that the prolonged exposures to ABA or osmoticum induce a dormancy or quiescence that the embryos cannot overcome when moved to germination conditions. We have tested a number of treatments designed to break this quiescence, including culture on agar-water for 1 week, 1-hour shake-rinses in distilled water or gibberellic acid (GA) at 10^{-6} – 10^{-4} M, and cold treatment at 6 °C for up to 3 months. Various modifications of the basal medium, including addition of 2 g/l activated charcoal, 10^{-5} or 10^{-6} M GA, or 10^{-5} or 10^{-6} M BA, also had no impact on embryo conversion frequency.

Due to these problems with conversion of yellow-poplar somatic embryos following prolonged culture in liquid medium, another method employing fractionation of embryogenic suspension cultures was tested, in which the objective was to obtain large numbers of roughly synchronous embryos from PEMs on semisolid medium rather than in liquid (Figure 2). In this protocol, 1 g of PEMs is transferred from liquid induction medium to liquid basal medium and sieved on a 140 μm screen. The fraction passing this screen is resieved on a 38 μm screen and the fraction remaining on the 38 μm screen is backwashed from the screen onto a single layer of filter paper in a Buchner funnel. PEMs on the filter paper are rinsed with liquid basal medium using a Pasteur pipet, and excess liquid medium is removed from the filter paper using low vacuum. Then the filter paper with PEMs is placed directly onto a 60mm plastic Petri dish containing 10 ml of semisolid basal medium. PEMs are incubated at 30° C under a 16 hour photoperiod with cool white fluorescent light. This procedure results in a population of 60–70% synchronous embroyos (Figure 3A). Similar results can be obtained by substituting 120 μm pore size nylon mesh for filter paper, for collection and plating of PEMs (Figure 3F).

Sieving followed by plating on filter paper eliminates the effects of having embryos in liquid medium. The procedure enriches the culture for embryogenic cells, and may almost completely eliminate nonembryogenic cells, since embryos appear to develop derectly from PEMs, with no underlying bed of callus. Another apparent effect of employing the PEM fraction between 38 and 140μm is that this fraction of PEMs yields 1 somatic embryo per PEM, resulting in approximately 90% singularized embryos, versus fused clusters of embryos (Figure 3B).

Somatic embryos produced using the fractionation/plating protocol usually reach the globular stage within 8 days following plating of PEMs, and appear mature within 14 days following plating (Fifure 3C). Depending on the clonal line used, up to 950 embryos per gram of fractionated PEMs can be obtained, and up to 720 of these will reach a developmental stage resembling a mature zygotic embryo (i.e. 1–2 mm long with 2–3 distinct cotyledons). When mature embryos abtained by this method are germinated on semisolid basal medium without CH (Figure 3D, 3E), followed by transfer to plantlet development medium (see above), up to 100% conversion can be obtained, again depending on clonal line.

3.4. *Plantlet acclimatization: nursery and field handling*

Following 8–10 weeks of growth in plantlet development medium, yellow-poplar plantlets derived from somatic embryos usually have produced 4–5 leaves and a 2–3 cm long tap root. At this stage, they can safely be transferred to potting mix and placed in a humidifying chamber for acclimatization. Before plantlets are removed from culture, book-design 'Hillson' root trainers (Spencer-Lamaire) are filled with Promix (Fafard), a peat-vermiculite

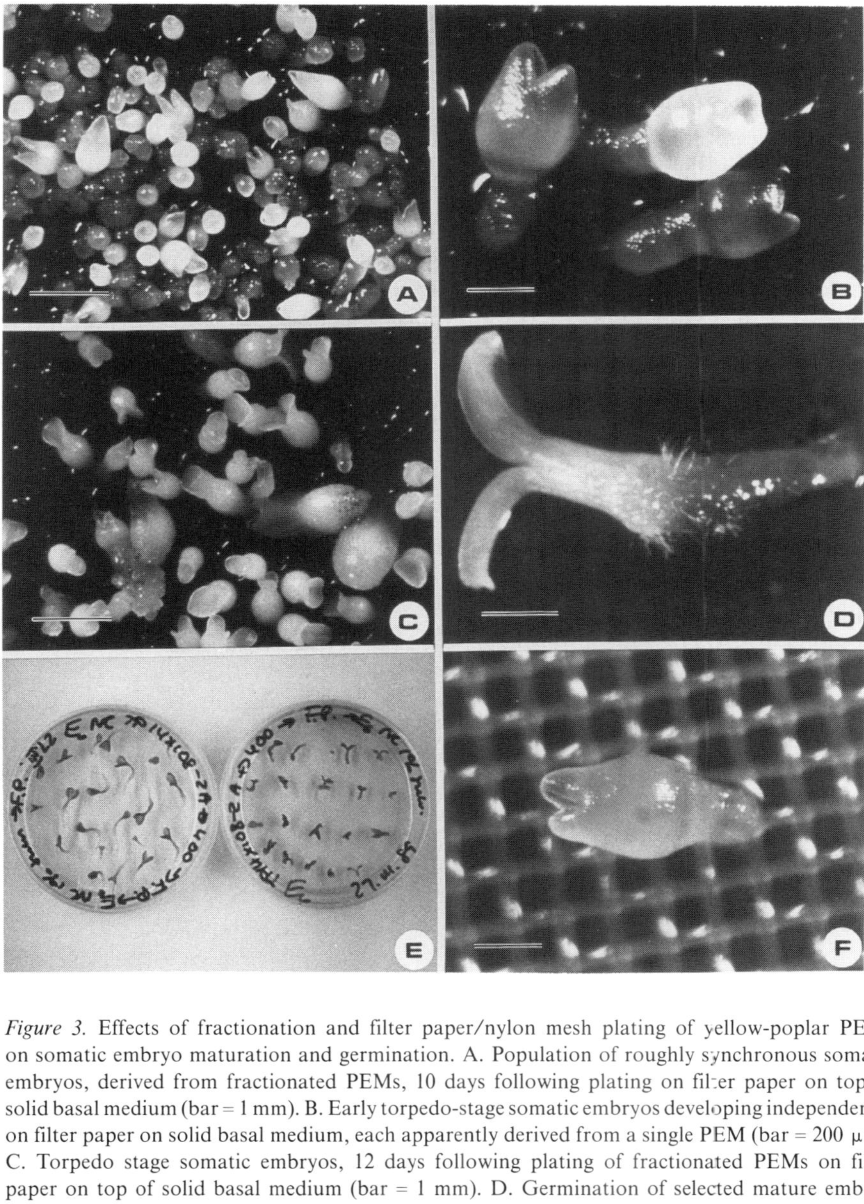

Figure 3. Effects of fractionation and filter paper/nylon mesh plating of yellow-poplar PEMs on somatic embryo maturation and germination. A. Population of roughly synchronous somatic embryos, derived from fractionated PEMs, 10 days following plating on filter paper on top of solid basal medium (bar = 1 mm). B. Early torpedo-stage somatic embryos developing independently on filter paper on solid basal medium, each apparently derived from a single PEM (bar = 200 μm). C. Torpedo stage somatic embryos, 12 days following plating of fractionated PEMs on filter paper on top of solid basal medium (bar = 1 mm). D. Germination of selected mature embryo on basal medium without CH, 6 days following transfer from filter paper (bar = 1 mm). E. Effects of 2 secondary media on cotyledon expansion and germination of somatic embryos derived from fractionated PEMs cultured on filter paper. Embryos in the left plate were cultured on basal medium without CH, while those in the right plate were cultured on basal medium with 0.1% CH. F. Early torpedo stage somatic embryo derived from fractionated PEM plated on nylon mesh on top of solid basal medium (bar = 200 μm). (From Merkle *et al.* [29]).

mixture, and allowed to become saturated with water for 3–4 days. Plantlets are removed from test tubes or Magenta boxes, and agar-solidified medium is gently and completely washed from the roots. Then the root trainers are opened and one plantlet is planted in each cell of the 'book', with the tap root laid down the center. As each 'book' is filled, is immediately placed in the humidifying chamber. Using an atomizer, relative humidity in the chamber is maintained at 100%. While in the chamber, plantlets are fertilized with a modified Hoaglands solution, supplemented with MS minor salts and MS iron, at the rate of 1 ml per plantlet per week. After 6–8 weeks, when plantlets have produced 2–3 additional leaves and a net-like root system, the atomizer is turned off and the humidity in the chamber is slowly lowered to ambient by gradually opening the doors of the chamber wider and wider over the course of 3 weeks. Although some older leaves are damaged by the drop in humidity, new leaves produced during the final weeks expand rapidly and less than 5% of plantlets are lost during acclimatization.

Somatic embryo-derived yellow-poplar plantlets have als been acclimatized in a greenhouse using a misting system to maintain high relative humidity. This system may ultimately be of advantage because of the large numbers of plantlets that can be accomodated. However its operation has yet to be optimized and early plantlet growth in the misting greenhouse has been disappointing (personal communication, S.E. Schlarbaum, Department of Forestry, Wildlife and Fisheries, University of Tennessee, Knoxville, TN).

Following acclimatization, yellow-poplar plantlets are moved to the greenhouse, where they are grown for 2–6 months in the same containers in which they were hardened off. Greenhouse-grown plantlets have been grown in raised concrete block nursery beds, where they displayed excellent survival and growth. Plantlets set winter buds and dropped their leaves in late autumn (November), as do yellow-poplar seedlings. They remained dormant throughout the winter, breaking bud and resuming growth in spring (March). In March, 1989, over 200 somatic embryo-derived yellow-poplar trees from 20–50 cm in height were transplanted from the nursery to the field at the University of Georgia School of Forest Resources' Whitehall Forest, near Athens, GA. Although early field growth and survival were promising, the trees were severely browsed by deer, killing many of them. Following erection of a deer fence around the site in 1990, several of the trees resprouted from the root collars.

3.5. *Studies of embryogenic potential and markers*

Although most research involving the yellow-poplar embryogenic system has been of an applied nature, primarily aimed at increasing plantlet production, a number of experiments have been directed at gaining a better understanding of the factors controlling embryogenesis in this species, including genetic and physiological factors. The goals of these experiments have included explant optimization using morphological and biochemical markers, and biochemical comparisons of embryogenic and nonembryogenic cultures.

Early observations with embryogenic yellow-poplar cultures indicated that one of the key factors involved in the initiation of embryogenic cultures was the stage of development of the explanted zygotic embryos. Merkle and Sommer [27] noted that only immature zygotic embryo explants from late August and early September produced embryogenic cultures. However, in that study all explants were from a single tree and sampling was not systematic throughout the period of seed maturation. Therefore, a study was executed to characterize the optimal stage of zygotic embryo to explant for initiation of embryogenic cultures, using morphological measurements and polypeptide profiles of the embryos [43].

In this study, developing seeds from six full-sib families were collected every 2 weeks from 4 weeks postpollination until seed maturity (18 weeks post-pollination). The seeds from each family were divided into two subsamples for each collection date. One subsample was surface-sterilized and dissected, and the embryos were used to initiate tissue cultures. The percentage of explanted embryos which ultimately gave rise to PEMs following 2 months in culture, defined as embryogenic potential, was calculated for each family/sampling date combination. The other subsample was dissected and the embryos were first measured for total length, cotyledon length and hypocotyl thickness. Then, total soluble proteins were extracted from these embryos and analyzed using sodium dodecylsulfate polyacrylamide gel electrophoresis (SDS-PAGE). We then attempted to relate changes in embryo size and quantitative or qualitative changes in soluble embryo proteins to changes in embryogenic potential of the explanted embryos.

Average embryogenic potential for embryos from the six families rose from near 0% for cultures initiated 4 weeks postpollination to a peak of 28% for those initiated 8 weeks postpollination. Embryogenic potential fell to 0% for cultures initiated from mature embryos at 16 weeks postpollination. Zygotic embryos at 8 weeks postpollination were at the globular stage and averaged 0.2 mm or less in diameter. The rapid decline in embryogenic potential displayed by embryos collected after 8 weeks postpollination appeared to coincide with rapid increases in embryo size, as the embryos developed from the globular to the torpedo stage. SDS-PAGE analysis of protein extracts from zygotic embryos collected from 10–18 weeks postpollination revealed some semi-quantitative differences in polypeptide patterns with increasing age of the embryos. The most conspicuous difference among embryos from the different sampling dates was a polypeptide of approximately 55 kD, which appeared to increase with later sampling dates (and therefore with declining embryogenic potential), finally accounting for approximately 25% of the total embryo protein by 18 weeks postpollination. Its accumulation as the embryos matured indicated that this polypeptide may be a storage protein. Although some qualitative differences in polypeptides were also found among embryos from different dates, these were not consistent among all the families, and thus were not judged to be useful markers.

As mentioned in Section 3.2. (above) explanted immature yellow-poplar embryos often produce both PEMs and nonembryogenic callus. Thus pairs of isogenic culture lines were available to be assayed for biochemical differences associated with the embryogenic versus the nonembryogenic state. Such comparisons were executed on three pairs of cultures using SDS-PAGE, isozyme analysis and ethylene evolution rates [42]. Polypeptide profiles obtained by SDS-PAGE analysis of PEMs and nonembryogenic callus from the same genetic source were almost identical, with the exception of a 24 kD polypeptide, which was present in the nonembryogenic samples but not detectable in the PEMs. No polypeptides were consistently present in PEMs that were not also in nonembryogenic callus. Isozyme gels for these comparisons were prepared in the same manner as SDS-PAGE gels, except that SDS was omitted from the gels and buffers so that proteins retained their activities. Following preliminary testing of stains for a number of isozymes, five isozymes were found to display polymorphic patterns among different embryogenic and nonembryogenic lines. These were acid phosphatase (ACP), alcohol dehydrogenase (ADH), glutamate dehydrogenase (GDH), glutamate oxaloacetate transaminase (GOT) and esterase (EST). However, of these five isozymes, only GOT was useful in detecting variation in banding patterns between PEMs and nonembryogenic callus derived from the same zygotic embryo. For the three lines tested, PEMs displayed two isozyme bands, while nonembryogenic calli exhibited only one band, which migrated at the same position as the slower moving band from the PEM samples.

To compare ethylene evolution rates of PEMs and nonembryogenic callus, equal amounts of PEMs and nonembryogenic callus from the three culture lines were transferred to culture tubes containing induction medium and grown for 5 days. The cap on each tube was then replaced with a rubber septum, and after 24 hours, atmosphere samples were taken from the tubes by syringe and injected into a gas chromatograph. After the sample was taken, the callus was removed and weighed to determine the increase in fresh weight over the six-day period. The ethylene evolution rate was determined and expressed in units of nl/mg/day. PEMs from all three culture lines produced ethylene at a higher rate than the nonembryogenic calli from the same culture line, although ethylene evolution also appeared to vary with cell line.

3.6. *Genetic stability of cultures and regenerants*

Although thousands of somatic embryo-derived yellow-poplar plantlets have been regenerated from at least 30 independent embryogenic culture lines, to date there has been no evidence of somaclonal variation among the plantlets produced by a particular line. Even embryogenic cultures that are over 3 years old produce plantlets with phenotypes that appear to display high clonal fidelity, although this observation has not been analyzed quantitatively. However, as cultures age, their ability to produce large numbers of embryos capable of conversion to plantlets declines, which might indicate the accumulation of deleterious genetic changes.

Chou *et al.* [6] reported within-clone phenotypic variation in PEM color and morphology among suspension cultured sublines of three embryogenic yellow-poplar clones. Restriction fragment length polymorphism (RFLP) analysis of the highly repeated fraction of DNA isolated from PEMs of the sublines revealed no variation among sublines of a given clone. It should be noted that only a single restriction enzyme (BamH1) was employed in this study. However, the technique was sensitive enough to detect polymorphisms among the three parent lines, which were all derived from embryo explants from the same half-sib family. Merkle *et al.* [24] were similarly unable to detect polymorphisms in BamH1-restricted highly-repeated DNA of yellow-poplar plantlets regenerated from a given embryogenic line, although polymorphisms were found among plantlets from different lines (again from the same half-sib family).

A more complete idea of the level of clonal fidelity will be obtained during the next decade, as data become available from large-scale field tests of somatic embryo-derived yellow-poplar trees, which are currently being initiated [26].

4. Protoplast isolation and culture

Yellow-poplar is one of a handful of hardwood forest species to be regenerated from protoplasts. Originally, protoplast research with this species was undertaken with the goal of providing a vehicle for direct gene transfer via electroporation or PEG-mediated DNA uptake (see section on gene transfer). The key to the ability to successfully culture yellow-poplar protoplasts has been the use of embryogenic yellow-poplar suspension cultures. Protoplasts isolated from PEMs can be induced to rebuild cell walls and divide to produce new PEMs, somatic embryos and eventually plantlets [28].

Suspension cultures to be used for protoplast isolation are grown in the light, since dark-grown cultures yield very low numbers of protoplasts. In the protocol published by Merkle and Sommer [28], before protoplast isolation, cultures were put on an accelerated transfer schedule, with about 5 g of PEMs being transferred to fresh induction medium every 6 days. However, it was later found that protoplast yields could be greatly increased by using 20 day old cultures [50]. In our current protoplast isolation protocol, approximately 1 g of PEMs are removed from suspension and placed in 10 ml of filter-sterilized digestion medium containing the same major salts, minor salts and iron as induction medium, plus 500 mg/l $CaCl_2 \cdot 2H_2O$, 100 mg/l MES (2[*N*-morpholino]ethanesulfonic acid), 0.5 M mannitol, 1% Cellulysin (Calbiochem-Behring Corp.), 0.5% Macerase (Calbiochem-Behring Corp.) and 100 mg/l bovine albumin (Sigma Chemical Co.). PEMs are incubated in digestion medium at 30 °C with agitation from a gyratory shaker run at 50 rpm. Following 24 hr of digestion, the suspension is filtered through two layers of Miracloth, then through a 25 μm pore-size, stainless steel sieve. The filtrate is then centrifuged at 100 × g for 5 min. The supernatant is discarded and the pellet resuspended in 5 ml of wash medium, which is of the same composition as digestion medium, but without enzymes or bovine albumin. Centrifugation and resuspension with fresh wash medium are repeated twice more. Although we have attempted various flotation techniques to purify yellow-poplar protoplasts, they are extremely dense and none of the flotation techniques could effectively separate them from undigested cells. However, the very long digestion time we employ (24 hr) digests virtually all single cells in the suspension to protoplasts, and undigested clumps are removed by the filtering and sieving process described above. Using 20-day-old embryogenic suspension cultures as starting material, an average of 1.7×10^7 protoplasts per gram of PEMs can be obtained.

Following the final wash step and centrifugation, the supernatant is removed and the pellet is resuspended in protoplast regeneration medium, which is of the same composition as induction medium, but supplemented with 500 mg/l $CaCl_2 \cdot 2H_2O$, 250 mg/l xylose and 250 mg/l fructose. In addition, instead of 40 g/l sucrose, protoplast regeneration medium has only 250 mg/l sucrose, but 0.5M glucose is used as osmoticum. Also, 2,4-D concentration is reduced from 2 mg/l to 1 mg/l. We employ a high density protoplast plating technique after Binding and Kollman [2]. In this procedure, protoplasts are diluted to $2–4 \times 10^4$ protoplasts per ml, and 0.2 ml drops of protoplast suspension are pipetted into plastic Petri dishes (1 drop per dish). There, the suspension is mixed with an equal volume of autoclaved regeneration medium containing 2.5% SeaPlaque low melting point agarose (FMC Corp.), which has been cooled to 38 °C. The mixture is immediately taken up with a pipette and redistributed as 0.05 ml droplets around the perimeter of the plate. When the droplets have solidified, the plate is flooded with approximately 2 ml of liquid regeneration medium. Petri dishes are sealed with Parafilm, stacked inside of inverted Magenta boxes to slow desiccation and incubated in darkness at 30 °C.

Cultured protoplasts usually regenerate cell walls within 3 days and begin dividing within 4–5 days. Divisions continue rapidly and eight-cell colonies are visible by 8 days. Plating efficiencies (i.e. percentage of protoplasts forming cell colonies) have ranged as high as 30%. Microcalli (or very early stage PEMs) are visible without a microscope within 3 weeks, and at this time, the regeneration medium is diluted with an equal of induction medium, which contains no osmoticum. One month later, agarose droplets containing PEMs can be transferred to induction medium in Erlenmeyer flasks on a gyratory shaker, and the PEMs will expand out of the droplets to form an embryogenic suspension culture. Alternatively, agarose droplets with PEMs can be placed on semisolid induction medium in plastic Petri dishes, and PEMs will proliferate from the droplets onto the surface of the induction medium. Protoplast-derived PEMs transferred to basal medium produce somatic embryos and plantlets in the same manner as described above for regular PEMs. Apparently, the embryogenic potential of the cells composing yellow-poplar PEMs is not altered by protoplasting or the subsequent manipulations employed to induce regeneration.

5. Gene transfer

Yellow-poplar is one of the few woody plants that have been transformed and regenerated. Two features of the yellow-poplar suspension culture system were exploited for gene transfer experiments: (1) protoplast regeneration and (2) size-fractionation of PEMs. Foreign DNA was introduced into protoplasts by electroporation and polyethylene glycol (PEG) treatment [50] and into PEMs by microprojectile bombardment [51]. These methods of direct DNA transformation were necessary to circumvent the host-range limitation of *Agrobacterium tumefaciens*.

Bacterial marker genes encoding β-glucuronidase (GUS) and neomycin phosphotransferase (NPT II), carried by the plasmid pBI121 [17], were used to monitor protoplast transformation. Transient expression of GUS could be detected in yellow-poplar protoplasts within 24 hours after a brief exposure to either a high intensity electric field (electroporation) or to 20% PEG-6000 in the presence of pBI121. GUS expression in the absence of pBI121 was negligible in fluorometric assays. Electroporation parameters were adjusted to give a pulse decay constant of 12 ms and protoplast viability of 53%. Transient GUS expression was found to be influenced by the length of the interval between subculture and exposure to DNA. The level of GUS activity in protoplasts isolated from cells 20 days after subculture was 5.3 and 2.6 times higher than in protoplasts from 5-day and 15-day old cells, respectively.

After electroporation or PEG treatment, a fraction of the protoplasts was plated in agarose droplets by the protocol described in Merkle and Sommer [28]. PEG-mediated transformation was less deleterious to yellow-poplar protoplasts than electroporation, as evidenced by improved colony production. Protoplast-derived cells were selected after 2 weeks with a level of the antibiotic kanamycin (100 μg/ml) sufficient to kill cells not exposed to pBI121. Some colonies derived from plasmid-treated protoplasts survived for up to three months under antibiotic selection, but failed to proliferate like the nonselected controls. Thus, the introduction of plasmid DNA into yellow-poplar protoplasts did not result in stably transformed callus.

Foreign DNA could be delivered directly into intact yellow-poplar cells by bombardment of PEMs with plasmid-coated microprojectiles according to the procedure of Klein *et al.* [20]. Approximately 0.5 g of a two-week old suspension culture was spread evenly on a 4.25 cm filter paper disc and bombarded with pBI121-coated tungsten particles using a Biolistic particle gun (DuPont). Transient GUS expression was detected by a histochemical assay of the cells 2 days after bombardment. However, selection of bombarded cells with kanamycin did not yield antibiotic-resistant callus, indicating that

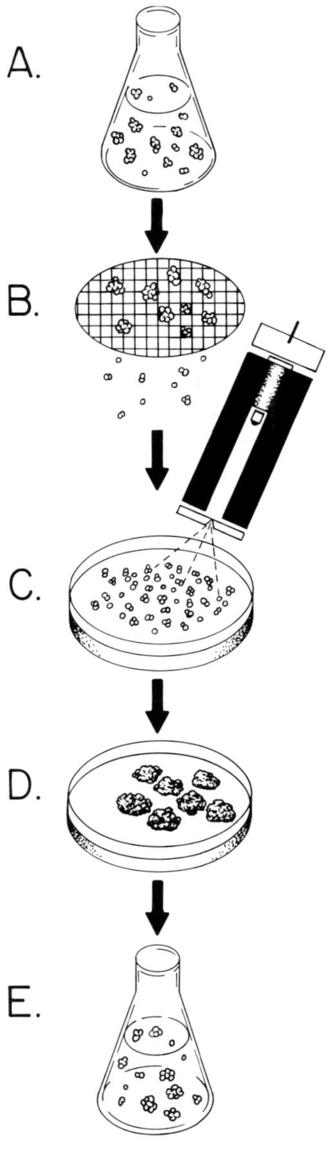

Figure 4. Transformation of embryogenic yellow-poplar. A. Embryogenic suspension cultures are grown for two weeks in liquid induction medium. B. Culture is fractionated and single cells and cell clusters less than 140 μm in diameter are collected on filter paper. C. Target cells are placed on solidified induction medium, bombarded twice with plasmid-coated microprojectiles, and incubated for two days. D. Bombarded cells are transferred to semisolid medium containing kanamycin. Transformed, antibiotic-resistant microcalli are visible after 5–6 weeks. E. Antibiotic-resistant microcalli are propagated individually and then subcultured into kanamycin-containing liquid medium. Transgenic plants can be regenerated from suspension cultures through somatic embryogenesis.

the plasmid DNA had not integrated at a frequency sufficient to permit recovery of transformed cells.

Stably transformed yellow-poplar callus could be obtained after the target for gene transfer was refined [52]. As shown in Figure 4, fractionation of suspension cultures through 140 μm mesh enriched for single cells and small cell clusters (PEMs). Approximately 10 000 cell clusters from this fraction were dispersed evenly on a filter paper disc and bombarded with pBI121-coated tungsten particles (Figure 4C). After two days of incubation, the filters were transferred onto semisolid induction medium containing 100 μg/ml kanamycin. Antibiotic-resistant microcalli, which appeared 5–6 weeks after bombardment (Figure 4D), were used to initiate suspension cultures in liquid medium containing 50 μg/ml kanamycin (Figure 4E). There was an average of four transformants per filter which, assuming 10 cells per cluster, resulted in a transformation frequency of 0.004%.

The number of copies of the GUS gene in independently transformed callus lines ranged from approximately 3–30. An ELISA for NPT II and a fluorometric assay for GUS showed that the expression of both enzymes varied by less than four-fold among callus lines. Transgenic plants were regenerated following previously established procedures [29]. Somatic embryos induced from transformed cell cultures were found to be uniformly GUS-positive by histochemical analysis. No somatic embryos developed from a non-transformed yellow-poplar culture under antibiotic selection. Somatic embryos from transformed lines converted into plantlets which grew vigorously in soil. All transgenic plants sampled expressed the two marker genes in both root and shoot tissues. A combination of small target size, early antibiotic selection, and propagation in suspension under selection pressure favored the recovery of transformed, nonchimeric yellow-poplar plants.

6. Summary

Yellow-poplar is a North American hardwood forest tree of substantial importance in terms of distribution and economics. Although it can be propagated using seeds and conventional vegetative propagation techniques, these methods have limitations which make *in vitro* propagation a desirable alternative. The only successful *in vitro* technique for mass propagation of yellow-poplar has been somatic embryogenesis. Embryogenic cultures are initiated from immature zygotic embryos, which display their highest embryogenic potential during the globular stage of development. When embryos at the proper stage are explanted on an auxin-supplemented medium, they will produce proembryogenic masses (PEMs). PEMs transferred to medium lacking auxin form somatic embryos from which plantlets can be produced. Hundreds of somatic embryo-derived yellow-poplar plantlets have been acclimatized and passed through greenhouse and nursery phases. Large scale testing of these plantlets is currently being initiated. PEMs can be maintained in suspension culture, which has facilitated a number of useful manipulations, including somatic embryo synchronization, protoplast isolation and culture, and gene transfer via microprojectile bombardment. Somatic embryo development can be synchronized by fractionating PEMs on sieves. When the proper fraction is cultured in liquid basal medium supplemented with ABA, synchronous populations of well-formed embryos can be produced, but these embryos fail to convert to plantlets. PEMs fractionated and immediately plated on a layer of filter paper overlayed on semisolid basal medium produce roughly synchronous populations of embryos which mature in 2 weeks and convert to plantlets at a very high frequency. Yellow-poplar protoplasts isolated from PEMs and cultured at high densities in agarose droplets will reform cell walls and divide to produce new PEMs. Thus, somatic embryos and plantlets of protoplast origin can be produced. Using PEMs which have been fractionated and plated on filter paper as target tissue, foreign DNA has been inserted into yellow-poplar cells via microprojectile bombardment. PEMs, somatic embryos and plantlets carrying and expressing genes for neomycin phosphotransferase and β-glucuronidase have been produced using this technique.

References

1. Belanger RP (1976) Grafting produces rootable cuttings from mature yellow-poplar trees. Plant Propagator 22(3):12–14.
2. Binding H and Kollman R (1985) Regeneration of protoplasts. In: Schafer-Menhur A (Ed.) *In Vitro* Techniques: Propagation and Long-Term Storage. Martinus Nijhoff/Dr. W. Junk Publishers, Dordrecht. pp. 93–99.
3. Bonner FT and Russell TE (19744) *Liriodendron tulipifera* L. In: Seeds of Woody Plants of the United States. USDA Forest Service, Washington, D.C. pp. 508–511.
4. Brotschol JV, Roberds JH, and Namkoong G (1986) Allozyme variation among North Carolina populations of *Liriodendron tulipifera* L. Silvae Genet 35(4):131–138.
5. Carpenter IW and Guard AT (1950) Some effects of cross-pollination on seed production and hybrid vigor of tulip tree. J For 48:852–855.
6. Chou PL, Merkle SA, and Sommer HE (1987) Repetitive sequences in the DNA of the genome of *Liriodendron tulipifera* tissue cultures. In: Proceedings of the 1986 TAPPI Research and Development Conference. pp. 95–97.
7. Chu CD (1979) *Liriodendron.* In: Technical Consultation on Fast-Growing Plantation Broadleaved Trees for Mediterranean and Temperate Zones. FAO Publication No. FO: FGB-79-8/14. pp. 381–383.
8. Churchwell NR (1965) Vegetative stage of stock and scion are important factors when field grafting yellow-poplar. Tree Planters Notes 70:27–29.
9. Donovan DM (1976) A list of plants regenerating from root cuttings. Plant Propagator 22(1):7–8.
10. Farmer RE, Russell T, and Krinard RM (1967) Sixth-year results from a yellow-poplar provenance test. In: Proceedings of the Ninth Southern Conference on Forest Tree Improvement. pp. 65–68.
11. Farmer RE Jr., Barnett PE, Thor E, and Rennie JC (1983) Heritability estimates for height growth of Tennessee yellow-poplar. Silvae Genet 32(1–2):15–18.
12. Fowells HA (1965) Silvics of Forest Trees of the United States. Agriculture Handbook No. 271. USDA Forest Service, Washington, D.C. 762 pp.

13. Funk DT (1962) Bud grafting yellow-poplar. In: Proceedings of the Third Central States Forest Tree Improvement Conference. pp. 19–21.
14. Furmanowa M and Rzedowski M (1979) *Liriodendron tulipifera* L. in tissue culture. Acta Polon Pharm 36(2):261–262.
15. Gresshoff PM and Doy CH (1972) Development and differentiation of haploid *Lycopersicon esculentum* (tomato). Planta 107:161–170.
16. Harlow WM, Harrar ES, and White FM (1979) Textbook of Dendrology, Sixth Edition. McGraw-Hill, New York. 510 pp.
17. Jefferson RA, Kavanaugh TA, and Bevan MV (1987) GUS fusions: β-glucuronidase as a versatile gene fusion marker in higher plants. EMBO J 6:3901–3907.
18. Kaeiser M and Boyce SG (1962) X-ray negatives for sorting yellow-poplar samaras with filled and empty seeds. J For 60:410–411.
19. Kellison RC (1968) Natural variation of yellow-poplar in North Carolina. In: Proceedings of the Fifteenth Northeastern Forest Tree Improvement Conference. pp. 7–14.
20. Klein TM, Gradziel T, Fromm ME, and Sanford JC (1988) Factors influencing gene delivery into *Zea mays* cells by high-velocity microprojectiles. Bio/technology 6:559–563.
21. Kormanick PP and Porterfield EJ (1966) Rooting yellow-poplar cuttings. Forest Farmer 26(24):41–42.
22. McAlpine RG (1964) A method for producing clones of yellow-poplar. J For 62:115–116.
23. McAlpine RG and Kormanick PP (1971) Rooted cuttings from select yellow-poplar trees. In: Proceedings of the Eleventh Conference on Southern Forest Tree Improvement. pp. 241–243.
24. Merkle SA, Chou PL, and Sommer HE (1988) Stability of highly repeated sequences in the DNA of embryogenic cultures of yellow-poplar. In: Cheliak WM and Yapa AC (Eds.) Molecular Genetics of Forest Trees. Petawawa National Forestry Institute Information Report PI-X-80. pp. 85–88.
25. Merkle SA, Parrott WA, and Williams EG (1990) Applications of somatic embryogenesis and embryo cloning. In: Bhojwani SS (Ed.) Plant Tissue Culture: Applications and Limitations. Elsevier, Amsterdam. pp. 67–101.
26. Merkle SA, Schlarbaum SE, Cox RA, and Wiecko AT (1990) Large scale production of yellow-poplar (*Liriodendron tulipifera* L.) somatic embryos for field testing [Abstract]. In: 11th North American Forest Biology Workshop, Program and Abstracts. pp. 61–62.
27. Merkle SA and Sommer HE (1986) Somatic embryogenesis in tissue cultures of *Liriodendron tulipifera*. Can J For Res 16:420–422.
28. Merkle SA and Sommer HE (1987) Regeneration of *Liriodendron tulipifera* (family Magnoliaceae) from protoplast culture. Amer J Bot 74(8):1317–1321.
29. Merkle SA, Wiecko AT, Sotak RJ, and Sommer HE (1990) Maturation and conversion of *Liriodendron tulipifera* somatic embryos. In Vitro Cell Dev Biol 26:1086–1093.
30. Murashige T and Skoog F (1962) A revised medium for rapid growth and bioassays with tobacco tissue culture. Physiol Plant 15:473–497.
31. Nitsch JP and Nitsch C (1969) Haploid plants from pollen grains. Science 163:85–87.
32. Parks CR, Miller NG, Wendel JF, and McDougall KM (1983) Genetic divergence within the genus *Liriodendron* (Magnoliaceae). Ann Missouri Bot Gard 70:658–666.
33. Parks CR and Wendel JF (1990) Molecular divergence between Asian and North American species of *Liriodendron* (Magnoliaceae) with implications for interpretation of fossil floras Amer J Bot 77(10):1243–1256.
34. Parks CR, Wendel JF, Sewell MM, and Qiu YL (1990) Genetic control of isozyme variation in the genus *Liriodendron* L. (Magnoliaceae). J Heredity 81:317–322.
35. Risser PG and White PR (1964) Nutritional requirements of spruce tumor cells *in vitro*. Physiol Plant 15:620–635.
36. Russell TE (1977) Planting yellow-poplar – where we stand today. USDA Forest Service Gen Tech Rep SO-17. South For Exp Stn, New Orleans, LA. 8 pp.
37. Rzedowski M (1981) Alkaloidy w kulturze tkantowej *Liriodendron tulipifera* L. PhD thesis. Medical Academy, Warsaw. 145 pp.

38. Santamour FS Jr (1972) Interspecific hybrids in *Liriodendron* and their chemical verification. For Sci 18(3):233–236.
39. Santamour FS Jr and McArdle AJ (1984) Cultivar checklist for *Liquidambar* and *Liriodendron*. J Arboriculture 10(11):309–312.
40. Sluder ER (1960) Early results from a geographic seed source study of yellow-poplar. USDA Forest Service Res Note SE-150, Southeast For Exp Stn, Asheville, NC. 2 pp.
41. Sommer HE and Brown CL (1980) Embryogenesis in tissue cultures of sweetgum. For Sci 26(2):257–260.
42. Sotak RJ (1989) Characterization of aspects of somatic embryogenesis in tissue cultures of yellow-poplar (*Liriodendron tulipifera*). MS thesis, University of Georgia, Athens GA. 182 pp.
43. Sotak RJ, Sommer HE, and Merkle SA (1991) Relation of the developmental stage of zygotic embryos of yellow-poplar to their somatic embryogenic potential. Plant Cell Rep. 10: 175–178.
44. Stefaniak B and Wozny A (1983) Investigations on the ultrastructure of the callus tissue of *Liriodendron tulipifera* L. developed *in vitro*. Acta Societatis Botanicorum Poloniae 82(1):3–8.
45. Taft KA (1962) The effect of controlled pollination and honeybees on seed quality of yellow-poplar (*Liriodendron tulipifera* L.) as assessed by X-ray photography. School of Forestry, NC State College. Tech Rep No 13. 21 pp.
46. Taft KA (1966) An investigation of the genetics of seedling characteristics of yellow poplar (*Liriodendron tulipifera* L.) by means of a diallel crossing scheme. PhD dissertation, North Carolina State University, Raleigh, NC. 59 pp.
47. Taylor FE (1964) Differences in the wood of yellow-poplar. Forest Farmer 24:8–9.
48. Thorbjornsen E (1961) Variation in density and fiber length in wood of yellow-poplar. J Tech Assoc Pulp and Paper Ind 44(3):192–195.
49. Wilcox JR and Taft KA (1969) Genetics of yellow-poplar. USDA Forest Service Res Pap WO-6. 12 pp.
50. Wilde HD, Meagher RB, and Merkle SA (1989) Introduction of foreign DNA into yellow-poplar protoplasts by electroporation. In: Proceedings of the 20th Southern Forest Tree Improvement Conference. pp. 389–394.
51. Wilde HD, Merkle SA, and Meagher RB (1991) Transfer of foreign genes into yellow-poplar (*Liriodendron tulipifera*). In: Ahuja MR (Ed.) Woody Plant Biotechnology. Plenum Press, New York pp. 227–232.
52. Wilde HD, Meagher RB, and Merkle SA (1992) Expression of foreign genes in transgenic yellow-poplar plants. Plant Physiol. 98:114–120.
53. Witham FH, Blaydes DF, and Devlin RM (1971) Experiments in Plant Physiology. Van Nostrand-Reinhold Co., New York. 245 pp.

18. Micropropagation of nitrogen-fixing trees

VIBHA DHAWAN
Tata Energy Research Institute, 90 Jor Bagh, New Delhi 110 003, India

1. Introduction

The increase in population and unit energy consumption has resulted in overexploitation of our natural resources, especially forests. The vicious cycles of drought and flood already visible in many parts of the world are probably a reflection of imbalance in nature due to indiscriminate destruction of forests. It is now realized that for the welfare of the human race, it is vital to protect the remaining forest cover from further damage and to reforest the presently denuded areas. Fast growing, hardy, robust, leguminous trees, that enrich the soil with nitrogenous bio-fertilizers, are the prime candidates for reclaiming and revegetating vast wastelands. The nitrogen fixing species not only survive under harsh conditions, but also, over a period of time, they improve the nutritional status of the soil, thus making it suitable for the growth of other species. Nitrogen fixation in higher plants is achieved through a symbiotic association of their roots either with bacteria belonging to the genus *Rhizobium* or actinomycetes belonging to the group *Frankia*.

Micropropagation is a useful technique for the rapid cloning of elite genotypes under aseptic conditions. This article reviews the available literature on *in vitro* propagation of nitrogen fixing tree species and, in a few cases, nodulation studies.

2. Actinomycetes-*Frankia* associations

Over 160 angiosperm species are known to form nitrogen-fixing root nodules with the actinomycete *Frankia* [2,30]. Elaeagnaceae tree species are particularly interesting because of their rapid growth and their capacity to grow in a wide range of edapho-climatic conditions. The available literature on tissue culture of the three genera (viz. *Alnus*, *Elaegnus* and *Hippophae*) is summarized below:

2.1. Alnus *spp.*

Alder species are known for soil improvement and high biomass production

M. R. Ahuja (Ed.), Micropropagation of Woody Plants, 303–315.

Thus, for a long time, studies have been conducted on the selection of superior genotypes, followed by the development of methods of cloning. Clonal propagation of *Alnus* species from mature plants by conventional methods of rooting of cuttings has been restricted to only a few species viz. *Alnus cordata*, *A. crispa*, *A. glutinosa*, *A. incana*, *A. rubra* and *A. rugosa*. However, the rates of multiplication are far from satisfactory due to the large size of the cutting and the particular season when cuttings could be rooted. *In vitro* propagation methods have been developed for many *Alnus* species [5,18,42,43, 53,56].

In *Alnus*, cultures were initiated from the buds of aseptically raised seedlings [42], 2-month old seedlings [18], 8-month and 2-year old plants growing in the field [55,56]. While for most species, there have not been many problems in raising aseptic cultures, in some species, such as *Alnus cordata*, severe bacterial contamination was observed which was largely due to the infestation of axillary buds with insects and insect larvae. Further losses were recorded due to oxidation of phenols that leached out from the cut ends of the explant; it was impossible to obtain healthy cultures from nodal segments of *Alnus cordata* [5]. However, 80% aseptic cultures could be obtained from shoot-tip explants (dissected from sterilized dormant axillary buds) but no growth was obtained due to vitrification. The hardwood stem cuttings of this species could, however, be rooted with 40% success and actively growing shoots from these served as the source of healthy explant [5].

No generalization can be made about the media requirement of different *Alnus* species. Both high salt medium e.g. MS – Murashige and Skoog [38] and low salt medium e.g. WPM – Lloyd and McCown [31] have been succesfully used in the micropropagation of *Alnus* spp. For *A. crispa* and *A. glutinosa*, WPM was superior to MS as it produced less callus and resulted in higher multiplication rates. Even the sugar requirements of different *Alnus* species vary. While some species respond better to sucrose, others do well with glucose. Tremblay and Lalonde [55] used a two-step procedure (i.e. different media for culture initiation and shoot multiplication) either by increasing sucrose concentration or replacing sucrose by glucose. A one-step shoot production system has been reported for *A. glutinosa*, *A. crispa* and *A. rubra* [18,47], but in these references, no comparison has been made for different sugars at the initiation and shoot multiplication stage. Tremblay and Lalonde [55] reported that *A. crispa* and *A. rubra* preferred glucose, while *A. glutinosa* preferred sucrose for shoot multiplication (Table 1). Barghchi [5] reported that although glucose is preferred by *A. cordata* for initiation, after six regeneration passages, the growth rates on sucrose and glucose supplemented media were comparable.

Perinet and Lalonde [42], for the first time reported *in vitro* nodulation of tissue culture propagated plants. Three clones of *Alnus glutinosa* (AG-2, AG-4 and AG-8), propagated by tissue culture, were inoculated with four *Frankia* strains (ARgN 22d, ARgP^{AC}, ACN14a and TN18^{AC}). AG-2 and AG-8 performed better than clone AG-4 thus proving that plant genotype is

Table 1. Micropropagation of adult material of *Alnus* species.

Species	Shoot Multiplication				Rooting				Reference
	Medium	Sugar	%	Rate of multiplication	Medium	Sugar	%	Percent rooting	
Alnus cordata	MS + BAP (1.1μM)	G	6	4–5 fold every 4–5 wk	MS1/2 + IBA (1.2–4.9 μM)	S	3	98%	5
A. crispa	WPM + BAP (0.1–1 μM)	S	2	+	*In vivo* rooting	–		95%	56
A. crispa	MS + BAP (2.5–5 μM)	S	6	10–12 fold every	MS1/2 + IBA (1μM)	G	1.5	60–100%	56
		G	3	3 wk		S	3		
A. glutinosa	WPM + BAP (1 μM)	S	2	3–5 fold in 3 wk	*In vivo* rooting	–		95%	18
A. glutinosa	MS + BAP (0.5–5 μM)	S	3	8–16 fold in 3 wk	MS1/23 + IBA (0–5 μM)	S or	3–8		
						G	1.5–3		
A. incana	MS + BAP (1 μM)	G	1.5	3–5 fold every 3 wk	MS1/2 + IBA (1 μM)	G	3	80–100%	55
A. japonica	MS + BAP (1 μM)	G	1.5	3–5 fold every 3 wk	MS1/2 + IBA (1 μM)	G	3	80–100%	55
A. rubra	MS + BAP (25 μM)	G	1.5	3–5 fold in 3 wk	MS1/2 + IBA (1 μM)	S	3–6	100%	55
						G	1.5–3		
A. sinuata	MS + BAP (2.5 μM)	G	3	5–7 fold every 3 wk	MS1/2 + IBA (1 μM)	G	1.5–3	50%	55
A. viridis	MS + BAP (2.5 μM)	G	3	5–7 fold every	MS1/2 + IBA (1 μM)	G	3	10–100%	55

G = Glucose, S = Sucrose, +, + = multiplication rates not mentioned.

responsible for the differences in growth rate between different clones inoculated with pure cultures of *Frankia* [42].

2.2. *Elaegnus angustifolia*

Of the many actinorrhizal species used for land reclamation, the Russian Olive (*E. angustifolia*) appeals because of its rapid juvenile growth and wide edapho-climatic adaptability. It is frost hardy and highly tolerant to alkaline and saline soils. With tissue culture, rapid year-round production of large number of uniform plants of desirable genotypes would become possible [6].

Six month old seedling plantlets have been propagated by the adventitious bud formation method. The maximum number of adventitious buds were formed with the cytokinin 2ip, followed by BAP and Kinetin, but the development of adventitious shoots was inversely proportional to the number of buds, i.e. maximum growth of shoots was obtained on kinetin supplemented medium. MS basal medium with 100 mM sucrose and 5 mM kinetin proved to be the best medium, giving a multiplication rate of 5–10 fold every three weeks. Cent percent rooting was obtained under *in vitro* conditions on a number of media compositions and about 50% plants formed roots *in vivo*. All rooted plantlets *in vitro* survived transfer to the artificial substrate. All plantlets formed nodules within 4–8 weeks after inoculation with *Frankia* regardless of the clone or the strain used.

2.3. *Hippophae rhamnoides*

Hippophae rhamnoides is a deciduous shrub commonly found in mountainous habitats, river banks and coastal dunes. This calciphic species has a high tolerance to low temperature and high pH and salt. Seedlings of *Hippophae rhamnoides* could be propagated at a rate of 3–5 fold every four weeks on MS medium supplemented with 10^{-6} M BAP [36]. The response varied with the clone. Clones from North America responded better than the clones from Holland. Also, growth was better in tubes than in Mason jars or GA7 vessels (from SIGMA).

At the multiplication stage some of the shoots showed signs of vitrification. These, however, could be converted to normal type by reducing macroelements of MS to half strength and increasing agar concentration to 1.5%. The optimal medium for rooting consists of quarter strength MS with 29.2 mM surcrose on which 60% rooting was obtained. All plantlets survived upon transplantation.

3. Legume-*Rhizobium* associations

Rhizobium infection of leguminoceae members leads to root nodule formation

– the site for atmospheric nitrogen fixation. With their inherent capacity to fix atmospheric nitrogen, they improve the soil nutrient status and pave the way for the growth of other species. The growth of tree legumes is determined not only by the site location and genotype, but also by the availability of suitable *Rhizobium* strains. *Rhizobium* also helps in the establishment phase. Our work with *Leucaena* showed that inoculation of micropropagated plants with an efficient strain of *Rhizobium* during the hardening phase considerably improved transplantation success [17]. For tree legumes it is important to infect the tissue culture propagated plants with *Rhizobium*. This is because the symbionts normally infect the seedling at early stages of development. In contrast, aseptically grown plantlets lack any associated bacteria. Thus when they are transferred to soil, their establishment is affected, adversely.

3.1. *Acacia species*

Acacias are prolific seed setters and are generally propagated from seed. They also coppice well and several species spread by root suckers. The trees are heavy seeders but being outbreeders there is a large degree of variation in the seedling progeny. There is a need to develop propagation methods for rapid clonal multiplication of elite individuals.

The first report on tissue culture of *Acacia* was from Professor Bonner's laboratory as long ago as 1942 [7]. The root culture of *A. melanoxylon* was maintained in liquid for over a year. The callus from seedlings stem was obtained by McLean *et al.* from *A. harpophylla* [34] and in *A. mearnsii* by Zeijlemaker [60]. Root differentiation from the cotyledonary callus of *Acacia senegal* was achieved by Kathju and Tewari [25]. Later, Agarwal and Prasad [1] obtained complete plantlets from the cultured stem tissue of *A. senegal*. Reports on successful propagation and transplantation of *A. koa* to the field was by Skolmen and Mapes [48–50].

A number of authors later succeeded in propagating other *Acacia* species namely *A. nilotica* [33], *A. mangium* and *A. stenophylla* [9]. *Acacia mangium* and *Acacia stenophylla* have been regenerated using seedling explants. Shoot multiplication was achieved on MS (half strength basal) with one micromole each of BAP and NAA. *In vitro* rooting in a moist vermiculite/perlite mixture was followed by hardening in plastic pots. The plants were later successfully transferred to the field.

In *Acacia nilotica*, the variety *cupressiformis* has been earmarked as ideal for agroforestry purposes. Attempts are being made to stimulate axillary shoot production from adult stem nodal explant. Initiation and multiplication methods are being standardized while attempts for rooting are underway. Regeneration using stem nodal explants on MS medium supplemented with IAA was reported by Mathur and Chandra [33], but we could not reproduce these results in our laboratory with seedling or adult tissue of *A. nilotica* var. cupressiformis. A micropropagation protocol is being standardized for

Table 2. Micropropagation of exotic *Acacias*.

Species	Medium for shoot multiplication	Rate of propagation every 4 weeks	Medium for rooting	Rooting percentage
Acacia bivenosa	MS + 3×10^{-6}M BAP	8–10 fold	MS1/2 + 6×10^{-6}M IAA	80
A. maconochieana	MS + 3×10^{-6}M BAP + 3×10^{-6} NAA	6 fold	MS + 3×10^{-6}M IAA + 10^{-6} NAA	100
Acacia sclerosperma	MS + 6×10^{-6}M BAP + 10^{-6}M NAA	8–10 fold	MS1/2 + 6×10^{-6}M IAA	80–90
A. victoriae	MS + 6×10^{-6}M BAP	9 fold	MS + 3×10^{-6}M IBA or MS1/2 + 10^{-6}M NAA	100

the same, starting with adult tissue. Over 60% of the initiates showed bud break. However, multiplication of *in vitro* developed shoots is not very consistent and very few cultures survived upto the 8th passage with average multiplication of two fold every four weeks (Nair, unpublished). In our laboratory, attempts are being made to culture a number of Australian *Acacia* species namely *A. ampliceps*, *A. auriculiformis*, *A. bivenosa*, *A. holosericea*, *A. ligulata*, *A. maconochieana*, *A. salicina*, *A. sclerosperma*, *A. stenophylla* and *A. victoriae*. Work was initiated with the seedling material (seeds obtained from CSIRO, Australia) and shoot multiplication was achieved on various media for four species (Dhawan, unpublished) (Table 2).

We are in the process of further optimizing the media and other physical factors for micropropagation. So far no attempts has been made to transfer them to the field or nodulate them. We are working on selecting *Rhizobium* strains for different *Acacia* species by using seed raised plants and will be later utilizing them for nodulation of tissue culture raised plants.

3.2. *Albizzia species*

A. lebbeck, or East Indian Walnut as it is known in India, provides fodder, timber and tannins, as well as host for the lac insect. It is adapted to a wide range of soil types. *A. lebbeck* has been regenerated from seedling tissues viz. root, hypocotyl, cotyledon and leaf. The method described by different authors [19,28,57] involves a callus phase followed by shoot bud differentiation or somatic embryogenesis.

Tomar and Gupta [54] obtained regeneration in three *Albizzia* species, namely *A. amara*, *A. lucida* and *A. richardiana*. In *A. richardiana* [53], they also obtained somatic embryogenesis. However, these studies are restricted to seedling tissues involving a callusing phase. For clonal propagation, the protocols with adult tissue still needs to be established.

3.3. *Dalbergia species*

Dalbergia is an important timber-yielding genus. For timber, good tree forms are highly desirable but due to cross pollination, the seed progeny gives rise to heterogenous population. Thus, there is a need to develop large scale cloning of desired genotypes.

Dalbergia lanceolaria [3], *D. latifolia* [29,39,45,46] and *D. sissoo* [11,37,51] have been propagated from seedlings as well as adult tissues.

Of these species, *D. sissoo* was the first to be regenerated; this being done from seedling root explants [37]. Shoots regenerated directly on the explants or from callus, were successfully rooted and transplanted. Sewal *et al.* [5] achieved 10–15 fold multiplication in every 8 weeks, from cotyledonary node explants. The *in vitro* multiplied shoots could be treated as microcuttings and rooted under *in vivo* conditions with 85% transplantation success. Datta *et al.* [1] reported multiple shoot formation from nodal explants of 30 year old tree and were also successful in inducing rooting, indicating the potential of the species to regenerate under culture condition. Datta and Datta [11] followed up their earlier work and reported callus and plantlet formation in same auxin combinations. Multiple shoots were obtained which were successfully rooted under *in vitro* conditions. In this report, however, there is no mention of *in vitro* cycles of multiplication or transplantation. Organogenic differentiation in tissue culture of *D. lanceolaria* was achieved by Anand and Bir [3].

Nataraja and Sudhadevi [40] successfully established callus from seedling explants (hypocotyl, stem and root) of *Dalbergia latifolia* followed by shoot bud differentiation. The shoots could be rooted and entire plants were raised. The studies were later extended to adult tissue by Lakshmi Sita and her co-workers [29]. Nodal and internodal explants from 5-year old trees produced callus followed by shoot bud differentiation. These shoot buds could be rooted but rooting was always accompanied by leaf drop. However, new leaves developed after the plant got established in the soil.

Most trees can be evaluated for the desired traits only after their half rotation. Unfortunately, the adult tissue is more recalcitrant to tissue culture techniques and many modifications in the protocol developed for seedling tissue are to be made. For *D. latifolia* successful protocols were developed from shoot-tips and shoot segments of 50-year old trees [45]. However, all these reports are on regeneration of shoot buds from callus and their clonal uniformity is to be documented by field trials.

3.4. *Leucaena species*

One of the most promising species of tropical legumes with versatility of uses for reclaiming and vegetating denuded soils is *Leucaena leucocephala*. Since 1979 several workers have studied the *in vitro* responses of seedling explants of *Leucaena*. Peasly and Collins restricted the scope of their work to callus growth [41]. Venketeswaran and Gandhi [58] and Glovak and Greatbatch [21] succeeded in regenerating plantlets from seedlings explants of *L. leucocephala*. Similarly, Ravishankar *et al.* [44] obtained plantlets from shoot tips of 4 day old seedling of *L. leucocephala*. Nagamani and Venketeswaran [39] cultured seedling explants of three species of *Leucaena* of which only *L. diversifolia* formed immature plantlets through somatic embryogenesis. In the cultures of *L. leucocephala* and *L. retusa* only leaf shoots were formed.

With the adult material Datta and Datta [10], were able to induce bud-break in single node segments and these shoots were directly rooted without attempting *in vitro* shoot multiplication. Also, there is no mention of the transfer of the plants to the field. Goyal *et al.* [23] described a full micro-propagation method for adult trees of cultivar K-67 and achieved 22.54-fold multiplication in approximately 21 weeks. The shoots were rooted in cultures with up to 90% success and transferred to soil with 80% survival. We [16,17] developed an efficient micropropagation protocol applicable to seedlings as well as adult trees of cultivars K-8 and K-28. The rate of multiplication was 5–6 fold every three weeks, thus resulting in approximately 280 000 shoots in 21 weeks starting from a single explant. The plantlets produced in cultures were nodulated with *Rhizobium* strain NGR-8 prior to their field transfer. The plants which were later transferred to the field, were clonally uniform and their growth could be correlated with nodulation pattern (Dhawan, unpublished).

3.5. *Mimosa pudica*

Gharyal and Maheshwari [20] aimed at achieving regeneration from various seedling explants of ‘touch me not’ – *Mimosa pudica*. Shoot regeneration followed by rooting were obtained but transplantation was not attempted.

3.6. *Ougeinia dalbergioides*

Ougeinia dalbergioides is an important tree of the hilly areas. The leaves provide nutritious fodder and the wood is used for making untensils, agricultural implements etc. Due to severe lopping, the seed set in this species is very poor. We initiated some experiments with the seedling explants. Cotyledonary and single node segments from three week old aseptically raised seedlings of *Ougeinia dalbergioides* could be multiplied 4–5 fold every 4 weeks on MS + BAP (3×10^{-6}M) + Kn (3×10^{-6}M) in our laboratory. There is lot of basal callusing and any attempt to reduce the callusing also resulted in stunted shoot growth. For rooting various media compositions were tried, but limited rooting (8%) has been achieved with NAA (5×10^{-6}). No attempt has been made to transfer these plants to the field (Dhawan, unpublished).

3.7. *Prosopis species*

Genus *Prosopis* consists of hardy, robust trees which produce useful wood and forage. Most *Prosopis* species are proving recalcitrant to tissue culture. However, success so far has been restricted either to seedling material or to bud break on explants from adult tissue. Tissue culture of *P. alba* [52], *P. chilensis* [24], *P. cineraria* [22] and *P. tamarugo* [24] has been attempted with seedling tissues. Success has been reported in *P. cineraria* starting with nodal segments from adult trees and plantlets were transferred to the soil. Bud break was obtained on nodal segments of *P. alba* [52] and *P. juliflora* [59]. The shoots, however, failed to attain a height of more than 2 cm and started turning brown from the top after 3 weeks. No multiplication has been achieved in *in vitro* formed explants of these species. We are also working on *P. juliflora* and irrespective of the media combinations, temperature range, etc. being tested, no success has been achieved so far in getting multiplication from shoots formed *in vitro* (Dhawan, unpublished).

3.8. *Robinia pseudoacacia*

Robinia pseudoacacia commonly called black locust, is extensively cultivated for timber and fodder, has proven useful for shelterbelts and in erosion control. The clonal propagation of *Robinia pseudoacacia* was first achieved by Chalupa and modified later by Davis' group [4,8,13–15]. Bargchi [4] developed a protocol using either stem or root cutting followed by enhanced axillary branching from nodal explants. A propagation rate of 5-fold every 4 weeks was achieved on MS supplemented with 0.25–1.0 mg l^{-1} BAP. Over 95% shoots formed roots on MS liquid medium supplemented with 2 mg l^{-1} IBA. The plants could be easily established in potting mix. Davis and Keathley [13], based on their work with five mature black locust trees, suggested that physiological/genetic factors inherent in the explants could make some clones more amenable than others to tissue culture. Merkle and Wiecko [35], later developed embryogenic callus from the immature seeds. The somatic embryos were well formed, with two distinct cotyledons. Direct and secondary embryogenesis was observed. Embryos germinated precociously, producing plantlets that were initially weak but later gained vigor and resembled seedlings. The clonal uniformity of these plants remains to be determined.

3.9. *Sesbania species*

Sesbania spp., improves soil and are potentially useful as ground cover to prevent soil erosion. Certain species are used for providing shade, production of paper pulp and fiber, food, fodder, medicinal purposes and as ornamental.

Khatter and Mohan Ram [26,27], succeeded in regenerating shoot buds from seedling explants, viz. cotyledon and hypocotyl of *Sesbania grandiflora* and *S. sesban*. No field trials are reported for these species and one is not sure of the clonal uniformity.

A protocol for organogenesis from different *S. rostrata* explants (cotyledons, hypocotyl, immature embryos) was established [59]. *S. rostrata* is a unique plant as it prolifically nodulates on the stem by *Azorhizobium sesbaniae*. Vlachova and his co-workers studied the susceptibility of *S. rostrata* to *Agrobacterium* mediated transformation using different wild type *A. tumefaciens* and *A. rhizogenes* strains. They reported that extensive systemic infection in *S. rostrata* by the agrobacterial strain occurs via spread of bacteria in the vascular bundle.

3.10. *Tamarindus indica*

While every part of the tamarind tree is useful, it is the fruit which is widely used for souring curries, and sauces. The average yields from a mature tree are in the range of 30–50 kg/annum but some trees yield as much as 180–225 kg. Also, the pulp quality varies from tree to tree. Since it is a cross pollinated species, the seed progeny gives rise to hetrogenous population and it is desirable to develop asexual methods for cloning of desirable individuals. Mascarenhas *et al.* [32] developed a method of cloning *T. indica* from various seedling explants. Hypocotyl, shoot tips and nodal explants on MS medium supplemented with Kn (0.2 mg l^{-1}), BAP (0.5 mg l^{-1}), calcium pantothenate (0.1 mg l^{-1}) and biotin (0.1 mg l^{-1}) gave rise to 1–5 shoots per explant within one month. Subculturing of shoots every 6 weeks onto the same medium gave 3–4 fold multiplication. Sixty to seventy percent of the shoots could be rooted in liquid White's medium supplemented with IAA, IBA, IPA, NAA (0.1 mg l^{-1}). Transplantation success of 50–60% was achieved in the field. The tissue culture propagated plants flowered in the 4th year as compared to 13–15 years in conventional seed raised plants. The yields are comparable to the seed raised plants with the added advantage of shortening the breeding cycle.

4. Conclusions

Fast growing trees belonging to family Leguminocae and Elaeagnaceae fix atmospheric nitrogen in symbiotic association of either *Rhizobium* or *Frankia*. These species are prime candidates for reclaiming the large wastelands existing all over the world. A programme should be launched for selecting superior trees, developing methods for their cloning, and selecting suitable *Frankia/Rhizobium* strains for different agroclimatic regions. This type of selection of macro- and micro-symbiont would lead to higher biomass production.

5. Summary

Considerable progress has been made over the last couple of decades in the development of aseptic techniques for multiplying superior and novel plants. While micropropagation methods are being widely exploited on a commercial scale for many ornamental and horticulture species their application in forestry continues to be meager. With increasing demand from our dwindling forest sources there is an urgent need to increase biomass production per unit area. This could be readily achieved by mass cloning of superior genotypes, either for raising seed orchards or for raising plantations. Since, for wastelands, nitrogen fixing trees are the prime candidates, it is important to develop tissue culture methods for cloning and wherever possible select the associated micro-organisms. The state of tissue culture of nitrogen fixing trees has been highlighted in this paper.

References

1. Agarwal SC and Prasad BN (Eds.) (1980) 3rd All India Bot Conference Dec 28–30. Abstr J Ind Bot Soc 59:657.
2. Akkermans ADL and van Dijk C (1981) Non-leguminous root-nodule symbiosis with actinomycetes and *Rhizobium*. In: Broughton WJ (Ed.) Nitrogen Fixation Vol. I. Ecology. Oxford Univ Press, New York. pp. 57–103.
3. Anand M and Bir SS (1984) Organogenetic differentiation in tissue cultures of *Dalbergia lanceolaria*. Curr Sci 53:1305–1307.
4. Barghchi M (1987) Mass clonal propagation in vitro of *Robinia pseudoacacia* L. (black locust) cv. 'Jaszkiseri'. Plant Sci 53:183–189.
5. Barghchi M (1988) Micropropagation of *Alnus cordata* (Loisel.) Loisel. Plant Cell Tissue Organ Cult 15:233–244.
6. Bertrand LJ and Lalonde M (1985) *In vitro* propagation and nodulation by *Frankia* of actinorhizal Russian olive (*Elaeagnus angustifolia* L.). Plant Soil 87:143–152.
7. Bonner J (1942) Culture of isolated roots of *Acacia melanoxylon*. Bull Torrey Bot Cl 69:130–133.
8. Chalupa V (1983) *In vitro* propagation of willows (*Salix* spp.), European mountain ash (*Sorbus aucuparia* L.) and black locust (*Robinia pseudoacacia* L.). Biol Plant 25:305–307.
9. Crawford DF and Hartney VJ (1986) Micropropagation of *Acacia mangium* and *Acacia stenophylla*. In: Turnbull JW (Ed.) Austr Acacias in Developing Countries. ACIAR Proc No 16. Proc Intl Workshop, Queensland, Australia. pp. 64–65.
10. Datta K and Datta SK (1985) Auxin + KNO_3 induced regeneration of leguminous trees – *Leucaena leucocephala* through tissue culture. Curr Sci 54:248–250.
11. Datta SK and Datta K (1983) Auxin induced regeneration of forest tree – *Dalbergia sissoo* Roxb. through tissue culture. Curr Sci 52:434–436.
12. Datta SK, Datta K, and Pramanik L (1982) *In vitro* clonal multiplication of mature trees of *Dalbergia sissoo* Roxb. Plant Cell Tissue Organ Cult 2:15–20.
13. Davis JM and Keathley DE (1985) Regeneration of shoots from leaf disk explants of black locust, *Robinia pseudoacacia*. In: Guries RP (Ed.) Proc Fourth North Central Tree Improvement Conf, East Lansing, MI, Aug 12–14, 1985. University of Wisconsin, Madison, pp. 29–34.
14. Davis JM (1987a) Toward efficient clonal propagation of mature black locust trees using tissue culture. Nitrogen Fixing tree Res Rep 5:57–58.
15. Davis JM (1987b) Differential responses to *in vitro* bud culture in mature *Robinia pseudoacacia* L. (black locust). Plant Cell Rep 6:431–434.

16. Dhawan V and Bhojwani SS (1985) *In vitro* vegetative propagation of *Leucaena leucocephala* (Lam.) de Wit. Plant Cell Rep 4:315–318.
17. Dhawan V and Bhojwani SS (1987) *In vitro* nodulation of micropropagated plants of *Leucaena leucocephala* by *Rhizobium*. Plant Soil 103:274–276.
18. Garton S, Hotser MA, Read PE, and Farnham RS (1981) *In vitro* propagation of *Alnus glutinosa* Gaertn. HortScience 16:758–759.
19. Gharyal PK and Maheshwari SC (1981) *In vitro* differentiation of somatic embryoids in a leguminous tree – *Albizzia lebbeck* L. Natuurwissenschaften 67:379.
20. Gharyal PK and Maheshwari SC (1982) Plantlet formation in tissue cultures of sensitive plant *Mimosa pudica* L. Z Pflanzenphysiol 105:179–182.
21. Glovak L, Greatbatch W (1982) Successful tissue culture of *Leucaena*. Leucaena Res Rep 3:81–82.
22. Goyal Y, Arya HC (1984) Tissue culture of desert trees I. Clonal multiplication of *Prosopis cineraria* by bud culture. J Plant Physiol 115:183–189.
23. Goyal Y, Binghan RL, and Felker P (1985) Propagation of the tropical tree, *Leucaena leucocephala* K67, by *in vitro* bud culture. Plant Cell Tissue Organ Cult 4:3–10.
24. Jordan M (1987) *In vitro* culture of *Prosopis* species. In: Bonga JM and Durzan DJ (Eds.) Cell and Tissue Culture in Forestry, Vol. 3, Case Histories: Gymnosperms, Angiosperms and Palms, Martinus Nijhoff, Dordrecht. pp. 370–384.
25. Kathjo S and Tewari MN (1979) Development of the root from the cotyledonary callus of *Acacia senegal*. Labd. J Sci Tech PB Life Sci 11:84–85.
26. Khattar S and Mohan Ram HY (1982) Organogenesis in the cultured tissues of *Sesbania sesban*, a leguminous shrub. Indian J Exp Biol 20:216–219.
27. Khattar S and Mohan Ram HY (1983) Organogenesis and plantlet *in vitro* in *Sesbania grandiflora* (L.) Pers. Indian J Exp Biol 21:251–253.
28. Lakshmana Rao PV and De DN (1987) Tissue culture propagation of tree legumes *Albizzia lebbeck* (L.) Benth. Plant Sci 51:263–267.
29. Lakshmi Sita G, Chattopadhyay S, and Tejovathi DH (1986) Plant regeneration from shoot callus of *rosewood*. Plant Cell Rep 5:266–268.
30. Lechevalier MP (1983) Cataloging *Frankia* strains. Can J Bot 61:2964–2967.
31. Lloyd G and McCown B (1980) Commercially-feasible micropropagation of Mountain Plant laurel, *Kalmia latifolia*, by use of shoot tip culture. Proc Int Prop Soc 30:421–427.
32. Mascarenhas AF, Nair S, Kulkarni VM, Agarwal DC, Khuspe SS, and Mehta UJ (1987) Tamarind. In: Bonga JM and Durzan DJ (Eds.) Cell and Tissue Culture in Forestry, Vol. 3, Case Histories: Gymnosperms, Angiosperms and Palms, Martinus Nijhoff, Dordrecht. pp. 316–325.
33. Mathur I and Chandra N (1983) Induced regeneration in stem explants of *Acacia nilotica*. Curr Sci 52:882–883.
34. McClean GD, Atherton JG, and Coaldrake JE (1966) Tissue Culture from an Australian acacia, *Acacia harpophylla* F.V.M. (brigalow). Phyton 23:101.
35. Merkle SA and Wiecko AT (1989) Regeneration of *Robinia pseudoacacia* via somatic embryogenesis. Can J For Res 19:285–288.
36. Montpetit D and Lalonde M (1988) *In vitro* propagation and subsequent nodulation of the actinorhizal *Hippophae rhamnoides* L. Plant Cell Tissue Organ Cult 15:189–199.
37. Mukhopadhyay A, Mohan Ram HY (1981) Regeneration of plantlets from excited roots of *Dalbergia sissoo* Indian J Exp Bioll 19:1113–1115.
38. Murashigee T and Skoog F (1962) A revised medium for rapid growth and bioassays that tobacco tissue cultures. Physiol Plant 15:473–497.
39. Nagamani R and Venketeswaran S (1983) *In vitro* culture of hypocotyl and cotyledon segments of *Leucaena*. Leucaena Res Rep 4:88–89.
40. Nataraja K and Sudhadevi AM (1985) Induction of plantlets from seedling explants of *Dalbergia latifolia* Roxb. *in vitro*. Beitr Biol Pflanz 59:341–350.
41. Peasly EL and Collins GB (1980) Development of an *in vitro* culture system for *Leucaena*. Leucaena News Lett 1:54.

42. Perinet P and Lalonde M (1983) *In vitro* propagation and nodulation of the actinorhizal host plant *Alnus glutinosa* (l.) Gaertn Plant Sci Lett 29:9–17.
43. Perinet P, Tremblay FM, Filion C, and Chatarpaul L (1987) Commercial miocropropagation of five *Alnus* species. Acta Hortic 212:551.
44. Ravishankar GA, Wali A, and Grewal S (1983) Plantlet formation through tissue cultures of *Leucaena leucocephala*. Leucaena Res Rep 4:37.
45. Ravishankar Rai V and Jagadish Chandra KS (1988) *In vitro* regeneration of plantlets from shoot callus of mature trees of *Dalbergia latifolia*. Plant Cell Tissue Organ Cult 18:77–83.
46. Ravishankar Rai V and Jagadish Chandra KS (1989) Micropropagation of Indian Rosewood by tissue culture. Ann Bot 64:43–46.
47. Read PE, Garton S, Louis KA, and Zimmerman ES (1982) *In vitro* propagation of species for bioenergy plantations. In: Proc Int Cong Plant Tissue Cell Cult, Tokyo. pp. 757–758.
48. Skolmen RG (1986) Acacia (*Acacia koa Gray*). In: Bajaj YPS (Ed.) Biotechnology in Agriculture and Forestry, Vol. 1. Trees I. Springer-Verlag, Berlin. pp. 375–384.
49. Skolmen RG and Mapres MO (1976) *Acacia koa* Gray plantlets from somatic callus tissue. J Hered 67:114–115.
50. Skolmen RG and Mapes MO (1978) After care procedures required for field survival of tissue culture propagated *Acacia koa*. Proc Intl Plant Prop Soc 28:156–164.
51. Suwal B, Karki A, and Rajbhandary SB (1988) The *in vitro* proliferation of forest trees. Silvae Genetica 37:26–28.
52. Tobone TJ, Felker P, Bingham RL, Reyes I, and Loughrey S (1986) Techniques in the shoot multiplication of the leguminous tree *Prosopis alba* and clones B–V50. In: Felker P (Ed.) Tree Planting in Semi-Arid Regions, Elsevier Science Publishers, Amsterdam. pp. 191–200.
53. Tomar UK and Gupta SC (1988) Somatic embryogenesis and organogenesis in callus cultures of a tree legume – *Albizzia richardiana* King. Plant Cell Rep 7:70–73.
54. Tomar UK and Gupta SC (1988) *In vitro* regeneration of leguminous trees (*Albizzia* spp). Plant Cell Rep 7:385–388.
55. Tremblay FM and Lalonde M (1984) Requirements for *in vitro* propagation of seven nitrogen fixing *Alnus* species. Plant Cell Tissue Organ Cult 3:189–199.
56. Tremblay FM, Nesme X, and Lalonde M (1984) Selection and micropropagation of nodulating and non-nodulating clones of *Alnus crispa* (Ait) Pursch. Plant Soil 78:171–k179.
57. Upadhyaya S and Chandra N (1983) Shoot and plantlet formation in organ and callus cultures of *Albizzia lebbeck* Benth. Ann Bot (Lond) 52:421–424.
58. Venketeswaran S and Gandhi V (1979) Mass propagation of two selected tree genera for biomass production *in vitro*. In: Proc 32nd Meeting Tissue Culture Assoc, Washington, D.C. pp. 118.
59. Vlachova M, Metz BA, Schell J, and de Bruijn FJ (1987) The tropical legume *Sesobania rostrata*: Tissue culture, plant regeneration and infection with *Agrobacterium tumefaciens* and *Rhizogenes* strains. Plant Sci 50:213–223.
60. Wainwright H and England N (1987) The micropropagation of *Prosopis juliflora* (Swartz) DC: Establishment *in vitro*, Acta Hortic 212:49–53.
61. Zeijlemaker FCJ (1972) Rep 1971–1972 On Plant physiology and pathology. Wattle Res Inst, Pietermaritzburg, S Afr Univ Natal. pp. 49.

19. Micropropagation of sandalwood (*Santalum album* L.) and mulberry (*Morus indica* L.)

P.S. RAO and V.A. BAPAT
Plant Biotechnology Section, Bhabha Atomic Research Centre, Trombay, Bombay 400 085, India

Sandalwood (*Santalum album.* L.)

1. Introduction

1.1. *Plant history and importance*

The Santalaceae are a family of some 250 species of mostly semi-root parasites. The genus *Santalum* comprises species of woody perenial plants widely distributed in tropical and subtropical regions of the world. Like many santalaceous parasites, *Santalum* attacks a wide range of host plants and even roots of the same species. Most members of the genus are evergreen trees and produce large quantities of seeds. The sandalwood tree reaches a height of 18 m and 2.4 m girth. The leaves are opposite and decussate. The shape and size of the leaves vary considerably. The bark is reddish brown to dark brown. Sandalwood starts flowering at an early age (3–4 years); however, good seeds can be obtained from seeds over 20 years old. The plant occurs mainly in India, Indonesia, South America, New Zealand, Hawaii and Australia and flourishes well from sea level upto 1800 m altitude, but the formation of the heartwood seems to be the best between 600 and 900 m. It grows very well in the rainfall from 60 to 160 cm. The plant requires special conditions of climate, altitude, rainfall and soil. In addition to all these, another factor which conditions its growth is abundant surrounding vegetation since sandalwood is a semi-root parasite and can thrive well in the midst of hospitable host plants [34].

The economic importance of *Santalum* are significant and will certainly continue to assume greater importance with passage of time. The sweet scent and medicinal properties of the plant have been known since ancient times and have come to be associated with culture and civilization through its use in the temples and religious functions. The volatile oil originating from the roots and heartwood of sandalwood is a pale yellowish, slightly viscous fluid having a peculiar sweet and a very lasting odour. The main constituent of sandal oil forming the higher boiling fractions are α and β-santals; both are

M. R. Ahuja (Ed.), Micropropagation of Woody Plants, 317–345.

sesquiterpene alcohols. The other ingredients, viz. acids, aldehydes, ketones, form only a comparatively minor part. The oil has medicinal properties and is used in ayurvedic as well as allopahtic systems of medicines. The oil is the main base for making sandal soaps, talcum powder, perfumes and incense. The paste of sandal is used as a cooling agent. The wood is hard and close-grained, like ivory and ebony, and is used for idols, furniture, combs, chess pieces and cosmetic boxes. Constituents of the benzene extract of sandalwood bark were reported to exhibit an excellent insect growth inhibiting property and chemosterilant activity [33,34].

Since sandalwood is an evergreen tree which survive for many years in the forest it represents an ideal habitat for disease causing agents. The most lethal and frequent disease is 'spike disease' caused by mycoplasma. Trees of all ages and sizes are affected by this disease and the tree dies within few years after the onset of the disease. Finding a spike disease resistant strain of sandal and biological control of the insect vector responsible for the transmission of spike disease are the permanent solution for the problem of sandal spike disease [24,32].

1.2. *Need for* in vitro *technology for sandalwood propagation*

The immense economic potential of *Santalum* demands a rapid micropropagation system. Generally the plant is multiplied by seeds. Normally seeds remain dormant for two months and gradually viability is decreased after 9–10 months period. Many adverse factors, such as attack by micro-organisms, heat, grazing by animals, excessive water etc. considerably reduces the survival of the seeds. The other methods of vegetative propagation – by cutting or by grafting – are of limited use in sandal since these methods are time consuming and are not always successful.

The relation between host and parasite has not been clearly understood in *Santalum* despite numerous conventional studies. Tissue culture possibly would be a better technique to deal with these problems. Isolation of disease-

Table 1. Highlights of *in vitro* investigations on sandalwood.

Nature of work	Reference
Adventive embryos from embryo callus and tissue cultures from endosperm	Rangaswamy and Rao (1963); Rao and Rangaswamy (1971)
Shoot buds and somatic embryos from hypocotyl	Rao and Bapat (1978); Bapat and Rao (1979)
Somatic embryos from stem callus	Lakshmi Sita *et al.* (1979)
Triploid plantlets from endosperm	Lakshmi Sita *et al.* (1980); Rao and Raghava Ram (1983)
Somatic embryos from protoplasts	Bapat *et al.* (1985); Rao and Ozias-Akins (1985)
Plantlets from synthetic seeds	Bapat and Rao (1988)

resistant cell lines and subsequent plantlet regeneration can help to obtain disease resistant plants. Cultures established from healthy, well characterized plant would certainly augment the stock of desired types of *Santalum*. Considering all these factors *in vitro* technology for the multiplication of sandal assumes greater significance (Table 1).

2. *In vitro* strategies for sandalwood micropropagation

2.1. *Organogenesis*

2.1.1. *Regeneration of shoots from hypocotyl and nodal explants*

Seeds were collected from a known elite tree. Following surface sterilization in $HgCl_2$ for 5 minutes the seeds were washed six times in autoclaved water and were sown on the basal medium. Hypocotyl segments (5 mm) from *in vitro* grown 4 week old seedling were excised and cultured aseptically on a basal medium. The basal medium (BM) [27] was supplemented with pyto-hormones as and when required either singly or in combinations. The pH of the medium was adjusted to 5.8 before adding agar to the medium. The cultures were grown under conitinuous light (1000 lux) at 25 ± 2° C and in rh of 55–60%.

Differentiation of shoot buds was readily achieved from hypocotyl cultures on BM supplemented with various auxins and cytokinins generally within 4 weeks of culture. Regeneration of buds was intense and exuberant on cytokinins compared with auxins. Among the cytokinins, 6-benzylaminopurine at 1 or 2 mg/l induced the maximum number of buds. Approximately 90% of the explants taken from the basal portion of the hypocotyl exhibited a higher potential for bud regeneration. This potential decreased to 80 and 60% for those explants obtained from middle and apical regions respectively. Data on the effect of age of seedling on the bud formation revealed that there was not much variation in bud regenerating capacity of explants excised from seedling of different ages. Position of the explants on the medium has a significant effect on bud forming capacity of the explants. Bud formation was 100% when only root end of the explant was either dipped in the medium or even kept touching the surface of the medium. In an opposite position i.e., shoot end in the medium bud formation was reduced to about 10%. When the entire explant was embedded inside the agar medium explants turned brown, however, if kept in liquid medium about 20% of the explants regenerated buds. In 60% of the cultures, explants when split longitudinally and kept in contact with the medium, shoot buds regenerated. Thin slices of explants produced shoot buds from all sides in 70% of cultures. However, if discs were placed in agitated liquid medium regeneration of shoot buds were observed which was poor.

Nodal segments bearing two axillary buds produced numerous shoot buds on MS + benzylaminopurine BAP (0.5 and 1 mg/l). Incorporation of other two cytokinins viz. kinetin or zeatin (0.1, 0.5 and 1 mg/l each) did not induce regeneration of shoot buds. Nodes nearer from the apex had more tendency for regeneration.

2.1.2. *Development of rooted plantlets*

Regenerated shoots buds were carefully isolated and transferred to a variety of root inducing media. After transfer, buds developed further and attained a height of 18 to 20 cms in all hormonal combinations tested. The plantlets grew on all auxin supplemented, agar or liquid medium on filter paper bridges. All the rooted plantlets were transferred to paper cups with vermiculite and were grown further for three weeks before transfer to the soil.

2.2. *Somatic embryogenesis*

Tissue cultures of sandalwood showed a great propensity for somatic embryo formation. The phenomenon was observed in tissue cultures of stem, endosperm, cell suspension cultures and protoplasts derived from stem callus and cell suspension cultures.

2.2.1. *Establishment and somatic embryogenesis from hypocotyl and stem callus*

Embryogenic callus was induced from hypocotyl and stem explants of *Santalum* on MS [17] + 2,4-dichlorophenoxyacetic acid (2,4-D 1 mg/l) or on MS + 2,4-D (1 mg/l) + kinetin (0,2 mg/l). Although different combinations of auxin-cytokinins induced embryos in the callus, high regeneration of somatic embryos was observed on MS + indoleacetic acid (IAA) + BAP (Table 4). The cultures on media with IAA, napthalene acetic acid, NAA with indolebutyric acid

Table 2. Effect of different combinations of IAA with various nitrogen sources on callus development and embryo differentiation.

Medium	Intensity of callus development	Intensity of embryo differentiation	% cultures producing embryos
BM + IAA	+++	+++	60
BM + IAA + YE	++	++++	70
BM + IAA + Casamino acid	+++	+++	70
BM + IAA + Peptone	+++	++	50
BM + IAA + CH	+++	++++	80

Data scored at the end of 4 weeks, 12 replicate per treatment. ++ moderate, +++ high, ++++ intense. Concentration of auxins 1 mg/l; nitrogen source 400 mg/l. (after Bapat and Rao, 1984).

(IBA) and gibberellic acid (GA) formed embryos that developed from the globular stage to mature embryos which occasionally produced adventitious buds, leaf like structures and roots. Well developed plants grew from these cultures after having been cultured over a long period of time. On the other hand, most of embryos were restricted to the globular stage on media containing 2,4-D. Addition of yeast extract, casein hydrolysate or casamino acid stimulated embryo differentiation on addition to MS + IAA (Table 2).

2.2.2. *Parameters controlling somatic embryogenesis*

In hypocotyl callus incorporation of BAP (1 or 2 mg/l) to the basal medium promoted shoot bud formation but affected embryo differentiation. An examination of the suitability of several mono, di and polysaccharide showed that sucrose was the best source for high frequency embryogenesis, maltose was second best followed by glucose and fructose. A correlation was also observed between sucrose concentration and embryo differentiation as shown in Table 3.

The development of sandalwood plants from somatic embryos was attempted by transferring the embryogenic tissue/single embryo to medium with lower concentration of sucrose and a low concentration of IAA and GA. In many cultures the embryo, instead of developing into plantlet, dedifferentiated again and produced a callus mass which again showed intense regeneration of somatic embryos. Continuously regenerating callus cultures were maintained by subculturing the tissue every 3 weeks.

2.2.3. *Somatic embryogenesis in cell suspension cultures*

From the undifferentiated callus tissue on MS + 2,4-D (1 mg/l) good suspensions were raised which consisted predominently of small cell aggregates.

Table 3. Effect of different concentrations of sucrose on callus development and embryo differentiation in *Santalum album*.

Sucrose concentration (%)	Intensity of callus development	Intensity of embryo differentiation	% cultures producing embryos
0	–	–	–
1	+	+	10
2	+++	+++	70
4	++++	++++	80
8	++	++	40
12	++	+	20
16	++	–	–
20	+	–	–

Data scored at the end of 4 weeks, 12 replicates per treatment, – nil, + low, ++ moderate, +++ high, ++++ intense.

Table 4. Effect of different concentrations of IAA and BAP on callus development and embryo differentiation in *Santalum album*.

IAA/BAP (mg/l)	Intensity of callus development		Intensity of embryo differentiation		% cultures producing embryos	
	IAA	BAP	IAA	BAP	IAA	BAP
0	–	–	–	–	–	–
0.5	++	++	++++	+	70	20
1	++	+++	++++	++	70	30
2	+++	+++	+++	++	60	50
4	+++	++++	++++	+++	60	50
8	+++	++	++	–	50	20
10	++	++	+	+	20	20
12	+	+	+	+	20	10

Data scored at the end of 4 weeks, 10 replicates per treatment, – nil, + low, ++ moderate, +++ high, ++++ intense.

The suspensions rapidly senesced at the end of the growth period (10 weeks). Therefore suspension cultures were transferred periodically to maintain active growth and embryogenic potential. As is well known from other plants and for sandalwood, too, a particular density is required for active growth of the cultured suspension cells. During the growth phase thousands of cells per ml were produced by rapidly dividing embryogenic cells. Hence, the population of cells comprised many highly vacuolated cells with a variety of size ranges and clusters of small, highly cytoplasmic cells. The small cytoplasmic cells and clusters were highly meristematic and as cell division proceeded smaller clusters and individual cells broke off. A large number of organized embryos were obtained in liquid medium when 2,4-D was replaced with IAA and BAP (0.5 mg/l) (Table 4). Such embryos in liquid medium were independent of each other. Mature, organized embryos developed into plantlets on transfer of these embryos from liquid to solid medium.

Different studies were conducted in order to obtain some degree of uniformity in embryogenic development. The embryogenic suspensions were passed

Table 5. Effect of ABA on embryogenesis in *Santalum album*[a].

MS + ABA	No. of normal embryos/g callus	No. of abnormal embryos/g callus
0	80	30
0.5	95	38
1	115	26
3	140	28

[a] Data scored at the end of 8 weeks. ABA: abscissic acid.

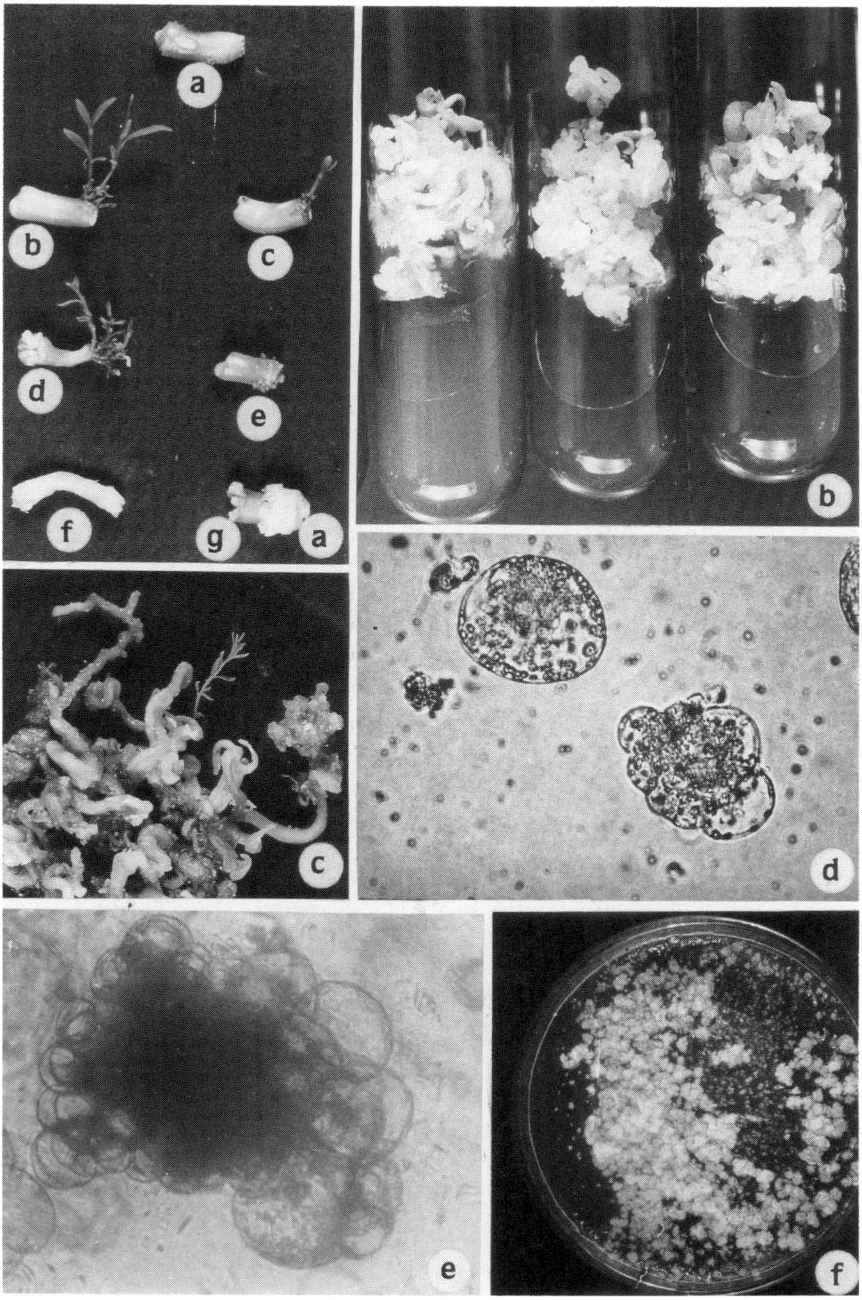

Figure 1. Sandalwood. **a.** Direct shoot bud regeneration from hypocotyl explants on auxin media. (a) control (b) BM + IAA (1 mg/l) (c) BM + IBA (1 mg/l) (d) BM + NAA (1 mg/l) (e) BM + NOA (1 mg/l) (f) BM + pCPA (1 mg/l) (g) BM + 2,4-D (1 mg/l).

b. Callus showing differentiation of embryos.

c. Embryos differentiated on medium containing maltose.

d, e, f. Protoplast division and colony and callus formation.

through the sieve (100 μ) which was followed by centrifugation in 16% ficoll containing 2% sucrose. The isolated cell clusters were cultured on MS + IAA + BAP. However, it was difficult to isolate purely embryogenic cells from non embryogenic cell population.

Abscisic acid (ABA) is known to check the development of abnormal somatic embryos in carrot and caraway [1]. It selectively inhibited additional growth centres leading to multiple embryos and shoot formation and did not interfere with the normal development. ABA was filter-sterilized and was added to MS + IAA + BAP in different concentrations (0.5, to 3 mg/l). The addition of ABA influenced greatly and increased the population of normal somatic embryos in embryogenic callus. In liquid medium also ABA markedly enhanced the percentage of normal embryos (Table 5).

3. *In vitro* studies on host-parasite relationship

It is well known that sandalwood is an obligatory root parasite and draws part of its nutrient from soil and part of the nutrient from the host through the haustorial connection. The exact nature and degree of its dependence upon the host has not been precisely elucidated. The parasitic behaviour of *Santalum* has been described by Iyengar [12]. The young hostless sandal seedling struggling to grow for haustorisation with host has repeatedly been observed. The seedlings exhibit a sudden and spectacular burst of vegetative activity when they are provided with the host. In conventional technique, a germinated seedling of *Santalum* is placed near a host plant. It has been observed that legumes, particularly *Cajanus*, is the best host plant in the seedling stage of sandal. Hence plantlets obtained from somatic embryos were placed on a medium containing 1/2 MS along with a seedling of *Cajanus cajan*. After 8–10 weeks, *in vitro* sandalwood plantlet produces the haustoria and connections are established with the roots of *Cajanus*. Thereafter, sandal plant grows very rapidly into robust and healthy trees.

In order to investigate the effect of host on callus growth and embryogenesis, seed extract of germinating seeds of *Cajanus* and *Glycine* was added to the

Table 6. Effect of *Cajanus* seed extract on callus growth and somatic embryogenesis of *Santalum album*.

Seed extract (%)	Intensity of callus development	Intensity of embryo differentiation	% cultures producing embryos
0	++	++	30
0.5	+++	++	50
1	+++	+++	50
2	+++	++	60

Data scored after 6 weeks, control contains only basal medium, 12 replicates per treatment, – nil, + low, ++ moderate, +++ high, ++++ intense.

MS basal medium. Ten grams of just germinating seeds were extracted in water with a mixer. The extract was diluted to one litre. From this 0.5, 1 and 2% were added to the MS medium. At 0.5 and 1% no significant change was noticed. Intense and extensive proliferation and differentiation was seen at 2% concentration of seed extract. a three fold increase over the control was noticed with the seed extract of *Cajanus* (Table 6).

Cell suspensions were plated on MS + seed extract medium. Multicellular colonies leading to callus development and subsequent embryo differentiation was noticed on seed extract medium indicating the affinity of extract even at cellular level.

4. Cryopreservation of somatic embryos

Preliminary investigations were carried out to assess the capability of somatic embryos as a source for preservation of elite germplasm of sandalwood. Somatic embryos were isolated from callus and were treated with DMSO + glycol + sucrose prepared in MS medium. The embryos were kept in this solution for 2 h at 4° C. Afterwards the embryos were blot dried to remove any traces of superficial moisture by using sterile filter paper. Embryos were then carefully transferred to screw cop polypropylene ampoules. Approximately 30 embryos were kept in each ampoule. The ampoules were stored at −70° C overnight and were transferred to cryocan at −196° C containing liquid nitrogen. For the thawing procedure and recovery of embryos, ampoules were transferred to water bath at 35° C. After 10 minutes, ampoules were opened and embryos were cultured on MS + IAA + BAP. Some of the frozen embryos (10%) proliferated and developed numerous secondary embryos. Further work is in progress.

5. Production of synthetic seeds

The consistent and high frequency regeneration of somatic embryos in cultures makes it an ideal system for preparation of synthetic seeds. Somatic embryos were carefully selected from callus or liquid cultures. Embryos were blot dried on a filter paper and were dipped for a few seconds in a mixture of 100 ml gel of sodium alginate (Sigma) prepared in MS basal medium. Subsequently, they were picked by forceps and were dropped into the solution of $CaCl_2 \cdot 2H_2O$ (1.036/150 ml). The drops (beads), each containing a single embryo were kept in this solution for 40 minutes on a gyratory shaker (80 rpm) in light (950 lux). After the incubation period the beads were recovered by decanting the $CaCl_2 \cdot 2H_2O$ solution and washing the beads 3–4 times with MS basal medium. Such encapsulated embryos were then cultured on the nutrient medium. One set of encapsulated embryos was stored at 4° C for 45 days on a filter paper soaked in MS medium in an aseptic condition. After the storage the beads

Table 7. Effect of different concentration of alginate on germination of synthetic seeds of *Santalum album*.

Alginate (%)	No. of seeds cultured	No. of seeds germinated	% Germination
2	60	5	8.3
3	60	6	10
4	60	3	8.3

Data scored at the end of 6 weeks on MS basal medium. (From Bapat and Rao, 1988. Reproduced with permission.)

were cultured on the nutrient medium.

Like isolated embryos, embryogenic cell suspensions were also encapsulated in an alginate matrix. First the suspensions were filtered on a nylon net and were washed twice in MS basal medium. Afterwards suspensions were centrifuged and cells were mixed with 2.5% Na-alginate solution in 1 : 1 proportion. The mixture was pipetted dropwise into 50 ml of MS liquid medium containing $CaCl_2 \cdot 2H_2O$ (1.036/150 ml). After incubation and washing, the beads were suspended in 25 ml medium in 100 flasks (10 beads/flask) and were shaken in continuous light (Ca 950 lux) at 25 ± 2° C. One set of beads was stored at 4° C in a petridish on filter paper soaked in MS basal medium.

At the end of 8 weeks germination ensued and the root portion of the embryos pierced through the matrix in 10 to 16% of beads. Subsequently, a shoot portion emerged and roots established contact with the medium. On transfer of germinated embryos to MS + IAA (1 mg/l) + IBA (0.5 mg/l) + sucrose (5%), the shoot and root portion elongated. A well developed plantlet was obtained at the end of 16 weeks. As shown in the following table, 3% alginate was the better concentration for germination (Table 7).

The encapsulated embryos stored at 4° C for 45 days germinated on transfer to MS basal medium at 25 ± 2° C. Different nutrient sources were tried to

Table 8. Effect of nutrient medium on germination of synthetic seeds of *Santalum album*.

Medium	No. of seeds cultured	No. of seeds germinated	% Germination and recovery of plantlets
MS	48	8	16.6
½ MS	48	9	18.7
WM	48	6	12.5
½ WM	48	6	12.5
TW	48	2	4.2

Data scored at the end of 8 weeks; MS, Murashige and Skoog; WM, White's medium; TW, tapwater. (From Bapat and Rao, 1988. Reproduced with permission.)

Table 9. Effect of maltose on germination of encapsulated embryos of *Santalum*[a].

MS + maltose n	No. of embryos encapsulated	No. of embryos germinated	% Germination
0 (+ sucrose)	50	8	16
0.5	50	4	8
1	50	6	12
3	50	17	34
5	50	15	30

[a] Data scored at the end of 8 weeks.

increase the germination rate of synthetic seeds and maximum germination was observed on MS basal medium (Table 8).

Addition of maltose as carbon source to the basal medium improved the germination rate of encapsulated beads (Table 9).

On MS + IAA (0.5 mg/l) + BAP (0.5 mg/l), encapsulated embryos started producing secondary embryos instead of showing germination. Such secondary embryos, on separation from the beads, developed into plantlets. This capacity was retained in the beads stored at 4° C for 45 days.

Cell suspensions encapsulated in alginate as beads and stored at 4° C for 45 days started liberating cells into surrounding liquid medium within 8–10 days when brought to the room temperature. Cells from suspensions on plating on agar medium grew into microcolonies and callus and finally into embryos. One of the drawbacks of alginate is its deformability, which limits its use in high volumetric scale reactions. Silica gel, on the other hand, is much harder but cannot be used to entrap cells. A composite of silica and alginate, however, combines the properties of both gels. Composite gels were prepared by mixing 50% gel-H with 4% alginate in 1 : 2 proportion. Embryos were dipped in this mixture and dropped in 5% $CaCl_2 \cdot 2H_2O$. Beads were kept in this solution for 2 h and afterwards were repeatedly washed in autoclaved water. The beads were cultured on MS + IAA (1 mg/l) for germination. Like alginate beads embryos in composite gels also germinated and formed plantlets (Table 10).

Table 10. Use of different matrices on germination of synthetic seeds of *Santalum*.

Matrix	MS	Cajanus seed extract
Alginate	18[a]	17[a]
Silica gel + alginate	22	20
Silica gel + alginate + charcoal	24	18

Data scored at the end of 8 weeks, 50 beads per treatment.
[a] No. of beads germinated in %.

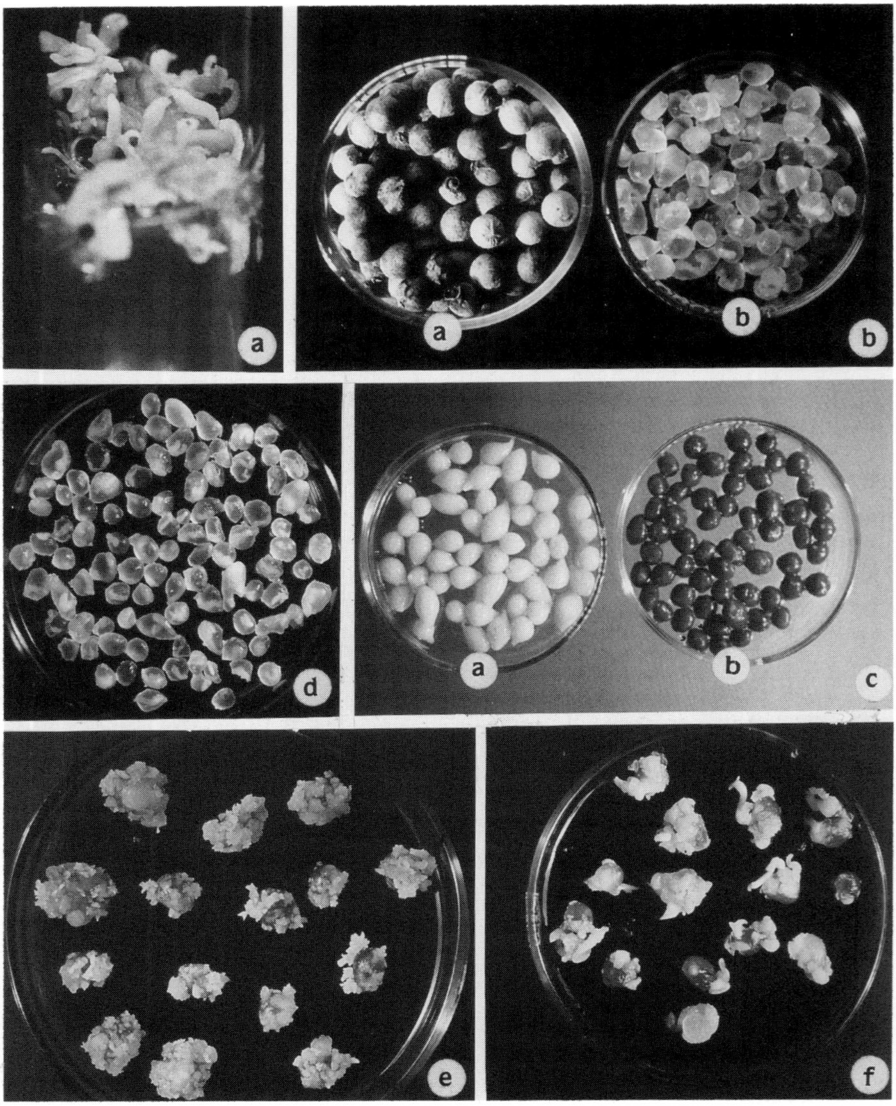

Figure 2. Sandalwood. **a.** Embryos selected for encapsulation.
b. (a) Natural seeds.
(b) Embryos encapsulated in alginate.
c. (a) Embryos encapsulated in silica gel + alginate.
(b) Embryos encapsulated in silica gel + alginate + charcoal.
d. Encapsulated embryos on MS + IAA (0.5 mg/l).**e, f.** Germinating embryos from alginate beads.

6. Protoplast culture

6.1. *Isolation of protoplasts from various sources*

Protoplasts were isolated from various sources such as hypocotyl and stem callus, cell suspensions and leaf mesophyll cells. Different enzyme mixtures containing combinations of cellulase (R-10, Onozuka), hemicellulase (Sigma), pectinase (Sigma) and macerozyme (Kinki Yakult) were employed. The callus from hypocotyl and stem weighing one gram was gently macerated in 10 ml nutrient medium and centrifuged to obtain pellet. For suspension cells, nutrient medium was removed by centrifugation followed by incubation in enzyme mixture. Young leaves from mature trees were macerated in 0.3% mannitol and then were incubated in enzyme mixture. The pH of the enzyme mixture was adjusted to 5.8. Following the incubation, the suspension was passed through a nylon mesh (μ 100) and centrifuged at 40 rpm for 3 min. The enzyme was decanted and the pellet was washed twice with the nutrient medium. The protoplasts were separated from the debris by floating the pellet on 20% sucrose followed by centrifugation.

6.2. *Somatic embryogenesis from protoplasts*

Protoplasts were isolated from stem callus by incubating 1 g of callus tissue in an enzyme mixture of Cellulase 1% (Onozuka) + macerozyme 0.5% (Kinki Yakult) with sorbitol or mannitol (0.6 M) as an osmoticum. The pH of the enzyme solution was adjusted to 5.8. The protoplasts were purified first by sieve and later by 20.5% sucrose by centrifugation. The protoplasts were repeatedly washed with nutrient solution to remove the traces of enzyme and sucrose. Afterwards protoplasts were cultured in the liquid medium or on agar jelled medium at a concentration of 4 to 5×10^5/ml.

The cultures were grown under fluorescent light (1000 lux) at $25 \pm 2°$ C and a relative humidity of 55 to 60%. After a week the cultures were transferred to fresh liquid medium containing reduced osmoticum. After the formation of microcolonies 2 ml of 0.4% agar containing nutrient medium was added.

Maximum yield of protoplasts (8.73×10^6 gram of tissue) was obtained after 16 h of incubation of the tissue in the enzyme mixture. Non-differtentiating tissue on MS + 2,4-D gave better yield of the protoplasts than the embryogenic tissue. Regeneration of cell wall in the cultured protoplast was visible with 18 h, detected by asymmetrical shape. This was followed by increase in size and formation of prominent nuclei and cytoplasmic strands. First division occurred between the 4th and 5th days. About 60–70% protoplasts divided and approximately 50% of such divided protoplasts showed sustained divisions leading to colonies. The optimum medium for protoplast division was as decribed in Table 11.

The other tested media, such as V47, B5 and VKM, were not found to

Table 11. Composition of nutrient medium for culture of protoplasts of *Santalum album* isolated from stem callus.

Major elements (mg/l)		*Iron and minor elements (mg/l)*	
NH_4NO_3	1444	$FeSO_4 \cdot 7H_2O$	27.85
KNO_3	1480	Na_2EDTA	37.25
$MgSO_4 \cdot 7H_2O$	984	KI	0.83
$CaCl_2 \cdot 2H_2O$	735	H_3BO_3	6.2
KH_2PO_4	68	$MnSO_4 \cdot 4H_2O$	22.3
		$ZnSO_4 \cdot 7H_2O$	8.6
Organic addenda		$Na_2MO_4 \cdot 2H_2O$	0.25
Thiamine HCl	1	$CuSO_4 \cdot 5H_2O$	0.025
Nicotinamide HCl	1	$CoCl_2 \cdot 6H_2O$	0.25
Riboflavin	0.1		
Ca pantothenate	1	*Carbon source*	
Biotin	1	Sucrose	20%
Choline chloride	1	Glucose	20%
Folic acid	1		
Pyridoxine HCl	1	*Hormones*	
Inositol	100	NAA	1
		2,4-D	1
		BAP	1
		Others	
		Charcoal	0.1%

[a] Mannitol/sorbitol was added as an osmoticum and an osmometer was used to adjust the proper osmotic value of the medium (700 to 750 mOsm), pH was 5.9. (From Bapat *et al.* 1985. Reproduced with permission.)

be suitable for sustained divisions of protoplasts. The microcolonies formed numerous somatic embryos when transferred to MS + IAA (1 mg/l) + BAP (1 mg/l).

7. Plantlet regeneration from somatic embryos

The ultimate success of any tissue culture programme depends upon the ability of the tissue culture-raised plants to grow in the soil. Although sandalwood has a very high frequency of somatic embryogenesis, the corresponding survival of plantlet is not very high and plantlets do not adapt easily to greenhouse conditions. The leaves desiccate very rapidly and respond to a changed environment by quickly wilting when plantlets are removed from the tubes. However, sandal plantlets can be acclimatized by keeping the culture tubes open for 3–4 days. Desiccation of leaves in the culture tube may still occur but this does not seem to affect the survival of the plantlets. At this stage good growth of roots appears to be more important. After removal from the tubes the plantlets are kept in a mist house with highly humid environment (70–80%) and this improves the survival rate. Plants which are difficult to establish in pots should be transferred for 10 days to an auxin-free liquid medium. This promotes rooting.

8. Discussion

The results presented here have demonstrated the possibility of micropropagation of sandal mainly through somatic embryogenesis. The *in vitro* production of plants could become a very useful tool for large scale propagation as has been shown in other forest trees [10], plantation crops [19], cereals [8] and legumes [9,15]. This would depend, of course, on the somatic embryos producing genetically stable plants that would be faithful copies of the source clone. Even if this is not the case variations among the plants could be used as tool for the genetic improvement of screening the seedlings for disease resistance and for other important silvicultural characters.

The advantage of cell suspensions or protoplast in tree tissue culture is that trees could be reduced to a convenient size for the experimentation and manipulation of cells or protoplasts in a way that a microbiologist is able to handle with microorganisms. Success in protoplasts culture in sandalwood as presented above is very significant as protoplast work in trees is very much restricted. This could be the starting point for the parasexual hybridization and genetic engineering programmes. Studies could be conducted on somatic hybridization of disease resistant and high yielding but susceptible trees. It would be possible to expose the protoplasts to chemical mutagens or ionizing radiation and then rear them to plants for introducing genetic variation. Protoplasts or cells plated on different concentrations of salts or disease extract may possibly give salt tolerant or disease resistant lines respectively. The barriers that remain do not appear to be insurmountable since cell suspensions or protoplasts have the capacity to regenerate plants. High rate of conversion of somatic cells to embryos as seen in sandalwood offers the possibility for commercial production of somatic embryos in large scale bioreactors. The use of either batch cultures or continuous cultures have to be employed for harvesting the embryos periodically. A new design has to be planned for such studies.

During the past two decades noteworthy progress has been made in understanding the biology of parasitic angiosperms [18]. A host stimulant requirement for root parasite has been known for a long time. In a few root parasites, haustoria have been successfully induced in the absence of host contact. Root exudates of many host plants have been added to induce haustorial induction. In *Santalum*, extract of germinating seeds of a host plant was added to gain an understanding of its effect on *in vitro* cultures. *Cajanus* is used in conventional methods as a host to young sandal seedling. The enhanced growth of *Santalum* embryogenic tissue on the extract of *Cajanus* seedlings shows the positive affinity of the host on the root parasitic nature of sandalwood. Likewise, further investigations into the mechanisms by which

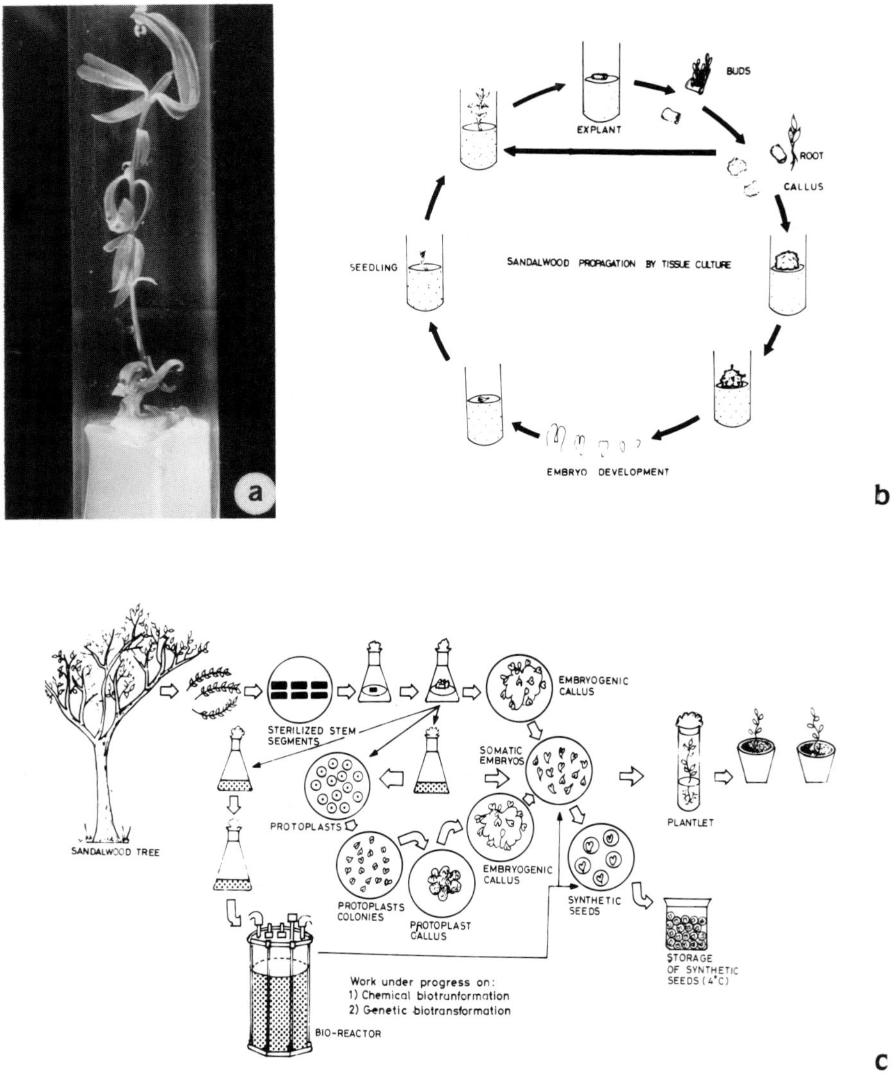

Figure 3. Sandalwood. **a.** Plantlet regenerating from somatic embryo.
b. Sandalwood propagation by tissue culture.
c. Biotechnological work on sandalwood.

these parasitic plants affect their host plant may provide a means for alleviation of the effects of parasitism under field conditions. The attachment of *in vitro* regenerated plantlet with the host shows that the requirement for the host has been retained in tissue culture raised plants. In addition, studies on host parasite interaction, development of haustorial connection and coordination between the host and the parasite, and host recognition and specificity could be easily studied through tissue culture technique.

The regeneration of plantlets from encapsulated somatic embryos in *Santalum* is significant observation since currently there are relatively few forest trees where predictable embryo regeneration and plantlet production are achievable. According to Redenbaugh *et al.* [30] artificial seeds will be useful for recovery of hybrids, elite germplasm and genetically engineered plants particularly those with sterile or unstable genotypes plus a potential delivery system. In addition, acclimatization steps would be avoided if direct sowing of encapsulated somatic embryos is achieved. However, this is a crucial stage and requires detailed experimentation. The size of the synthetic seeds is advantageous for several functions such as storage, handling, shipping and planting [30]. The matrix which forms a coating may be biologically inert but several other growth nutrient, biofertilizers, pesticides and other beneficial adjuvants can be incorporated to the encapsulating matrix. Encapsulation of cells (immobilization) has been carried out mainly for studies on biocatalyst, biotransformations and for protoplast immobilization [5,11]. The major advantage of encapsulation of cells is that alginate beads reduce changes or strong gradients in the environment. This protection against fluctuations in temperature, pH, osmolarity and nutrient conditions may therefore give both improved survival and simplified handling in general. In case of *Santalum*, regeneration of somatic embryos form encapsulated cells indicate that this could be an effective tool for high regeneration and preservation of germplasm.

There seems to be great promise for the future in a new line of attack on all the problems associated with sandalwood. Tissue culturing of sandalwood material combined with concentrated efforts to delineate the detailed physiology and genetics of sandalwood could make it possible to produce plants with superior characters and yet resistant to disease, climatic conditions and other adverse factors. This is admittedly a tall order, but it is certainly a goal worth pursuing.

Mulberry (*Morus indica* L.)

1. Introduction

1.1. *Crop history*

The genus *Morus* consists of trees and shrubs and is distributed in the temperate and subtropical regions of the northern hemisphere. Commonly known as mulberry, a few of the *Morus* species are valued for the their foliage, which constitutes the chief feed for mulberry silkworms. Some species are grown for their edible fruits and timber. The species *M. indica* is a monoecious, occasionally dioecious shrub or moderately sized tree. The plant regenerates itself naturally by seeds or can be propagated by cutting or grafting. Numerous mulberry types are under cultivation in various silk producing countries of the world and the types differ in their adaptability to various soils and climates, resistance to various diseases, food value of the leaf crop for silkworms and suitability for use as stock or scion in grafting. The plant is attacked by various pathogens. Mildew attacks leaves and causes spots on the leaves; other type of fungus grow in the main stem. Shownia disease affects plantations, reducing leaf output. Many pests also attack the plants posing serious problems for mulberry cultivation. Beetle, scale insects, borers, aphides, thripps and white-grubs generally reduces mulberry cultivation.

1.2. *Need for* in vitro *technology for mulberry propagation*

For successful micropropagation of mulberry through tissue culture three main steps must be accomplished.

(1) establishment of axillary buds in culture;
(2) achievement of high multiplication rate through multiple bud formation;
(3) acclimatization and good rooting of propagules after transfer to the soil.

In mulberry, results reported so far indicate the feasiability of using tissue culture technology for rapid micropropagation independent of several natural factors. It would be possible to produce unlimited plants per variety. Success if achievable, in anther culture and protoplast culture could speed up breeding work and be useful in systematic hybridization and realization of possible crosses in wild and cultivated species.

Only a limited number of reports have appeared on tissue culture of mulberry. Oka and Ohyama (1981) reviewed the reported work on mulberry. Ohyama [20] and Yamamoto and Kamata [37] reported the establishment of callus from hypocotyl and embryo. However, regeneration of either shoot buds or

Table 12. In vitro studies on *Morus* (Mulberry) species.

Nature of the work	Reference
Suspension cultures from hypocotyl and embryos	Ohyama (1970); Yamamoto and Kamata (1971); Yamada and Okamoto (1977)
Rooting from callus	Ohyama (1970); Seki *et al.* (1971)
Callus from stem, root	Yamada and Okamoto (1977)
Buds from hypocotyl	Ohyama and Oka (1982)
Plantlets from axillary buds	
Buds from leaf	Oka and Ohyama (1981)
Single plantlets from axillary buds, male and female flowers	Patel *et al.* (1983)
Multiple buds from axillary buds and leaf	Mhatre *et al.* (1985)
Plantlets from encapsulated axillary buds	Bapat *et al.* (1987)
Plantlets from meristems stored in liquid nitrogen	Yakuwa and Oka (1988)

somatic embryos from callus have so far not been reported. Regeneration of single and multiple plantlets from axillary bud has been reported [16,22]. Recently Bapat *et al.* (1987) used encapsulated shoot buds in studies connected with the micropropagation of mulberry (Table 12).

2. *In vitro* work on mulberry

2.1. *Shoot multiplication*

2.1.1. *Culture of stem, leaf and petiole*

Axillary buds were excised from a 10 year old, sexually mature tree of *Morus indica*. The buds were washed several times with liquid soap followed by rinsing in tap water and dipping in 70% alcohol. The material was finally surface-sterilized with 0.1% $HgCl_2$ for 5 min and washed several times in autoclaved water. The explants were aseptically placed on MS + IAA (1 mg/l) medium containing 3% sucrose. The buds developed into plantlet. From such shoot cultures, stem internodes, leaf, petiole and axillary buds were cultured. According to the requirement of the experiment various hormones were added to the basal medium. The pH of the medium was adjusted to 5.8 before jellying the medium with 0.8% agar. All cultures were maintained in culture room at 25 ± 2 °C at a relative humidity of 50–60% and were exposed to continuous fluorescent light (ca. 1000 lux). Each treatment consisted of 24 replicates and the experiments were repeated three times.

Leaf discs, stem and petiole segments did not show any response on MS alone. However, explants proliferated into callus tissue on MS provided with IAA, IBA, NAA or 2,4-D at 0.5, 1 or 2 mg/l within 4 weeks. From the primary explants, callus tissues were isolated and repeatedly subcultured on the fresh medium of the same composition. No organogenesis or embryogenesis was obtained from the callus tissue. The callus was green and friable on 2,4-D whereas on other auxins callus was compact and brownish. Similarly none of the cytokinins tested (Zea, Kin, BAP, 2ip) promoted shoot differentiation from callus.

2.1.2. *Culture of axillary buds*

The axillary buds cultured on MS basal medium sprouted a little but did not show further growth. However, buds cultured on MS + IAA or NAA or 2,4-D produced plantlets with roots. Abundant rooting was observed on MS + IAA which was sparse on NAA or 2,4-D. On different cytokinins (BAP, Kin, 2 ip or Zea) rooting was totally inhibited and buds developed into shoots. The results are presented in Table 13.

On various auxin-cytokinin combinations, mainly on MS+NAA+BA or Kin, sprouting of buds occurred within three days and in another ten days shoots

Table 13. Morphogenetic response of axillary buds of *Morus indica* to various growth substances.

Medium	No. of surviving ex-plants ± SE	No. of explants showing regeneration of shoots	No. of explants showing inflorescence	No. of plants regenera
MS+Kn(1)	21.0±0.58	16.0±0.58	16.0±0.58	–
MS+BAP(1)	19.7±1.20	18.0±0.58	10.3±1.86	–
MS+NAA(1)	12.0±1.16	6.0±0.58	5.0±1.00	5.0±1.0
MS+NAA(0.5)+Kn(1)	18.0±1.53	16.0±1.16	12.0±2.52	12.0±2.5
MS+NAA(1)+Kn(0.5)	9.0±1.00	4.3±0.67	3.0±0.00	2.0±0.5
MS+NAA(1)+Kn(1)	18.0±1.53	12.0±1.53	12.0±1.53	12.0±1.5
MS+NAA(0.5)+BAP(1)	22.7±1.33	16.0±1.53	12.0±3.56	–
MS+NAA(1)+BAP(0.5)	10.0±1.16	5.0±0.58	3.0±0.58	–
MS+NAA(1)+BAP(1)	18.3±0.88	10.0±1.16	4.0±1.00	10.0±1.1
MS+IBA(1)+Kn(1)	12.0±2.00	12.0±2.00	4.0±0.58	12.0±2.0
MS+IBA(1)+BAP(1)	12.0±1.73	10.7±1.33	–	–
MS+NAA(1)+CM(10%)	6.0±1.16	4.0±0.00	4.0±0.00	4.0±0.0
MS+NAA(1)+Kn(1)+CH(0.1%)	18.3±0.88	16.0±1.00	1.0±0.00	16.0±1.0

Data based on 24 explants after 4 weeks of growth period, conc. of growth substances mg/l. (After F *et al.*, 1983 Reproduced with permission.)

elongated and developed 4–5 robust, alternate leaves. The basal portion of the explanted bud invariably produced a friable, brownish callus. A single plantlet developed from each explanted bud. Rooting occurred form the shoots on the same medium resulting into the plantlet development on MS+IBA+Kin. In some of the cultures, development of male and female inflorescences took place. Ripe fruits were developed from the female inflorescences within 40 days.

2.1.3. *Pretreatment of explants*

For the experiments relating to pretreatment of the explants, nodal and internodal segments, sessile and leaves with petiole were excised from shoot cultures, were soaked individually in MS liquid medium for 72 h. To this medium cytokinins such as BA, Kin, or Zea at three concentrations (0.5, 1 and 2 mg/l) were added. Following the pretreatment the explants were cultured on solid medium of the same composition. Pretreated leaves sessile and with petiole, were cultured in an upright position with only half its lower portion dipped in the medium. Nodal portions were also cultured in the upright position whereas internodal segments were placed horizontally on the surface of the medium.

The presoaked axillary buds produced multiple shoot buds on a cytokinin provided MS medium. On an average 6–8 buds from single axillary bud were formed on MS+BAP (0.5, 1 and 2 mg/l) medium. The other two cytokinins used (Kin, Zea) were less effective in promoting multiple buds from axillary

Table 14. Response of axillary buds and leaves of *Morus indica* to various cytokinins.

Addition to MS medium	No. of explants cultured	No. of explants showing multiple bud induction	
		Leaves	Axillary buds
BA (0.5)	48	14	34
BA (1)	48	18	36
BA (2)	48	20	40
Kn (0.5)	48	nil	nil
Kn (1)	48	nil	nil
Kn (2)	48	nil	nil
Z (0.5)	48	nil	nil
Z (1)	48	nil	nil
Z (2)	48	nil	nil

The explants were pretreated with corresponding cytokinin for 48 hrs, concentration; mg/l; On Kn and Z only single bud gegeneration. (From Mhatre *et al.* 1987. Reproduced with permission.)

buds. Plantlets were regenerated from the multiple buds on excision and transfer of the shoots to the hormone free MS basal medium (Table 14).

The maximum induction of shoot buds occurred from leaf explant cultured in an upright position in the cultured medium. The basal portion of the explant (petiole) enlarged and formed primordia which developed into green leafy shoots after 5–7 weeks of culture. Sessile leaves also showed differentiation of numerous shoot buds under similar conditions. Generally 3–4 shoots developed per explant. This response was observed only with leaves soaked in BA and not with those soaked either in Kin or in Zea. Non-soaked leaves did not show multiple bud induction and necrosed after 6–8 weeks. Callus development occurred at the base of the leaf along with the differentiation of shoot buds. On separation of the callus from the explant, numerous shoot buds were differentiated which when subcultured on the hormone free medium developed into plantlets.

3. Encapsulation of buds

For encapsulation, axillary buds from established shoot cultures were isolated along with the nodal region and a portion of the stem (approx. 0.5 cm length). These were then blot dried on the filter paper for a few seconds and then mixed with a gel of sodium alginate (Sigma) prepared in MS basal medium. The explants were mixed well in the gel, picked up by means of forceps and dropped into a solution of $CaCl_2 \cdot 2H_2O$ (1.036/150 ml). The drops were allowed to remain in this solution for 30 minutes on a gyratory shaker (80 rpm) in light (950 lux). After the incubation period the beads were recovered by decanting off the $CaCl_2 \cdot 2H_2O$ solution and washing the beads 3–4 times with autoclaved washing medium (MS salts). Subsequently the beads were cultured on the MS basal medium.

3.1. *Encapsulation in alginate*

The axillary buds were encapsulated in alginate matrix of different concentration to determine the suitable concentration of alginate. The alginate matrix was prepared either in MS basal medium or in distilled water. After few days encapsulated buds germinated and subsequently the developing shoots and roots broke through the alginate matrix and developed into plantlets within 3 weeks. As shown in Table 15 there was no striking difference when the encapsulated buds were kept at 4 °C for 45 days and then cultured for germination. The plantlets were robust and had a very good shoot and root growth.

3.2. *Encapsulation in agar*

Axillary buds were coated in two concentrations of different types of agar (Sisco, Bombay and Difco). Encapsulation was done either in agar prepared in MS basal or in distilled water. In both types of blocks the buds germinated and formed plantlets at 2 and 3% agar concentration. However, when the blocks were stored at 4 °C for 45 days and subsequently cultured their germination was drastically affected being reduced to 10% (Table 16).

3.3. *Encapsulated buds in different substrates*

To simplify the methods for the germination of encapsulated buds, the axillary buds encapsulated in 4% alginate were cultured on different substrates as

Table 15. Percent germination response of axillary buds of mulberry encapsulated in sodium alginate prepared in MS medium and in distilled water.

Na-alginate (%)	MS		Distilled water	
	4 °C	Room temp.	4 °C	Room temp.
2	50%	60%	50%	40%
3	50%	40%	60%	50%
4	80%	80%	90%	50%

Beads were stored at 4 °C for 45 days; 12 beads per treatment. (From Bapat *et al.*, 1987. Reproduced with permission.)

Table 16. Percent germination responses of axillary buds of mulberry encapsulated in agar blocks.

Matrix used		No. of beads germinated (%)
Distilled water + Difco agar	(2%)	0
Distilled water + Difco agar	(3%)	0
Distilled water + Sisco agar	(2%)	30
Distilled water + Sisco agar	(3%)	20
MS + Difco agar	(2%)	20
MS + Difco agar	(3%)	0
MS + Sisco agar	(2%)	50
MS + Sisco agar	(3%)	10

No. of beads cultured in each treatment 10. (From Bapat *et al.*, 1987. Reproduced with permission.)

Table 17. Germination response of encapsulated axillary buds (beads) of mulberry to various treatments.

Type of substrate used for germination	% Germination response of beads prepared in	
	MS	Distilled water
(1) Vermiculate + tap water	100%	0
(2) Agar + tap water	100%	10%
(3) Soil + tap water	0	0
(4) Filter paper + tap water	30%	10%

24 beads cultured per treatment, data scored after 70 days. (From Bapat *et al.*, 1987. Reproduced with permission.)

shown in Table 17. The beads germinated and formed plantlets on filter paper or vermiculate or agar with tap water as sole source of nutrient. Germination rate was drastically reduced when beads were prepared in distilled water.

3.4. *Pretreatment with cytokinin*

Benzylaminopurine was shown to induce multiple bud formation in mulberry when the explants were soaked in a solution. The explants were pretreated for 48 h in BAP medium and subsequently encapsulated in 4% sodium alginate matrix. It was observed that the capacity of the explants to form multiple buds did not alter after encapsulation in alginate matrix. They readily formed shoots (4–5 per bead) within 6 weeks. The shoots emerged from the matrix and developed into plantlets.

The major hurdle in the development of synthetic seed technology is the germination of encapsulated shoot buds or embryos under non-aseptic conditions. In case of mulberry, addition of a proper fungicide to the alginate matrix prevents the beads from the attack of microorganisms and allows germination in an non-aseptic environment.

Under non-aseptic condition, buds were encapsulated in an autoclaved alginate matrix containing a fungicide (Carbendenzum, Benomyl) and Bavastin). These fungicides were added separately at a concentration of 50 mg/l and 100 mg/l. The alginate matrix consisted of MS basal medium containing sucrose (3%). The soil was autoclaved in petri dishes. After the encapsulation in each Petri dish 10–12 beads were kept. Every alternate day soil was irrigated with 1/2 strength MS basal medium without sucrose. All petri dishes were initially covered and sealed with Parafilm. The petri dishes were kept in the light for 16 h at 25 + 2 °C. After the emergence of shoots from the beads the lids on the petri dishes were removed. The fungicide carbendenzum proved better and more effective in checking the attack of microorganisms (Table 18). On an average 30–40% encapsulated beads shrivelled and showed no growth. Some of the beads turned brown after 10–15 days but survival was apparent within a few days by the appearance of the roots. The roots emerged and penetrated the soil. Once roots were established in the soil, rapid growth of the plantlet took place subsequently. The plantlets with good root and shoot growth were transferred to paper cups.

When the regenerated plantlets having a good root system attained a height of 6″, cotton plugs were removed from the tubes and to the agar medium 1/2 strength MS solution devoid of sugar was added. In this condition plantlets were kept for two days. Afterwards plantlets were repeatedly washed with water and were transferred to paper cups with autoclaved soil. They were irrigated with tap water and were kept in illuminated shelves covered from all sides with polythene sheets. At the end of three weeks the plantlets were transferred to the pots and were eventually established in soil where they displayed vigorous growth and reached a height of 10 feet within a period of a year. In some plants variation in leaf form was noticed. Leaves lobed in contrast to normal simple leaf shape, sometimes leaves were with 4 to 5 lobes. Variation in branching was also noticed. Some plants were profusely branched and some were poorly branched and taller compared to other plants. However, most of the plants were of uniform growth as shown in figure 4.

Table 18. Effect of various fungicides added in encapsulated matrix on *in vivo* germination of mulberry shoot buds.

Type fungicide (50 mg/l)	No. of beads encapsulated	No. of beads germinated	% Germination
Carbendazum	150	95	63.3
Benomyl	150	45	30
Bavastin	150	32	21.3

Data scored after 6 weeks.

Figure 4. Mulberry. **a.** Plantlet regenerated from axillary bud.
b. Regenerating plantlet with development of flowers.
c. Axillary buds encapsulated in alginate.
d. Multiple shoot buds from axillary bud.
e. *In vivo* germination of encapsulated axillary buds on soil.
f. Germinating encapsulated axillary buds.
g–h.. Establishment of plants in the field.

Although variation in leaf morphology is seen in the natural population the variation in the tissue cultured plants could be attributed to the juvenile and mature phases of the regenerated plants.

4. Potential and prospects of *in vitro* technology for large scale propagation

The results achieved in mulberry are encouraging since axillary and leaf explants have the capacity to form multiple shoots in large numbers. The mortality rate of tissue cultured raised plant is also low in mulberry. This provides a strong basis for micropropagation of desirable varieties. The major hurdle for the adaptation of more culture methods such as protoplast and anther culture and cell suspension is the inability of mulberry callus to undergo differentiation.

Soaking of explants prior to culture was effective for inducing multiple bud formation from the explants. Possibly the pretreatment confers a physiological status on the explants that is favourable for shoot bud formation. In the case of leaves, shoot buds were regenerated from leaf base directly as well as from callus. These observations are significant and could be employed for high multiplication rate of elite species of mulberry.

The distribution of tissue culture material and survival rate of plantlets depend on the period of transportation, conditions during transit and handling after transplanting. A prolonged period during transportation and unfavourable conditions during transit often lead to the deterioration of the plant material. Hence, recently synthetic seeds have been prepared from somatic embryos in carrot, alfalfa, celery and sandalwood. [7,30] as a unit of delivery. However, this technique is not possible to use in plants like mulberry which have not produced somatic embryos in culture so far. In such cases axillary buds can be encapsulated to produce synthetic seeds. This would be an easy transport delivery system as compared to large parcels of cuttings from one place to another. In case of mulberry only 30–40% cutting survive the time between pruning, transportation and final transplantation whereas the encapsulated buds could be easily packed in bottles (vials) thus limiting space and ensure the viability and survival rate.

Among the encapsulated matrices alginate was superior and could protect the enclosing tissue from dessication. The other advantage of alginate is that it is soluble at room temperature and forms a gel without heating with $CaCl_2{\cdot}2H_2O$. The encapsulated buds did not lose the capacity to germinate even after storing for 45 days at 4 °C. The potentiality to form multiple buds would be of advantage for storing propagating elite clones of mulberry. Germination of encapsulated buds on tap water stresses the simplicity of the system.

The results also showed that alginate matrix containing a fungicide protects the tissue from the attack of microorganisms and could supply the nutrients to the embedded tissue upto a certain time *in vivo*. Rough handling of the beads in non-sterile environment did not kill the encapsulated buds; hence it offers a system acceptable at farm level.

The work reported and discussed here so far illustrates the potentiality of the technique for rapid micropropagation of mulberry through axillary bud culture.

5. Summary

Demands for forest trees are continually increasing throughout the world due to large scale afforestation programmes. The latest advances in biotechnology and particularly in plant tissue culture have potential for large scale application of this tehcnology for rapid multiplication of elite clones of forest trees. Towards this goal work presented in this chapter shows the possibility of *in vitro* multiplication of *Santalum album*. Although direct induction of shoot buds was achieved from hypocotyl segments, rooting from exiced shoot buds was not easy. However, an actively growing callus which yielded somatic embryos was established from both hypocotyl and stem segments. Plantlets were regenerated from somatic embryos and were established in the field. Somatic embryos were also regenerated from cell suspensiosn and protoplasts isolated from various sources. Somatic embryos encapsulated either in alginate or in composite gel of alginate + silica gel retained the capacity of germination. The concept of 'synthetic seed' is a new idea and could be an effective tool for micropropagation of forest plants.

Mulberry is an important plant in sericulture. Mulberry is propagated by asexual means, such as stem cuttings and root grafts. The technique of tissue culture offers an additional approach to the vegetative multiplication of difficult-to-root, elite cultivars of mulberry. In the present chapter work on the use of leaves and axillary buds for micropropagation of mulberry has been presented. Pretretament of axillary buds and leaves with cytokinin prior to culture improved the morphogenetic response. The plants raised *in vitro* showed certain characteristics which could be useful for obtaining more yield of leaves per plant, such as more branching, short internodes, and variations in phyllotaxy. The encapsulation of axillary buds provide an easy and novel propagation system in mulberry which could be easily packed in to bottles, thus limiting space and ensure the viability and survival rate.

References

1. Amirato PV and Styer DJ (1985) Strategies for large scale manipulation of somatic embryos in suspension cultures. Biotechnology in plant science. In: Zaitlin M, Day P, and Hollaender A (Eds.) Academic Press, New York. pp. 161–178.
2. Bapat VA and Rao PS (1979) Somatic embryogenesis and plantlet formation in tissue cultures of sandalwood (*Santalum album* L). Ann Bot 44:629–630.

3. Bapat VA and Rao PS (1984) Regulatory factors for *in vitro* multiplication of sandalwood tree (*Santalum album* L). I. Shoot bud regeneration and somatic embryogenesis in hypocotyl cultures. Proc Indian Acad Sci 93:19–27.
4. Bapat VA, Gill R, and Rao PS (1985). Regeneration of somatic embryos and plantlets from stem callus protoplasts of sandalwood tree (*Santalum album* L).Curr Sci 54:978–982.
5. Bapat VA, Rao PS, and Heble MR (1986) Immobilization of cells and protoplasts of *Catharanthus roseus*. Proc Indian Acad Sci 96:413–418.
6. Bapat VA, Mhatre M, and Rao PS (1987) Propagation of *Morus indica* L. (mulberry) by encapsulated shoot buds. Plant Cell Rep 6:393–395.
7. Bapat VA and Rao PS (1988) Sandalwood plantlets from 'Synthetic Seeds'. Plant Cell Rep 7:434–436.
8. Bajaj YPS (1986) Biotechnology in Agriculture and Forestry, Vol. II, Springer-Verlag, Berlin.
9. Barwale UB, Kerns HR, and Widholm JM (1986) Plant regeneration from callus cultures of several soybean genotypes via embryogenesis and organogenesis. Planta 167:473–481.
10. Bonga JM and Durzan DJ (1987) Cell and Tissue Culture in Forestry. Vol 3. Case Histories: Gymnosperm, Angiosperm and Palms, Martinus/Nijhoff, pp. 416.
11. Brodelius P and Nilsson K (1980) Entrapment of plant cells in different matrices. FEBS Lett 122:312–316.
12. Iyengar AV (1965) The physiology of root parasitism in sandal (*Santalum album* L). Indian Forester 91:246–258.
13. Lakshmi Sita G, Raghava Ram NV, and Vaidyanathan CS (1979) Differentiation of embryoids and plantlets from shoot callus of sandalwood. Plant Sci Lett 15:265–270.
14. Lakshmi Sita G, Raghava Ram NV, and Vaidyanathan CS (1980) Triploid plants from endosperm cultures of sandalwood by experimental embryogenesis. Plant Sci Lett 20: 63–69.
15. Lazzeri PA, Hilderbrand DF, and Collins GB (1987) Soybean somatic embryogenesis: effects of hormones and culture manipulation. Plant Cell, Tissue and Organ Culture 10:197–208.
16. Mhatre MV, Bapat VA, and Rao PS (1985) Regeneration of plants from the cultures of leaves and axillary buds in mulberry (*Morus indica* L). Plant Cell Rep 4:78–80.
17. Murashige T and Skoog F (1962) A revised medium for rapid growth and bioassays with tobacco tissue culture. Physiol Plant 15, 473–497.
18. Musselman LJ (1980) The biology of *Striga* and Orobanche and other parasitic weeds. Ann Rev Phytopathol 18:463–489.
19. Novak FJ, Afza R, Von Duren M, Perea-Dallos M, Conger BV, and Xiaolang T (1989) Somatic embryogenesis and plantlet regeneration in suspension cultures of dessert (AA and AAA) and cooking (ABB) bananas (*Musa* sp.). Biotechnology 7:154–159.
20. Ohyama (1970) Tissue culture in mulberry tree. Japan Agric Res 5:30–34.
21. Ohyama K and Oka S (1982) Multiple shoot formation from mulberry (*Morus indica*) hypocotyls by *N*-(2-chloro-4-pyridyl)-*N*-phenylurea. In: Fujiwara A (Ed.) Plant tissue and cell culture. Maruzen, Tokyo. pp. 149–150.
22. Oka S and Ohyama K (1981) *In vitro* initiation of adventitious buds and its modifications by high concentrations of bezyladenine in leaf tissue of mulberry (*Morus alba* L). Can J Bot 59:68–74.
23. Patel G, Bapat VA, and Rao PS (1983) *In vitro* cultures of organ explants of *Morus indica*. Plant regeneration and fruit formation in axillary bud cultures. Z Pflanzenphysiol 111:465–468.
24. Parthasarathi K and Venkatesan KR (1982) Sandal spike disease. Curr Sci 51:225–230.
25. Rangaswamy NS and Rao PS (1963) Experimental studies on *Santalum album* L. establishment of tissue cultures of endosperm. Phytomorphology 13:450–454.
26. Rao PS and Rangaswamy NS (1971) Morphogenic studies in tissue cultures of the parasite *Santalum album* L. Biol Plant (Praha) 13:200–206.
27. Rao PS and Bapat VA (1978) Vegetative propagation of sandalwood plants through tissue cultures. Can J Bot 56:1153–1156.

28. Rao PS and Raghava Ram NV (1983) Propagation of sandalwood (*Santalum album* L) using tissue and organ culture technique. In: Sen SK and Giles KL (Eds.) Plant Cell Culture in Crop Improvement. Plenum Press, New York. pp. 119–124.
29. Rao PS and Ozias Akins P (1985) Plant regeneration through somatic embryogenesis in protoplast cultures of sandalwood (*Santalum album* L). Protoplasma. 124:80–86.
30. Redenbaugh K, Paasch BD, Nichol JW, Kossler ME, Viss PR, and Waikar KA (1986) Somatic seeds: Encapsulation of asexual plant embryos. Biotechnology 4:797–801.
31. Seki H, Takeda M, Tsutsumi KM and Ushiki Y (1971) Studies on the callus cultures of the mulberry tree (I), the effects of concentration of auxin and kinetin on the callus cultures of mulberry stem. J Seric Sci Japan 40:81–85.
32. Sen-Sarma PK (1977) Sandalwood – its cultivation and utilization of medicinal and aromatic plants. In: Atal CK and Kapur BM (Eds,) Reg Res Laboratory, Jamnu Tawi, India. pp. 287–297.
33. Shankaranarayana KH, Aiyar KS, and Krishna Rao GS (1980) Insect growth inhibitor from the bark of *Santalum album* . Phytochem 19, 1239–1240.
34. The Wealth of India (1972): Raw Materials, Vol IX, In: Chadha YR (Ed.) Publication and Inofrmation Directorate, New Dehli. pp. 208–224.
35. Yakuwa H and Oka S (1988) Plant regeneration through meristem culture from vegetative buds of mulberry (*Morus bombycis* Koidz) stored in liquid nitrogen. Ann Bot 62:79–82.
36. Yamada Y and Okamoto A (1977) A basic study of mulberry cell culture for silkworm feeding. A gric Biol Chem 41:2261–2267.
37. Yamamoto A and Kamata M (1971) Studies on the tissue cultures of mulberry tree. Part 1. Dedifferentiation of tissue of mulberry tree and culture of callus. Bull Fac Tex Sci, Kyoto University, Ind Art Text Fibers. 6:108–121.

20. Micropropagation of *Pinus sylvestris*

JANA ŽEL
Institute of Biology, Aškerčeva 12, 61 000 Ljubljana, Slovenia

1. Introduction

Pinus sylvestris is one of the most widespread conifers in Europe and Asia. Its geographic boundaries lie between 70°20′ northern latitude in Scandinavia and 38°41′ south in Central Turkey, and the species survives at altitudes of up to 2000m [52]. It is a very variable species that grows on different types of soils.

The vegetative propagation of the Scots pine, especially that of mature trees, is still problematic. The difficulty lies in the low percentage of rooting of cuttings [59]. Therefore, the propagation of *Pinus sylvestris* using tissue culture methods could be very helpful in vegetative propagation. A review of the methods of embryo culture, tissue culture and micropropagation of European pines was presented in 1987 by Jelaska [28]. For an understanding of physiology of *Pinus sylvestris in vitro* different types of cultures as root [16,17,41-43], embryo [40,48], callus [3,8,25,26,38] and cell cultures [11,22,23,35,36] are very helpful. The results of such studies can therefore also be useful also in establishing a micropropagation process.

In the process of mass propagation of the Scots pine *in vitro*, rooting is still the limiting step. From experiments on the rooting of hypocotyl cuttings more can be learned about the behaviour of the explant under rooting treatment *in vitro* [18,19]. The response of explants of mature trees introduced in tissue culture is difficult. The first encouraging results in establishing callus cultures from explants of mature 10- to 40-year-old Scots pine trees have been achieved recently [25,26].

This chapter is a recapitulation of experience on the Scots pine cultured *in vitro* with special reference to recent results and perspectives.

2. Micropropagation from embryos

Embryos (Table 1) and seedlings aged from 2 days to 5 months are the starting material used for the micropropagation of *Pinus sylvestris*.

A precise description of adventitious bud formation on embryos and few days old seedlings was presented by Tranvan [50,51]. The concentrations of BA needed for formation of adventitious buds increased with the physiological

M. R. Ahuja (Ed.), Micropropagation of Woody Plants, 347–365.

Table 1. Development of adventitious buds on embryonal explants.

Explant	Age of donor day (d) week (w)	Basal medium	Hormones tested (μM)	Optimal hormone concentration (μM)	Adventitious buds	References
embryo		Lin&Staba[31]	+0.5–100BA	2.5–25 BA	+	[50]
		Nitsch&Nitsch				
seedling	2,5d	(vitamins)[32]			+	
		2–4% sucrose				
cotyledon	2,10d	0.8% agar			+	[51]
embryo		mod.LP[35]		10 BA	+	[54]
		3.4% sucrose				
		0.7% agar				
embryo		½ MS[37]	0.5–25 BA		+	[15]
			0.5–25 2iP			
			0.5–25 Z			
embryo		½ MS[37]	0.5–100 BA	0.5–10 BA	+	Žel, Gogala
		mod.Cheng[39]				unpublished
		3% sucrose				
		1% agar				

age of the explant. Buds formed at two different sites: near the shoot apical meristem they resulted from the partition of the apex or from the axillary buds of the cotyledons, as well as from the formation of adventitious vegetative primordia following the dedifferentiation of groups of internodal cells. At the subterminal tip of the cotyledon, only a single bud arose from a zone of dedifferentiated tissue under the epidermis situated on the apical side of the cotyledon's procambium. At the tip of the cotyledon so-called 'touffes du feuilles' – adventitious leaves, not organized in bud form – sometimes formed. The formation of buds was also influenced by correlative growth between organs. Total or partial removal of cotyledons caused a decreased number of adventitious buds.

There are some other reports on adventitious bud formation on embryos of Scots pine [15,33,49]. Some interesting results were achieved with embryos isolated from seeds obtained after different controlled crossings, results that showed variability in the capacity of adventitious bud formation [54]. Žel and Gogala, working with embryos isolated after 2 days imbibition of the seeds in sterile, double distilled water (J. Žel, N. Gogala, unpublished), found that the medium for adventitious bud formation used by Tranvan [50] did not give the expected results. This can probably be explained by different genetic and environmental origins of the seed [61] and thus presumably with different requirements for nutrients [34]. Half strength MS medium [31] modified by Cheng [10] was used with different BA concentrations. Sites of adventitious bud formation were found to be the same as those reported by Tranvan. We noticed that at lower BA (1μM) concentrations adventitious buds formed mainly on the tips of elongated cotyledons (Figure 1). Callus

Figure 1. Adventitious buds formed only on the elongated cotyledon tips, when cultured on a low BA medium.

formed at high concentrations (50μM) but at the very highest concentrations used (100μM), embryos only swelled and rarely formed buds (Figure 2).

The position of the embryos on the medium was also important for the formation of buds. Embryos placed in a vertical position formed twice the number of buds when compared to embryos placed horizontally. In the latter ones cotyledons in touch with the medium formed groups of meristematic tissue, whereas cotyledons which were not in direct contact with the medium formed only single buds (Figure 3). Similar effects were noticed for *Pinus radiata* embryos [2,37].

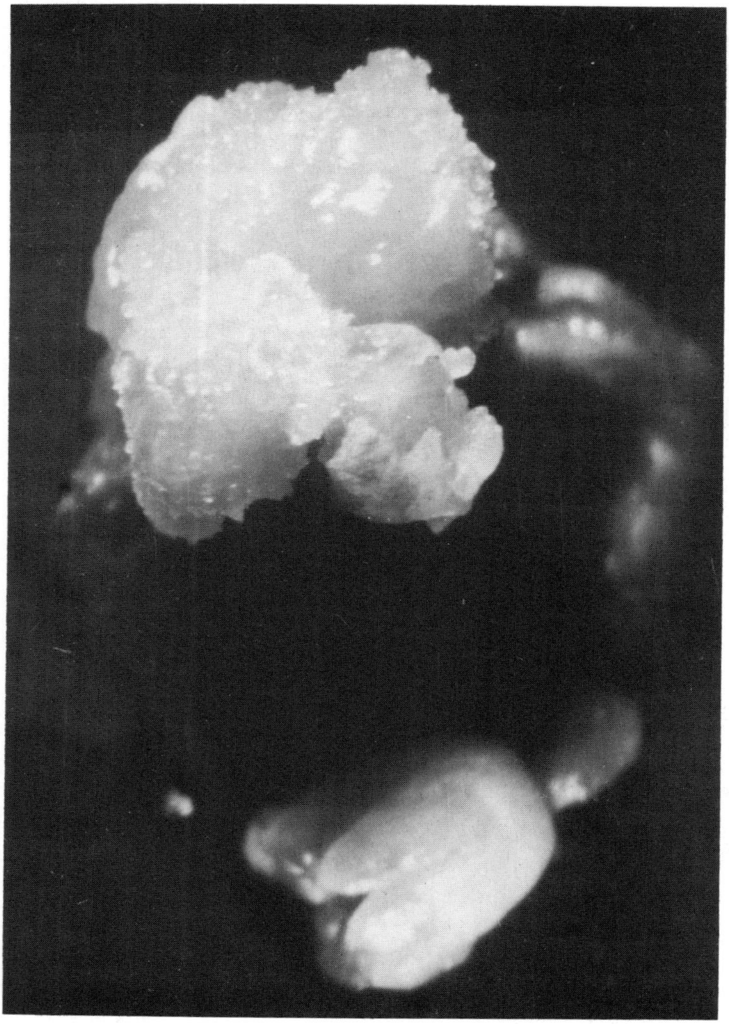

Figure 2. Embryos only swollen or form a callus on a high BA medium.

3. Micropropagation from seedling explants

Explants of seedlings have consisted of isolated cotyledons, brachyblasts and shoot apices (Table 2). The most successful micropropagation seems to be from axillary buds formed on the seedling apex.

Bornman, whose work was done mostly on *Picea abies*, also applied the

Figure 3. Influence of position of embryo on the medium: cotyledons in contact with medium formed a mass of meristematic tissue; cotyledons not directly in contact with the medium formed only single buds.

results to *Pinus sylvestris* [5–8]. He followed organogenesis and plantlet regeneration from different explants. Cotyledons and seedling apices of 2-week-old plants formed adventitious buds, while on 6- to 8-week-old seedling apices, mostly axillary buds developed. Development of suppressed buds of brachyblasts was also obtained. Bornman compared the influence of different BA applications on bud induction. This period could be shortened by more than half by pulse treatment with plant growth regulators of high concentration or by infusion and pre-incubation of the entire hypocotyl explants. Both treatments were more successful than incorporation of the hormones in the medium [6,7]. A positive effect was also noticed when 8-week-old seedling explants were pretreated with BA, applied as a foliar spray before explanting, and culturing on a medium free of plant growth regulator [6].

Observations of ^{14}C BA uptake in explants of *Pinus sylvestris* and *Picea abies* showed a passive mode of uptake [53]. It was directly proportional to the external concentretion.

Different types of cytokinin applications can have an influence on the site

Table 2. Micropropagation from seedling explants.

Explant	Age of donor day (d) week (w) month (m)	Basal medium	Optimal hormone concentration (μM)	Adventitious buds	Axillary buds	Plantlet regeneration	References
brachyblast, apical slices	20w (frosthardened)	Sommer *et al.*[17] mod. de Fossard[18] 3mM myo-inositol 3% sucrose 0.6–0.7% agar	5–10 BA& 5–10 2iP or 20 BA	+ +			[8]
plumule	14d		10 coumarin				[8]
cotyledon	12–21d	MCM[46] 0.5–3% sucrose 0.25–0.7% agar	2.5–10 BA& 2.5–10 2iP or 125 BA 3h or infusion and preincubation 5 BA 24h	+		+ elongation: 1% activated charcoal, irrad. far red light, rooting: 120mgl^{-1}auxin 24h, in vivo	[5,7]
brachyblast,	5m (frosthardened)	MCM[46] 0.5–3% sucrose 0.25–0.7% agar	pretreatment of plant: 670(1250)BA sprayed in 2d intervals 2w (10d)	+			[5]
	24w (frosthardened)					+ 625μM NAA, 24h (rare)	[6]
seedling apex	2w	MCM[46]	5 BA or 125 BA 2h	+		+ 625μM IBA 12h, in vivo	[6]
seedling apex	6w	MCM[46]	pretreatment of plant: 1250 BA sprayed in 2d intervals 10d	+	+	+ 625μM IBA 12h, in vivo	[6]
seedling apex	8w	MCM[46]	125 BA 24h	+	+	+ 625μM IBA 12h, in vivo	[6]

Table 2. (Continued).

Explant	Age of donor day (d) week (w) month (m)	Basal medium	Optimal hormone concentration (μM)	Adventitious buds	Axillary buds	Plantlet regeneration	References
seedling apex	3,6,9w	½Hoagland's solution ⅛MS[37] mod.Cheng[39] 3% sucrose 1% agar	625 BA 2d 111 BA 2h	+ +	+ +	 + elongation: irrad. far red light 1w, rooting: 53.8μM NAA, 0.6% agar 24 h, 1/16 MS mod. Cheng, 1% sucrose	[6] [59]
seedling apex	45–105d	½MS[37]–½Bourgin& Nitsch[50] (organic substances) 1% sucrose 1% agar	22–44 BA or 88 BA&4.9 IBA	+	+	+ 9.8μM IBA 4d, perlite, Knop mycorrhiza formation	[45–47]

of origin of adventitious buds. Histological observations on their origins from cotyledonary needles of *Picea abies*, showed that cell divisions in cotyledons on media incorporating BA appeared to be initiated mainly in cells of the epidermis, while those in pulse-treated or infused cotyledons commenced mainly in sub-epidermal tissues [4].

Uptake as well as leaching of substances at the time of soaking proceeds simultaneously. Soaking of explants of *Eucalyptus grandis* in distilled water at least two hours before treatment had a positive influence on organogenesis [12]. Phenols and other substances, which were inhibitory for further development of explant, were leached in this way. Elongation of buds was achieved on hormone-free medium with 1% activated charcoal by irradiation with far-red light (1–2 weeks), followed by irradiation with light of high intensity (30–40 W m^{-2}) for 3–4 weeks.

The influence of coumarin on rooting *in vitro* was tested. A relatively high rooting frequency in plumule explants of 14-day-old seedlings was achieved [8].

Regeneration of plantlets from adventitious buds induced on cotyledons, or from the activation of suppressed buds that arose from brachyblasts, occurred only occasionally. Activation of the brachyblast meristem can be achieved after rooting, and conversely, rooting can be induced after activation of the suppresed bud. In both cases the terminal step is very difficult to achieve [7].

Regeneration of plantlets from 2-week-old explants took place exclusively via adventitious buds, and in the case of 8-week-old explants, predominantly from axillary buds [6]. Shoots were usually rooted *in vivo* by, firstly, an overnight treatment with auxin, followed by placement in a mistbed [6,7].

Žel *et al.* used about 1-cm-long seedling apices (roots, a part of the hypocotyl and the distal parts of cotyledons were removed) from 3-, 6- and 9-week-

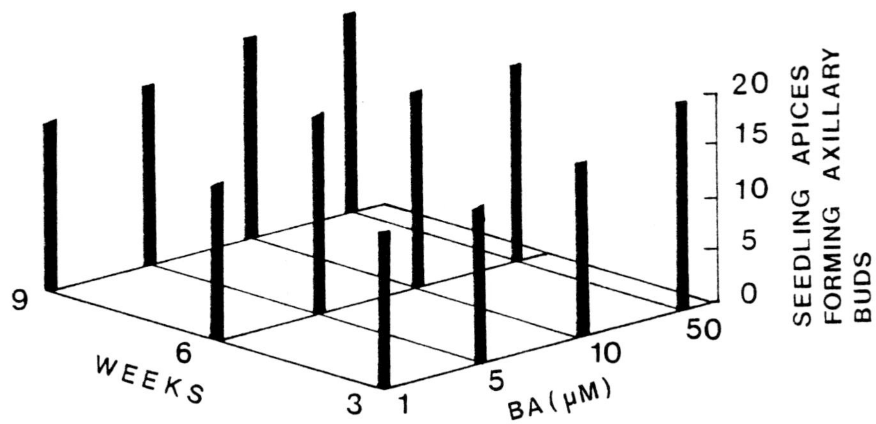

Figure 4. The influence of the age of seedlings and different BA concentrations on the number of seedling apices that respond.

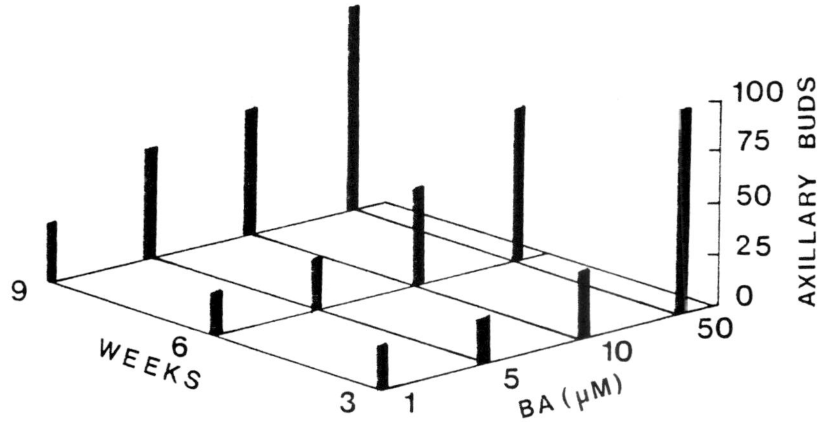

Figure 5. The influence of the age of seedlings and different BA concentrations on the number of induced axillary buds. The sum of axillary buds induced on 20 seedling apices is given.

old seedlings [60]. The explants of various ages did not display a wide variability in their response (Figures 4–6).

Pinus sylvestris is sensitive to high salt concentrations [27]. Media used for tissue cultures are usually developed for fast growing tissues. These media have to be diluted when used for slow growing tissues such as conifers [82]. Žel *et al.* [60] found that dilution of the medium to one-eighth strength was beneficial for growth.

BA appears to be the cytokinin of choice for induction of buds in *Pinus sylvestris* cultures. Higher BA concentrations in media have been found to increase the number of buds but to decrease their length. Pulse treatments gave better results, similar to those mentioned earlier [47], which were better expressed in longer buds (Figure 9). Besides axillary buds, adventitious ones were also formed, especially with higher BA concentrations in the medium and after soaking (Figure 7). Further growth of adventitious buds was reduced

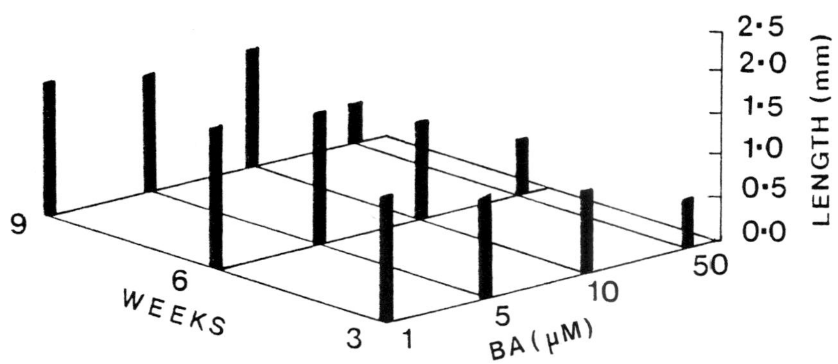

Figure 6. The influence of the age of seedlings and different BA concentrations on the average length of axillary buds.

Figure 7. Formation of axillary buds.

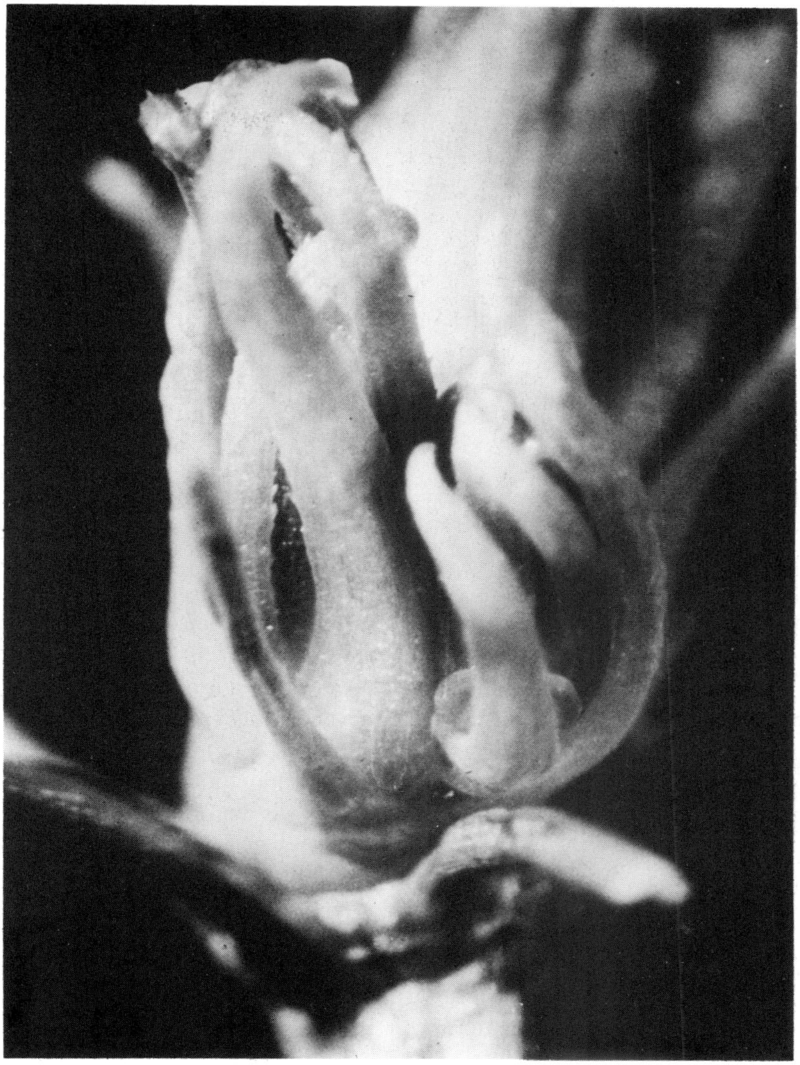

Figure 8. Formation of secondary axillary buds.

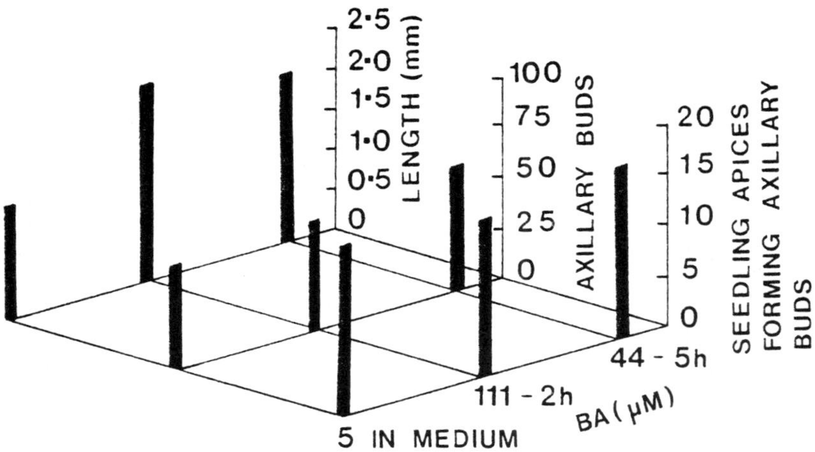

Figure 9. The influence of soaking seedling apices in BA on the number of seedling apices that respond, on the sum of axillary buds induced on 20 seedling apices, and on their average length.

because of the rapid development of axillary buds. Secondary and tertiary buds were spontaneously formed on axillary ones and it is here that the possibility of continual *in vitro* propagation exists (Figure 8).

Elongation of separated buds was achieved with exposure to far-red light and by reducing sucrose in hormone-free medium (Figures 10,11). Thereby very small buds also elongated and could thus be included in the propagation process. Of the different auxins tested, NAA was most successful. A 24-h treatment with NAA in 0.6% agar, followed by transferral to one-sixteenth strength medium with 1% sucrose, gave a rooting frequency of 64% (Figure 12).

Variability in results is often mentioned in papers concerning the *in vitro* tissue culture of *Pinus sylvestris* [6,7,54,60]. The reason for this is to be sought in many factors. Seeds of Scots pine are open-pollinated, so the genotype of each seed is different, and consequently it may be expected that the ratios of endogenous hormones may also be variable. Environmental conditions such as a photoperiod and temperature affect the endogenous hormonal status of a plant [39]. The season can also strongly influence results [3,20,25]. There are seasonal changes in the cellular structure and metabolism of plants [24]. Designing suitable culture conditions would be facilitated were the physiological state of the explant better understood. Varying treatments may also contribute to the differences. It is important to adopt a uniform procedure for the disinfection of seeds. Treatment with H_2O_2 promotes germination. Seedlings from seeds which undergo different aseptic procedures can develop differently.

An efficient rooting system is one of the primary conditions for successful transferral of plantlets to *in vivo* conditions. Therefore, the inducement of mycorrhizae in the period before transplantation of the plants to conditions

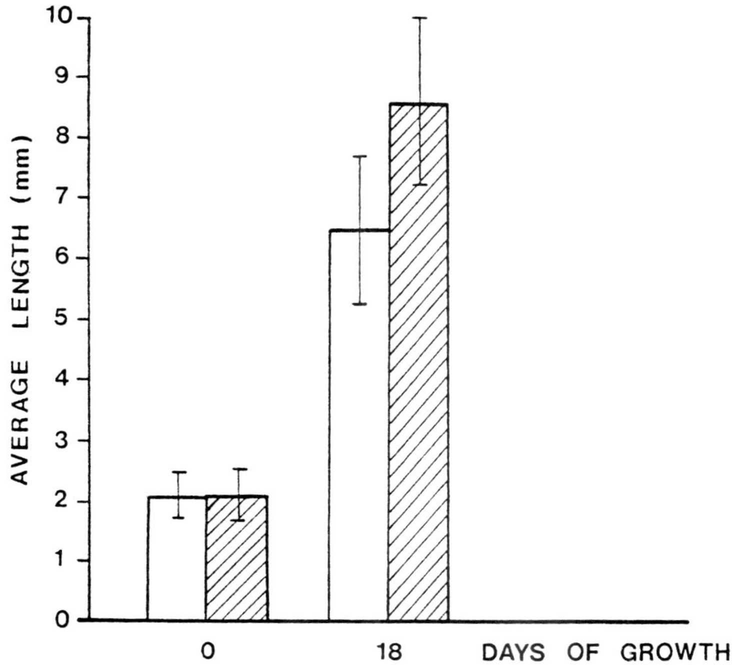

Figure 10. The influence of far-red light on the average length of buds. 0 = length at the beginning of 8 days of treatment. Vertical lines represent 2 × SE. Untreated buds (not hatched), and buds treated with far-red light (hatched).

in vivo can contribute to the survival of plantlets. This was clearly demonstrated on *Pinus pinaster* shoots [13]. Agar culture permitted intense meristem initiation, but prevented active root elongation. In the presence of the mycorrhizal fungi *Pisolithus tinctorius* and *Hebeloma cylindrosporum* grown on perlite, roots resumed growth and short, lateral roots were stimulated.

Recently, succesful mycorrhization was also demonstrated in *Pinus sylvestris* plantlets [45,47]. Axillary and adventitious buds were obtained on seedling apices. The formation of adventitious buds was also achieved on the leaves of *Pinus sylvestris* from remultiplied axillary buds (90-day-old) on modified half-strength MS-BN medium (S. Suprianto, R. Rohr, personal communication).

Rooting of axillary shoots was achieved by treatment with IBA and transferal into Knop's medium on perlite. Adventitious roots developed 14 days later. The inoculum of mycorrhizal fungi (*Hebeloma cylindrosporum* and *Pisolithus tinctorius*) was first grown on the perlite substrate also containing Knop medium and 10 g/l sucrose. The mycorrhizal fungus was inoculated on the plantlet medium when the root system was well developed (2 cm in length) by placing 2–3 pieces of inoculated perlite near the roots. The dichotomous root branches

Figure 11. Elongated axillary shoots before rooting.

Figure 12. Plantlet formation in culture.

appeared 2 weeks after inoculation. Such mycorrhizal plantlets are easily able to adapt to autotrophic conditions and are much more drought resistant than non-mycorrhized ones.

4. Conclusions

Although *Pinus sylvestris* is one of the most recalcitrant species to be used in tissue culture, recent results show that progress is being made, as with numerous other succesfully propagated trees. Successful micropropagation method [60] and the mycorrhization of plantlets [45] are a good basis for mass propagation and the eventual incorporation of *Pinus sylvestris* tissue culture in vegetative propagation. First results with mature trees [25] indicate, that in the future the propagation of elite trees might be done *in vitro*.

The method of protoplast fusion [57] as well as induction of tumours [1] are first steps towards including methods for the genetic improvement of *Pinus sylvestris in vitro*.

Perhaps it should be mentioned that, at least in some countries in Europe, *Pinus sylvestris* is one of the coniferous species most threatened by air pollution and acid rain. The techniques of micropropagation and cryopreservation would be helpful in the selection of clones to ameliorate these effects.

5. Summary

The vegetative propagation of *Pinus sylvestris*, especially from mature trees, is still limited. Micropropagation of Scots pine has recently become better understood and the possibility of including this species among those in mass propagation is therefore coming closer to realisation. In this chapter the tissue culture of *Pinus sylvestris* has been reviewed, with special attention to recent results.

Acknowledgements

I thank Professor Dr C. Bornman for critically reading the manuscript and Professor Dr N. Gogala and Dr M. Ravnikar, for providing helpful suggestions.

References

1. Ahuja MR (1988) Gene transfer in forest trees. In: Genetic manipulation of woody plants. Basic Life Sciences, Vol. 44, Plenum Press, New York. pp. 25–41.
2. Aitken J, Horgan KJ, and Thorpe TA (1981) Influence of explant selection on the shoot-

forming capacity of juvenile tissue of *Pinus radiata*. Can J For Res 11:112–117.
3. Bogdanović M (1968) The influence of different factors on stem tissue cultures of *Pinus silvestris* L. Zemljište i biljka 17:135–145.
4. Bornman CH (1983) Possibilities and constraints in the regeneration of trees from cotyledonary needles of *Picea abies in vitro*. Physiol Plant 57:5–16.
5. Bornman CH (1981) *In vitro* regeneration potential of the conifer phyllomorph. Symp on Clonal Forestry, Uppsala, Sweden. pp. 43–56.
6. Bornman CH (1984) In: Proceedings of the International Symposium of Recent Advances in Forest Biotechnology, Traverse City, Michigan. pp. 178–194.
7. Bornman CH and Jansson E (1982) Regeneration of plants from the conifer leaf with special reference to *Picea abies* and *Pinus sylvestris*. In: Colloque international sur la culture *in vitro* des essences forestiers, IUFRO, Fontainebleu, AFOCEL, Nangis, France. pp. 41–53.
8. Bornman CH and Jansson E (1980) Organogenesis in cultured *Pinus sylvestris* tissue. Z Pflanzenphysiol 96:1–6.
9. Bourgin JP and Nitsch JP (1967) Obtention de *Nicotiana* haploides à partir d'étamines cultivées *in vitro*. Ann Physiol Vég 9:377–382.
10. Cheng TY (1975) Adventitious bud formation in culture of Douglas fir (*Pseudotsuga menziesii* (Mirb.) Franco). Plant Sci Lett 5:97–102.
11. Christersson L (1978) The influence of photoperiod and temperature on the development of frost hardiness in seedlings of *Pinus sylvestris* and *Picea abies*. Physiol Plant 44:288–294.
12. Cresswell R and Nitsch C (1975) Organ culture of *Eucalyptus grandis* L. Planta 125:87–90.
13. David A, Faye M, and Rancillac M (1983) Influence of auxin and mycorrhizal fungi on the *in vitro* formation and growth of *Pinus pinaster* roots, Plant Soil 71:501–505.
14. De Fossard RA (1976) Tissue Culture for Plant Propagators. The University of New England Printery, Armidale, Australia.
15. Dencso I (1985) Possible ways of *in vitro* morphogenesis in tissue cultures of some Pinus species II. Organogenesis in embryo cultures. Erdeszeti es Faipari 1–2:153–162.
16. Fortin JA (1966) Synthesis of mycorrhizae on explants of the root hypocotyl of *Pinus sylvestris* L, Can J Bot 44:1087–1092.
17. Gogala N (1971) Growth substances in mycorrhiza of the fungus *Boletus pinicola* Vitt. and the pine-tree *Pinus sylvestris* L, Razprave (Dissertationes) 14/5:123–202.
18. Grönroos R and von Arnold S (1985) Initiation and development of wound tissue and roots on hypocotyl cuttings of *Pinus sylvestris in vitro*. Physiol Plant 64:393–401.
19. Grönroos R (1987) Rooting of *Pinus sylvestris* and *Pinus contorta* hypocotyl cuttings *in vitro*. Acta Univ Ups 88:1–26.
20. Hansen J and Erstsen A (1982) Seasonal changes in adventitious root formation in hypocotyl cuttings of *Pinus sylvestris*: Influence of photoperiod during stock plant growth and of indolebutyric acid treatment of cuttings. Physiol Plant 54:99–106.
21. Heller R (1953) Recherches sur la nutrition minérale des tissues végetaux cultivés *in vitro*. Ann Sci Nat Bot Biol Veg 14:1–223.
22. Hellergren J (1981) Frost hardiness development in response to fertilization. Physiol Plant 52:297–301.
23. Hellergren J (1983) Cold acclimatization of suspension cultures of *Pinus sylvestris* in response to light and temperature treatments. Plant Physiol 72:992–995.
24. Hohtola A (1988) Seasonal phenomena in natural and cultured bud tissues from mature Scots pine. Acta Univ Oulo Ser A 204:1–54.
25. Hohtola A (1988) Seasonal changes in explant viability and contamination of tissue cultures from mature Scots pine. Plant Cell Tiss Org Cult 15:211–222.
26. Hohtola A and Pitkänen R (1988) Browning of Scots pine tissue cultures. J Ult Mol S 98:336.
27. Ingestad T (1979) Mineral nutrient requirements of *Pinus sylvestris* and *Picea abies* seedlings. Physiol Plant 45:373–380.
28. Jelaska S (1987) European pines. In: Cell and Tissue Culture in Forestry, Martinus Nijhoff

Publishers, Boston. Vol. 3, pp. 42–60.
29. Lin ML and Staba J (1961) Peppermint and spearmint tissue culture. I. Callus formation in submerged culture. Lloydia 24:139–145.
30. Linsmaier EM and Skoog F (1965) Organic growth factor requirement of tobacco tissue cultures. Physiol Plant 18:100–127.
31. Murashige T and Skoog F (1962) A revised medium for rapid growth and bioassays with tobacco tissue cultures. Physiol Plant 15:473–497.
32. Nitsch JP and Nitsch C (1965) Néoformation de fleurs *in vitro* chez une espèce de jours courts: *Plumbago indica*. Ann Physiol Vég 7:251–256.
33. Paasisalo S (1988) *In vitro* formation and development of multiple shoots from embryos of *Pinus sylvestris* L. Physiol Plant 73:22A.
34. Patel KR and Thorpe TA (1984) *In vitro* differentiation of plantlets from embryonic explants of logepole pine (*Pinus contorta* Dougl. ex Loud.). Plant Cell Tiss Org Cult 3:131–142.
35. Ramsden L and Northcote DH (1987) Glucomannan synthase from suspension cultures of *Pinus sylvestris*. Phytochemistry 26:2679–2684.
36. Ramsden L and Northcote DH (1987) Tracheid formation in cultures of pine *Pinus sylvestris*, J Cell Sci 88:467–474.
37. Reilly K and Washer J (1977) Vegetative propagation of Radiata pine by tissue culture: plantlet formation from embryonic tissue. NZ J For Sci 7:199–206.
38. Rogozinska JH (1970) Culture of Scots pine callus and its nutritional requirements, Acta Soc Bot Pol 39:151–160.
39. Sandberg G and Oden PC (1982) Effects of a short-day treatment on pool size, synthesis, degradation and transport of 3-indoleacetic acid in Scots pine (*Pinus sylvestris* L.) seedlings. Physiol Plant 55:309–314.
40. Schmidt A (1924) Über die Chlorophyllbild in Koniferenembryo Bot Arch 5:260–282.
41. Slankis V (1948) Verschiedene Zuckerarten als Kohlenhydratquelle für isolierte Wurzeln von *Pinus silvestris*. Physiol Plant 1:278–289.
42. Slankis V (1948) Einfluss von Exudaten von *Boletus variegatus* auf die dichotomische Verzweigung isolierter Kieferwurzeln. Physiol Plant 1:390–400.
43. Slankis V (1950) Effect of α-naphthaleneacetic acid on dichotomous branching of isolated roots of *Pinus silvestris*. Physiol Plant 3:40–44.
44. Sommer HE, Brown CL and Kormanik PP (1975)) Differentiation of plantlets in longleaf pine (*Pinus palustris* Mill) tissue cultured *in vitro*. Bot Gaz 136:196–200.
45. Supriyanto S and Rohr R (1988) Micropropagation and *in vitro* mycorrhization of *Pinus sylvestris* L. 6th FESPP Congress, Split, Yugoslavia. 14/34.
46. Supriyanto S and Rohr R (1987) Bourgeonnement massif des microboutures de Pin noir et de Pin sylvestre cultivees en conditions aseptiques sur milieu syntehtique. Actes de 2e Colloque Sciences et Industries du Bois, Nancy. pp. 291–295.
47. Supriyanto S and Rohr R (1988) Use of selected mycorrhizal fungi to facilitate the transfer to soil of Scots pine axenic plantlets. Intern Symp on For Tree Physiol, INRA, IUFRO, Nancy, France.
48. Thomas MJ (1972) Comportement des embryons de trois espèces de pins (*Pinus mugo* Turra, *Pinus silvestris* L. et *Pinus nigra* Arn.), isolés au moment de leur clivage et cultivés *in vitro*, en présence de cultures-nourrices. CR Acad Sc, Serie D. 274:2655–2658.
49. Toribio M and Pardos JA (1981) *In vitro* organogenesis of *Pinus sylvestris* L. tissues. Coll Int sur la culture *in vitro* des essences forestieres, AFOCEL. pp. 143–147.
50. Tranvan H (1979) *In vitro* adventitious bud formation on isolated seedlings of *Pinus silvestris* L. Biol Plant 21:230–233.
51. Tranvan H and Thomas MJ (1981) Aptitudes morphogénétiques des plantules et des fragments de plantules du *Pinus silvestris* L. en culture *in vitro*. Bull Soc Bot Fr. 128:151–163.
52. Vidaković M (1982) Četinjače – morfologija i variabilnost, JAZU, Zagreb.
53. Vogelmann TC, Bornman CH, and Nissen P (1984) Uptake of benzyladenine in explants of *Picea abies* and *Pinus sylvestris*. Physiol Plant 61:513–517.
54. Von Arnold S and Eriksson T (1981) Production of adventitious plants from spruce and

pine. Symp on Clonal Forestry, Uppsala. pp. 7–31.

55. Von Arnold S and Eriksson T (1981) *In vitro* studies of adventitious shoot formation in *Pinus contorta*. Can J Bot 59:870–874.
56. Von Kirsten U, Jacob HE, Tesche M, and Kluge S (1986) Isolierung und Elektrofusion von Koniferenprotoplasten, Silvae Genet 35:5–6.
57. White PR (1943) A Handbook of Plant Culture, The Jacques Catell Press Inc., Tempe, Ariz.
58. Whitehill SJ and Schwabe WW (1975) Vegetative propagation of *Pinus sylvestris*. Physiol Plant 35:66–71.
59. Žel J, Gogala N, and Camloh M (1988) Micropropagation of *Pinus sylvestris*. Plant Cell Tiss Org Cult 14:169–175.
60. Želavski W, Gowin T, and Lotocki A (1972) Preliminary study of the ecotype differentiation on three plantations of Scots pine *Pinus silvestris* L. in Poland. Ekologia Polska 20:763–770.

21. Improved aseptic germination and controlled growth for micropropagation of Douglas fir

MICHELLE J. HUTZELL and DON J. DURZAN
Department of Environmental Horticulture, University of california, Davis, CA 95616 U.S.A.

1. Introduction

Genetic gains may be captured by micropropagation using improved seed and aseptically germinated seedlings as explant sources [1,10]. Currently, excised embryos, hypocotyl segments, and cotyledons are major sources of explants for a number of conifers, e.g., *Picea abies* [3], *Pseudotsuga menziesii* [4,6,7], *Pinus radiata* [9] and *Picea glauca* [2], to name a few. These sources are easy to handle and tissues are more responsive to growth regulators than explants from mature trees.

This paper describes improvements in early stages of a micropropagation system using seedlings germinated in a novel way to produce more vigorous explant material. The system uses an improved substrate and a controlled humid atmosphere under aseptic germination conditions for several coniferous species. We have often observed that the viability of surface-sterilized seeds on Petri dishes has not always been as optimal as one would expect. The traditional method of germinating seed on moist filter paper inhibited the growth of seedlings when compared to our systems employing other substrates, and controlled humidity in a sealed polypropylene bag.

In this study we also explore local microenvironmental gaseous factors that contribute to improved viability and growth. Even though seedlings are germinated in sealed polypropylene bags permeable to oxygen and carbon dioxide, the effects of humidity and removing organic volatiles could be studied.

2. Materials and methods

Douglas fir [*Pseudotsuga menziesii* (Mirb.) Franco.] seeds (Douglas-fir seed lot #202-302.30-05-58-2-1003-82) were obtained from the USDA Forest Service nursery in Placerville, CA (lat. 38° N 44.2′, long. 120° W 44.5′, elev. 4000′). Seeds were surface-sterilized using 30-minute washes in each of the following solutions: (1) 1% Linbro, (2) 1% H_2O_2, and (3) 1% sodium hypochlorite. The seeds were rinsed three times in sterile distilled water prior to imbibition in sterile water in a refrigator overnight. Following this, the seeds were re-sterilized briefly in 1% (w/v) sodium hypochlorite and rinsed several times

M. R. Ahuja (Ed.), Micropropagation of Woody Plants, 367–372.

in sterile distilled water.

In order to illustrate the importance of substrate and atmosphere, the seeds were germinated aseptically on filter paper in Petri dishes and compared to germinating seeds in sterile, sealed polypropylene bags (Clavies™). Petri dishes containing 2 g seed on moist filter paper were incubated under continuous light (1.0, nil, 0.75 $\mu W\ cm^{-2}$ in red, far red, blue spectrum) for a period of two weeks. Polypropylene bags each containing 30 ml damp, sterile vermiculite plus 2 g seed were incubated under the same conditions. While the polypropylene bags may be permeable to oxygen and carbon dioxide, they are impermeable to water vapor en ethylene.

Several modifications of the bag method were tested to control local gas exchange and atmospheric influences upon seed germination. Modifications included additions of either KOH pellets to absorb CO_2, or $KMnO_4$ pellets (Purafil™) to oxidize ethylene and other organic volatiles, or a combination of the two types of pellets. The bags were divided into compartments to separate the pellets from the seeds, thus preventing chemical mixing and toxicity. Bags containing 30 ml damp vermiculite were autoclaved prior to use. Approximately 2 g seed was added to each bag, and 2 g of either KOH or $KMnO_4$, or a combination of the two were added to separate compartments in each bag (Figure 1). The bags were sealed (using a Seal-a-Meal™) and placed upright in a hot-pac walk-in growth chamber, 23° C, 60% humidity, continuous/diffuse

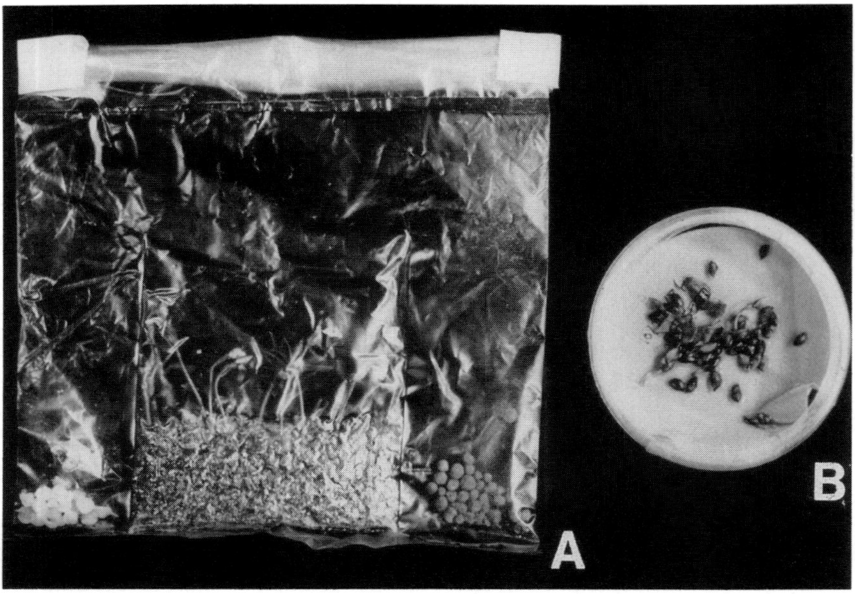

Figure 1. Comparison of two systems for seed germination and vigor in Douglas fir after 14 days. (A) Clavies polypropylene bag +30 ml damp, sterile vermiculite +2 g seed. Equivalent weights of KOH (left) and/or $KMnO_4$ pellets (right) are added to either side of the bag. (B) Surface-sterilized seeds germinated on moist filter paper. See Figure 2 for growth comparisons between seedlings.

Table 1. Comparison of mean seedling height and germination after 14 days.

Treatment	Height (cm)	% Germination
Petris Dishes	3.2	8.3
Polypropylene		
Clavies		
A. Control	11.4	58.7
B. KOH Pellets	11.9	37.0
C. $KMnO_4$ Pellets	14.2	49.3
D. KOH + $KMnO_4$	12.8	54.0

light (1.0, nil, 0.75 $\mu W\ cm^{-2}$ in red, far red, and blue spectrum, respectively). After two weeks, measurements of seedling height and germination percent (based on all seeds) were recorded.

3. Results and observations

Seedling height measurements of the ten most vigorous seedlings for each treatment, and germination percentage are shown in Table 1. The seedlings in the bag treatments were much more vigorous, and germination rates much higher than those in Petri dishes (Figure 2). Statistical analysis comparing the two methods showed significant differences between the Petri dishes and the bags, but there was no significant difference among the various bag treatments (Table 1).

The suitability of seeds germinated in polypropylene bags for micropropagation was tested in the following way. Cotyledon and hypocotyl explants of the Douglas-fir seedlings were cultured on DCR medium in agar Petri plates (Figure 3). Within two weeks, these explants showed multiple, adven-

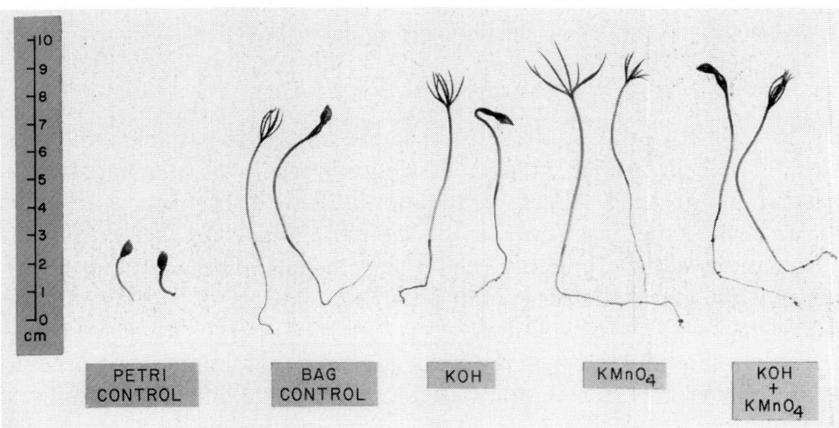

Figure 2. Comparison of 14-day-old Douglas-fir seedling height among treatments.

Figure 3. Multiple adventitious buds induced on cotyledons after 9 days exposure to 2 mg l^{-1} BAP and 0.5 mg l^{-1} kinetin at 23 to 25 °C. Douglas-fir cotyledon (C) on DCR medium [6].

titious bud formation on both cotyledons and hypocotyls within two weeks. From these buds, rooted plantlets have been obtained, which are now hardened-off and maintained under greenhouse conditions (Figure 4).

4. Discussion

The results from this study show that the bag germination method yields, within 2 weeeks, a high number of vigorous seedlings, as compared to the conventional method of seed germination in Petri dishes. Increased humidity due to sealing the bags seems to be the most important factor in increased germination, not the presence of CO_2 or ethylene. This is supported by the fact that the control bags without additions of KOH or $KMnO_4$ did just as well as bags with those additions.

Further modifications of the seed sterilization method may need to be incorporated into this system for seed lots which are difficult to sterilize. For example, a wash with mercuric chloride or a similar compound may be needed [6,7]. The stratification time of the seeds may be avoided by using the H_2O_2

Figure 4. Rooted Douglas-fir adventitious buds after 3–4 weeks. Micropropagated shoots were exposed to 0.5 DCR, NAA (2.0 mg l^{-1} and 1% sucrose for 6 days (2 in darkness followed by 4 in light) and then transferred to 0.5 DCR and 1% sucrose for 3–4 weeks in light. Rooting after week 4 was 40% [7].

wash [5]. Since the seed material may have different requirements or uses, modifications of seed density/bag, and possible nutrient or hormone additions to the bags could be employed. In this way, the germination percent and seedling vigor could be increased even more than presented in our preliminary studies.

Advantages of our method of aseptic seed germination for *in vitro* culture include reducing the time necessary for preparing the seedlings for culture, especially the elimination of the need to excise the embryos from seeds, which is quite labor-intensive. The bag method reduces the damage which may occur during excision, and which may have a detrimental effects on the seedling explant prior to culture. Furthermore, a direct estimate of the quality and vigor of seedlings in the seed lot is obtained. This is important where micropropagation of elite seeds is goal of tree improvement. We have found the bag method especially useful under dry laboratory conditions encountered in hot and cold dry climates.

Another use for this type of system could be the conversion or germination of somatic embryos from cell cultures. Based on experience with encapsulated zygotic embryos and seeds (e.g. [8], more work will be needed to improve the recovery of viable, clonally propagated and encapsulated embryos. Our bag method provides an improved environment for the germination of these embryos. In fact, environments could be selected which represent conditions at high elevations for trees destined to such locations.

5. Summary

The quality of explant material from Douglas-fir seedlings has been improved by developing a novel method for seed germination under controlled atmospheres. Explant quality is evidenced by a significant improvement in the vigor and rate of growth and development in seedlings over a 2-week period. Improvement in germination under aseptic conditions in sealed polypropylene bags was due mainly to the control of humidity and by provision of a suitable substrate. These observations were supported by non-significant differences in the humid atmosphere of sealed bags when KOH or permanganate were added to trap carbon dioxide and to oxidize organic volatiles during germination. Explants from seedlings in polypropylene bags responded within 9 days to N^6-benzyladenine by forming multiple adventitious buds. This result shortens the time by several days to attain cytokinin-responsive explant material in comparison to the filter paper/Petri dish method.

References

1. Bonga JM and Durzan DJ (Eds.) (1982) Tissue Culture in Forestry. Martinus Nijhoff/Dr. W. Junk, The Hague, The Netherlands.
2. Campbell RA and Durzan DJ (1976) Vegetative propagation in *Pica glauca* by tissue culture. Can J Forest Res 6:240–243.
3. Chalupa V (1977) Organogenesis in Norway spruce and Douglas fir buds *in vitro*. Commun Inst Forestry Czech 10–79.
4. Cheng TY 16975) Adventitious bud formation in culture of Douglas fir (*Pseudotsuga menzeisii* (Mirb.) Franco), Plant Science Letters 5:97–102.
5. Ching TM and Parker MC (1958) Hydrogen peroxide for rapid viability tests of some coniferous tree seeds. Forestry Scinece 4(2):128–133.
6. Gupta PK and Durzan DJ (1985a) Short multiplication from mature trees of Douglas fir (*Pseudotsuga menziesii*) and sugar pine (*Pinus lambertiana*). Plant Cell Reports 4:177–179.
7. Gupta PK and Durzan DJ (1985b) *In vitro* establishment and multiplication of juvenile and mature Douglas-fir and sugar pine. In: proceedings from the Symposium on *In Vitro* Problems Related to Mass Propagation of Horticultural Plants, September 16–20, Gembloux, Belgium. p. 18.
8. Linden T (1984) Planting seeds for the future. Encapsulation. Western Grower and Shipper. January. pp. 12–14.
9. Reilly K and Washer J (1977) Vegetative propagation of Radiata pine by tissue culture. Plantlet formation from embryonic tissue. N Z J Forest Science 7:199–206.
10. Timmis R and Ritchie GA (1984) Progress in Douglas fir tissue culture. In: Proceedings of the International Symposium of Recent Advances in Forest Biotechnology, June 10–13, Michigan Biotechnology Institute, Traverse City, Michigan. pp. 37–46.

22. Micropropagation of *Larix* species and hybrids

DAVID F. KARNOSKY, YINGHUA HUANG and DONG-ILL SHIN
School of Forestry, Michigan Technological University, Houghton, MI 49931, U.S.A.

1. Introduction

Larch (*Larix*) species and hybrids are important conifers that are grown for lumber, poles and pulp as well as ornamental purposes in northern temperate zone forests across North America, Europe and Asia. Among the principal commercial larch species are tamarack (*Larix laricina)*, Western larch (*L. occidentalis*), European larch (*L. decidua*), Japanese larch (*L. leptolepis*), Dahurian larch (*L. gmelinii*) and Siberian larch (*L. siberica*). Hybrid larches have shown heterosis in growth, stem form and wood mechanical properties as compared to the parental plants [36].

Traditional breeding and large-scale production of hybrids for larch are hindered by the poor seed yields obtained. Empty seeds often account for two thirds or more of the seed obtained following controlled pollinations [29,17]. The relatively small cone sizes and the early seasonal flowering phenology which often leads to low-temperaturee injury to flowers also contribute to the difficulty in breeding these trees. Therefore, *Larix* species and hybrids are good candidates for vegetative propagation via micropropagation [21].

While conifers have generally been more difficult to handle *in vitro* than have hardwood trees, larches are a unique set of conifers in that they are highly amenable to micropropagation systems. Larches have been regenerated from juvenile and mature tissues via organogenesis, caulogenesis and embryogenesis. Haploid plants can be obtained by culturing female gametophyte tissue [34,39] and whole plants have been obtained from larch protoplasts [27]. In this chapter, we will describe some significant breakthrough in larch micropropagation and then discuss how we see micropropagation being used in larch tree improvement programs.

2. Micropropagation from juvenile tissues

The first reported *in vitro* culture of larch was that of immature *Larix decidua* embryos by Sterling [38]. Over 30 years later, the first organogenesis from callus obtained from seedling apical meristems was reported [32]. Then in the early 1980s, systems were developed for micropropagation via adventitious

M. R. Ahuja (Ed.), Micropropagation of Woody Plants, 373–382.

bud production from juvenile tissues of *L. laricina* [6] and *L. decidua* [23]. Significant improvements in these systems developed in our laboratory [12,24,25] have resulted in initial multiplication rates of 70 new shoots per initial larch seedling explant. The resultant shoots can serve as source tissues for axillary bud production in a continuous shoot culture system developed by Ellis *et al.* [13].

Buds can be initiated on cut cotyledons or hypocotyls or on intact seedling explants by adding BAP (10 mg/l) to the initiation medium for two to three weeks. Buds are elongated by transferring them to hormone-free media. Shoots that stop elongating and that set bud can generally be induced to reflush by placing them on media supplemented with 18 mg/l BAP for two to three weeks before transferring buds to hormone-free medium. Shoots grown *in vitro* can be induced to root by supplementing the basal medium with 10 mg/l NAA or they can be rooted by dipping them in Rootone F rooting powder and placing them in a potting-soil mix under intermittent mist in the greenhouse. Spontaneous rooting does occur occasionally both *in vitro* and in the greenhouse, but root-system quality is enhanced by the addition of the rooting hormone (Figure 1). Plantlets obtained by these methods transplant readily to pots for greenhouse culture or to the field for plantation culture. While detailed studies have not yet been done to compare seedlings and plantlings with respect to growth and form, our experience suggests that plantlings tend to look very much like rooted cuttings for the first year; that

Figure 1. Comparison of *Larix decidua* root systems of a seedling (far left) and tissue culture plantlings rooted in the greenhouse without hormone treatment (left) and with a dip in Rootone (right) and *in vitro* with 1/2 ppm NAA added to the medium for a seven-day exposure (far right).

is, they tend to have less branched root systems and tops than do seedlings (Figure 2). However, after a few years in the field, plantlings and seedlings are indistinguishable (Figure 3).

Embryogenesis in juvenile larch tissues was first reported by Nagmani and Bonga [34]. They obtained somatic embryos from subcultured callus derived from *Larix decidua* gametophyte tissue. Embryogenesis was prolific through six to eight successive transfers. Many somatic embryoids formed plantlets which later died in the greenhouse. These cultures were subsequently found to be haploid [39] and the authors speculated that unmarked deleterious lethal genes probably hindered plantlet survival.

Recently, Klimaszewska [28] has obtained somatic embryos via embryogenesis of larch calli initiated from developing zygotic embryos. In what may be the first regeneration of a conifer from protoplasts (Figure 4), she has recovered somatic embryos from protoplasts [27] derived from her embryogenic callus lines. While these embryogenesis results are preliminary in nature and

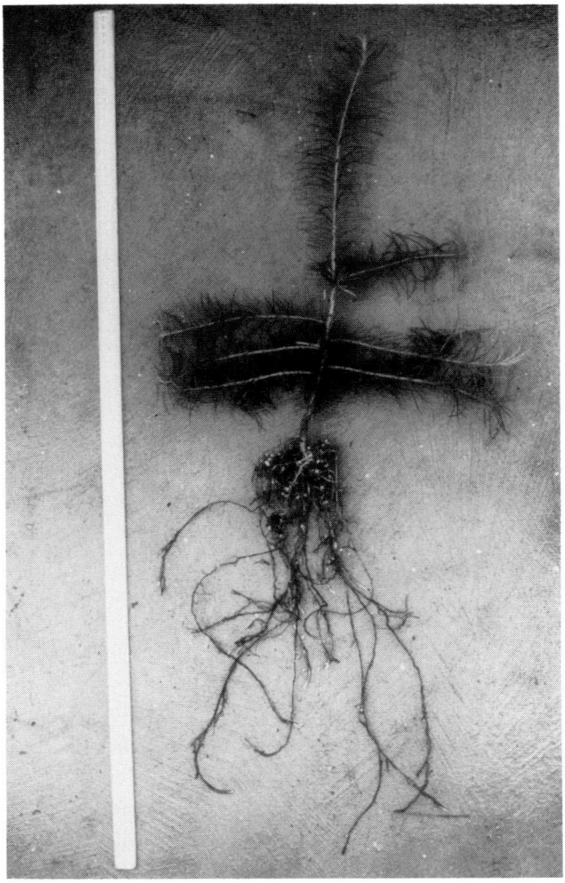

Figure 2. One-year-old European larch tissue culture plantling. The white bar is a meter stick.

Figure 3. Four-year-old *Larix decidua* plantling in an experimental planting near Houghton, Michigan.

have not yet resulted in plants suitable for field testing, they suggest that juvenile larch tissues are highly amenable to the newly developing micropropagation technologies.

3. Micropropagation from mature trees

The earliest studies of micropropagation from mature larch tissues resulted in callus proliferation but no further development [8,18]. The next major event in the micropropagation from mature larch tissues occurred when Bonga [3] obtained shoot formation via caulogenesis in callus derived from female strobili of a 25-year-old *Larix decidua* tree. In repeating this research, Bonga [4,5] obtained additional shoots, some of which rooted to form plantlets, from

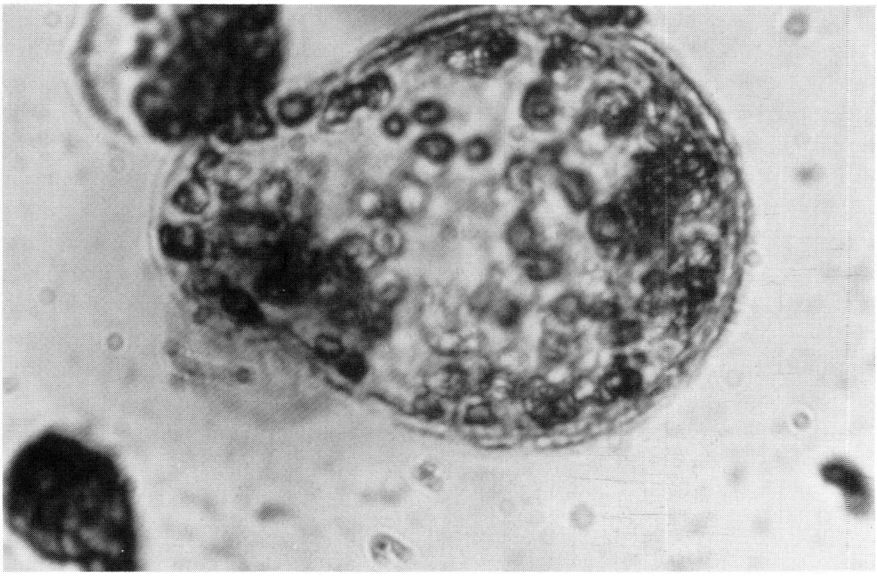

Figure 4. One-week-old *Larix decidua* protoplast that is just starting to reform its cell wall.

the bracts and scales of female strobili. None of these plants survived transfer to soil.

Initiation of shoot buds that later elongated rootable shoots was obtained from vegetative buds taken from 30-year-old *Larix decidua* trees both by adventitious bud production from the surface of the apical dome and by caulogenesis of callus initiated from the apical dome region [7]. Sustained caulogenesis in callus cultures initiated from short shoot buds of a 12-year-old *Larix* × *eurolepis* was obtained by Laliberte and Lalonde [30]. Their cultures were still very productive after 32 months in culture. Up to 22% of their shoots rooted spontaneously *in vitro* and one quarter of these survived transfer to greenhouse conditions. Normal, orthotropic growth was exhibited by the plantlings grown in the greenhouse. To the best of our knowledge, embryogenic callus has not yet been obtained from mature larch tissues.

4. Potential uses of and research needs for micropropagation

4.1. *Vegetative propagation*

Strategies for the genetic improvement of *Larix* species and hybrids generally include a component of clonal propagation through mass propagation of juvenile cuttings from a limited quantity of genetically improved seed [15,33,37]. It is our opinion that the bulking up of planting stock from a limited number of genetically improved seed will eventually involve a micropropagation component, either alone or in combination with the rooting of cuttings. Currently, this micropropagation phase could involve adventitious bud production from seedling explants. After an initial multiplication *in vitro*, planting stock could be additionally increased via rooting of cuttings taken from the plantlings.

In the future, the bulk up of planting stock could employ an embryogenesis system. However, before that is possible, a number of research questions must be addressed. Methods are needed to improve the frequency of embryogenesis in culture, to increase the conversion rate of somatic embryos to plantlings and to compare the fidelity and performance of somatic embryogenesis derived plantlings with seedlings. The development of artificial seeds using encapsulation of somatic seeds would be particularly useful in larch where traditional seeds generally have a very low viability.

Additional research is needed for micropropagation from mature trees; the recent successes to date in propagating from mature larch trees are far from being commercially feasible. Methods of deploying clonal larch forests to compromise between desirable growth, form and hardiness characteristics and maintenance of genetic diversity are needed. Thus, the desired end product (i.e. the planted forest) should in some ways drive developing micropropagation technology.

4.2. *Germplasm preservation*

Some larch species such as *Larix decidua* occupy very small natural ranges and portions of these natural ranges are being threatened by extinction. For example, the Tatra and Sudeten Mountain populations of *Larix decidua* are being threatened by the air-pollution loading from the Eastern Block countries in this area. Seed collected from these two populations have consistently been among the best performers in *Larix decidua* provenance tests around the world [2,16,20]. A program to collect and preserve a sizable sample of these two populations is needed [22]. The existing technologies of cryopreservation would be a relatively low-cost method to preserve this valuable germplasm. However, experiments are needed to determine the best methods for larch cryopreservation.

4.3. *Disease screening*

Abdul Rahman *et al.* [1] have described a micropropagation system for testing the response of larch species to scleroderris canker. The authors suggested this system could be a valuable initial screening step in testing the scleroderris canker resistance of various seedlots. While there are a number of practical limitations to *in vitro* disease screening [10], these systems can provide useful preliminary information and can be used for basic research studies of host-parasite interactions.

4.4. *Gene transfer*

Larch species appear to be amenable to *Agrobacterium*-mediated gene transfer [9,11,26,19]. In our laboratory, we have shown opine production and improved kanamycin resistance of transformed larch callus, documented *Agrobacterium* attachment to larch cells via electron microscopy, we have demonstrated stable plasmid DNA incorporation in the larch genome, and regenerated transgenic plants [19].

There remain many research needs with this newly developing technology. First, the expression of a transferred gene in a transformed larch plantlet has not been demonstrated. More efficient systems of plantlet recovery and screening for transformed plantlets are needed. Regeneration of shoots from transformed roots would be particularly useful as prolific multiple roots are readily obtained following inoculation of each seedling with several strains of *A. rhizogenes*. No one has yet reported adventitious shoot introduction from larch roots.

To facilitate field testing and plantation establishment of transformed larch plantings, a means of preventing pollen transport into native larch populations is needed. Thus, the isolation of male sterile genes and the incorporation of these genes into the plasmid of choice would be very helpful. Single genes already available that could be used in larch improvement would be those for insect and herbicide resistance. The Bt toxin genes could be expected to increase larch's resistance to larch casebearer. Resistance to the herbicide 'Roundup' is conveyed through the products of genes like the *aroA* gene [14]. Such a gene in larch could enable foresters to have a chemical to release larch from competing vegetation. Since larches are notoriously susceptible to herbicides [35], an herbicide-resistant larch would be valuable. While developing herbicide-resistant plants is considered environmentally unwise in some parts of the world, we feel that glyphosate (Roundup) resistant larch trees could actually reduce the use of more toxic environmentally damaging herbicides that are commonly used to prepare larch planting sites. Since it is now possible to regenerate somatic embryos from larch protoplasts [27], the usefulness of protoplasts for gene transfer via electroporation, particle bombardment, or laser beam should be examined.

4.5. *Somatic hybridization*

Larch hybrids have shown a high degree of heterosis as compared to their parental species. However, as has been described previously in this paper, larch hybrids are difficult to produce in large numbers. Since haploid gametophyte tissue can be cultured in larch, this material would be a good donor tissue for protoplast isolation and fusion. Haploid protoplasts of the species of interest could be fused to create a diploid fusion hybrid product. Protoplast fusion has not yet been attempted with larch so that research is needed in this area to indentify suitable fusion conditions and to select and regenerate the hybrid fusion products. This type of research has been hampered by the lack of suitable selection systems. However, hybrid cell selection on the basis of antibiotic or herbicide resistance genes brought by the two parents of fusion as has been recently described for potato by Masson *et al.* [31] should be possible with larch as gene transfer of the selection gene with larch is possible [26].

5. Summary

Larch (*Larix*) species and hybrids are commercially important forest trees in northern temperate forests around the world. Because of their small cone size, low numbers of seed per cone, early seasonal flowering phenology, and high degree of embryo abortion, the number of viable seeds obtained from control pollinations and from hybrid seed orchards are severely limited. Thus, these species are exellent candidates for asexual production via micropropagation. In contrast to most conifers, larches are highly amenable to *in vitro* culture and manipulation. Whole plants have been regenerated from juvenile and mature larch tissues via organogenesis, caulogenesis and embryogenesis. Larches are the first conifer genus in which whole plants have been obtained from haploid and from protoplasts. In addition to the bulking up of plants from superior seedlots and hybrids, micropropagation will be used for preserving superior and endangered genotypes, *in vitro* disease screening, gene transfer, and basic genetic studies such as the study of the behaviour of homozygous diploid plants.

References

1. Abdul Rahman NN, Diner AM, Skilling DD, and Karnosky DF (1987) *In vitro* responses of conifer adventitious shoots and calli inoculated with *Gremmeniella abietina*. For Sci 33:1047–1053.
2. Barnes BV (1977) The international larch provenance test in south-eastern Michigan, USA. Silv Gen 26:145–148.
3. Bonga JM (1982) Shoot formation in callus from the stalks of young female strobili of *Larix decidua*. Can J Bot 60:1357–1359.
4. Bonga JM (1983) *In vitro* propagation of conifers. Proc Nineteenth Can Tree Improv Assoc Meeting. Toronto, Canada. pp. 75–83.
5. Bonga JM (1984) Adventitious shoot formation in cultures of immature female strobili of *Larix decidua*. Physiol Plant 62:416–421.
6. Bonga JM and McInnis AH (1983) Origin and early development of roots in plantlets derived from embryo sections of *Larix laricina in vitro*. Canadian Forestry Service Research Notes 3(2):12–14.
7. Bonga JM and von Aderkas (1988) Attempts to micropropagate mature *Larix decidua* Mill. In: Ahuja MR (Ed.) Somatic Cell Genetics of Woody Plants. Kluwer Academic Publishers, Dordrecht. pp. 155–168.
8. Bychenkova EA (1963) An investigation of callus formation in certain trees and shrubs by the method of tissue culture *in vitro*. Dokklady Akademii Nauk USSR 151:732–736.

9. Diner AM and Karnosky DF (1986) *Agrobacterium*: Potential for genetic engineering. Proc TAPPI Research and Development Conf Technology Park, Atlanta, Georgia. pp. 93–94.
10. Diner AM and Karnosky DF (1987) Tissue culture application to forest pathology and pest control. In: Bonga JM and Durzan DJ (Eds.) Cell and Tissue Culture in Forestry. Vol. 2, Martinus Nijhoff Publishers, Boston. pp. 351–373.
11. Diner AM and Karnosky DF (1987) Differential responses of two conifers to *in vitro* inoculation with *Agrobacterium rhizogenes.* Eur J For Path 17:211–216.
12. Diner AM, Strickler A, and Karnosky DF (1986) Initiation elongation and remultiplication of *Larix decidua* micropropagules. New Zealand J For Sci 16:306–318.
13. Ellis D, McCown B, Skilling D, Barker M, Serres R, and Ostry M (1989) Establishment of a shoot culture system for *Larix decidua.* Proc IUFRO Working Party for Somatic Cell Genetics. Workshop in Woody Plant Biotechnology. Placerville, California. (In press.)
14. Fillatti JJ, McGown BH, Sellmer J, and Haissig B (1986). The introduction and expression of a gene conferring tolerance to the herbicide glyphosate in *Populus* NC5339. Proc TAPPI Research and Development Conf. pp. 83–85.
15. Fowler DP (1986) Strategy for the genetic improvement of *Larix* species and hybrids in the Maritimes. In: 1986 Larch Workshop. The New Brunswick Forest Research Committee. Fredericton, New Brunswick. pp. 73–78.
16. Giertych M (1979) Summary of results on European larch (*Larix decidua* Mill.) height growth in the IUFRO 1944 provenance experiment. Silv Gen 28:244–256.
17. Hall JP and Brown IR (1977) Embryo development and yield of seed in *Larix*. Silvae Gen 26:77–84.
18. Harvey AE and Grasham JL (1969) Procedures and media for obtaining tissue cultures of 12 conifer species. Can J Bot 47:547–549.
19. Huang Y, Diner AM and Karnosky DF (1991) *Agrobacterium rhizogenes*-mediated genetic transformation and regeneration at a conifer; *Larix decidua.* In Vitro Cell Dev Biol (Plant) 3:201–207.
20. Jeffers RM and Isebrands JG (1972) Larch potential in the North-central States. Proc Eighth Central States For Tree Imp Conf Columbia, Missouri. pp. 80–85.
21. Karnosky DF (1981) Potential for forest tree improvement via tissue culture. BioScience 31:114–120.
22. Karnosky DF (1989) Ecological genetics and changes in atmospheric chemistry: The application of knowledge. In: Clegg MT, Pitelka LF, and Taylor Jr GE (Eds.) Ecological Genetics, Terrestrial Vegetation and Antropogenic Changes in the Atmosphere. Springer Verlag (In Press).
23. Karnosky DF and Diner AM (1984) A cotyledon culture system for cloning *Larix decidua* and *Pinus banksiana*. Proc. TAPPI Research and Development Conf. Technology Park, Atlanta, Georgia. pp. 13–15.
24. Karnosky DF and Diner AM (1986) Techniques for increasing clone sizes in *Larix*. Proc VI Int Congr of Plant Tissue and Cell Culture. Univ Minnesota, Minneapolis, Minnesota. p. 10.
25. Karnosky DF and Mulcahey A (1988) Explant orientation and media manipulation affect *Larix decidua* adventitious bud quality and quantity. In Vitro 24(3) Part II p. 51A (Abstract).
26. Karnosky DF, Diner AM, and Barnes WM (1988) A model system for gene transfer in conifers: European larch and *Agrobacterium*. In: Ahuja MR (Ed.) Somatic Cell Genetics of Woody Plants. Kluwer Academic Publishers, Dordrecht. pp. 55–63.
27. Klimaszewska K (1988) Recovery of somatic embryos from protoplast cultures of hybrid larch. Fourth International Conifer Tissue Culture Work Group. Saskatoon, Saskatchewan, Canada. p. 46 (Abstract).
28. Klimaszewska K (1988) Somatic embryogenesis and plantlet formation from immature zygotic embryos of hybrid larch. Fourth International Conifer Tissue Culture Work Group. Saskatoon, Saskatchewan, Canada. p. 74 (Abstract).
29. Kosinski G (1987) Empty seed production in European larch (*Larix decidua*). For Ecol Management 19:241–246.

30. Laliberte S and Lalonde M (1988) Sustained caulogenesis in callus cultures of *Larix* × *eurolepsis* initiated from short shoot buds of a 12-year-old tree. Amer J Bot 75:767–777.
31. Masson J, Lancelin D, Bellini C, Lecerf M, Guerche P, and Pelletier G (1989) Selection of somatic hybrids between diploid clones of potato (*Solanum tuberosum* L.) transformed by direct gene transfer. Theor Appl Genet 78:153–159.
32. Momot TS (1976) Organogenesis of leaf tissues of *Larix dahurica* Turez, cultured *in vitro*. Lesnop Zhurnal 5:27–29.
33. Morgenstern EK (1987) Methods for rooting of larch cuttings and application in clonal selection. For Chron 63:174–178.
34. Nagmani R and Bonga JM (1985) Embryogenesis in subcultured callus of *Larix decidua*. Can J For Res 15:1088–1091.
35. Netzer DA (1984) Herbicide trials on European larch in northern Wisconsin. USDA Forest Service North Central Forest Experiment Stn. Research Note NC-318. 2 pp.
36. Paques LE (1989) A critical review of larch hybirdization and its incidence on breeding strategies. Ann Sci For 46:141–153.
37. Rauter RM and Graham BJ (1983) The genetics and tree improvement potential of larches in Ontario. Proc Larch Symposium. Ontario Ministry of National Resources and Faculty of Forestry. University of Toronto. pp. 11–26.
38. Sterling C (1944) Preliminary attempts in larch embryo culture. Bot Gaz 111:90–94.
39. Von Aderkas P and Bonga JM (1988) Formation of haploid embryoids of *Larix decidua*: Early embryogenesis. Am J Bot 75:619–628.

23. Commercial micropropagation of radiata pine

BEVERLEY J. NAIRN
Forest Nursery and Tree Improvement Centre, Tasman Forestry Ltd., P.O. Box 149, Te Teko, New Zealand

1. Introduction

The incentive for Tasman Forestry Ltd. (T.F.L.) entering a programme of clonal forestry is the prospect that some clones will produce a *large* increase in value over unimproved stock. Some of the most genetically advanced control pollinated families are anticipated to yield an approximate 20% increase, over unimproved seed. Major benefits are sought in uniformity of growth and form, plus the opportunity to capture correlation breaking individuals.

The best route currently available to maximise these gains is micropropagation. Other propagation methods such as embryogenesis and serial stool bed derived cuttings both promise lower costs, but embryogenesis on a commercial basis may be still some time away, and stool bed cuttings offer limited numbers on a clonal basis because of the relatively short productive life of the stools.

Thus Tasman Forestry established a pilot laboratory in 1984 and moved into the present building with a capacity of 3 million plantlets per year in December 1988.

1.1. *Present position*

By the close of the planting season 1991, TFL will have in the order of 1000 hectares of clonal plantings, originating from micropropagated stock. The earliest clonal plantings, though small in number, were made in 1986 in Tarawera Forest, originating from specific control pollinated crosses. A larger planting in 1988 has now undergone first lift pruning and trees measure approximately 4m on average.

Clonal tests replicated on both field and forest sites have been established each year since 1986 with a total of 700 clones under analysis. The 1992 clonal test program will contain a further 900 clones. Results from these early tests will be examined to disclose:

- minimum number of ramets/clone rerquired;
- comparison of field and forest sites;
- clone to clone variation within families;
- clone to clone variation across families;

M. R. Ahuja (Ed.), Micropropagation of Woody Plants, 383–394.

- variation within clone;
- *in vitro* ageing;
- value of 'reference' clones;
- performance of adventitious compared with axillary shoots;
- comparison of *in vitro* and *ex vitro* 'vigour'.

Replicates of all clones under test are maintained in cool storage, the method is described Section 2.3. The earliest stock has now been successfully stored for three years.

Adventitious and axillary derived shoots are represented in existing clonal tests originating from different clones. Under incubation currently, and due for clonal testing in 1993, are adventitious and axillary shoots derived from the same embryo.

Micropropagated stock at 2.0 million plantlets/year will form no more than 30% of forest planting.

1.2. *Program*

A seven part program is currently in place, the functions of each group are described below.

1.2.1. *Production clones*

Of the stock in hand, 80% is worked to produce 2.5 million plantlets to the glasshouse each year. An 80% throughout from the glasshouse and 95% success rate of nursery lined out stock results in 2.0 milion plants being dispatched to the forest. On a weekly basis this is 48 000 plantlets for acclimatization.

The breakdown of clonal numbers within this total is flexible. The low productivity of early methods indicated we would require 1000 clones, each yielding 2500 ramets/year. Growth and multiplication improvements have allowed a much greater yield, resulting in a 30-fold reduction in the number of clones required and an associated 30-fold increase in the number of ramets per clone. In progress at the moment is the regime of 32 production clones each yielding 78 000 ramets/year for acclimatization.

A flowchart showing production steps for this regime appears in Figure 4.

1.2.2. *Replacements clones*

Ten clones are maintained in active growth in readiness to replace any of the 32 production clones should they be rejected for any reason. Four percent of stock in hand falls into this category.

1.2.3. *Test and store clones*

Under the current strategy, every three years a further 4000 clones will be initiated into tissue culture. Thirty ramets will be sampled for clonal testing and 100 shoots are to be cool stored until field information is forthcoming. Incubation numbers of these vary from year to year but currently account for approximately 4% of working stock. Selections from this field tested stock will be bulked up for future production clones.

1.2.4. *Experimental clones*

Clones which are difficult to propagate, become vitrified readily or are reluctant to root, are maintained as rapid indicators in experimental work. The effects of changes to culture medium are often not disclosed fully for six months and 'sensitive' clones are valuable markers. Generally improvements are seen to be common to all clones, the main variation *between* clones is their tolerance to inappropriate treatment.

1.2.5. *Dormant bud culture*

The goal of being able to replicate superior, mature, individual trees remains paramount, however, maturation problems have not been overcome, and this avenue offers access to early superior trees for a limited time only. Resting buds are collected in winter months from young adolescent trees – success being greatest in the 5–10 year age range. Large scale production from this source has not been achieved to date, but culture improvements made in the last 12 months have seen a rapid change in multiplication rates, and we now have clones of 5000 shoots 18 months after initiation. The majority of clones however, are difficult to multiply rapidly and yields are closer to 200–500.

1.2.6. *Other species*

Development programs encompass Douglas fir, Sequoia, Loblolly pine and five Eucalyptus species, namely: globulus, nitens, globulus-nitens hybrid, regnans and saligna. Clone tests will be established in 1992 for Douglas fir, Sequoia and some Eucalyptus clones.

1.2.7. *Gene reserve*

There are currently 2500 clones in cool storage. Deletions are made periodically as performance information comes to hand. Storage is expected to be four years in duration before reliable information identifies superior clones for production purposes. Protocols are discussed in Section 2.3.

2. Protocols

2.1. *Growth rooms*

Four independent, computer-controlled growth rooms provide incubation space for 160 000 containers. Sub-floor cooling fan coils supply a vigorous vertical air flow up through mesh shelves and out ceiling vents. Computer-controlled dampers vent this air to waste or recirculate it, depending on the heating or cooling demand. The vertical cool air flow minimises condensation on container lids, which is a potential source of loss through vitrification and contamination. Aeration has been demonstrated to be important and reduces ethylene and other gas concentrations. Plants are grown under cool white fluorescent lighting with a 16 h, 40 $\mu E\ m^{-2}5^{-1}$ photoperiod, and 25/20 °C day/night temperatures. Culture pots are stacked for space efficiency without significant growth loss.

2.2. *Media preparation*

Large volume requirements of each media type allows bulk processing. We have had custom built, a continuous flow sterilizer with the capacity to sterilize and dispense 300 litres of media into pre-sterilized pots in six to seven hours. This includes one hour of set up time, one hour of cleaning down and four to five hours of dispensing at a rate of approximately 1500 pots per hour. Labour costs are $< 3^c$ per media unit. Polyethylene food containers are gamma ray sterilized and bulk packed prior to use. Heat and time parameters were established from Patashnic's [9] principles of sterilization, and have achieved the heat curve below (Figure 1). In short, eight minutes at 121° C is sufficient for sterilization, however, we use a 10 minute treatment time which allows for minor temperature and flow rate fluctuations.

As with UHT (ultra high temperature) treatments there is very little heat damage to the medium. The pH after sterilization shows no drop, maintaining at 5.7; whereas during conventional autoclaving, it falls to 5.4 ± 0.1. We have seen an improvement of growth and, in particular, *in vitro* rooting with medium sterilized by this method. Media storage is on large trolleys (1200 pots each) which, after a seven day holding period, are transported to the transfer rooms for use.

2.3. *Culture protocols*

Methods of Aitken-Christie, Horgan, Smith and Thorpe [1,3,4,11,12] were put in place in 1984 and formed the basis for operations. Over the years extensive adaptations have been made in order to reduce costs, optimize

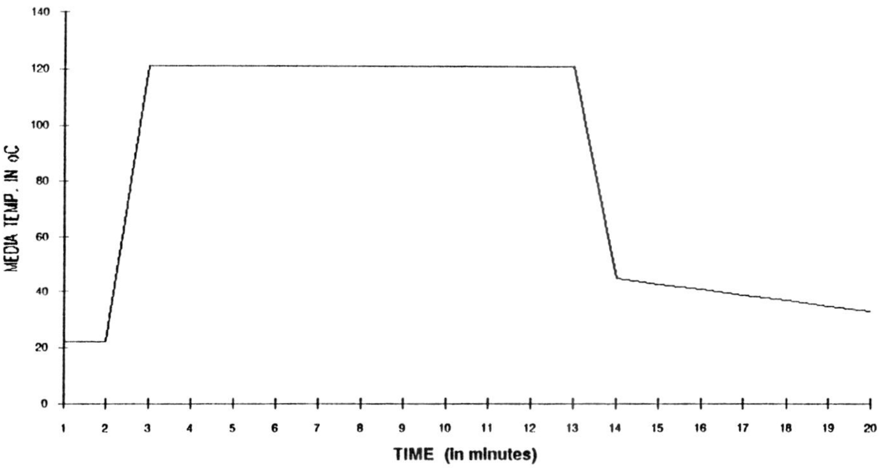

Figure 1. Temperature curve of T.F.L. sterilizing unit.

growth, increase multiplication, improve rooting and expand the range of genotypes being represented in culture.

2.3.1. *Mineral salt solutions*

Modified Quoirin and le Poivre's [10] (QLP) medium forms the base of growth media, though we have found satisfactory growth to be maintained on Gresshoff and Doy medium [13] and McGown and Lloyd's [5] Woody Plant medium. Murashige and Scoog [6] medium gives undesirable results – yielding shorter, yellow shoots, secondary needles and more vitrification. For *some* elements it would appear absolute levels are not critical as manipulations to the base minerals in order of double, half or even quarter strength have had little effect on growth rates.

Boron levels are more critical. Foliar levels of >70 ppm are believed to be almost certainly toxic. (G. Will, personal communication). Foliar analysis of shoots exhibiting extensive dieback have revealed boron levels of 96, 80 and 84 μg/g. Assay of three lots of Difco Bacto agar showed levels of 167, 122 and 70 μg/g dried weight.

Thus:	QLP boron supplied as H_3BO_3	1.09 mg/l^{-1}
plus	Agar boron (8g) at 167 μg/g DW	1.10 mg/l^{-1}
	Total boron:	**2.19 mg/l^{-1}**

In some cases there was *more* boron being added by the agar than the mineral salt solution, and this may have contributed in part to the powerful inhibitory effects of agar (see gelling agents below). Our experience shows foliar levels of boron, with time, mirror the levels of the medium they are grown in (unpublished data). QLP stock boron has been reduced by 25%, and the appropriate selection of gelling agents has seen the foliar boron levels fall to 30 μg/g.

2.3.2. *Gelling agents*

Above *all* media components, gelling agents have exerted the most powerful controlling influence on successful micropropagation. No commercially available source of agar sampled to date would support the growth and viability, long term, of *P. radiata* tissues, without the addition of activated charcoal. Substitution of agar with Gelrite removed the toxicity but resulted in vitrification, so a series of blends were devised to minimise toxicity and vitrification losses [7]. The restrictions of this management practice have finally been overcome in 1991 with the refinement of a non-gelling, non-toxic antivitrifying polysaccharide (AVF).

Improvements are seen in all facets of micropropagation using the AVF: i.e. more genotypes represented in culture, less vitrification in culture stock generally, higher multiplication rates, e.g. some poor performing clones showed a 4 fold growth improvement, less ‘cytokin associated’ death: it also opens the avenue of liquid culture. (Meristematic modules and elongating shoots have now grown in a liquid medium for eight months without vitrification.) A patent is being sought for this product (NZ Application No. 242308).

Overcoming the gel agent induced inhibition, along with the ‘stem section’ method of multiplication (see below) are the two main developments which allow the large numbers per clone required in the Tasman Forestry program.

2.3.3. *Multiplication methods*

A base stock of a few dozen to a few hundred adventitious cotyledonary shoots per clone were established by initial exposure to benzyl adenine (BA) and subsequent outgrowth. Secondary multiplication by topping and growing on the bases proved to be slow and unreliable, presenting a major bottleneck to producing the very large numbers required per clone. During 1988 a rapid multiplication ‘stem section’ method was developed at Tasman Forestry [8]. It become clear that apical dominance was suppressing the flushing of the numerous axillary meristems along the shoot stem. By isolating meristems from one another, through cutting small (4 mm) ‘stem sections’, the multiplication from a single shoot tripled. The gel agent in the culture medium was adjusted to compensate for the added vitrification encountered by using tiny explants. Where previously a 12 mm shoot base yield 2.5 shoots, if cut into 3 × 4 mm pieces it yielded 7.5 shoots in the same time.

The use of BA has been suspended as a production tool for the last few years as losses in some families were unacceptable. The death appeared to be a result of the vitrification/toxicity/agar complex, which is exacerbated by the use of cytokinins. The advent of this AVF re-opens this avenue and work continues to seek the best option.

FAMILY	0-48 SHOOTS	48-80 SHOOTS	81-96 SHOOTS	>96 SHOOTS	TOTAL % > 48 SHOOTS	SHOOT NO. FOR HIGHEST CLONE
1	87	2	1	10	13	304
2	80	4	4	12	20	304
3	89	0	3	8	11	384
4	94	1	1	4	6	143
5	93	3	0	4	7	110
6	95	1	1	3	5	150
7	93	2	3	2	7	81
8	100	0	0	0	0	48
9	91	0	2	7	9	176
10	96	0	1	3	4	128
11	99	0	1	0	1	96
12	98	1	1	0	2	96
13	99	1	0	0	1	80
14	97	2	0	1	3	464
15	100	0	0	0	0	48
16	99	1	0	0	1	64
MEAN	94	1	1	3	6	N/A

EXPRESSED AS % OF CLONES FALLING INTO EACH BRACKET OF SHOOT NUMBERS AT 12 MONTHS.

NB: % DOES NOT ADD EXACTLY BECAUSE OF ROUNDING

Figure 2. Shoot counts per family with BA in initiation stage (a).

The following charts compare the number of shoots yielded per clone after 12 months of culture by (a) cotyledonary explants on BA and outgrown (Figure 2), and (b) simple germination followed by 'stem section' axillary bud multiplication (Figure 3). (These results were collated prior to the introduction of the AVF and are therefore no longer representative of newer methods. Information on new clones using AVF is not yet available.)

Clones which yield fewer than 48 shoots within 12 months are culled from the program as commercially unattractive. This cull rate has reduced from 94% using BA, to 72% without BA. Likewise, the use of BA resulted in 6 of 16 families failing to yield more than 2% commercially useful clones, compared with nil without BA. Thus family representation in tissue culture has risen from 62% to 100%, and representation overall from 6% to 28%. (Using AVF representation looks to be in the order of 65%). The average shoot number for the 'acceptable' clones by the axillary method is 90 each at 12 months. High performance clones have been found in all families, but with differing frequency.

The establishment path for 32 clones for an annual rootable shoot yield of 2.5 million plantlets is shown in Figure 4.

FAMILY	0-48 SHOOTS	48-80 SHOOTS	81-96 SHOOTS	>96 SHOOTS	TOTAL % > 48 SHOOTS	SHOOT NO. FOR HIGHEST CLONE
1	84	11	2	3	16	152
2	86	3	3	8	14	192
3	78	11	2	9	22	192
4	69	17	4	10	31	144
5	53	26	5	16	47	224
6	97	3	0	0	3	80
7	65	17	2	16	35	416
8	47	28	10	15	53	336
9	78	15	1	6	22	208
10	93	2	2	3	7	176
11	64	18	4	14	36	144
12	79	8	3	10	21	192
13	96	0	1	3	4	176
14	37	26	4	33	63	608
15	48	12	16	24	52	320
16	82	11	3	4	18	160
MEAN	72	13	4	11	28	N/A
EXPRESSED AS % OF CLONES FALLING INTO EACH BRACKET OF SHOOT NUMBERS AT 12 MONTHS.						

Figure 3. Shoot counts per family without BA in initiation stage (b) but multiplied by axillary stem sections.

2.3.4. *Acclimatization*

Material for rooting is collected 10–15 mm in height and is elongated for rooting simultaneously with root induction over a five week *in vitro* transfer period. The resulting 30 mm plantlets with multiple root initials are slowly acclimatized in the glasshouse, with six gradually reducing stages of protection until a rooted, 45 mm plantlet is ready for growing on in the nursery.

Step 1: The lids are removed in the first glasshouse environment, but left in the gelled medium for three days under a wet sheet prior to washing and pricking out.

Step 2: Plantlets are pricked out under mist into a potting mix of 50% peat and 50% pumice with slow release fertilizer added. The mixture is drenched with fungicide prior to use.

Step 3: Plantlets are kept in the first acclimatization environment, fogged to 95% humidity at 23° C for 5–10 days.

Steps 4–6: The mobile trolleys carrying the shoots are advanced through a flexible barrier into the second bay which has fogging to 90% humidity, and the third at 80%. From this stage, 21 days onwards, the plants

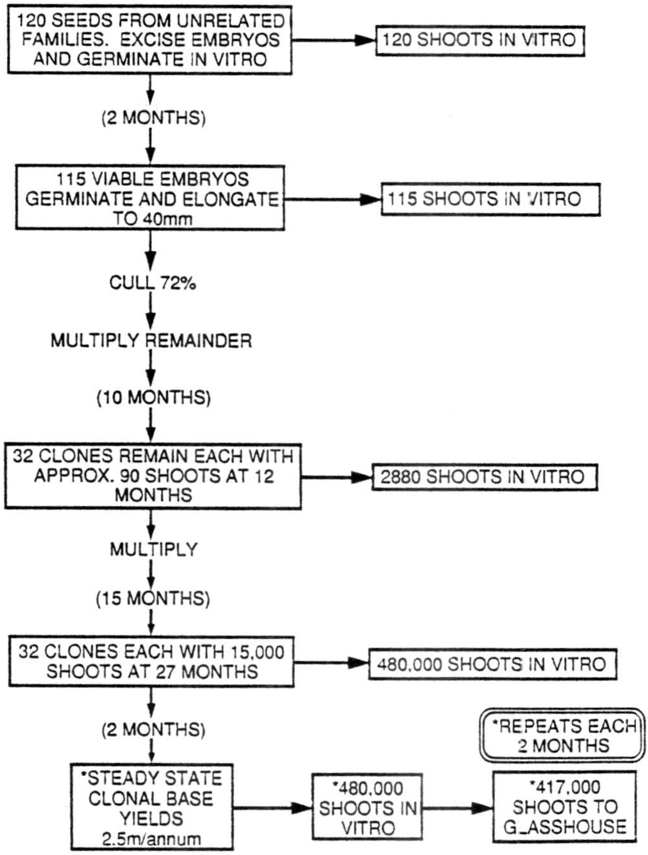

Figure 4. Establishment of clonal base stock (to yield 2.5 million shoots/annum).

are irrigated and liquid fed by an overhead automatic system. The final glasshouse stage is a shadehouse only, for final hardening before being lined out under shade in the nursery beds. Forest establishment takes place over four winter months.

Ex vitro rooting achieved a range of responses across clones from 5% to 85% with a mean of 45%. Shoot survival was 95% or greater, but shoots which had not rooted by 10 weeks seldom subsequently rooted.

In vitro rooting achieves a range from 55% to 100% with a mean of 88%. Though *in vitro* rooting of single shoots is thought to be expensive, in fact the final elongation from 15–30 mm happens simultaneously with root induction, resulting in a superior plantlet in the same time period.

Seasonal changes see a small drop in viability over winter months; and the period for adequate top growth can extend from 6–10 weeks.

2.3.5. *Cool storage*

Representatives of all clones in the testing programme and under production are held semi-dormant under cool storage. Root establishment and field testing takes approximately five years. Currently we use a combination of cool temperature and nutrient deprivation to perpetuate a very slow growth phase, refreshing the medium every 18 months. Stock under storage for three years has maintained good health and has resumed growth on re-introduction to the growth rooms. We encounter a 10% loss due to contamination and mortality under long term storage.

3. Costs

Differences in clonal performance with regard to multiplication rates and root initiation, exert powerful effects on cost. Certain families and clones are more costly to represent but are maintained to ensure the genetic diversity of the clonal selection. At a coarse level, however, efficiency screening removes approximately 70% of all clones sampled. This proportion promises to be changed radically by new protocols, alluded to in Section 2. Transfer labour in the laminar flow cabinets is the major laboratory cost at approximately 65% of direct costs. Shoot transfer rates of 2500 per person/day compare favourably with many other commercial operations and may leave little room for improvement with traditional methods. Incentive bonus schemes are in place and have resulted in a 6% labour cost reduction.

The greatest potential cost savings include:
(i) automation of sterile transfers;
(ii) raising *in vitro* rooting of all clones to 100%;
(iii) further improvement to multiplication rates;
(iv) meristematic liquid culture;
(v) improved glasshouse protocols;
(vi) cryopreservation of genetic material.

Some indication of comparative cost trends is given below (Figure 5) where seedlings and tissue culture plantlets are shown. (Tissue culture plantlets now cost less than the market price of field cuttings).

4. Future aspects

4.1. *Clonal screening*

Early screening methods will be invaluable for cost savings. Currently the early performance data of clones which prove to grow well in the field are being examined for early indicators of good performance. Criteria being studied include, *in vitro* multiplication rates, shoot dry weights, plant form, accli-

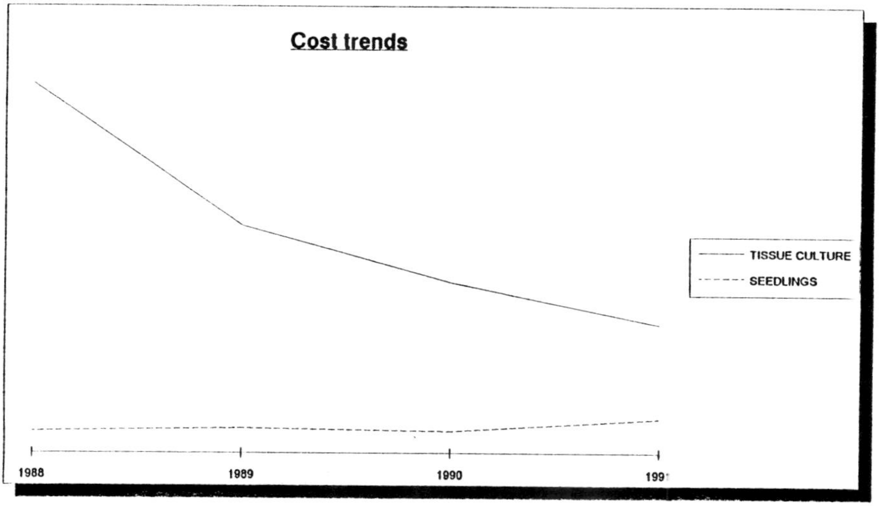

Figure 5. Cost trends 1988–1991.

matization success, and root number and mass. It is apparent, while handling clones, that their response to different environmental and nutritional treatments vary enormously. The opportunity exists to screen clones in a glasshouse or nursery situation to a variety of soil mineral conditions, with a view to planting tolerant and adaptive clones to a given forest site.

4.2. *Meristematic tissue*

Whilst we find axillary multiplication to be the most productive route at the moment, the multiplication and subsequent elongation of meristematic tissue [2] promises great savings. Work will continue to optimise this method to represent a greater proportion of genotypes, speed outgrowth rates and improve shoot types. Conversion to a liquid based system will bypass expensive manual multiplication stages. This method may also provide the vehicle for genetic manipulation of clonal tissues.

5. Summary

Regimes to produce 2.0 million plantlets annually from 32 clones are discussed, highlighting factors which have had a major impact on productivity. It is apparent that further improvements over the current method are possible with the development of a liquid meristematic tissue system for initiation, genetic engineering and possibly cryopreservation, but also over more traditional methods by further increasing growth rates and improving acclimatization success.

Acknowledgements

Judith Redington and Dianne Patangata are thanked for their technical assistance.

References

1. Aitken J, Horgan KJ, and Thorpe TH (1981) Influence of explant selection on the shoot forming capacity of juvenile tissues *Pinus radiata*. Can J For Res 11:112–117.
2. Aitken-Christie J, Davies H, and Singh AP (1988) Multiplication of meristematic tissue: A new tissue culture system for radiata pine. In: Hanover JW and Keathley DE (Eds.) Genetic Manipulation of Woody Plants.
3. Aitken-Christie J and Thorpe TA (1984) Clonal propagation. Gymnosperms. In: Vasil IK (Ed.) Cell Culture and Somatic Cell Genetics of Plants. Academic Press, Florida, Vol. 1. pp. 82–95.
4. Horgan K and Aitken J (1981) Reliable plantlet formation from embryos and seedling shoot tips of radiata pine. Physiol Plant 53:170–175.
5. Lloyd G and McCown BH (1981) Commercially feasible micropropagation on mountain laurel (Kalmia latiflora) by use of shoot tip culture. Proc Int Plant Propagators Soc 30:421–427.
6. Murashige T and Skoog F (1962) A revised medium for rapid growth tobacco tissue cultures. Physiol Plant 15:473–497.
7. Nairn BJ (1988) Significance of gelling agents in a production tissue culture laboratory. IPPS Combined Proceedings. Vol 37:200–205.
8. Nairn BJ (1990) gel agent manipulation and the rapid multiplication of micropropagated *Pinus radiata* via stem sections. Oral Paper. Abstract ICBWG Meeting 1990 in Sittingboom (submitted for publication).
9. Patashnic M (1953) A simplified procedure for thermal process evaluation. Food Technol 7:1–6.
10. Quoirin M and Lepoivre P (1977) Etudes de milieux adaptes aux cultures *in vitro* de *Prunus*. Acta Hort 78:437–442.
11. Reilly K and Washer J (1977) Vegetative propagation of radiata pine by tissue culture: Plantlet formation from embryonic tissue. NZ J For Sci 7:199–206.
12. Smith DR, Horgan KJ, and Aitken-Christie J (1982) Micropropagation of *Pinus radiata* for afforestation. Fujiwara A (Ed.() Proc 5th Intl Cong Plant Tissue and Cell Culture. pp. 723–724.
13. Sommer HE, Brown CL, and Kormanik PP (1975) Differentiation of plantlets in long leaf pine (*Pinus palustris* Mill) tissue cultured *in vitro*. Bot Gaz 136:196–200.

24. Micropropagation of palms

P. S. RAO and T. R. GANAPATHI
Plant Biotechnology Section, Bhabha Atomic Research Centre, Trombay, Bombay 400 085, India

1. Coconut (*Cocos nucifera* L.)

Since time immemorial, palms have provided the people of tropical and subtropical areas with food, drink, fuel and shelter. The coconut palm in particular is popularly known as 'Tree of life' or 'Kalpavriksha'. They are extensively grown in the tropics and millions of people depend on this crop for their livelihood.

1.1. *Distribution*

India is one of the largest coconut producing countries of the World besides Philippines and Indonesia. In India it is cultivated in an area of about 1.2 million hectares with an annual production of 5871 million nuts [99]. Coconut is also cultivated in Sri Lanka, Thailand, Malaysia, Papua New Guinea, Fiji, Zanzibar, Trinidad and Tobago, and Western Samoa. The fruits yield a valuable oil, which is used as margarine, cooking oil and in the manufacture of soaps, detergents, cosmetics, pharmaceutical products and also as a fuel. The productivity of the coconut tree is generally low, despite the fact that many efforts have been made to improve its yield potential. The long life span and heterozygosity seem to be the limiting factor in coconut improvement through breeding. Thus production of inbred lines for this purpose is difficult and time consuming. Unconventional methods must therefore be evolved as a means of shortening the inbreeding cycle [100].

1.2. *Conventional method of propagation*

Coconut can be propagated naturally only from seed. Since the palms do not breed 'true to type', the offspring differ greatly in vigour, productivity and resistance to disease. These traits are often not expressed until the palms are 10 years old, and the commercial life of a coconut palm may be as long as 60–70 years. Therefore selecting the parent palms and breeding new hybrids can take decades. The variable quality of palms obtained from seeds could therefore affect many smallholders throughout the tropics by reduction in

M. R. Ahuja (Ed.), Micropropagation of Woody Plants, 395–421.

yield through diseases etc. [13]. Several attempts at the vegetative propagation of coconut have been made by airlayering of tall trees [26]. The rooted upper portion can be cut and replanted but it does not help in the multiplication of palm. The spadixes occasionally revert from generative to vegetative phase resulting into bulbil shoots. Although these can be rooted by air layering and planted, such mutant palms produce only bulbils and no nuts [95]. There is a great deal of variability in seed nuts. Some plantations may contain palms producing as few as ten nuts a year whilst others produce over a hundred. The average does not exceed 30–40 nuts per palm per year, whereas individual palms have been recorded which produce upto 400 nuts per year [44]. Therefore, to produce identical clones in large numbers from such high yielding palms, the technique of tissue culture is the best alternative.

1.3. In vitro *studies*

Advances in coconut propagation through tissue culture have been relatively slow. The progress of the work conducted so far has been reviewed [11,12,101]. Earlier investigations on coconut tissue culture were based on the assumption that the coconut would respond in a similar manner to that of oil palm: i.e., that a slow-growing callus would be produced which could be multiplied and induced to differentiate somatic embryos, eventually to produce plantlets. However, such expectations have not been fulfilled and it has proved difficult to obtain plantlets in coconut tissue cultures and the production of callus and organogenesis in callus is somewhat different from oil palm. It has been observed that coconut tissue cultures do show a tendency towards somatic embryogenesis; however, they do not multiply further and generally follow an unbalanced growth resulting in occasional production of shoots and rarely a complete plantlet [14]. To date there is only one documented example of survival and establishment in soil of tissue culture regenerated coconut plant [91]. This was accomplished in Unilever Labs and was planted in the Solomon Islands in 1984. Germination *in vitro* of zygotic embryos has been possible. These are slow growing and require a long period of nursing before they can be established in soil. In general, the process of germination tends to be one year behind in their development, compared with the direct germination of embryos from the nuts. In view of the considerable time and cost involved in raising and establishing coconut plants through zygotic embryo culture it is unlikely to become an inexpensive large scale process in comparison with oil palm, embryo culture would useful in the propagation of elite parent palms for hybrid seed production.

1.3.1. *Types of explants used*

Branton and Blake [13,15] used primarily inflorescence tissue, whereas Pannetier and Buffard-Morel [68] used leaf tissue. Other types of explants such as secondary or tertiary roots, immature flowers, spear leaves, apical meristem, leaves and leaf bases, male flowers and anthers, rachilla, zygotic embryo and endosperm have also been used to initiate the cultures by various investigators [70,4].

1.3.2. *Mode of regeneration*

Regeneration of coconut plants reported so far is through the development of somatic embryos. No reports are available on the differentiation of shoot buds either directly or through callus. In the majority of the investigations, the initial explants proliferate to form a nodular callus. Nyman *et al.* [54] suggested the term 'calloid' for the callus they observed; it is more organised than a callus and appears to resemble orchid protocorm initiation. In general, these nodules produces only roots [69,38]. However, Blake [12] reported the formation of embryo-like bodies in the calloid structures under certain conditions, and observed that these may develop comparably with a zygotic embryo. However, a variety of embryo-like and abnormal structures are also frequently observed [69] and these include leaf-like structures, often with leaves fused, as well as the development of haustorial type tissue, with or without roots. Interestingly, direct differentiation of embryos from leaf cultures of coconut has been reported by Raju *et al.* [81]. However, details about the protocol used are not given and no further details have been published. In general, the protocol for embryo and plantlet development has not yet been defined, but Blake [12] reported in the recent review that the gradual lowering of 2,4-D level is associated with embryo production. Complete elimination of auxin results in a haustorial type of development with or without roots. The role of cytokinins has not been determined but it appears that a moderate level is usually incorporated throughout the culture period.

Coconut embryo culture offers enough scope for basic and application oriented research and would be especially advantageous in the cultivation of hybrid embryos to maturity. Transportation of coconut germplasm by embryos than nuts would be easier, relatively inexpensive and safer. The culture of zygotic embryos and their establishment in soil should provide information which can be related to establishment of clonal plants.

1.3.3. *Zygotic embryo culture*

Extensive studies on zygotic embryo culture using the Makapuno variety has originated from de Guzman's laboratory in the Phillipines. In recent times several laboratories, including those in India, have also published work on the culture of zygotic embryos of other varieties [10,30,38,45,66]. While normal development of the zygotic embryos can be achieved, no laboratory has reported consistent success through all the stages from excision of the embryo to establishment of plants in the nursery [12]. Several investigators have obtained proliferation of the callus or calloid on zygotic embryos, but the results are not consistent.

Considerable variation in success with the culturing of zygotic embryos has been reported and a wide variety of media and additives have been used. Gupta *et al.* [38] harvested mature and immature embryos of 'West Coast Tall', which were cultured on Eeuwens [35] Y_3 liquid medium with variations in auxins, cytokinins and coconut water. They obtained normal germination but without root elongation. Such cultures upon transfer to Y_3 medium without additives showed elongation of roots and some produced plantlets which could be transplanted to soil : sand : compost (3 : 3 : 1) mixture. Immature embryos could not be cultured through callus to the plantlet stage but 60% of the embryos of 7 × 4 mm size yielded plantlets. However, none of the plantlets survived after 4 weeks. Padmanabhan [66] successfully cultured embryos of West Coast Tall on Y_3 basal medium with NAA, BA and GA_3 and ascorbic acid. The author, however, has not reported the establishment of plantlet in soil. Iyer [44] reported the use of MS medium for successful embryo culture but failed to obtain successful establishment. Pannetier and Buffard-Morel [69] quote unpublished results of some success with embryos of the hybrid PB121 (Malayan Yellow Dwarf × West African Tall). Dwarf coconut embryos have been cultured by Karunaratne and his coworkers [52].

The most comprehensive published work on embryo culture has been carried out by Assy Bah [4], who used mature embryos (11–12 months after fertilization) of 20 varieties. The basic medium was MS and germination was obtained only in the presence of agar and activated charcoal. Germination was often slow, especially on liquid medium or agar medium without charcoal. Shoot development occurred first and some development of the haustorium was observed with both ages of embryo. Root development was obtained by adding 20 mg/1 of NAA. After 7 months in culture, the plants were successfully established in soil. All varieties which were tested showed germination but embryo from Tall varieties were slower in development than Dwarf varieties. Blake [12] reported Watanayathin's unpublished data on the embryo culture of Malayan Dwarf, Thai Dwarf and Malayan Dwarf × East African Tall. In their study initial growth during the first few weeks was often good but later it was difficult to maintain growth, especially that of roots. Further they have observed a few embryos in each batch formed good plantlets in less than 6 months. The medium used for the culture of embryos was MS or Y_3 [14] either liquid or with agar at 0.5% and 4–12% of sucrose. Growth adjuvants used included auxins, cytokinins, gibberlic acid, coconut water, hydrolysed casein and aminoacids like glutamine, arginine and aspargine, potassium chloride and activated charcoal. Establishment of plantlets derived from embryo is feasible in glass house in a mixture of coir, sand and loam if the plantlets had two good leaves and more than 15 lateral roots.

Recently the optimum conditions for the *in vitro* culture of zygotic embryos has been standardised [5]. Conversion of embryos into plantlets was obtained on a single medium with the removal of haustorium after 3 months. Further the authors have reported that 73 percent of the plantlets obtained by their method has been transferred to natural conditions; the survival rate after

6 months was 92 percent and the leaf emission rate was comparable to seed nut germination (5 leaves in 6 months).

The culture of makapuno embryos is an example of the technique of embryo rescue [86]. The makapuno variety of the Philippines has a jelly endosperm instead of the normal endosperm and solid copra and water. Although a normal looking embryo is present it never germinates.

1.3.4. *Regeneration of plantlets through zygotic embryo callus*

The earliest record on the successful production of callus was that of de Guzman *et al.* [40]. The use of 2,4-D in the culture medium gave a variety of development from the cotyledonary sheath tissue which included nodular callus, root like structures and a spongy tissue which was suggestive of haustorial tissue. The authors suggested that the callus nodules resembled the protocorms of orchids.

Jagadeesan and Padmanabhan [50] cultured sliced sections of the twenty-day-old germinated embryos on Y_3 medium. The portion containing the cotyledonary sheath was the only one to survive and in the presence of NAA and BAP produced callus which formed roots and pneumatophores. Tahardi [96] also used Y_3 medium with activated charcoal and 2,4-D or 2,4,5-T and sliced mature embryos, before culturing. Nodular proliferations were obtained which were thought to be similar to orchid protocorms as has been observed by D'Souza [29]. Such proliferations when transferred to a medium with reduced auxin level, callus or further nodules resembling shoots were produced. Bhalla Sarin *et al.* [9] cultured immature embryos (8–10 months from pollination) of West Coast Tall variety, on Gamborg's B_5 medium with a variety of additions. They obtained a white, compact nodular callus, which produced a few plants.

1.3.5. *Culture of anther, endosperm and protoplasts*

Naturally occurring haploid embryos in coconut palm has been reported with authentic evidence of its haploid chromosome number [59]. Thanh-Tuyen and de Guzman [100] and Monfort [57] reported the formation of pollen embryos from cultured anthers. Iyer [44] was able to obtain multicelled proembryoids within the exine of coconut microspores in cultured anthers. Than-Tuyen and de Guzman [100] also obtained the various stages in the development of pollen embryos. Similarly the emergence of an embryo from cultured anthers has been described by Monfort [57]. Although a few reports are available on the development of pollen embryos, the regeneration of plantlets still remains a challenge. More recently Verdeil and Pannetier [114] studied the optimum conditions for *in vitro* germination of pollen grains and a defined medium for pollen germination was reported. Prakash Kumar et al. [70] obtained callusing from cultured endosperm of coconut on Eeuwens Y_3 [35] medium with a higher level of 2,4-D. Microcallus has been obtained from the protoplasts derived from tender tissue from the rachillae enclosed in the protective sheaths of inflorescences [41].

1.4. *Conclusions*

One of the most serious problems encountered in coconut tissue culture is the variation in culture responses from tree to tree or batch of collections. It has been observed that the desired growth proliferation is not always obtained in different batches of culture which have been subjected to exactly similar cultural conditions [39]. Hence further investigations should be directed towards the understanding of the factors responsible for the uniform growth, differentiation and the regeneration of plantlets.

Despite the difficulties, research on tissue culture of coconut should be augmented and it is desirable that more number of laboratories should get involved. If successful and reproducible protocols are established, coconut palms with outstanding traits could be propagated. Multiplication of productive clones as for example from trees bearing more than 400 nuts per annum would have a significant impact on coconut oil and copra yield. It would also have a bearing on the husks and fibres around the nuts which constitute an underexploited source of energy. Furthermore around fifty years are required to select a productive hybrid though conventional breeding programmes compared to a few years of sustained efforts to isolate and produce palms through *in vitro* techniques. In addition embryos of interesting hybrids may be isolated and preserved rather than whole nuts, and plants could be retrived by embryo culture technique.

2. Date palm (*Phoenix dactylifera* L.)

The Date palm (*Phoenix dactylifera* L.) belongs to the Palmaceae. There are 12 species in the genus *Phoenix*. Among them *P. dactylifera* L. and *P. sylvestris* Roxb. (Wild date palm) are important. The former date palm is cultivated for its edible fruit and the latter is cultivated in India as a source of sugar. Although date palms are primarily cultivated for their fruits, many other parts of the palm can be used for various purposes.

2.1. *Distribution*

The date palm is subtropical in origin and cultivation. Records dating back to 3000 B.C. suggest Mesopotomia (Iraq) as possible site of origin of this crop [16]. Currently date palms are widely distributed throughout North Africa, the Middle East, Central America and parts of South America. Approximately 90 million date palms are in existence and about 6 million palms per year are required for replanting [16].

2.2. *Conventional method of propagation*

Date palm can be propagated through the seed. However, since the date palm is dioecious approximately one half of the progeny will be males and as it is heterozygous there will be lot of variation within the progeny and even some of the desirable characters may be lost. Date palm can be vegetatively propagated through the offshoots, but the number of offshoots produced from mother palm is limited [76]. In a normal life span of about 100 years a date palm produces anywhere between 30–40 shoots. Consequently, the clonal multiplication rate is slow and the cost of production is high. Furthermore, a serious threat to date palm plantations is a disease called 'bayoud' caused by *Fusarium oxysporum*. More than 10 million date palms have been wiped out by this disease during the last century and in Morocco about 5 million trees are threatened with extinction by the spread of this disease [88]. To overcome all these problems micropropagation through tissue culture techniques is the only way of multiplying this species. Elite palms which are high yielding and resistant to fungal disease would form the donor stock for multiplication.

2.3. In vitro *studies*

In vitro studies on date palm have been carried out in many laboratories and these include research groups working in Algeria, France, Morocco and Tunisia. In India work on date palm has been carried out at Hissar [90]. For details see reviews [108,17,16]. Many reports are available on successful production of plantlets and suggests that date palm cloning is feasible. Some of the commercial tissue culture laboratories have already adopted this technique for large scale production of elite palms [93]. *In vitro* techniques will not only provide clonal material in large numbers but it can also be used as a supplement to plant breeding programmes. Although many reports are available on *in vitro* studies on date palm leading to regeneration of plantlets, only a few published reports are available on the establishment of plantlets in soil (Table 1).

2.3.1. *Types of explants used*

A variety of explants have been employed successfully and these include exised zygotic embryos, rapidly dividing meristematic tissues such as meristems, shoot tips and axillary buds and stems, roots and leaves from juvenile material. From mature palms, however, only inflorscences have been used.

2.3.2. *Initiation of cultures*

Cultures were initiated from various explants after sterilization with mercuric

Table 1. Present status of Date palm tissue culture*.

Variety	Source of explant	Type of development	Author*
Deglet Noar Desert Nugget Medjoot, Honey	Embryo, seedling young leaf	Callus, roots plantlets	Schroeder [89]
Deglet Noar	Cotyledon	Callus, plantlets	Amar & Benbadis [1]
Bou-Feggous Bour-Sthammi	Apical meristem or shoot tip	Axillary shoot, plantlets	Rhiss *et al.* [85]
–	Embryo, stem, axillary meristem apical meristem	Callus, embryoid, plantlets	Tisserat [104]
Medjool	Axillary meristem	Callus, embryoid, Axillary shoot development into plantlets	Tisserat [106]
Barhee, Halawy	Mature embryo	Callus, embryoid, plantlets	Mater [56]
Kenta	Axillary meristem	Adventitious buds, plantlets	Drira [27]
Bou Feggous Bou Skri Bou Stammi Noir cvs.	Axillary meristem	Development of plantlets	Beauchesne [7]
Deglet Noor	Embryo	Callus, floral plantlet development	Ammar & Benbadis [2]
–	Apical meristem or shoot tip	Axillary shoot plantlets	Tisserat [107]
Thoorly Khalasaa	Apical meristem or shoot tip	Callus, embryoid plantlets	Gabr & Tisserat [36]
Khadrawy	Axillary shoot	Callus, embryoid plantlets	Kackar *et al.* [51]

* Only reports of field establishment of *in vitro* plants are given (Adapted from Branton and Blake, 1989).

chloride [90,51] or sodium hypochlorite [107]. A quick dip in alcohol has often been used in addition to the treatment with a sterilant [16]. To check the browning of the explants vaious antioxidants like citric acid and ascorbic acid have been used [121]. Sharma *et al.* [90] used a solution of ascorbic acid (22mg/1) and cysteine (25mg/1). Kackar *et al.* [51] used citric acid (100mg/1) and then the explants were treated with carbendazim solution for five minutes before treating the explants with mercuric chloride.

2.3.3. *Media composition*

The majority of investigators have used modified versions of Murashige and Skoog's [58] medium with agar, activated charcoal and slightly higher level of phosphate. For the initiation of callus 2,4-D and 2-ip have been used as adjuvants to modified MS medium [84,104]. The differentiation of embryoids from callus has been observed on a medium lacking both the growth regulators. Similar medium has also been used by Sharma *et al.* [90], but with increased levels of phosphate and thiamine and instead of 2-ip, BA was incorporated to obtain somatic embryogenesis and plantlet development. In general the growth regulator combinations used are auxins and cytokinins, such as IAA, NAA or IBA and cytokinins- KN, BA or 2-ip. In most of the cases for the initiation of callus a higher level of auxin, usually 2,4.D, with a cytokinin, either 2-ip or BA, has been used, and for the regeneration the callus was transferred to hormone-free medium. Callus induction was achieved within just two weeks on a modified MS medium with 2,4-D and BA [51]. For the development of plantlets normally auxins alone (IAA, NAA or IBA) or along with low concentrations of cytokinins has been found to be effective.

2.3.4. *Mode of regeneration of plantlets*

The regeneration of plantlets in date palm can be broadly categorised into three types:
1. regeneration from callus;
2. regeneration from axillary buds;
3. regeneration from inflorscence tissue.

Regeneration from callus is usually through the somatic embryogenesis pathway [51,90,108,110]. There has been no report on shoot morphogenesis from callus. Somatic embryos obtained from callus developed into normal plantlets. Continuous production of plantlets is possible through multiplication of embryogenic tissue through process of secondary embryogenesis.

The production of shoots from axillary bud cultures may be achieved by proliferation of axillary buds or by adventitious bud development, currently explored as a type of micropropagation. Histological studies carried out on young plantlets derived from tissue cultures by Tisserat and De Masson [111,112] have revealed that the shoots were derived from a common shoot axis via lateral bud proliferations. The authors, however, do not make it clear whether any of the media constituents have a controlling effect on axillary bud proliferation. They have also reported precocious inflorscence production on lateral buds developed on excised zygotic embryos in culture which they attributed to the presence of higher levels of cytokinins in the medium. Similarly, Ammar *et al.* [3] have reported precocious flowering after five months in culture on MS medium with BAP, IAA and glucose.

Shoot production via inflorscence tissue through reversal of the generative meristem to a vegetative form has been reported [83]. Reversion of floral primordia invloved organisation of leaf like structures from the interior of the flowers as well as from the axils of the sepals [28]. The stage of development of inflorscence at the time of culture was important and it appears that portions of inflorescence (0.8–4.5 cm long) with young flowers showing stages from sepal to carpel initiation were most successful. When shoots were produced amongst the carpels, there seemed to be a proliferation comparable to the production of shoots at the base of adult plants. Plantlets were produced in nine months from the time of taking the explants [16].

2.3.5. *Establishment of regenerated plantlets soil*

Once plantlets have been obtained *in vitro*, whatever their source of origin, the mode of their further development follows a similar pattern. Hence all plantlets can be treated alike for the purpose of producing palms [109].

Establishment of date palm plantlets in soil requires a high humidity levels followed by a gradual lowering of humidity. This is accomplished in high humidity chambers or with fogging systems or mist beds. Kackar *et al* [51] hardened the *in vitro* plantlets by raising the temperature to 32 °C for two weeks before transferring the plantlets to pots and treatment with 0.2% carbendazin and 0.1% streptocyclin for 30 seconds to avoid fungal and bacterial infections. The soil medium consisted of peat moss and vermiculite 1 : 1 ratio, although a variety of other soil mixtures have been employed without any notable difference in growth responses [105]. The plants are initially covered with a polythene bag and incubated in an environmentally controlled chamber for about two weeks and then transferred to a shaded portion of a green house. The plantlets are gradually acclimatized to the green house humidity conditions by punching holes in plastic covers. After two months the covers are removed and the plants are treated as normal palm seedlings.

2.4. *Conclusions*

Although many reports are available on the successful regeneration of date palm plantlets, either through somatic embryogenesis or axillary shoot development, the protocol for commercial propagation and the cost factors have to be worked out. No studies have been made, either, on the genetic stability of the tissue cultured plants, large scale field planting and the evaluation of yield, resistance to diseases etc. Elucidating the mechanism of lateral bud differentiation on demand and maximizing plantlet production with minimum labour requirements are the challenging areas for future research on date palm.

3. Oil palm (*Elaeis guineensis* Jacq.)

The oil palm (*Elaeis guineensis* Jacq.) is an extremely important oil yielding crop belonging to the Palmaceae. There are two species in the genus *Elaeis, E. guineensis* (African oil palm) and *E. oleifera* (South American oil palm).

3.1. *Distribution*

E. guineensis, a native of West Africa which has been cultivated in plantations since the beginning of the 20th century, is now grown extensively in Malaysia, Nigeria, Indonesia, Republic of Zaire and Ivory Coast, and to a limited extent in India, Ghana as well as in South and Central America. It was first introduced as an ornamental palm [43].

3.2. *Morphological characteristics of oil palm*

Oil palm plantations are generally set with 143 trees per hectare, the space between each palm being around 9 meters. The young trees are planted after one year of being in a nursery and they start producing fruits two to three years later [61]. Oil palm grows well in localities where the annual temperature varies between 20–35 °C, and a moist deep loamy soil rich in humus with good water permeability suits palm best. The palm has only one growing point. The stem is an erect and fairly uniform column having persistent leaf basis. The crown is compact with a large number of long leaves arranged in a spiral fashion. Oil palm is monoecious, bearing male and female flowers in separate bunches in the same palm. In the axil of every leaf there is a bud which may develop into a male, female or occasionally bisexual inflorescence. The inflorescence is branched and the flowers are arranged spirally around the branches. The portion of male, female or bisexual inflorescences may not show any regularity. The female inflorescence is a spadice and bears anywhere between 500 and 3000 fruits and weighs 50 kg when bearing ripe fruit [61].

3.3. *Palm oil and its uses*

Palm oil is extracted from the mesocarp of the fruit, whilst palm kernel oil is extracted from the seed enclosed in the endocarp of the fruit [103]. Three oil palm varieties have been indentified, based on the fruit structure. They are Dura, Pisifera and Tenera. Dura has a thick shell with low to medium mesocarp content. Pisifera is characterised by a shell-less fruit and pea-like kernel inside: the kernel is also often absent. Tenera is a hybrid form obtained by crossing Dura (female) and Pisifera (male). It has a thin shell and a medium to high mescocarp content of about 60–95%. This is the widely cultivated type all over the world due to the higher mesocarp content and resultant oil outturn. Under favourable environmental conditions the yield of oil varies between 4 and 5 tonnes per hectare per annum and can go up to 6 or 7 tonnes per hectare per annum for selected high yielding palms. Once the bearing sets in it generally lasts for 30 years. Palm oil is used mainly in the manufacture of cooking oil, vanaspati (vegetable ghee), margarine, soaps and detergents, shortenings, confectionary fats and ice cream [8]. Palm kernel oil contains lauric (48%), myristic acid (16%), and resembles coconut oil. It is used in the manufacture of soap, methyl esters, fatty alcohols, confectionary fats, imitation dairy products, margarines and cooking oils [53].

Palm oil has been in use for many years and it is now consumed throughout the World. Palm oil contains mainly 40% oleic acid, 10% linoleic acid, 44% palmitic acid and 5% stearic acid and its fatty acid content is equivalent to human milk [77]. Like any other vegetable oil, palm oil is cholesterol-free and studies conducted in various countries reveal that a palm oil enriched diet did not raise the blood cholesterol. According to Temple and Basu [98] unrefined palm oil contains beta-carotene, which is an anti-cancer agent, and it is also a rich source of vitamin E, tocopherols and tocotrienols.

3.4. *Diseases of oil palm*

Most of the diseases of oil palm are caused by fungal pathogens. The important disease in South Asia and the South Pacific is basal stem rot of mature palms caused by *Ganoderma*, which can result in high losses if left unchecked [113]. In Africa, *Fusarium* or vascular wilt can cause heavy losses of palms in oil palm plantations [82]. The incidence of these diseases can be overcome by the application of relevant fungicides, improved nursery growing techniques and good estate practice.

As the oil palm is an out crossing species, the individual palms are highly heterozygous, and consequently there is considerable variation in the oil yield. In Indonesia under highly favourable environmental conditions, the average oil production per tree per annum was 44.3 kg (6.3 tonnes per hectare per annum) but the individual production varied between 21 kg and 77 kg or between 3 and 11 tonnes per hectare per annum [61]. The sexual breeding cycle in oil palm is very long; as a result the performance and oil yield of the progeny cannot be predicted before 15–20 years. Oil palm is a non-suckering palm and has only one growing point. Conventional methods of vegetative propagation, such as rooting of cuttings, grafting etc., are not possible. Therefore the clonal propagation of the oil palm is advantageous, not only for the multiplication of individuals with exceptional agronomic traits but also for the multiplication of trees resulting from interspecific crosses. Tissue culture has thus provided a means whereby selected high yielding palms from the field can be vegetatively propagated without destroying the mother palm.

This enables the production of uniform plantlets from any selected elite palm. The variation due to genetic differences could be avoided if oil palm was propagated clonally. Moreover it is estimated that higher yields and improved oil quality would be obtained by cloning palms possessing the desired characters. Yield increases of the order of 30% are feasible through cloning [23,117]. Clonal propagation also has the advantage that secondary characteristics such as reduced palm height and superior quality, found in selected individual palms, could be uniformly expressed in all individuals or ramets of the respective clones [73].

3.5. In vitro *studies*

Methods for the large scale clonal propagation of oil palm are now available in several laboratories but due to its commercial application all the details regarding its production have not been published. However the advantages of producing clones of an individual selected palm using tissue cultures have been described by several authors [20,46,60] and the available literature on palm tissue culture has been reviewed [11,19,47,49,71,72,83,118,119]. In the early 1970s, oil palm tissue culture was dominated by two groups working independently – the British [20,46,92] and the French [78–80]. At present, due to its commercial potential, several groups have set up tissue culture laboratories to specifically clone oil palms. About a dozen such laboratories are in existence and about ten are to be found in Malaysia alone [119]. However, the publications from these commercial laboratories are not available.

3.5.1. *Types of explants used to initiate the cultures and the surface sterilization*

The various explants used to initiate cultures include: meristem tissue from 2–6 year old palms [94]; seedling embryos [78]; stem apex and leaf base from 15-month old palms [79]; seedling roots, shoot apex from aseptically germinated embryos [46]; leaf tissues [80]; young leaf sections from 2-year-old palm [54]; roots from mature palms [120]; young leaf sections from nursery and mature palms [42]; roots and young leaf from aseptically grown seedlings and also tertiary roots, inflorescence and young leaves from mature palms [74]; leaf and roots from aseptic seedlings as well as from nursery seedlings of pisifera [63] and young leaves and meristematic leaf bases from 6-month-old palm [102].

The first successful report of regeneration of plantlets from leaf tissues came from the work of Rabechault and Martin in 1976 [80]. Following this there were refinements in the techniques and Corley *et al.* [22] and Hanower and Pannetier [42] obtained similar results. In these experiments the explants from shoot apex region of nursery or adult trees were removed without damaging the apex and used for culture. The various explant sources, the available media composition and the type of development are summarised recently by Wooi [119].

Table 2. Developments and the current status of oil palm tissue culture*

Nature of work	Author*
Leaves, roots and callus nodules from meristematic tissue (from 2–6 year-old-palm)	Staritsky [94]
Nodular cell colonies from seedling embryo	Rabechault *et al.* [78]
Callus, roots and shoots from stem apex and leaf bases excised from 15-month-old palms	Rabechault *et al.* [79]
Callus and suspension cultures from mature palm roots and apical tissues	Smith & Thomas [92]
Callus and embryoids from seedling roots, shoot apex from aseptically germinated embryos	Jones [46]
Callus, embryoids and rooted shoots from leaf tissues	Rabechault & Martin [80]
First report of the establishment of regenerated plantlets in polybag nursery and field planting of the plantlets in Malaysia	Corley *et al.* [20]
Performance of four clones compared	Corley *et al.* [23]
Callus, embryoids and plantlets from roots of mature palms	Wooi *et al.* [120]
First report of isolation of protoplasts	Vouyouklis [115]
Slow growing callus, embryoids and plantlets from leaf sections from nursery and mature palms	Hanower & Pannetier [42]
Callus, embryoids and plantlets from roots from aseptically grown seedlings, and tertiary roots, inflorscence and young leaves from mature palms	Paranjothy & Oathman [74]
Cryopreservation and recovery of embryos	Grout *et al.* [37]
Use of charcoal in media and plants regenerated from pisifera	Nwankwo & Krikorian [63]
Cultures from anthers	Odewale [65]
Oil composition of clonal palms described	Jones [48]
Cell colonies developed from protoplasts	Bass & Hughes [6]
IRHO and ORSTOM concluded two aggrements with Indonesian companies for large scale production of plantlets	Noiret *et al.* [61]
Callus, embryoids, shoots and rooted shoots from young leaves and meristematic leaf bases from 6-month-old palm	Thomas & Rao [102]
Reported abnormal flowering in clonal oil palms	Corley *et al.* [21]
Successful cryopreservation of embryoids	Engelmann & Duval [33]
Unilever has built up a unit at UK for producing 500,000 plantlets a year for export	Sasson [88]
Commercial production of *in vitro* plants was programmed from 1991 onwards	Sasson [88]

* Source Sasson, 1988; Jones & Hughes, 1989; Wooi, 1990. Reference Nos in brackets.

For surface sterilization of the explants, majority of the investigators have used mercuric chloride or sodium or calcium hypochlorite either alone or both. Normally root explants from field grown palms are difficult to decontaminate [92]. Staritsky [94] used 2–6 year old green house palms and immersed the whole 'cabbage' consisting of apical meristem and very young leaf primordia in a saturated solution of calcium hypochlorite for 1–2 hours and then the explants were taken after removing the outermost layers. Even with this procedure the percentage of contamination was high, except for vigorously growing palms.

Rabechault *et al.* [79] used 0.1% mercuric chloride and followed a similar protocol. However, young leaves still enclosed by leaf bases and inflorescence still enclosed within spathes need no sterilization. These can be sampled without damaging the apex. Swabbing the outer surface of leaf bases or spathes with 70% alcohol and careful removal of inner tissues under aseptic conditions is sufficient to establish sterile explants.

3.5.2. *Media composition*

Exact media formulations are not available because of intense commercial interests in oil palm tissue culture. However, data published so far indicate that the majority of them have used modified Murashige and Skoog's medium. Although it is apparently clear that different protocols are used by the various laboratories to produce oil palm clones, it is generally observed that, at least for the initiation of callus from different sources, auxins are essential, and also for the rooting of the shoots. Most of the investigators have used 2,4-D alone or in combination with a cytokinin for the initiation of callus and NAA for the induction of rooting of shoots.

3.5.2. *Initiation of callus*

Smith and Thomas [92] induced callus from roots of aseptically grown seedlings within three weeks of culture on media containing 2,4-D or NAA. Similarly, Wooi *et al.* [120] reported induction of callus from the root segments of mature palms in two months. The growth rate of callus derived from root segments have been reported to be significantly variable among the different genotypes studied [120]. Callus appears to originate from the pericycle and central stele region of the explants [47,74]. For initiation on the explants usually dark incubation at 28±1 °C has been found to be essential.

Paranjothy [72] described that the younger and vigorously growing palm leaves produced callus easily. A similar phenomenon was also observed for palms grown under green house conditions [63]. Callus tissues from leaves originated from the veins of young leaf tissues [67]. Induction of callus in oil palm is dependent on a variety of factors, viz.: the genotype of the palm, the distance tissues are located from the meristem in case of young leaf tissues and also the type of auxin used [31,32,54,79]. The presence of NAA tends to give a rooty callus and 2,4-D induces nodular type [46,92]. Browning of explant and the medium is often a common phenomenon in oil palm cultures. Many investigators have added activated charcoal to the medium to check browning. But Jones [46] stated that browning need not necessarily be harmful, as successful cultures have been grown from explants which had turned almost black. A reduction in the level of auxin is necessary for the maintenance of callus cultures [74,80,92,102].

3.4.5. *Induction of embryogenesis*

There is no published report on direct embryogenesis from explant cultures in oil palm; however, Wooi [119] reported, on the basis of some personal communications, that it is occasionally possible to obtain embryos formed directly on leaf explants. Somatic embryos have been obtained from parts of aseptically grown seedlings [46], roots [20] and young leaves of selected field palms [80]. Similarly Wooi [119] reported the unpublished results on somatic embryo formation from inflorescence explants, their multiplication and regeneration of plantlets. Most of the published reports reveal that the primary explant produces callus and the embryos are formed on the secondary callus. Observations of embryos formed have led to the conclusion that these are somatic embryos [67]. Normally embryos are observed when callus cultures on nutrient medium rich in auxin are transferred to medium with decreasing amounts of auxins or on to a medium devoid of it. Paranjothy and Oathman [74] have obtained embryos by lowering the auxin levels plus the incorporation of cytokinins in the culture medium. Nwankwo and Krikorian [63] reported that lowering the NAA concentration from 20 to 0 mg/l in presence of activated charcoal gives rise to embryos. Such embryos produced shoos and roots in the presence of NAA (20mg/l) and GA_3 (1mg/). Thomas and Rao [102] noted the nodular callus formed in the presencee of 2,4-D became embryogenic on media containing 0.01–0.5 mg/l of 2-ip. Loiret [54] reported that the whitish, friable or fast growing callus was more highly embryogenic.

Clearly there are two distinct phases in clonal multiplication of oil palm, the first corresponds to the formation of calli and the differentiation of somatic embryos, (generally 8–10 months), the second phase involves the multiplication of embryos and their conversion into plantlets (4–6 months). In many instances the embryogenic calli have been grown on a medium devoid of growth hormones and the multiplication of embryos occurred through adventive embryogenesis [42,74]. Callus initiation and embryogenesis are dependant on age of the palm, type of explants and the genotype. The explants from younger palms [3–5 years] are more prone to produce callus compared with the explants from older palms (more than 4- years). Seed embryos from some genotypes are an excellent source of embryogenic callus.

3.5.5. *Embryo culture*

Early studies in oil palm tissue culture have utilized embryos in fundamental studies to formulate defined media and culture conditions. Callus cultures are easily established from zygotic embryos, and aseptically germinated seedlings provide a ready source of explants for *in vitro* sudies [119]. Paranjothy [71] has reviewed the progress on embryo culture. Cui and Gong [24] have reported the induction of callus from seed embryos and have mentioned that the frequency of callus induction could be markedly improved with heat pretreatment of seeds. Heat treatmant was given by placing the cleaned seeds at 38–40 °C. The swollen embryos were split and grown into callus within 30 days on medium with 2,4-D or 2,4-D in combination with NAA at 3 : 4 to 1 : 2 ratio. The highest frequency of callusing obtained was 95%. The calli obtained may differentiate into embryos on differentiation media with or without KN and with reduced auxin concentration. High proliferation frequency of embryos was obtained on media with 2.2μM BA or 2.3μM zeatin. Maximum proliferation of embryos was observed on media with BA 2.2μM, and 3mg/l Royal jelly. For the conversion of embryos into plantlets, Cui and Gong [24] used reduced macro elements of MS medium and micro elements of H medium and the medium was supplemented with low concentration of auxin. It was not difficult to induce plantlets from most of the well developed embryos.

Embryo culture has been used for the germination of the shell-less seeds of pisifera and the thin-shelled from tenera palms and also for the conservation of germplasm. It might also facilitate the movement of sterile cultures internationally thereby minimizing the risk of the introduction of diseases [75]. Methods have been worked out for storage oof zygotic embryos as well as somatic embryos [34,75,97]. Besides germplasm preservation, the cryopreservation of somatic embryos can be used to maintain cultures for clonal propagation [73]. French workers at CNRS have successfully stored embryos at −196 °C and retrieved normal plants from such freeze dried embryos [33].

3.5.6. *Rooting of shoots and development of plants*

Most of the investigators have used NAA for the induction of rooting [55]. According to Wooi [119] rooting of shoots does not seem to pose any difficulty. Further it is stated that the efficacy varies with the genotype and short exposure of shoots to a relatively high concentration of NAA and then to basal medium or a continuos exposure of shoots to low levels of an auxin is adequate for the successful rooting. Nwankwo and Krikorian [63] have used NAA (20mg/1) in combination with GA_3(1mg/1) for the development of both shoot and roots. They have suggested that the addition of a low dose of gibberellic acid helps embryonal structures to grow further and develop into plantlets, especially for the emergence of both shoot and roots. Similarly Thomas and

Rao [102] have used 1mg/l each of NAA and GA_3 for rooting of shoots.

de Touchet *et al.* [117] reported plantlet regeneration from embryogenic suspension cultures. Embryogenic suspensions were transferred to hormone-free liquid medium for sychronous maturation and shoot and root regeneration was achieved upon planting.

3.5.7. *Transplantation of regenerated plantlets to soil*

Once a balanced shoot and root system is formed, the establishment of *in vitro* plantlets to soil does not appear to be problematic [20,54,74,80]. According to Wardle *et al.* [116] tissue cultured plantlets do not develop a normal cuticle and hence require gradual weaning from the *in vitro* conditions. Therefore maintanance of high humidity around newly transplanted plantlets is by far the most commonly practiced method for hardening of plantlets and success rates above 90% have been reported and good survival was obtained by transplanting into non-sterile soil [120]. The growth rate of clonal plantlets in a growth cabinet are generally comparable to those of seedlings in a polybag nursery [20]. The plantlets are normally grown for about one year in the nursery before field planting.

3.5.8. *Protoplast culture*

The isolation of protoplasts from oil palm was first reported by Vouyouklis [115] using leaves and roots as the source material. Bass and Hughes [6] described the isolation and culture of protoplasts from oil palm suspension cultures. Paranjothy [73] reported that the protoplasts were cultured using a 'nurse culture' of oil palm cells and at PORIM (Palm Oil Research Institute of Malaysia) protoplasts were routinely isolated from a range of oil palm tissues. Further Paranjothy [73], based on Ismail and Ali and Sambanthamurthi's unpublished data, reported that the protoplasts prepared from leaves and inflorescences have been found to regenerate cell walls and undergo limited cell division to form micro colonies. Sambanthamurthi *et al.* [87] described the isolation of protoplasts from embryogenic cultures and the mesocarp of oil palm. However, till todate there is no report on the successful regeneration of plantlets from cultured protoplasts.

3.5.9. *Genetic stability*

Genetic stability in oil palm plants obtained through tissue culture, is an important factor for cloning. It has been observed that in oil palm the plants are regenerated from callus. It is generally known that the callus is unstable [25]. Jones [47] recorded that plants regenerated from oilpalm tissue cultures were not indentical and included somatic variants. In general, such variations are of gentic origin and are induced by *in vitro* culture [18]. Corley *et al.* [23] reported the four abnormal palms derived from a single culture of a clone representing 1.8% of the plants produced were with unusual leaf morphology and occasionally branched rachises. Later Corley *et al.* [21] reported the abnormal flower production on clonal materials. Although the rate of abnormal mitosis was low, chromosome number variations still occurred in clonal progenies [24].

Paranjothy [73] reported the morphological abnormalities in the development of the individual flowers on both male and female inflorescences and the following types were recorded: (i) parthenocarpic development of the gynoecium; (ii) development of the normally vestigial stamens in the female inflorescence into a mantle of fleshy carpels surrounding the fruit; (iii) development of stamens in the male inflorescence into carpellary structures; and (iv) feminization of the male flowers in the female inflorescences. Further it is stated that parthenocarpy in oilpalm is induced by auxins, and growing the shoots in high concentrations of benzylaminopurine or other cytokinins results in the feminization of flowers. These abnormalities are caused by heritable changes in DNA and is an epigenetic phenomenon and further the unstable expression of the abnormalities and the reversion of abnormal clonal palms into normal ones supports the epigenetic mechanism [73]. Therefore studies on genetic stability are important for improvement of oil palm tissue culture and before taking up of large scale production of oilpalm through tissue culture, it is better to evaluate the genetic stability of the regenerants.

3.5.10. *Field trials*

The Unilever group carried out the first clonal trials derived from seedling materials in Malaysia during 1977 - 1980. In 1981 tissue cultured plantlets were planted in various agroclimatic conditions by the Institut de Recherches pour les Huiles et Oleagineux (IRHO) and Institut de francais de recherche scientifique pour le development en cooperation (ORSTOM) to test their clones and also to promote the clonal propagation of oil palm. Initially about 40,000 plantlets were produced in Cote D'Ivoire and were planted in Africa, South America and South East Asia and their performance was recorded [61]. These studies were conducted to compare the tissue cultured plantlets with that of donor parent plants and the selection of best clones based on their rate of growth, oil yield and resistance to pathogens etc. [88].

3.5.11. *Commercial propagation*

Commercial propagation of oil palm was taken up by IRHO and ORSTOM with an aggrement with Federal Land Development Authority in Malaysia to set up a pilot plant in Kuala Lumpur with a production capacity of one million plantlets per year and to establish one hundred clones. Similar aggrements were also made with Indonesian companies, (SOCFINDO– a mixed plantation company, and PNP– National Governmental Plantations Company) and Latin American and Far Eastern countries for large scale production of plantlets [62]. According to Noiret *et al.* [61] the commercial production of oil palm was programmed for 1991 and its expansion depends on the availability of new clones and their fidelity.

3.5.12. *Conclusions*

The oil palm tissue culture has become commercially important and is feasible, even if the production cost of tissue cultured plants is five to ten times higher than that of seeds, the increase in production expected and the increase in the exploitation period of the trees make the operation a profitable venture [61]. Additionally, it should provide a method for the propagation of any improved or elite palm. Future research on oil palm should be directed to improve the efficiency of the micropropagation process by shortening time in culture for production of embryos and plantlet regeneration and minimizing the use of plant growth regulators in culture [119]. Optimum medium or media combinations for various developmental stages has to be standardised with a suitable explant source, because various explants like leaf, root and inflorescence have been used as a starting material and opinion differs as to which explant source gives the best results. At present there are also no published data showing any one tissue to be superior.

4. Summary

The majority of palms are propagated by seeds, except date palm, which can be propagated vegetatively by offshoots. Despite the selection there is a great deal of variation between the seedlings. Coconut and oil palms are specially cultivated for the production of edible oil. The variation due to seed propagation affects the yield and consequently the oil production. Therefore, for the production of identical clones in large numbers from high-yielding palms, the technique of tissue culture offers a vast scope.

Considerable success in tissue culture has been achieved specially in date and oil palm. However, the progress in coconut tissue culture is rather slow. Early investigations assumed that coconut would also respond similarly to that of oil palm. This did not prove true. Initial results on the regeneration of plants from coconut leaf tissue cultures have been encouraging. Nevertheless, reproducible results have not yet been obtained in many laboratories working with coconut tissue culture and therefore the problem of regeneration of plants from callus culture still persists. Todate, the most serious problem encountered is the variation in the culture responses from tree to tree and also for different batch of collections subjected to similar cultural conditions. Recently, however, encouraging results have been obtained in coconut with embryo culture [5]. The technique standardized is advantageous specially for the transportation of coconut germplasm by embryos, which is easier, comparatively inexpensive and safer. The success obtained in the culture of makapuno embryos can be exploited for the cultivation of hybrid embryos and it is reasonable to assume that this aspect of work will be pursued further. Though a few reports are available on culture of anther, endosperm and protoplast, the regeneration of plantlets still remains a challenge.

Date palm is cultivated for its edible fruit, which is a rich source of carbohydrate. The micropropagation technique has been standardized for date palm and many reports are available on successful production of plantlets either through somatic embryogenesis from callus or through axillary shoot development. The later type is ideal for cloning the date palm. A few commercial laboratories have adapted this technique.

Unlike the other two palms, oil palm is a very important oil yielding crop and has gained prominence in recent years throughout the world for its edible oil, which is cholesterol free. Due to its heterozygous nature, there is a considerable variation in the oil yield. A protocol for the large scale multiplication is now available in many laboratories, but due to its commercial importance all the details are not available.

The technique of embryo culture in oil palm can be used for germplasm conservation, storage and exchange of cultures without the risk of introduction of diseases. The survey of available literature clearly indicates that the plantlets have been obtained through somatic embryogenesis from a piece of callus. One of the major concerns in oil palm micropropagation has been the genetic stability of the plants obtained *in vitro*. It has been observed that the plants are produced through callus phase and as the callus is generally unstable, the plants regenerated may not always be uniform. Field trials conducted so far have also indicated some abnormalities in tissue cultured palms. Hence investigations should be focussed towards the maintenance of the genetic stability by minimising the use of growth regulators and the duration of the culture period. The date and oil palm tissue culture is feasible and commercially viable. Even though the cost of production may be higher than that of seeds, the higher yield expected of tissue cultured clones should make it a profitable business.

References

1. Ammar S and Benbadis A (1977) Multiplication of vegetative du palmier datteier (*Phoenix dactylifera* L.) par la culture de tissus de jeunes plantes issues de semis. C R Acad Sci 284:1789–1792.
2. Ammar S and Benbadis A (1983) Vegetative propagation of date palm (*Phoenix dactylifera* L.) by *in vitro* culture. 1st Symp Date Palm 1982. King Faisal Univ Hofuf, Saudi Arabia. pp. 1580–166.
3. Ammar S, Benbadis A, and Tripathi BK (1987) Floral induction in date palm seedlings (*Phoenix dactylifera* var. Deglet Nour) cultured *in vitro*. Can J Bot 65:137–142.
4. Assay Bah B (1986) *In vitro* culture of coconut zygotic embryos. Oleagineux 41:321–328.
5. Assy Bah B, Durrand-Gosselin T, Engelmann F, and Pannetier C (1989) Culture *in vitro* d'embryons zygotiques de cocotier (*Cocos nucifera* L.) Methode, revisse et simplifies d'obtention de plants de cocotiers transferables au champ. Oleagineux 44:515–523.

6. Bass A and Hughes W (1984) Conditions for isolation and regeneration of viable protoplasts of oil palm (*Elaeis quineensis*). Plant Cell Rep 3:169–171.
7. Beauchesne G (1983) Vegetative propagation of date palm (*Phoenix dactylifera* L.) by *in vitro* culture. 1st Symp. Date palm 1982. King Faisal Univ Hofuf, Saudi Arabia. pp. 698–699.
8. Benger KG (1986) Food uses of palm oil. Occ Paper No 2. Palm Oil Res Inst Malaysia. (Revised from 1981.)
9. Bhalla-Sarin N, Bagga SS, Spory K, and Guha-Mukherjee S (1986) Induction and differentiation of callus from embryos of *Cocos nucifera* L. by IAA conjugates. Plant Cell Rep 5:322–324.
10. Bhalla-Sarin N and Suman Bagga (1988) *In vitro* culture of embryos and other parts of *Cocos nucifera*. In: Proceedings of the National seminar on plant tissue culture held on 2–4 March, 1983 at CPCRI, Kasargod, ICAR, New Delhi. pp. 132–139.
11. Blake J (1982) Tissue culture propagation of coconut, date and oil palm In: Dodds JH (Ed.) Tissue culture of Trees, Croom Helm, London. pp. 29–50.
12. Blake J (1990) Coconut (*Cocos nucifera* L.) Micropropagation. In: Bajaj YPS (Ed.) Biotechnology in Agriculture and Forestry Vol 10. Legumes and oil seed crops. Springer Verlag, Berlin, Heidelberg. pp. 538–554.
13. Branton RL and Blake J (1983a) A lovely clone of coconuts. New Sci 98:554–557.
14. Branton RL and Blake JJ (1983b) Development of organized structures in callus derived from explants of *Cocos nucifera* L. Ann Bot 52:673–678.
15. Branton RL and Blake J (1986) Clonal propagation of coconut In: Pushparajah E, Chew Poh Soon (Eds.) Cocoa and Coconuts: Progress and Outlook. Kuala Lumpur Inc Soc Plant pp. 771–780.
16. Branton RL and Blake J (1989) Date palm (*Phoenix dactylifera* L.) In: Bajaj YPS (Ed.) Biotechnology in Agriculture and Forestry Vol 5. Trees-II Springer-Verlag, Berlin, Heidelberg. pp. 161–175.
17. Brackpool AL, Branton RL, and Blake J (1986) Regeneration in palms In: Vasil IK (Ed.) Cell culture and somatic cell genetics of plants. Vol 3. Academic press, London New York. pp. 207–222.
18. Cocking EC, Davey MR, Pental D, and Power JB (1981) Aspects of plant genetic manipulation. Nature (London) 293(5830):265–270.
19. Corley RHV (1982) Clonal planting material for the oil palm industry. Planter 58:515–528.
20. Corley RHV, Barrett JN, and Jones LH (1977) Vegetative propagation of oil palm via tissue culture. Oil Palm News 22:2–7.
21. Corley RHV, Lee CH, Law IH, and Wong CY (1986) Abnormal flower development in oil palm clones. Planter 62:233–240.
22. Corley RHV, Wooi KC, and Wong CY (1979) Progress with vegetative propagation of oil palm. Planter 55:377–380.
23. Corley RHV, Wong CY, Wooi KC, and Jones LH (1981) Early results from the first oil palm clone trials. In: Pushparajah E, and Chew PS (Eds.) Oil Palm in the Agriculture in the Eighties. Inc Soc Plant Kuala Lumpur. pp. 173–176.
24. Cui Y and Gong Z (1990) Oil Palm. In: Chen Z, Evans DA, and Sharp WR (Eds.) Handbook of Plant Cell Culture. Macmillan, New York. pp. 440–452.
25. D'Amato F (1978) Chromosome number variation in cultured cells and regenerated plants. In: Thorpe TA (Ed.) Frontiers of Plant Tissue Culture. Int Assoc Plant Tissue Culture, Calgary. pp. 287–295.
26. Davis TA (1962) Rejuvenation of coconut palms. World Crops 14:254–259.
27. Drira N (1983) Multiplication vegetative de palmier dattier (*Phoenix dactylifera* L.) par la culture *in vitro* de bourgeons axillaires et de feuilles quien derivent. CR Acad Sci Paris 296 Ser III: 1077–1082.
28. Drira N and Benbadis A (1985) Multiplication vegetative du palmier dattier (*Phoenix dactylifera*) par reversion en culture *in vitro* d'ebauches florales de pieds femelles. J Plant Physiol 119:227–235.

29. D'Souza L (1982) Organogenesis in coconut embryo callus. Plant Tissue Culture 1982. Marzuen, Tokyo. pp. 179–180.
30. D'Souza L, Paily J, and Arokiadoss V (1988) Coconut embryo and tissue culture: Potentialities, problems and applications In: Proceedings of the National seminar on plant tissue culture held on 2–4 March 1983 at CPCRI, Kasargod, ICAR New Delhi. pp. 140–145.
31. Duval Y, Gasselin TD, Konan K, and Pannetier C (1987) *In vitro* vegetative micropropagation of oil palm *Elaeis guineensis* Jacq.): Choice of strategy and results. In: Hassan AH, Chew PS, Wood BJ, and Pushparajah E (Eds.) Int Oil Palm Conf Kuala Lumpur. pp. 191–196.
32. Duval Y, Durand Gasselin T, Konan K, and Pannetier C (1988) Multiplication vegetative du palmier a huile (*Elaeis guineensis* Jacq.) par culture *in vitro*. Strategie et resultats (1). Oleagineux 43:39–47.
33. Engelmann F and Duval Y (1986) Cryopreservation d'embryons somatiques de palmier a huile (*Elaeis guineensis* Jacq.): Resultats et perspectives d'application. Oleagineux 41:169–174.
34. Engelmann F, Duval Y, and Pannetier C (1987) Use of cryopreservation for setting up a bank of oil palm (*Elaeis guineensis* Jacq.) somatic embryos. In: Hassan AH, Chew PS, Wood BJ, and Pushparajah E (Eds.) Int Oil plalm Conf Progress and Prospects-Agriculture, PORIM, Kuala Lumpur. pp. 648–652.
35. Eeuwens CJ (1976) Mineral requirements for growth and callus initiation of tissue explants excised from mature coconut palms (*Cocos nucifera*) cultured *in vitro*. Physiol Plant 36:23–28.
36. Gabr H and Tisserat B (1985) Propagating palms *in vitro* with special emphasis on the date palm (*Phoenix dactylifera* L.). Sci Hortic 25:255–262.
37. Grout BWW, Shelton K, and Pritchard HW (1983) Orthodox behaviour of oil palm seed and cryopreservation of the excised embryo for genetic conservation. Ann Bot (London) 52:381–384.
38. Gupta PK, Kendurka SV, Kulkarni VD, Shirgurkar MV, and Mascarenhas AF (1984) Somatic embryogenesis and plants from zygotic embryos of coconut (*Cococ nucifera* L.) *in vitro*. Plant Cell Rep 3:222–225.
39. de Guzman EV (1980) *In vitro* culture studies on coconut: Progress and problems. Professorial Lecture Univ Philippines, Los Banos.
40. de Guzman EV, del Rosario AG, and Ubalde EM (1978) Proliferative growths and organogenesis in coconut embryo and tissue cultures. Philip. Coconut Stud 3:1–10.
41. Haibou TK and Kovoor A (1982) Regeneration of callus from coconut protoplasts In: Rao AN (Ed.) Proc COSTED Symp on Tissue Culture of Economically Important Plants. Singapore, 1981. pp. 149–151.
42. Hanower J and Pannetier C (1982) *In vitro* vegetative propagation of the oil palm (*Elaeis guineensis* Jacq.) In: Fujiwara A (Ed.) Plant Tissue and Cell Culture. 1982. Maruzen, Tokyo. pp. 745–746.
43. Hardon JJ (1984) The oil palm; Progress through plant breeding in Malaysia. Span 27:59–61.
44. Iyer RD (1982) Embryo and tissue culture for crop improvement, especially of perennials, germplasm conservation and exchange-relevance to developing countries. In: Rao AN (Ed.) Proc. COSTED Symp on Tissue culture of economically important plants, Singapore 1981. pp. 219–230.
45. Iyer RD, Raju CR, and Prakash Kumar P (1988) Palm tissue culture: Potential priorities and applications in coconut and oil palm In: Proceedings of the National seminar on plant tissue culture held on 2–4 March 1983 at CPCRI, Kasargod, ICAR New Dehli. pp. 129–131.
46. Jones LH (1974) Propagation of clonal oil plams by tissue culture. Oil Palm News 17: 1–8. Planter 50:374–381.
47. Jones LH (1983) The oil palm and its clonal propagation by tissue culture. Biologist 30:181–188.
48. Jones LH (1984) Novel palm oil from cloned palms. JAOCS 61:1717–1719.
49. Jones LH and Hughes WA (1989) Oil palm (*Elaeis guineensis* Jacq.) In: Bajaj YPS (Ed.) Biotechnology in Agriculture and Forestry. 5. Trees II Springer-Verlag, Berlin, Heidelberg New York Tokoyo. pp. 176–202.

50. Jagadeesan M and Padmanabhan D (1982) Induction of rooting in cotyledon callus of coconut. Curr Sci 51:567.
51. Kackar NL, Solanki KR, and Joshi SP (1989) Micropropagation of date palm (*Phoenix dactylifera* L.) cv. Khadrawy using tissue culture technique. Annals Arid Zone 28:137–141.
52. Karunaratne S, Kurukulaarachoh C, and Gamage C (1985) A report on the culture of embryos of dwarf coconut *Cocos nucifera* L. var. nana *in vitro*. Cocos 3:1–8.
53. Knaut J and Richtler HJ (1985) Trends in industrial uses of palm and lauric oils. J Am Oil Chem Soc 62:317–327.
54. Loiret C (1981) Vegetative propagation of oil palm by somatic embryogenesis. In: Pushparajah E and Chew PS (Eds.) Oil Palm in the Eighties. Inc Soc Plant Kuala Lumpur. pp. 163–172.
55. Martin JP and Rabechault H (1976) Procede de multiplication vegetative de vegetaux et plants ainsiobtenus. French Patent 2364611. Inst Nat Propriete Industrielle Paris.
56. Mater MA (1983) Plant regeneration from callus cultures of *Phoenix dactylifera* L. Date Palm J 2:57–77.
57. Monfort S (1985) Androgenesis of coconut embryos from anther culture. Z Pflanzenzuecht 94:251–254.
58. Murashige T and Skoog F (1962) A revised medium for rapid growth and bioassays with tobacco tissue cultures. Physiol Plant 15:473–497.
59. Ninan CA and Raveendranath TG (1965) A naturally occurring haploid embryo in the coconut palm (*Cocos nucifera* L.). Caryologia 18:619–623.
60. Noiret JM (1981) Application of *in vitro* culture to improvement and production of clonal material in the oil palm. Oleagineux 36:123–126.
61. Noiret JM, Gascon JP, and Pannetier C (1984–85) En France, les success de la recherche. 1. Production de palmier a huile par culture *in vitro*. Cahiers des Ingenieurs Agronomes (Paris). 382:57–64.
62. Noiret JM, Gascon JP, and Pannetier C (1985) La production de palmier a huile par culture *In vitro*. Oleagineux 543:365–372.
63. Nwankwo BA and Krikorian AD (1986) Morphogenetic potential of embryo and seedling derived callus of *Elaeis guineensis* Jacq. var. Pisifera Becc. Ann Bot 51:65–76.
64. Nyman LP, Gonzales CJ, and Arditti J (1983) Reversible structural changes associated with callus formation and plantlet development form aseptically cultured shoots of Taro (*Colocasia esculenta*) var. antiquorum). Ann Bot 51:279–286.
65. Odewale JO (1983) Development of tissue culture techniques and callus induction in the oil palm (*Elaeis guineensis* Jacq.) anthers. PhD Thesis Univ Ife. (Paranjothy K pers comm from Jones and Hughes 1989).
66. Padmanabhan D (1988) Potentials of embryo culture in *Cocos nucifera*. In: Proceedings of the National seminar on plant tissue culture held on 2–4 March 1983 at CPCRI Kasargod ICAR New Delhi. pp. 146–148.
67. Pannetier C, Arhuis P, and Lievoux D (1981) Neoformation of young *Elaeis guineensis* plants from primary callus obtained from leaf fragments cultured *in vitro*. Oleagineux 36: 119–122.
68. Pannetier C and Buffard-Morel J (1982) First results of somatic embryo production from leaf tissue of coconut *Cocos nucifera*. L. Oleagineux 37:349–354.
69. Pannetier C and Buffard-Morel J (1986) Coconut palm (*Cocos nucifera* L.) In: Bajaj YPS (Ed.) Biotechnology in Agriculture and Forestry Vol 1. Trees. Springer-Verlag, Berlin, Heidelberg New York Tokyo. pp. 430–450.
70. Prakash Kumar P, Raju CR, Chandramohan M, and Iyer RD (1985) Induction and maintenance of friable callus from the cellular endosperm of *Cocos nucifera* L. Plant Sci 40:203–207.
71. Paranjothy K (1984) Oil palm. In: Ammirato PV, Evans DA, Sharp WR, Yamada Y (Eds.) Handbook of Plant Cell Culture. Macmillan New York. pp. 591–605.

72. Paranjothy K (1986) Some aspects of oil palm tissue culture in relation to ortet selectrion and clonal evaluation In: Colloq Breeding and Selection for Clonal Oil Palms. ISOPB, Kuala Lumpur.
73. Paranjothy K (1989) Research strategies and advances in oil palm cell and tissue cultures presented at Commonwealth Science Council workshop, Pune, India February 1989. pp. 119–125.
74. Paranjothy K and Oathman R (1982) *In vitro* propagation of oil palm In: Fujiwara A (Ed.) Plant Tissue Culture 1982. Maruzen, Tokyo. pp. 747–748.
75. Paranjothy K, Rohani O, and Hashim AH (1985) Oil palm germplasm conservation: The possible use of *in vitro* methods. Proc Inter Workshop on Oil Palm Germplasm and Utilization. Kuala Lumpur. pp. 202.
76. Pareek OP (1984) Date palm growing potential of Indian arid zone. Indian Horticulture 29:2–8.
77. Patton S and Jensen PG (1975) Lipid metabolism and membrane functions of the mammary gland. Prog Chem Fats Other Lipids 14:167–277.
78. Rabechault H, Ahee J, and Guenin G (1970) Colonies cellulaires et formes embryos *in vitro* a partier de cultures d'embryons de palmier a huile (*Elaeis guineensis* Jacq. var. dura Becc.) CR Acad Sci Paris. 270:3067–3070.-
79. Rabechault H, Martin JP, and Cass (1972) Recherches sur la culture des tissues de palmier a huile (*Elaeis guineensis* Jacq.). Oleagineux 27:531–534.
80. Rabechault H and Martin JP (1976) Multiplication vegetative du palmier a huile (*Elaeis guineensis* Jacq.) a l'aide de cultures de tissue foliares. CR Acad Sci Paris Ser 283:1735–1737.
81. Raju CR, Kumar PP, Chandramohan M, and Iyer RD (1984) Coconut plantlets from leaf tissue cultures. J Plant Crops 12:75–78.
82. Renard JL (1976) Diseases in Africa and South America. In: Corley RHV, Hardon JJ, and Wood BJ (Eds.) Oil Palm Research. Elsevier, Amsterdam. pp. 447–466.
83. Reynolds JF (1982) Vegetative propagation of palm trees In: Bonga JM and Durzan DJ (Eds.) Tissue Culture in Forestry. Nijhoff, The Hague. pp. 182–207.
84. Reynolds JF and Murashige T (1979) Asexual embryogenesis in callus cultures of palms. *In vitro* 15:383–387.
85. Rhiss A, Poulain C, and Beauchesne G (1979) La culture *in vitro* appliquee a la multiplication vegetative du palmier-dattier (*Phoenix dactylifera* L.). Fruits 34:551–554
86. del Rosario AG and de Guzman EV (1982) The status of plant tissue culture in the Philippines. In: Rao AN (Ed.) Proc COSTED Symp on Tissue Culture of Economically Important plants, Singapore 1981. pp. 293–294.
87. Sambanthamurthy R, Oo KC, and Ong ASH (1987) Lipid metabolism in oil palm (*Elaeis guineensis* and *Elaeis oleifera*) protoplasts. Plant Science 51:97–103.
88. Sasson A (1988) Biotechnologies and development UNESCO, Paris, France. pp. 25–31.
89. Schroeder CA (1970) Tissue culture of date shoots and seedlings. Date Growers Inst Rep. 47:25–47.
90. Sharma DR, Dwara S, and Chowdhury JB (1984) Somatic embryogenesis and plant regeneration in date palm (*Phoenix dactylifera* L.) cv. Khadravi through tissue culture. Indian J Exp Biol 22:596–598.
91. Smith BG (1986) Tissue culture of *Cocos nucifera* – biochemical changes preceding embryogenesis. In: Pushparajah E, Chew Poh Soon (Eds.) Cocoa and Coconuts: Progress and outlook. Kuala Lumpur Inc Soc Plant pp. 781–786.
92. Smith WK and Thomas JA (1973) The isolation and *in vitro* cultivation of cells of *Elaeis guineensis*. Oleagineux 28:123–127.
93. Spurr JW (1985) Commercial date palm micropropagation at Twyfords. Agri Cell Rep 4:40.
94. Staritsky G (1970) Tissue culture of the oil palm (*Elaeis guineensis* Jacq.) as a tool for its vegetative propagation. Euphytica 19:288–292.

95. Sudarship H, Keat H, and Davis A (1978) Clonal propagation of the coconut via the bulbils. Philipp J Cocon Stud 3:5–15.
96. Tahardi S (1985) Organogenesis in hybrid coconut embryo and tissue cultures. Poster Abstract Symp. *In vitro* problems related to mass propagation of horticultural plants. Gembloux. Belgium 16–20, Sept.
97. Tarmizi AH and Paranjothy K (1988) *In vitro* storage of oil palm polyembryogenic cultures. Proceedings of Symposium Biologi Kebangsaan Ke–III.
98. Temple WJ and Basu TK (1988) Does beta carotene prevent cancer? A critical appraisal. Nutr Res 8:685–701.
99. Thampan PK (1986) Coconut – The veritable source of food and oil. In: Srivastava HC, Bhartendu Vatsya, Menon KKG (Eds.) Plantation crops opportunities and constraints. Vol. II. Proceedings of the symposium on plantation opportunities in India. Hindustan Lever Research Foundation. pp. 81–96.
100. Thanh-Tuyen NT and de Guzman EV (1983) Formation of pollen embryos in cultured anthers of coconut (*Cocos nucifera* L.) Plant Sci Lett 29:81–88.
101. Thanh-Tuyen NT (1990) Coconut (*Cocos nucifera* L.): Anther culture. In: Bajaj YPS (Ed.) Biotechnology in Agriculture and Forestry Vol. 10. Legumes and Oil seed crops. Springer-Verlag Berlin–Heidelberg. pp. 555–568.
102. Thomas V and Rao PS (1985) *In vitro* propagation of oil palm (*Elaeis guineensis* Jacq. var. Tenera) through somatic embryogenesis in leaf derived callus. Curr Sci 54:184–185.
103. Timms RE (1986) Processing of palm oil Fette Seifen. Anstrichmittel 8:294–300.
104. Tisserat B (1979) Propagation of date palm (*Phoenix dactylifera* L.) *in vitro*. J Exp Bot 90:1275–1283.
105. Tisserat B (1981) Production of free-living date palms through tissue culture. Date Palm J 1:43–54.
106. Tisserat B (1982) Factors involved in the production of plantlets from date palm callus cultures. Euphytica 31:201–214.
107. Tisserat B (1984a) Propagation of date palms by shoot tip cultures. Hort Sci 19:230–231.
108. Tisserat B (1984b) Date palm. In: Sharp WR, Evans DA, Ammirato PV, and Yamada Y (Eds.) Handbook of Plant Cell Cultures. Vol. 2, Macmillan, New York. pp. 505–545.
109. Tisserat B (1987) Palms. In: Bonga JM and Durzan DJ (Eds.) Cell and Tissue Culture in Forestry, Vol. 3. Martinus Nijhoff publishers. pp. 338–356.
110. Tisserat B and De Mason DA (1980) A histological study of development of adventive embryos in organ cultures of *Phoenix dactylifera* L. Ann Bot (London) 46:465–472.
111. Tisserat B and De Mason DA (1985) Occurence and histological structure of offshoots and inflorescences produced from *Phoenix dactylifera* L. plantlets *in vitro*. Bull Torrey Bot Club 12:35–42.
112. de Touchet B, Duval Y, and Pannetier C (1991) Plant regeneration from embryogenic suspension cultures of oil palm (*Elaeis guineensis* Jacq.). Plant Cell Rep 10:529–532.
113. Turner PD (1976) Oil palm diseases in South East Asia and the South Pacific. In: Corley RHV, Hardon JJ, and Wood BJ (Eds.) Oil palm Research. Elsevier, Amsterdam. pp. 427–445.
114. Verdeil JL and Pannetier C (1990) Optimisation des conditions de germination *in vitro* du pollen de cocotier (*Cocos nucifera* L.) pourlamise au point d'un test de viabilite. Oleagineux 45:175–181.
115. Vouyouklis GV (1981) Use methode d'isolement de protoplasts du palmier a huile *Elaeis guineensis* Jacq. Phyton 40:169–178.
116. Wardle K, Quinlan A, and Simpkins I (1979) Abscicic acid and the regulation of water loss in plantlets of *Brassica oleracea* L. var. botrytis regenerated through apical meristem culture. Ann Bot 43:745–752.
117. Wood BJ (1986) A Brief Guide to Oil Palm Science, Incorporated Society of Planters. Kuala Lumpur. pp. 39.
118. Wooi KC (1984) Palm tissue culture. In: George EF (Ed.) Micropropagation of Selected

Root Crops, Palms, Citrus and Ornamental Species. FAO plant production and protection paper 59, FAO, Rome. pp. 88–112.

119. Wooi KC (1990) Oil palm (*Elaeis guineensis* Jacq): Tissue culture and micropropagation. In: Bajaj YPS (Ed.) Biotechnology in Agriculture and Forestry. Vol. 10. Legumes and Oil Seed Crops I. Springer-Verlag Berlin Heidelberg. pp. 569–592.
120. Wooi KC, Wong CY, and Corley RHV (1981) Tissue culture of palms – a review. In: Rao AN (Ed.) Tissue Culture of Economically Important Plants. COSTED and ANBS, Singapore. pp. 138–144.
121. Zaid A and Tisserat B (1983) Morphogenetic responses obtained from a variety of somatic explant tissues of date palm. The Bot Mag (Tokyo) 96:67–73.

SECTION THREE

Tree improvement

25. Micropropagation and clonal options in forestry

W. J. LIBBY[1] and M. R. AHUJA[2]

[1] *Department of Forestry and Resource Management, University of California, Berkeley, CA 94720, USA*

[2] *Institute of Forest Genetics, Federal Research Centre for Forestry and Forest Products, Sieker, Landstrasse 2, 2070 Grosshansdorf, Germany*

1. Introduction

Micropropagation has been achieved in a number of woody plant species. That is only the beginning in the production process. Before clonal material is regenerated by tissue culture, it would be desirable to consult geneticists/breeders for inclusion of pedigreed material in the program. Production costs, genetic fidelity, and performance of *in vitro* regenerated plants will determine the future of micropropagation in clonal forestry. Although true-to-type plants are the desired goal of a clonal forestry program, micropropagation may not always yield high fidelity clonal regenerants. A spin-off from tissue culture is somaclonal variation, which offers opportunities for selection of new useful variants. In addition micropropagation technology has been exploited for manipulation of maturation states, germplasm preservation, and genetic engineering in plants. Thus, micropropagation has become useful in several different areas of biotechnology. In this chapter we examine to what extent micropropagation can contribute or may contribute to clonal options in forestry.

2. Definitions of micropropagation and clonal options

2.1. *Micropropagation*

This term refers to *in vitro* vegetative propagation of plants, and it is abundantly described in context in other chapters of this volume. The products of such micropropagation are often called plantlets, somatic embryos, or micropropagules. By contrast, macropropagation also implies vegetative propagation and refers to such techniques as rooting of detached shoots, grafting of twigs (scions) onto stock plants, inserting buds attached to a bit of stem tissue into slits in stock plants, and air-layering of attached shoots.

There is perhaps a grey area when shoots of tissue-culture origin are detached and then rooted in soil (substrate), and the qualifying criterion here may

M. R. Ahuja (Ed.), Micropropagation of Woody Plants, 425–442.

be whether the shoot originated *in vitro*, or from plants rooted in soil in a pot or in the field.

2.2. *Clonal options*

Opting for the use of clones implies that one can use clonally propagated (nearly) intact genotypes for various purposes, instead of using (usually) novel and unique genotypes that are the immediate offspring of sexual propagation. These uses of clones may be for research, or for production forestry. Within production forestry, clones may be used both within zygotic forestry and clonal forestry. Details of various uses will be given in Section 6, but first we define zygotic forestry and clonal forestry at this point [31].

Clonal forestry involves much more than the use of vegetative propagules in forestry. It has as qualifying criteria the deployment of relatively few well-tested and well-understood clones in ways that allow substantial increases in management efficiency and effectiveness. It also allows the deployment of clones that would rarely survive in a genetically heterogeneous mixture (such as short fat trees), or that would rarely occur in zygotic forestry (such as low-sex trees).

Whether the deployed propagules or naturally-reproduced seedlings are single-copy genotypes or vegetatively-multiplied propagules, the important criterion of zygotic forestry is that the individual genotypes are not characterized or well known. Performance knowledge focuses on the family, line, breed or population, but management can not take advantage of differences among individual clones, or of extensive knowledge about a particular clone.

It should be noted in this context that micropropagation can serve both clonal forestry and vegetative multiplication for zygotic forestry in various ways, discussed below.

3. The selection of genotypes for micropropagation

This topic is discussed in detail [5,7,30], and additional information and discussion appear in Section II of this volume. Briefly, the main criteria and principles to consider when selecting genotypes for micropropagation are as follows.

3.1. *Ease of propagation*

For some purposes, one might screen many genotypes to find some small number of them that propagate most easily. On the other hand, if one is working on increasing the effectiveness of the micropropagation technique, say for large scale operation, one might wish to include difficult-to-propagate genotypes to investigate their barriers to effective propagation.

3.2. *Use pedigreed clones*

While the genetic identity of a clone may not be important during early phases of some micropropagation research, it may become important later. It becomes important at the time of deployement and of possible further breeding. Breeders, regeneration foresters and silviculturists are going to want to know the provenance origin and immediate pedigree of micropropagated clones that reach the point of field deployment or of entry in a breeding program. Provenance origin identifies the location and environment in which the clone's recent ancestors evolved. Pedigree identifies the clone's parents, grandparents, etc., and often includes performance data on its relatives (from sibs and offspring as well as from those in the parental line).

3.3. *Employ known material*

A particularly desirable approach is to work with clones that are already well-known. When these clones are deployed to the field, those familiar with them prior to micropropagation can effectively spot differences in performance, whether due to the propagation technique or to some sort of genetic engineering done to purposefully modify them.

In short, when beginning a line of micropropagation research, first contact breeders working with the species, and then try to employ material that will be of interest to them, and in which they have some confidence.

4. The testing of mircropropagated clones

This topic is covered in recent publications [23,28,29,30]. The failure to adequately test micropropagated material is identified as one of the serious concerns about the field (see Section 7.5 below). There are at least three dimensions of testing that deserve attention.

4.1. *Genetic fidelity*

In any breeding program, there must be sustained quality control to ensure that clones, lines or cultivars have not been confused or contaminated. These problems are most commonly caused by errors in labelling, planting location or record-keeping. In these cases, the clones have not changed genetically, but their identities have been confused or are in error. Thus, this form of quality-control testing is most focused on whether the genetic entry is still true-to-type, and it is commonly performed by growing out samples that are inspected by knowledgeable experts. Biotechnology is now adding new tools for such maintenance of genetic fidelity, for example biochemical fingerprinting [19].

In any somatic line, somatic mutations occur that cause real but generally trivial genetic changes. These accumulate in somatic lineages and create differences among different sublines of a clone. In some cases the changes are not trivial; quality-control testing will seek to identify and remove those ramets with non-trivial deleterious changes, and occasionally useful ones will thus be found for further testing and deployment. Their classification is further discussed under Section 5, below.

Finally, there is substantial evidence that some types of tissue-culture are likely to produce high rates of somaclonal and other genetic changes [1]. These are further discussed under Section 5.2.3 below. Testing for variants produced during tissue culture will often focus first on traits being selected for desirable variants, then on the stability of such desirable variants once found, and on whether such variants will be transmitted through meiosis to sexual offspring. Testing in this regard will also need to be particularly sensitive to cryptic or uncharacterized genetic changes that are likely to occur in other parts of the selected tree's genome. This is a difficult task, as one does not know quite what one is looking for. This will be more effective if the clones being propagated are already well known by the investigators.

4.2. *Performance*

When some sort of genetic engineering or either qualitative or quantitative selection has preceded the propagation, then performance with respect to some trait or index of traits will be evaluated by testing.

It is also possible that the type of propagule produced has a general effect on performance. Differences between rooted cuttings and zygotic seedlings in subsequent field performances are well documented [26, 44] and these differences are often biologically and economically significant. It is possible that differences of similar or even greater importance will be found to exist between seedlings and (tissue-culture origin) plantlings [23], and between zygotic and somatic seedlings as well. Until we know otherwise, it will be important to adequately test for such differences between propagule types.

Testing for differences in propagule type and for selection response are both well served by having adequate and appropriate control plants for comparison. Two important principles are that the control plants be related to the test plants, and that the field design be efficieint (see 4.3 below) and unbiased. The most appropriate relative to use as a control plant is an unchanged member of the same clone. Where that is not possible, an outcrossed full-sib has important advantages over other kinds of relative. An important bit of advance planning is to store a supply of such full-sib seeds (or otherwise maintain such relatives) when beginning an experiment, just in case it works; if it does, it will then be important to go to field tests with the products.

4.3. *Cost*

Field tests of forest trees are costly in terms of space, time and money. Thus, tree breeders pay unusually close attention to the efficiency of field-test design and execution.

The simplest and most efficient field design is one that sensitively compares the tested genotypes (propagules) with the appropriate controls. Randomised pair tests are hard to beat. In this design, each test plant is first matched to an appropriate control plant with respect to size, health and physiological condition shortly before planting in the field test. Replications of such matched pairs are then planted in random orientation, usually in random locations within the general test area. The crictical measurements are of differences between the members of each pair. Fairly robust non-parametric statistics (sign tests, for example) can be used with such paired data, as can parametric statistics (paired *t*-tests, for example) when appropriate. Time costs can be lessened by putting the test on good animal-protected sites that are kept free of competing vegetation and watered if necessary. (One must consider possible genotype-by-environment interactions if performance on tougher sites is an important goal.) Space costs can be lessened by anticipating the size and age at which data are no longer of importance, and spacing the trees so that they are free-to-grow until they reach that size and age. If long-term productivity per unit area is a trait of importance, then large contiguous plots should be used instead of single-tree paired plots, with the clear recognition that such large-contiguous-plot experiments are exceedingly costly in terms of time, space and money.

5. The genetics of clones

Why clones are attractive for various production and practical purposes has been recently discussed and reviewed [9,10,11]. Briefly, clones provide substantial uniformity and reliability, and allow the selection and large-scale deployment of unusual combinations of additive-genetic, non-additive-genetic and developmental-genetic variation for traits of importance. Other advantages of clones in forestry have been discussed [31].

Why clones are not common in nature is reviewed and discussed by Libby and Ahuja [34]. Perhaps the dynamic adaptation to non-uniform and changing environments favors maintenance and repeated reconstitution of genetic variability within populations. This is well served by sexual recombination coupled with abundant reproduction, perhaps reinforced by an escape from parasites that is associated with meiosis.

We discuss below the occurrence and some possible uses of maturation-related developmental differences within clones (see Section 5.3), the occurrence and some possible uses of somaclonal variation (Section 5.2.3), and its problems (Section 7.3).

We also propose a conceptual hierarchy for thinking about clones and their production and use, under the provisional terminology 'macroclones, mini-clones, microclones, and cycloclones'. These topics will deal with genetic or developmental variability that occurs among the somatic offspring of a single zygote.

5.1. *Macroclones*

In terms of the large genetic library that comprises a zygotic genome, much more of it covaries (is transmitted in common) among the somatic lineages of a zygote than among lineages of its sexual offspring, even if the sexual lineages are produced by selfing. Thus, although various kinds of genetic variation can and do arise within clonal lineages, our experience affirms that these variations are of a substantially lower magnitude than are the variations arising as a result of sexual recombination, at least for normally outcrossing organisms such as trees. We thus define a 'macroclone' as a set of lineages that has been somatically propagated from a single zygote, with the majority of their genetic libraries varying in common as a result. Implied in this concept is that the different sublines of a macroclone have most or all traits of interest or importance in common, while differing from other macroclones in various subsets of this interesting and important suite of traits.

5.2. *Miniclones*

We suggest that a 'miniclone' is a subline of a macroclone that differs from other sublines of that macroclone in one or a few non-trivial traits. A good example of a miniclone is the semi-evergreen sport that first appeared on a branch of the Lombardy poplar (*Populus nigra italica*) clone in Chile before 1900 [43]. Although the gene for the semi-evergreen trait has been transferred sexually to other lines and species, the semi-evergreen miniclone has remained generally stable and similar to its progenitor Lombardy poplar for about a century. For example, the leaf shape, branch angle, crown form, and sex of the semi-evergreen miniclone are all identical or nearly so to that of the deciduous Lombardy poplar. These two lineages differ mainly in the time at which they shed their leaves, and also in their susceptibility to *Melampsora* leaf rust as a result of that difference. These, being non-trivial differences, merit recongnition, and calling these two lines miniclones of the same macroclone seems a useful level of distinction.

There are at least three ways to find or produce such miniclones.

5.2.1. *Sports and other somatic variations*

Somatic mutations are occurring at generally low rates for all genes of the genome. As these are noticed, often as branch sports on an otherwise 'normal' plant, they can be vegetatively propagated and form a miniclone. The semi-evergreen miniclone of the Lombardy poplar macroclone, described in Section 5.2 above, originated in this way. One may also notice that a whole ramet or whole subline of a clone is usefully different from the other members of that clone, and thus identify and name a miniclone. In this latter case, the somatic mutation(s) leading to the miniclone occurred one or more somatic generations before they were noticed.

5.2.2. *Genetic engineering*

By employing genetic engineering [2–4,46], it should be possible to take useful clones that are deficient in one or a few respects, and precisely change and improve the genetic mechanisms controlling those few defects. The derived lines would be part of their generally useful macroclones, but would differ in these few non-trivial, genetically engineered traits.

5.2.3. *Somaclonal variation*

This is defined as variation arising in tissue culture of somatic tissue, often at unusually high rates due to higher rates of gene mutation, transpositions, and various chromosomal abnormalities [20] (Skirvin *et al.*, this volume). A discussion of the effect of time in culture and details on likely causes of at least some kinds of somaclonal variation are given in a recent review [21]. In one sense it is an annoying kind and level of variation (see 7.3 below), but it is also a source of useful variation, particularly if the useful variant is stable. The stable variants are likely to be mutations of various kinds that have escaped repair and become incorporated in the genome.

The combination of a high rate of mitotic division in tissue culture, and the occurrence in tissue culture of perhaps unusual combinations of nutrients and hormones active in both the production and repair of mutational events, may explain at least in part the higher rate of mutations in tissue culture. There is substantial evidence that dividing cells are more at risk of mutations than are quiescent cells; a recent review provides evidence for this statement and six ways it can happen [12].

Somaclonal variation that is fixed as mutations may be essentially identical to sports and other somatic variations in terms of kinds and function of mutations, but it differs from them in two important respects. The first is in its (often) much higher rate of occurrence. The second is the opportunity to screen variants *in vitro*. In this latter case, miniclones may be produced as the direct products of tissue culture.

5.3. *Cycloclones*

Maturation may be viewed as a special case of developmental processes [26]. In this sense, different parts of the genetic library of a clone are being read in different tissue or at different times in the same tissue. Cyclophysis is defined as 'the process of maturation of the apical meristem' [22], while topophysis is defined as 'the phenomenon that scions, buddings and cuttings for some time after grafting, budding or rooting maintain the branch-like growth habit they had as shoots on the ortet'. In general, vegetative propagules shift from typophytic growth to normal or near-normal growth during a period of months to years after propagation, while those exhibiting cyclophysis may be different from (usually more juvenile) control plants for many years [16], and perhaps even for many cycles of vegetative propagation.

We have thus resisted the temptation to include 'topoclones' as forms of miniclones, but 'cycloclones' seem to qualify as a result of the non-trivial differences in performance that are associated with the maturation state of the clones that have mid-term and perhaps long-term stability. In this sense, one is selecting within-clone variation that is based on reading different parts (or the same parts with different intensity) of the same genetic library, rather than recombining and selecting or changing the DNA of a clone's genome.

In radiata pine, selecting cycloclones at a maturation state of 4–8 years, compared to using the same macroclones at more juvenile states, provides substantial advantages of both harvest index and disease resistance, somewhat offset by greater difficulty of propagation and slower diameter growth [25].

5.2. *Microclones*

On the one hand, mutation rates per locus for various organisms are commonly reported in the neighborhood of 1 in 100 000 (10^{-5}) to one in 1 000 000 (10^{-6}) per cell division. This leads to the prediction that about one cell in a hundred has a mutation at one of its loci. On the other hand, premutational events apparently occur at rates of 100 000 to a million per cell per day! Most types of such premutational DNA damage can be repaired by efficient enzyme systems, but not all repairs are completed, particularly when cells are in the process of division [13].

If one takes the view that a cell mutates about once in a hundred divisions, then small cell lines may have complete genetic fidelity and form true clones. But if one takes the view that each cell may have many mutations, then there is no such thing as a true clone, even at the cellular level, and each cell may be considered its own 'microclone'.

Under either view, in independent plant propagules that are 15 cm or so tall, each consisting of many millions of cells, it is likely that no two propagules are absolutely alike genetically at all of their loci, and thus each may be considered a 'microclone' with respect to all others of its macroclone. This is interesting intellectual detail, but if the differences are very small, that is, biologically or economically trivial, the microclone level is not a very useful level. We present it to differentiate it from the miniclone level, which may be useful both biologically and economically. It may be noted that, as perceptions or values change, what could or would have been classified as trivial microclonal variation may be upgraded to the level of miniclones.

We also mention that we have resisted the temptation to define a microclone as a product of micropropagation. That would have been a possibility. But when considered in the context of genetic covariance and genetic variation, the next logical level from a miniclone would be a microclone. Therefore, there might be some ambiguity in the use of the term microclone.

6. Uses

This section focuses on the uses of tissue culture for producing clones to be used for various options in forestry. The broader uses of clones in forestry have been presented or reviewed [9,31,37]. We take the view that clones produced by tissue culture in the narrow sense, namely the production of plantlets from callus or organ cultures, will remain labor intensive, and thus high-cost and specialized. But if somatic embryogenesis [15] (Gupta and Kreitinger, this volume) can be achieved and automated to mass produce artificial seeds, then these artificial seeds will serve the broad range of uses referenced above, and will in many cases do so better than grafts and rooted cuttings now do.

6.1. *For* in vitro *selection*

There are already examples of screening for and selecting useful or interesting variants in cell culture or tissue culture (see Michler, this volume), and then recovering trees with those traits by propagating plantlets from the selected lines. Fertile imaginations can create and have created lists of projects where such screening can be done more effectively in culture than in field trials of whole plants.

In vitro selection for salt or drought tolerance has been made in tissue cultures of colt cherry [40]. By challenging leaf tissues to glyphosate and sulfometuron, putative enhanced herbicide tolerance was observed in a *Populus* hybrid [35] (Michler, this volume). In addition, *in vitro* genetic selection has been made in *Populus* for tolerance against three diseases, *Septoria* leaf spot

[41], and *Melampsora* leaf rust [42], *Hypoxylon* canker [49]. All these are preliminary results, and the tolerant plants have not yet been field tested.

6.2. *Control of maturation state*

The fate of clonal forestry is not clear. It is clouded by three main uncertainties: cost, breeding effectiveness, and clonel maturation [37]. Of these uncertainties, cost seems the least serious. It will yield to a demonstration of sufficient benefits to justify the higher costs of cloning, and/or to experience-generated efficiencies and economies of scale that will lower the costs of clonal relative to zygotic production of propagules.

For many decades to come, there will be a contest between the values of excellent well-known clones and the values of even better zygotic lines. Its outcome will depend on two variables: (a) how much better the excellent clones are than the population from which they were selected; and (b) how long it will take aggressive breeders to produce zygotic lines that, on average, are better than the clones in use. These zygotic populations must be sufficiently better than the deployed clones to offset their advantages of uniformity and familiarity. Experience and knowledge about the performance of deployed clones throughout the harvest cycle develops management efficiency and confidence in those clones. If breeders can produce lines that substantially exceed the performance of the deployed clones in less than two or three harvest cycles, then many of the important advantages of clonal forestry cannot be realized [31]. Only time will tell, and the answer will probably not be the same for all situations.

The problem that many, perhaps most, clones change maturation state, and thus change their average performance in important traits, is currently the most serious one impeding the adoption of clonal forestry. The possibility of manipulating the maturation states of important clones in tissue culture is thus of great interest to all present and potential practitioners of clonal forestry.

In a practical sense, the manipulation of maturation state can have one or more of three goals: to maintain a particular maturation state, to rejuvenate a clone, or to rapidly mature a clone. The fundamental knowledge likely to be obtained while learning to do such maturation-state manipulation may also contribute in important ways to the broader understanding of the biology of aging and maturation [9,26].

6.2.1. *Maintenance*

If a clone has been tested at one or more maturation states, it can be responsibly deployed and used only at those states. Each maturation state of a clone can be considered a miniclone, or cycloclone, see above; to be true-to-type, the maturation state of the clonal donor tissue or plants must be maintained.

It is possible to do this crudely by hedging donor plants [33] or by serial propagation [48], but careful control of media and environment in tissue culture may do so more precisely.

Cryopreservation of tissue samples of a clone is an attractive option [6,8]. In this scheme, the maturation-state 'window' would be identified, with acceptable true-to-type performance by ramets at the maturation states defined by that window. Donor tissues or plants would be used as long as their maturation state was within the window. They would be replaced by new donors when they mature beyond the window. This will work if the replacement donors have their maturation state maintained at or below the low end of the window during cryopreservation, and if they then mature within it during their use as donors.

6.2.2. *Rejuvenation*

The ability to repeatedly return donor tissue to a fully juvenile state, or to some defined less-mature state, would make maintenance of maturation state less important. It would also allow outstanding trees or clones that are not identified until they are beyond juvenile to be returned to a juvenile state for testing and possible eventual deployment [18]. This is not yet reliable technology, although there are surely examples of the performance of tissue with respect to ease of propagation and subsequent growth being improved during tissue culture [38,45]. The development of reliable diagnostic biochemical markers to detect rejuvenated tissue in culture is an attractive goal [17]. It is both difficult and important to differentiate reinvigoration from rejuvenation, one difference being in the subsequent stability of the more juvenile performance.

6.2.3. *Maturation*

Of the three manipulation options, rapidly increasing maturation state during tissue culture seems the least important. Yet, it is not without merit, and there is some evidence that it has been done by mistake and thus can be done on purpose [23]. Shortening the time to reproductive competence may be the most frequent purposeful use of this technique; putting clones into test in an advanced maturation state may also allow them to be screened for such things as susceptibility to insects and diseases that do not normally attack juveniles.

6.3. *To produce hedges or other steckling donors*

It is possible to combine hedging and tissue culture in production programs. For example, in coast redwood, the maturation state of clones is reduced or maintained in tissue culture. Plantlets for donor hedges are repeatedly extracted from the tissue-culture lines in small numbers, hedged, and the hedges are used to achieve the multiplying power to produce sufficient cuttings to deploy about a million stecklings per year [45].

6.3.1. *Cost*

With some exceptions, the cost of a rooted cutting is substantially less than the cost of a plantlet, and, because the plantlet needs more intensive aftercare, the cost differential between a steckling and plantling is ever greater. Thus, the scheme described above is cost effective, with small numbers of high-cost plantlings being used to produce very large numbers of lower cost stecklings.

An alternative is to use 100 percent plantlings. This would require a large tissue-culture facility and a large intensive-care greenhouse (see 6.4.1 below). The redwood operation described above needs only a modest tissue-culture facility, about a hectare of land devoted to a hedge-orchard, and the rooting of the cuttings and production of stecklings is accomplished in the same (slightly modified) greenhouse facilities as had been used to produce redwood seedlings.

6.3.2. *To achieve desired cycloclonal state*

Suppose the optimal maturation state for deployment is the equivalent of 4–8-year-old seedlings. This in fact is the case for one radiata pine program in New Zealand [25]. Rather than trying to maintain tissue-culture stocks at this somewhat-more-difficult maturation state, it may be possible to maintain clones in a fully juvenile state (see 6.2.1 above), and then grow cutting donors until they enter the '4–8' maturation-state window (not necessary exactly four years after they come out of storage). There is then approximately a five-year period when each plant can be donating hundreds of cuttings per year, and as each donor matures through the '8' end of the window, new donors are entering the '4' beginning of the window.

6.4. *For direct deployment in production forestry*

By this is meant that the immediate products of tissue-culture will, after hardening-off and some growth in a nursery, be deployed as plantlings to practical forest plantations.

6.4.1. *Plantlings*

In spite of our expressed viewpoint in the introduction to Section 6 above, there is at least one conifer program that is moving to operational-level production of plantlings. These are produced as tissue-culture shoots that are excised and rooted to form plantlets, and that then go through a four-stage greenhouse (a series of three hardening sections, and a normal growing area) to produce plantation-ready plantlings. A large facility was completed in New Zealand in 1988, and it is commissioned to produce 2.5 million radiata pine plantlings per year (see Nairn, this volume; also [24]). The projected net difference between these tissue-culture clones and the seedling stock recently used for afforestation is about NZ $ 60 000 000 (ca. US $ 35 000 000) per year. Similarly, a large-scale production of aspen and hybrid aspen by tissue culture was achieved at the Kleinwanzlebener Saatzucht (KWS) in Germany [14]. There are also clonal forestry programs in other countries, including both macro- and micropropagation, for large scale production of forest tree species (see [47]).

6.4.2. *Germinated somatic embryos*

In spite of our being not quite right about production deployment of plantlings, it is also our view that production deployment of germinated somatic embryos through a nursery to plantations will be problematic. We base this opinion on the likelihood that handling germinated somatic embryos will continue for some time to be delicate and labor-intensive. They will serve to establish early test plantations, so that information and confidence about the performance of this class of propagules may be gained.

6.4.3. *Encapsulated somatic embryos*

At such time that the technology of automating the production and encapsulation of somatic embryos becomes available and reliable, then it is our view that clonal forestry will largely replace zygotic forestry in most forest plantations. This presupposes that control of maturation state has been achieved as part of the process of somatic embryogenesis.

6.5. *For genetic conservation*

Genetic conservation of our forest tree species is a good idea under any circumstances. For those of us involved in the replacement of natural systems with human-designed ones, which clonal forestry surely does, then particular attention to the needs of genetic conservation is ethically appropriate. It is fortunate that tissue culture contributes not only to the problem but also to the solution. The solutions range from contributing to the research that characterizes patterns of natural and domesticated genetic variability, to cryopreservation for long-term conservation, to recovery of live plants from conserved material for reintroduction in the forests [36].

6.5.1. *Biodiversity*

Conserving individual genotypes of important forest tree species is one thing. Conserving the biodiversity of natural ecosystems, or creating appropriate biodiversity in a reconstituted ecosystem, are more complex challenges.

Conserving the biodiversity of natural ecosystems seems best accomplished by allowing many ecosystems to continue to exist and evolve with minimum human impact. Tissue culture can contribute indirectly to this solution by increasing the productivity of plantations, thus reducing the need to enter such natural forests to satisfy demands for a renewable wood resource.

We know little about biodiversity development in reconstituted ecosystems. We are gaining experience, whether we want to or not. In general, a good principle is that nature abhors a vacuum, and that niches in a reconstituted ecosystem will sooner or later be filled. Perhaps if, through tissue culture or other means, we have conserved appropriate samples of the keystone species, then a healthy and a diverse reconstituted ecosystem is more likely to quickly develop.

7. Concerns

7.1. *Loss of genetic variability?*

This used to be discussed as a problem of clonal forestry. Our view is that it is not a problem of clonal forestry *per se*, but that clonal forestry technology does provide a new opportunity for the management error of deploying too few clones. By proper deployment of sufficient numbers of pedigreed clones, genetic variability can be maintained, better controlled, and even enhanced, compared to management options with zygotic forestry [27]. One avoids the all-too-possible management errors through effective education and regulation.

7.2. *Regulations*

However, regulations can themselves cause problems. Some of these problems are caused by imposing rules that unnecessarily raise costs, and by imposing rules that are counterproductive. An example of the latter is to require too many clones to be deployed to each region or planting unit [39], thus losing many of the advantages of clonal forestry while adding little to or detracting from the goal of defending against physical or biotic loss [27]. Partly at issue here is the question of the robustness and ecological stability of different clones. There is a developing literature that indicates that clones can be effectively selected for such stability and robustness, and that a clonal plantation of such clones is more stable than is a zygotic plantation of recombined and untested genotypes [50].

7.3. *Somaclonal variation*

The term 'cryptic and uncharacterized variability' creates nervousness in the conservative forester. While all forest populations contain such variability, if the process of tissue culture substantially increases it [20], then that is cause for concern. Substantial rigorous testing of trees originating from tissue culture will be needed both to assess the problem, and to gain the confidence of field foresters.

7.4. *Promises unfulfilled*

The problem of promises unfulfilled, perhaps leading to a withdrawal of funding or even of reprisals, is a concern of workers in the field of forest biotechnology [32]. A useful countermeasure for tissue culture programs is to take the extra steps to get plantlings originating from tissue culture into tests and demonstration plantations. The need to overpromise to remain competitive in the current climate for funding is understandable, but may be counterproductive.

7.5. *Insufficient testing*

Many of the promises made have included the possibility of greatly shortening the time until large-scale planting can capture the benefits and profits that will flow from investments in tissue culture and other fields of biotechnology. One of the main concerns is that this will lead to *premature deployment*, before the new releases have been adequately tested. This has particular force for tissue culture, where the possibility of unusual kinds or amounts of somaclonal variation is an additional concern.

8. Summary

This chapter addresses the rather specific topic of how micropropagation may contribute to the general fields of forest science and of operational forestry. Other propagation methods, such as production of zygotic seedlings, grafting, or rooting of cuttings, are mentioned as exemplary contrasts to micropropagation. The genetics of clones is considered in the framework of new terms 'macroclones, miniclones, cycloclones, and microclones', thus allowing us to discriminate between sub-lines of a clone.

Acknowledgements

We are pleased to acknowledge useful reviews by Thimmappa Anekonda, Iris Libby, and Deborah Rogers.

References

1. Ahuja MR (1987) Somcaclonal variation. In: Bonga JM and Durzan DJ (Eds.) Cell and Tissue Culture in Forestry. Vol 1. Martinus Nijhoff Publishers, Dordrecht. pp. 272–285.
2. Ahuja MR (1988) Gene transfer in forest trees. In: Hanover JW and Keathley DE (Eds.), Genetic Manipulation of Woody Plants. Plenum Press, New York. pp. 25–41.
3. Ahuja MR (1988) Gene transfer in woody plants: perspectives and limitations. In: Ahuja MR (Ed.) Somatic Cell Genetics of Woody Plants. Kluwer Academic Publishers, Dordrecht. pp. 83–101.
4. Ahuja MR (1988) Molecular genetics of transgenic plants. In: Hällgren JE (Ed.) Fras Kempe Symposium on Molecular Genetics of Forest Trees. Swedish University of Agric Sci Umeå. pp. 127–145.
5. Ahuja MR (Ed.) (1988) Somatic Cell Genetics of Woody Plants. Kluwer Academic Publishers, Dordrecht. 225 pp.
6. Ahuja MR (1989) Storage of forest tree germplasm at sub-zero temperatures. In: Dhawan V (Ed.) Application of Biotechnology in Forestry and Horticulture. Plenum Press, New York. pp. 215–228.
7. Ahuja MR (Ed.) (1991) Woody Plant Biotechnology. Plenum Press, New York. 373 pp.
8. Ahuja MR (1991) Application of biotechnoloty to preservation of forest tree germplasm. In: Ahuja MR (Ed.) Woody Plant Biotechnology. Plenum Press, New York. pp. 307–313.
9. Ahuja MR and Libby WJ (1993) Genetics, biotechnology and clonal forestry. In: Ahuja MR and Libby WJ (Eds.) Clonal Forestry, Vol 1, Genetics and Biotechnology. Springer Verlag, Berlin, Heidelberg (in press).
10. Ahuja MR and Libby WJ (Eds.) (1993) Clonal Forestry, Vol 1. Genetics and Biotechnology. Springer Verlag, Berlin, Heidelberg (in press).
11. Ahuja MR and Libby WJ (Eds.) (1993) Clonal Forestry, Vol 2. Conservation and Application. Springer Verlag, Berlin, Heidelberg (in press).
12. Ames BN and Gold LS (1990) Chemical carcinogenesis: Too many rodent carcinogens. Proc Nat Acad Sci USA 87:7772–7776.
13. Ames BN and Gold LS (1991) Endogenous mutagens and the causes of aging and cancer. Mutation Res (in press).
14. Barocka KH, Baus M, Lontke E, and Sievert F (1985) Tissue culture as a tool for *in vitro* mass propagation of aspen. Z Pflanzenzüchtung 94:340–343.
15. Becwar MR (1993) Conifer somatic embryogenesis and clonal forestry. In: Ahuja MR and Libby WJ (Eds.) Springer Verlag, Berlin, Heidelberg (in press).
16. Bolstad PV and Libby WJ (1982) Comparison of radiata pine cuttings of hedge and tree-form origin after seven growing seasons. Silvae Genet 31:9–13.
17. Bon M-C and Monteuuis O (1991) Rejuvenation of a 100-year-old *Sequoiandendron giganteum* through *in vitro* meristem culture. II. Biochemical arguments. Physiol Plantarum 81: 116–120.
18. Bonga JM and von Aderkas P (1993) Rejuvenation of tissues from mature conifers and its impact for propagation *in vitro*. In: Ahuja MR and Libby WJ (Eds.) Clonal Forestry, Vol 1. Genetics and Biotechnology. Springer Verlag, Berlin, Heidelberg (in press).
19. Cheliak, WM (1993) Clone identification. In: Ahuja MR and Libby WJ (Eds.) Clonal Forestry, Vol 1. Genetics and Biotechnology. Springer Verlag, Berlin, Heidelberg (in press).

20. Chen Zhenghua and Ahuja MR (1993) Regeneration and genetic variation in plant tissue cultures. In Ahuja MR and Libby WJ (Eds.) Clonal Forestry, Vol 1. Genetics and Biotechnology. Springer Verlag, Berlin, Heidelberg (in press).
21. Cullis CA (1990) DNA rearrangements in response to environmental stress. Adv. in Genetics 28:73–97.
22. Dodd RS and Power AB (1988) Clarification of the term topophysis. Silvae Genet 37:14–15.
23. Frampton LJ and Foster GS (1993) Field testing vegetative propagules. In: Ahuja MR and Libby WJ (Eds.) Clonal Forestry, Vol 1. Genetics and Biotechnology. Springer Verlag, Berlin, Heidelberg (in press).
24. Gleed J (1991) Towards clonal afforestation. In: Miller CI (Ed.) FRI/NZFP Forests Ltd Clonal Forestry Workshop, New Zealand Min For Bull 160. pp. 61–62.
25. Gleed J (1993) Development of plantlings and stecklings of radiata pine. In: Ahuja MR and Libby WJ (Eds.) Clonal Forestry, Vol 2. Conservation and Application. Springer Verlag, Berlin, Heidelberg (in press).
26. Greenwood MS and Hutchinson KW (1993) Maturation as a developmental process. In: Ahuja MR and Libby WJ (Eds.) Springer Verlag, Berlin, Heidelberg (in press).
27. Libby WJ (1982) What is a safe number of clones per plantation? In: Heybroek HM, Stephan BR, and von Weissenberg K (Eds.) Resistance to Diseases and Pests in Forest Trees. Purdoc, Wageningen. pp. 342–360.
28. Libby WJ (1987) Testing and deployment of genetically engineered trees. In: Bonga JM and Durzan DJ (Eds.) Cell and Tissue Culture in Forestry, Vol 1. Martinus Nijhoff Publishers, Dordrecht. pp. 167–197.
29. Libby WJ (1987) Testing for clonal forestry. Annales Forestales 13:69–75.
30. Libby WJ (1988) Testing and deployment of brave new plantlings. In: Ahuja MR (Ed.) Somatic Cell Genetics of Woody Plants. Kluwer Academic Publishers, Dordrecht. pp. 201–209.
31. Libby WJ (1990) Advantages of clonal forestry revisited. In: Proc Joint Meeting WFGA and IUFRO Working Parties S2.02–05, 06, 12 and 14. Weyerhaueuser, Tacoma, Sec 5.18. 13 pp.
32. Libby WJ (1991) The problem of biotechnological constipation. In: Ahuja MR (Ed.) Woody Plant Biotechnology. Plenum Press, New York. pp. 323–328.
33. Libby WJ, Brown AG, and Fielding JM (1972) Effects of hedging radiata pine on production, rooting, and early growth of cuttings. NZ J For Sci 2:263–283.
34. Libby WJ and Ahuja MR (1993) The genetics of clones. In: Ahuja MR and Libby WJ (Eds.) Clonal Forestry, Vol 1. Genetics and Biotechnology, Springer Verlag, Berlin, Heidelberg (in press).
35. Michler CH and Haissig BE (1988) Increased herbicide tolerance of *in vitro* selected hybrid poplar. In: Ahuja MR (Ed.) Somatic Cell Genetics of Woody Plants. Kluwer Academic Publishers, Dordrecht. pp. 183–189.
36. Millar CI (1993) Conservation of forest tree germplasm. In: Ahuja MR and Libby WJ (Eds.) Clonal Forestry, Vol 2. Conservation and Application, Springer Verlag, Berlin, Heidelberg (in press).
37. Miller JT (Ed.) (1991) Proc. FRI/NZFP Forests Ltd Clonal Forestry Workshop. Min For FRI Bull 160. 200 pp.
38. Monteuuis O (1991) Rejuvenation of 100-year-old *Sequoiadendron giganteum* through *in vitro* meristem culture. I. Organogenic and morphological arguments. Physiol. Plantarum 81: 111–115.
39. Muhs HJ (1993) Policies, regulation, and laws affecting clonal forestry. In: Ahuja MR and Libby WJ (Eds.) Clonal Forestry, Vol 2. Conservation and Application. Springer Verlag, Berlin, Heidelberg (in press).
40. Ochatt SJ and Power JB (1989) Selection for salt and drought tolerance in protoplast- and explant-derived tissue-culture of colt cherry (*Prunus aviatum* × *pseudoserasus*). Tree Physiol 5:259–266.
41. Ostry ME and Skilling DD (1988) Somatic variation in resistance of Populus to *Septoria musiva*. Plant Disease 72:724–727.

42. Prakash CS and Thielges BA (1989) Somaclonal variation in eastern cottonwood for race-specific partial resistance to leaf rust disease. Phytopathology 79:805–808.
43. Pryor LA and Willing RR (1965) The development of poplar clones suited to low altitudes. Silvae Genet 14:123–127.
44. Roulund H (1981) Problems of clonal forestry in spruce and their influence on breeding strategy. For Abstr 42:457–471.
45. Rydelius JA and Libby WJ (1993) Arguments for redwood clonal forestry. In: Ahuja MR and Libby WJ (Eds.) Clonal Forestry, Vol 2. Conservation and Application. Springer Verlag, Berlin, Heidelberg (in press).
46. Sederoff RR and Stomp A-M (1993) DNA transfer in conifers. In: Ahuja MR and Libby WJ (Eds.) Clonal Forestry, Vol 1. Genetics and Application. Springer Verlag, Berlin, Heidelberg (in press).
47. Talbert CB, Ritchie GA, and Gupta PK (1993) Conifer vegetative propagation: an overview from a commercial perspective. In: Ahuja MR and Libby WJ (Eds.) Clonal Forestry, Vol 1. Genetics and Biotechnology. Springer Verlag, Berlin, Heidelberg (in press).
48. St Clair JB, Kleinschmit J, and Svolba J (1985) Juvenility and serial vegetative propagation of Norway spruce clones (*Picea abies* Karst.). Silvae Genet 34:42–48.
49. Valentine F, Baker S, Belanger R, Manion P, and Griffin D (1988) Screening for resistance to *Hypoxylon mammatum* in *Populus tremuloides* callus and micropropagated plantlets. In: Ahuja MR (Ed.) Somatic Cell genetics of Woody Plants. Kluwer Academic Publishers, Dordrecht. pp. 181.
50. Zobel BJ (1993) Clonal forestry in the eucalypts. In: Ahuja MR and Libby WJ (Eds.) Clonal Forestry, Vol 2. Conservation and Application. Springer Verlag, Berlin, Heidelberg (in press).

26. *In vitro* genetic selection for woody plant improvement

C. H. MICHLER

USDA Forest Service, North Central Forest Experiment Station, Forestry Sciences Laboratory, P. O. Box 898, Rhinelander, WI 54501, U.S.A.

1. Introduction and scope

Larkin and Scowcroft [34] introduced the term somaclonal variation to describe variation produced from somatic cell and tissue culture procedures that resulted in phenotypic and genotypic variation among progeny. Individual plants with somaclonal variation were thus 'somaclones'. The definition has been extended to include any variation existing in genetically variable cell cultures whether it be pre-existing variation or variation induced by the cell culture process. The process of selecting for a specific variant trait *in vitro* has been termed 'somaclonal selection'.

Early plant tissue culture researchers theorized that living totipotent cells placed in culture would regenerate clones or exact copies of the parent plant material. Phenotypic variants that arose from tissue culture regeneration methods were thus assumed to have a non-genetic ('epigenetic') rather than genetic basis and to be caused by growth regulators used in culture. Skirvin and Janick [60,61] demonstrated the importance of this useful *in vitro* variation and selected a new horticultural cultivar, 'Velvet Rose' *Pelargonium*. Shepard *et al.* [59] performed a classic study with plants regenerated from potato protoplasts and demonstrated genetic variability that could result from *in vitro* manipulations of plant cells. Chromosome number variation that resulted in aneuploids and octoploids was found to be caused by endomitosis, nuclear fragmentation, and chromosome disjunction [19].

The term somaclonal selection can have several meanings. Selected plants may have either altered genotype or phenotype or both. Genotypic variants would be the most desirable, and many phenotypic variants often revert to normal pre-selected phenotypes after removal from the tissue culture environment. In the following discussion, the term '*in vitro* genetic selection' will be substituted for somaclonal selection to emphasize the importance of selection for genetic variation. *In vitro* genetic selection describes the process of selecting for genetic variation in tissue culture, validation of the genetic change through vegetative and sexual generations, and molecular and biochemical analysis of the altered genotype.

Of the plants regenerated *in vitro* having genetic variation, only a fraction

M. R. Ahuja (Ed.), Micropropagation of Woody Plants, 443–455.

have undergone the scrutiny of genetic analyses [13,15]. Most of these variants have resulted from chromosomal changes and not from altered single genes. Chromosomal changes that have been reported in cell culture regenerants of herbaceous plants include polyploidy [6,13,54,61,63] and aneuploidy [13,27, 31,56]. Altered traits resulting from *in vitro*-derived genetic variation include fruit color [14], seed color [18], disease resistance [16], flower color [23], and leaf color [11]. Evans and Bravo [13] reported heritable single gene variation in several tomato varieties in frequencies as high as 5% with genetic variation from both nuclear and chloroplast DNA origin. Single gene alteration was also reported in maize following molecular analysis which showed a single base pair change in the gene for alcohol dehydrogenase [5].

In vitro genetic selection has recently become a valuable tool for woody plant genetic improvement to speed selection for useful plant traits such as herbicide and disease tolerance [24,25,44–46,51]. Traditional breeding of woody plants has been long-term due to extended juvenile growth before flowering, coupled with multi-generational field evaluation.

In vitro genetic selection is possible because plant cells are known to be genetically variable under *in vitro* conditions, which has been recognized for at least 30 years [47]. Heinz and Mee [26] discovered that *in vitro* genetic variation could be passed on to regenerated sugar cane progeny and expressed. Since that study, numerous examples of genetic variation in tissue culture regenerants have been reported [24,27,50,61].

Early experimentation using *in vitro* genetic selection was initially aimed at creating useful variants for use in large-scale herbaceous breeding programs. *In vitro* genetic selection also applies to woody plant breeding, although many important woody species have long generation times. Thus, plants regenerated from *in vitro* selection experiments could also be useful for short-term tree improvement with progeny that may be routinely vegetatively propagated.

This paper will review previous research on *in vitro* genetic selection of woody plants to obtain useful genetic alterations, and it will describe applications of this new technology for woody plant genetic improvement. Cell culture-derived genetic and epigenetic variation in non-woody species has been reviewed in depth elsewhere [9,12,13,15,21,33,34,53,57,66,67] and will not be the focus here.

2. Possible bases for *in vitro* genetic variation

Important genetic traits of *in vitro* variants should be passed to progeny by both vegetative and sexual means. Initially, genetic stability of *in vitro* variants may be determined by successive vegetative generations to screen against epigenetic causes. These improved plants could be released and planted following commercial vegetative propagation. In most cases, it will also be necessary to sexually transmit new traits if we intend to introduce them into other important breeding lines without repeating the initial somaclonal selection procedure with each target genotype.

In vitro genetic selection may result in selection of cells with changes in DNA that may result in altered heritable characters. As previously reviewed [1,35], chromosome numbers can change during cell culture, resulting in increased or decreased ploidy levels and gross karyotypic changes among regenerants. DNA may be altered in nuclear, chloroplast, and mitochondrial genomes. Genes may become amplified; single genes can mutate; chromosomes may rearrange by reciprocal translocation, deletion, and inversion; and genes can recombine at increased rates. Somatic crossing over and sister chromatid exchange are known to cause further deletions and duplications. Transposable elements can cause expression of genes that are normally expressed, as well as deletions and rearrangements due to imprecise excision of transposable genetic elements.

Although not an important contributor, genetic variation can be attributed to culture media components. Synthetic plant regulators such as 2,4-D, NAA, and kinetin have been suspected as causal agents that led to regenerated progeny with such traits as albinism and altered leaf shape [53]. Plant growth regulators may be less important as agents to induce genetic variation; instead they may act to increase the rate of cell division thereby increasing the potential for genetic variability.

Variation in phenotype of plants resulting from tissue culture that cannot be explained by genetics or transmitted through meiosis is termed epigenetic [13]. Epigenetic changes can result from a change in environment or temporary morphogenetic response to components of the tissue culture growth medium. Cytokinin habituation [3,40–42], cycloheximide resistance [22,37], and picloram resistance [7] are examples of the latter. These epigenetic traits were stable through several cell cycles without the inducible factors, but were reversible. Epigenetic responses can temporarily alter the physiology of regenerated plants so that their response to biotic and abiotic stress is changed, sometimes due to temporary physiological adaptation to the stress. The unaltered genome of the epigenetic plants will eventually cause reversion to normal morphological, physiological, or biochemical response to the environment [9].

Epigenetic effects will continue to plague tissue culture researchers until biochemical markers are developed to gain a greater understanding of regeneration, and growth and developmental physiology *in vitro*. For example, in one study, plum rootstocks regenerated *in vitro* grew as rosettes instead of upright trees [30]. Further investigations showed that the variance was due to lack of chilling, which could be overcome by cold temperatures or gibberellic acid (GA_3) applications [28].

Practical use of *in vitro* genetic selection may be limited to single gene or simple genetic alterations [67]. Few, if any, reports have shown where increased ploidy levels or large changes in DNA have been useful. This would be expected because most beneficial horticultural gene mutations that have resulted in new varieties were single trait characteristics [67].

3. *In vitro* genetic selection and regeneration methods

Selection for genetic variability in cell culture, in theory, should be a straightforward process. Unfortunately, efficient selection strategies require numerous *in vitro* manipulations requiring an understanding of plant growth and development, and biochemistry for each step [67]. *In vitro* genetic selection can require: (1) maintenance of continuous cell and shoot cultures; (2) selection at the cellular, tissue, and whole plant level with both direct and indirect stress agents; (3) regeneration of plants from protoplasts (single and fused), cell suspensions, and organ explants by organogenesis or embryogenesis; (4) knowledge and use of mutagenic agents if applied in culture; (5) development of haploids for selection of recessive traits; (6) development of physical tissue culture techniques (i.e. nurse cultures, suspended artificial supports for cells and tissues, roller culture); and (7) testing of the biochemical and genetic bases for the new traits.

With woody plants, regeneration protocols usually are not transferable between species, and basic research is required to determine mechanisms for regeneration and plant development *in vitro*. Research with species such as *Populus* will further our basic understanding of selection and regeneration, although protocols may not be directly and easily applied to other woody species.

Callus and cell suspension culture allows populations of genetically variant cells to divide and multiple in a culture thereby increasing the possibility that plants can be regenerated with genetic variation [55]. Unfortunately, shoots are difficult to regenerate from callus cells, and regeneration potential decreases with increasing culture age. Regeneration of plants by somatic embryogenesis from cell suspensions may be an alternate means of producing genetic variants although it may not be easier than callus regeneration. Reports of successful regeneration of woody species by somatic embryogenesis are increasing [8,17,20,43,48,62].

Protoplasts, individual cells without cell walls, may produce the most progeny with genetic variation. Plants regenerated from protoplasts may result in clonal replicates of the initial protoplast that may contain an altered genome from the parent plant. The first reports of plants produced from protoplasts included genetic variants [64]. These plants were altered in morphology, fertility, and chromosome number.

The limiting factor for using protoplasts for *in vitro* genetic selection has been the lack of regeneration techniques for protoplasts of woody species, although successful regeneration has now been noted for more than 12 species [39]. In only one case has stable variation been noted for a woody species regenerated from protoplasts and that has been with hybrid poplar (B. McCown, personal communication). Altered traits noted for these protoplast-derived progeny following vegetative propagation and two seasons in field trials were increased leaf length, altered leaf pubescence, and dwarfism.

Haploid culture can be the step before *in vitro* genetic selection because it provides a more efficient method selecting for recessive traits [66]. Plants with important new traits would presumably result from genetic expression from a single dominant gene that before haploid culture had been recessive. Haploid culture allows for genes that may normally be recessive to be expressed after chromosome doubling and regeneration.

To maintain genetic changes that occur from *in vitro* genetic selection, clonal micropropagation techniques must be used to produce progeny of genetic variants. Shoots regenerated from either shoot meristems or axillary buds rarely produce genetic variants unless spontaneous mutations within those meristems pre-exist before culture [68]. Clonal fidelity can be maintained best with axillary shoot proliferation because meristem culture often results in callus formation at the base of the wounded shoot meristem. Genetic variation could result from adventitious regeneration initiated from this wound callus.

Selection and, to a greater extent, regeneration of useful genetically variant cells are rare. Thus, increased selection and regeneration efficiency is needed to improve the production frequency of important somaclones. Cellular genetic variation can often occur at frequencies of 1–10%, but resulting useful genetic variation for a specific trait is a fraction of this rate [21]. For example, if disease tolerance were the goal, the genetic variant would have to exist in culture, and more importantly, be regenerated. Because of the low probability of cellular regeneration with genetic variation, regeneration protocols must produce high frequency regeneration from existing cells. The variant cells may exist in an explant, or genetically variant cell populations must be enriched as in cell suspension culture to increase the number and thus the frequency of variant plant production. Frequent subculture of cell suspensions is one way to increase variation frequency by maintenance of continuous rapid cell division.

Physical parameters of cell culture can be optimized to increase rates of genetic variation [21]. Temperature affects rate of cell division, and faster cell division promotes DNA alteration. Heat shock is an example of a known agent that may influence genetic instability by causing spontaneous mutation [10]. Light effects may be less important due to dose levels of ultraviolet rays are not high enough to mutagenic. Light can affect gene amplification, but there is a lack of evidence to support any quantitative genomic changes associated with this parameter. The osmolarity of culture media also affects cell division rate that could likewise affect genetic variation rate.

Compared to physical parameters, control of chemical parameters in cell culture is more readily recognized as a means of generating genetic variability. Auxins and cytokinins can be manipulated to induce rapid, unorganized growth leading to genetic alteration. 2,4-D has been used most widely because it induces cell suspensions to produce cells that will not differentiate in its presence to form organized plant tissues.

4. Applications in tree improvement

Most of the available evidence for somatic genetic variation comes from horticultural literature. These genetic variants arise naturally, and many have been selected as superior genotypes that provide us with many of our important fruit and ornamental species. As evidence for this, 5000 of 8800 plant varieties cultivated in Europe in 1899 were the result of bud sports [58]. *In vitro* genetic selection techniques allow scientists to produce plants with somatic variation *in vitro* instead of relying only on visual observation of bud sports in field plantings. The following are examples of direct *in vitro* selection for tree improvement.

4.1. *Plant morphology*

Lester and Berbee [36] reported variation in poplar trees regenerated from callus culture. Trees derived from *Populus* × *euramericana* cultures had altered leaf morphology, branching habit, and height. Chromosome alteration was evidence for some of the phenotypic variation seen, but altered characteristics could have partially resulted from virus elimination due to heat treatment and the use of 2,4-D as a synthetic auxin source in the culture medium.

Leaf variegation is an ornamental feature that often results from using somaclonal selection techniques with many woody plants. Brand and Lineberger [4] reported three new leaf variegation patterns that resulted from adventitious shoot regeneration on leaf explants of *Liquidambar styraciflua* 'Variegata'. Although this was an example of somaclonal variation, it was also a chmieral rearrangement. Stable spontaneous leaf variegation has also been noted in adventitious shoot cultures of hybrid poplar (C. Michler, unpublished). Of three clones producing variegation, one resulted in a periclinal chimera. In another example, Marcotrigiano and Jagannathan [38] have commercially introduced a variegated cultivar of *Paulownia tomentosa* that arose as a variant from regenerated callus tissue.

4.2. *Herbicide tolerance*

The first putative enhanced chemical tolerance produced in a forest tree species through *in vitro* genetic selection techniques was achieved with hybrid poplar [44–46]. Glyphosate and sulfometuron methyl were used as herbicidal selective agents to screen for increased herbicide tolerance in adventitious shoots regenerated from leaf explants and in cell suspension callus. Somatic embryos were regenerated from herbicide tolerant callus and have yet to be tested as whole plants *ex vitro*. Herbicide tolerant shoots regenerated from leaf explants were sprayed with commercial formulations of the herbicides, and several variants were selected for field testing based upon useful tolerance levels expressed in greenhouse tests.

Enzyme activity assays were performed on trees with putative increased sulfometuron methyl tolerance to determine acetolactate synthase (ALS) activity. Results of those tests showed both increased ALS activity in some variants and other trees with equal ALS activity compared to non-selected parent clones. The results suggested several mechanisms of apparent herbicide tolerance.

4.3. *Salt and drought tolerance*

Citrus callus cell lines [2,32] have been selected *in vitro* for tolerance to salt and water stress, but the first woody tree species that was regenerated with salt or drought tolerance was Colt cherry (*Prunus avium* × *pseudocerasus*) [49]. Callus cells from leaf and root explants and from cell suspension protoplasts were selected on NaCl, Na_2SO_4, or KCL. Direct recurrent selection was used whereby the salt or osmotic agent was used only in alternating culture cycles. This procedure could possibly remove epigentic variant cell lines. These trees are currently in greenhouse trials. Callus cultures were re-initiated from these trees and successfully tested for stability of salt or drought tolerance in further *in vitro* tests.

4.4. *Disease resistance*

In vitro genetic selection has been used to produce *Populus* with putative resistance to two disease causal agents, Septoria leaf spot [51] and Melampsora leaf rust [52], and preliminary tolerance studies were attempted with a third agent, Hypoxylon canker [65]. All the foregoing studies are preliminary and lack proof of genetic variation at this time. Variants of hybrid poplar with increased resistance to Septoria leaf spot were regenerated from internodal stem callus. These variants were screened with a leaf disk bioassay by using a crude culture filtrate from the causal organism. Plants with increased resistance selected from susceptible clones are undergoing field evaluation.

Resistance to Melampsora leaf rust was studied using *Populus deltoides* shoots regenerated from callus. Tissue culture derived plants were tested in the greenhouse by inoculations with two races of *M. medusae* onto leaf disks. A differential response was seen; regenerates had both increased and decreased resistance to the pathogen.

A similar procedure was developed to screen and select aspen callus and micropropagated plantlets for resistance to Hypoxylon canker. Fungal culture filtrates were used to select callus lines, and ascospores were used to inoculate regenerated plants. Although no resistant plant have been reported to date, knowledge was gained about applying these screening methods for use in selection of resistant aspen genotypes *in vitro* as opposed to field trial selection.

Peach trees with putative enhanced disease resistance have been produced by using *in vitro* genetic selection techniques [25]. Embryo callus regenerants of two varieties, 'Sunhigh' and 'Redhaven', were screened at the whole plant level with a leaf bioassay for resistance to leaf spot (*Xanthomonas campestris* pv. *pruni*) and bacterial canker (*Pseudomonas synringae* pv. *synringae*). Trees were produced with enhanced resistance to both diseases and were successfully vegetatively propagated while retaining disease resistance. Epigenetic variants did occur for canker resistance. At this point, there is no other evidence for a genetic basis for resistance.

4.5. *Flowering*

Huhtinen [29] reported variation in *Betula pendula* for altered time of flowering. Plants regenerated from cambial callus of an early flowering cultivar had a mix of flowering and non-flowering branches on first year's growth compared to the normal trait of all flowering branches. Unfortunately, no further reports were made of vegetative propagation, long-term field observation, or genetic analysis of these plants.

5. Future applications

5.1. *Plant morphology*

Selection for altered plant morphology has implications for improvement in all commercial uses of woody plants. One important trait is branch angle, which is genetically controlled and highly heritable. In fruit tree production, much labor must be put into manual branch spreading techniques and pruning to improve light interception within the tree canopy for enhanced spur branch production. In forestry, maximum photosynthesis through canopy light interception and upright growth habit are important forest tree attributes. In woody ornamental plant improvement, compactness, dwarfism, and upright growth are some characteristics that allow use of landscape plants in smaller landscapes where space is limited. Other characteristics such as altered flower, leaf, fruit color and shape, and sterility could be important where desired morphological changes would increase commercial usage or plant productivity.

5.2. *Biotic and abiotic stress tolerance*

Biotic and abiotic stresses limit the productivity of many woody plants. These stresses include drought, frost, temperature extremes, salt and metal toxicity, insect and disease pests, and air pollutants. If used for *in vitro* genetic selection these stress agents can be directly applied to cell cultures or substitute chemicals are available that mimic the biochemical causal agent. For example, sodium chloride can be added to cell cultures for salt tolerance screening. Conversely, diseases and insects are very complex, and plants may need multiple mechanisms to resist stress from these agents. A greater basic understanding of tree physiology is necessary before tolerance to these stresses can be selected for and studied because we lack knowledge of biochemical markers, mechanisms of action, and cellular response to these agents.

5.3. *Post-harvest tree fruit quality*

In vitro genetic selection to improve post-harvest fruit quality is an unexplored avenue of fruit tree improvement. Although attention has not been paid to this subject, post-harvest fruit quality is a well-studied topic in plant physiology, and much is known about improving fruit quality through manipulation of senescence by alteration of the ethylene biosynthesis pathway and inhibition of peroxidases. The negative aspect of this research is that fruit trees have generation times of at least 3–5 years before results can be initially evaluated.

5.4. *Interface with plant breeding and molecular biology*

The successful use of genetic variant selection techniques will require a close relationship with plant breeding. Once problems are overcome with plant regeneration from cell culture, rapid plant genetic improvement can be made by using *in vitro* genetic variants as parent breeding material for future superior breeding lines.

Plant breeders will be interested in using *in vitro* genetic variants only after the variants have passed stringent tests of classical breeding programs [12]. Genetic changes must be transmitted to sexual and vegetative progeny. Changes must have agricultural value or scientific utility, and the traits must be unavailable in current parent germplasm. Selection of variants must be quicker than conventional breeding methods, and these variants should occur in important parent breeding lines to be rapidly introduced into commercial use.

In vitro genetic variants may also be genetically engineered with available genes to provide plants with multiple introduced traits. For example, poplars with glyphosate tolerance could be transformed to express insect resistance to cottonwood leaf beetle, thereby producing a commercially important tree with two important traits that would normally limit its use and productivity.

5.5. *Production of secondary metabolites*

Besides turpentine and resin, woody plants are an untapped source of secondary metabolites. Perhaps cell lines in bioreactors could be used to produce pigments, medicinals, and other useful byproducts. In reality, only high value bioreactor-generated products would be feasible as a substitute for known synthetic chemical production methods or when natural secondary production within the plant is the most efficient method.

6. Summary

Since the recognition of *in vitro* genetic variation as a useful tool for genetic improvement, several important applications have been made to woody plants to produce herbicide tolerant and disease resistant poplars, and disease resistant peach trees. The first and possibly most useful application has been to screen and select genetic variants at the cellular level, and at the whole plant level. Abiotic and biotic agents, i.e. chemicals, disease toxins, metals, etc. can be used to directly screen for genetically variant cells; the limiting factor to success is lack of adequate regeneration protocols for many woody species. A greater understanding of plant growth and development and the use of biochemical markers is needed to increase selection and regeneration frequencies. Future widespread use of biotechnology in plant improvement will include use of this elite germplasm in breeding programs and genetic engineering.

References

1. Bayliss M (1980) Chromosomal variation in plant tissues in culture. Intern Rev Cytol Suppl 11B:113–144.
2. Ben-Hayyim G (1987) Relationship between salt tolerance and resistance to polyethylene glycol-induced water stress in cultured *Citrus* cells. Plant Physiol 85:430–433.
3. Binns AN and Meins F (1980) Chromosome number and the degree of cytokinin habituation of cultured tobacco pith cells. Protoplasma 103:179–187.
4. Brand MH and Lineberger RD (1988) *In vitro* adventitious shoot formation on mature-phase leaves and petioles of *Liquidambar styraciflua* L. Plant Science 57:173–179.
5. Brettell RIS, Dennis ES, Scowcroft WR, and Peacock WJ (1986) Molecular analysis of a somaclonal variant of maize alcohol dehydrogenase. Mol Gen Genet 22:235–239.
6. Brossard D (1976) Influence of kinetin on formation and ploidy level of buds arising from *Nicotiana tabacum* pith tissue grown *in vitro*. Z Pflanzenphysiol 78:323–333.
7. Chaleff RS and Parsons MF (1978) Direct selection *in vitro* for herbicide-resistant mutants of *Nicotiana tabacum*. Proc Natl Acad Sci USA. 75:5104–5107.
8. Cheema GS (1989) Somatic embryogenesis and plant regeneration from cell suspension and tissue cultures of mature himalayan poplar (*Populus ciliata*). Plant Cell Rep 8:124–127.
9. Cullis CA and Creissen GP (1987) Genome organization and variation in higher plants. Ann Bot 60:(Suppl. 4)103–113.
10. Drake JW and McGuire J (1967) Characteristics of mutations appearing spontaneously in extracellular particles of bacteriophage T4. Genetics 55:387–398.
11. Dulieu H and Barbier M (1982) High frequencies of genetic variant plants regenerated from cotyledons of tobacco. In: Earle L and Demarly Y (Eds.) Variability in Plants Regenerated from Tissue Culture. New York, Praeger Press. pp. 211–229.

12. Earle ED and Gracen VE (1985) Somaclonal variation in progeny of plants from corn tissue cultures. In: Henke RR, Hughes KW, Constantin MJ, and Hollaender A (Eds.) Tissue Culture in Forestry and Agriculture. New York, NY. Plenum Press. pp. 139–152.
13. Evans DA and Bravo JE (1986) Phenotypic and genotypic stability of tissue cultured plants. In: Zimmerman RH *et al.* (Eds.) Tissue Culture as a Plant Production System for Horticultural Crops. Dordrecht, Martinus Nijhoff Pub. pp. 73–94.
14. Evans DA and Sharp WR (1983) Single gene mutations in tomato plant regenerated from tissue culture. Science 221:949–951.
15. Evans DA and Sharp WR (1986) Applications of somaclonal variation. Bio/Technology 4: 528–532.
16. Evans DA, Sharp WR, and Medina-Filho HP (1984) Somaclonal and gametoclonal variation. Amer J Bot 71:759–774.
17. Fukui H, Nishimoto K, Marase I, and Nakamura M (1988) Somatic embryogenesis from the leaf tissues of continuously sub-cultured shoots in Japanese persimmon (*Diospyros kaki* Thunb.). Japan J Breed 38:465–469.
18. George L and Rao PS (1983) Yellow seeded variants in *in vitro* regenerants of mustard (*Brassica juncea* Coss var. RAI-5). Plant Sci Lett 30:327–330.
19. Gill BS, Kam-Morgan LNW, and Shepard JF (1986) Origin of chromosomal and phenotypic variation in potato protoclones. J Hered 77:13–16.
20. Gonzalez ML, Veitez AM, and Veitez E (1985) Somatic embryogenesis from chestnut cotyledon tissue cultured *in vitro*. Sci Hort 27:97–103.
21. Gould AR (1986) Factors controlling generation of variability *in vitro*. In: Vasil IK (Ed.) Cell Culture and Somatic Cell Genetics of Plants, Vol. 3, Orlando, FL. Academic Press pp. 549–568.
22. Gresshoff PM (1979) Cycloheximide resistance in *Daucus carota* cell cultures. Theor Appl Genet 54:141–143.
23. Groose RW and Bingham ET (1984) Variation in plants regenerated from tissue culture of tetraploid alfalfa heterozygous for several traits. Crop Sci 24:655–658.
24. Hammerschlag FA (1988) Selection of peach cells for insensitivity to culture filtrates of *Xanthomonas campestris* pv. *pruni* and regeneration of resistant plants. Theor Appl Genet 76:865–869.
25. Hammerschlag FA and Ognjanov V (1989) Somaclonal variation in peach: screening for resistance to *Xanthomonas campestris* pv. *pruni* and *Pseudomonas syringae* pv. *synringae*. Acta Hort (in press).
26. Heinz D and Mee G (1969) Plant differentiation from callus tissue of *Saccharum* species. Crop Science 9:346–348.
27. Heinz D and Mee G (1971) Morphologic, cytogenic, and enzymatic variation in *Saccharum* species hybrid clones derived from callus tissue. Amer J Bot 58:257–262.
28. Howard BH and Oehl VH (1981) Improved establishment of in vitro-propagated plum micropropagules following treatment with GA_3 or prior chilling. J Hort Sci 56:1–7.
29. Huhtinen O (1976) Early flowering of birch and its maintenance in plants regenerated through tissue cultures. Acta Hort 56:243–249.
30. Jones OP and Hopgood ME (1979) The successful propagation in vitro of two rootstocks of *Prunus*: The plum rootstock Pixy (*P. institita*) and the cherry F 12/1 (*P. avium*). J Hort Sci 52:63–66.
31. Karp A and Maddock SE (1984) Chromosome variation in wheat plants regenerated from cultured immature embryos. Theor Appl Genet 67:249–255.
32. Kochba J, Ben-Hayyim G, Spiegel-Roy P, Saad S, and Neumann H (1982) Selection of stable salt-tolerant callus cell lines and embryos in *Citrus sinensis* and *C. aurantium*. Z Pflanzenphysiol 106:111–118.
33. Krikorian AD, O'Connor SA, and Fitter MS (1989) Chromosome number variation and kayotype stability in cultures and culture-derived plants. In: Evans DA, Sharp WR, Ammirato PV, and Yamada Y. Handbook of Plant Cell Culture. Vol. 1. Techniques for propagation and breeding, NY. MacMillan Pub. pp. 541–581.

34. Larkin P and Scowcroft W (1981) Somaclonal variation – a novel source of variability from cell cultures for plant imporvement. Theor Appl Genet 60:197–214.
35. Lee M and Phillips RL (1988) The chromosomal basis of somaclonal variation. Ann Rev Plant Physiol Plant Mol Biol 39:413–437.
36. Lester DT and Berbee JG (1977) Within-clone variation among black poplar trees derived from callus culture. Forest Sci 23:122–131.
37. Maliga P, Lazar G, Svab Z, and Nagy F (1976) Transient cycloheximide resistance in a tobacco cell line. Mol Gen Genet 149:267–271.
38. Marcotrigiano M and Jagannathan L (1988) *Paulownia tomentosa* 'Somaclonal Snowstorm'. HortScience 23(1):226–227.
39. McCown BH (1988) Recent advances in protoplast culture of horticultural crops: Ornamental trees and shrubs. Sci Hort 37:257–265.
40. Meins F and Binns A (1977) Epigenetic variation of cultured somatic cells: Evidence for gradual changes in the requirement for factors promoting cell division. Proc Natl Acad Sci 74:2928–2932.
41. Meins F Jr. and Lutz J (1979) Tissue specific variation in the cytokinin habituation of cultured tobacco cells. Differentiation 15:1–6.
42. Meins F, Lutz J, and Foster R (1981) Factors influencing the incidence of habituation for cytokinin on tobacco pith tissue in culture. Plants 150:264–268.
43. Michler C and Bauer E (1987) Somatic embryogenesis in plant cell cultures of *Populus. In Vitro* Cell & Dev Biol 23(3) Part II:46A.
44. Michler CH and Bauer E (1988) Selection for herbicide tolerance in *Populus* to sulfometuron methyl. In: Hanover W and Keathley DE (Eds.) Genetic Manipulation of Woody Plants. New York. Plenum Press. pp. 479–480.
45. Michler CH and Haissig BE (1988) Increased herbicide tolerance of *in vitro* selected hybrid poplar. In: Ahuja MR (Ed.) Somatic Cell Genetics of Woody Plants. Dordrecht. Kluwer Acad. Pub. pp. 183–189.
46. Michler CH, Voelker TM, and Bauer EO (1989) Production of putative herbicide tolerant plants from somatic embryos regenerated from *in vitro* selected callus. In Vitro Cell & Dev Biol 25(3) Part II:62A.
47. Mitra J and Steward F (1961) Growth induction in cultures of *Haplopappus gracilis*. II. The behavior of the nucleus. Amer J Bot 48:358–368.
48. Muralidharan EM, Gupta PK, and Mascarenhas AF (1989) Plantlet production through high frequency somatic embryogenesis in long term cultures of *Eucalyptus citriodora*. Plant Cell Rep 8:41–43.
49. Ochatt SJ and Power JB (1989) Selection for salt and drought tolerance in protoplast- and explant-derived tissue cultures of Colt cherry (*Prunus avium* × *pseudocerasus*). Tree Physiol 5:259–266.
50. Oinuma T and Yoshida T (1974) Genetic variation among doubled haploid lines of burley tobacco varieties. Jap J Breed 24:211–216.
51. Ostry ME and Skilling DD (1988) Somatic variation in resistance of *Populus* to *Septoria musiva*. Plant Disease 72:724–727.
52. Prakash CS and Thielges BA (1989) Somaclonal variation in eastern cottonwood for race-specific partial resistance to leaf rust disease. Phytopathology 79:805–808.
53. Reisch B (1983) Genetic variability in regenerated plants. In: Evans DA, Sharp WR, Ammirato PV, and Yamada Y (Eds.) Handbook of Plant Cell Culture. Vol. 1. Techniques for propagation and breeding, New York. Macmillan Pub. pp. 748–769.
54. Reisch B and Bingham ET (1981) Plants form ethionine-resistant alfalfa tissue cultures: variation in growth and morphological characteristics. Crop Sci 21:783–788.
55. Sacristan M (1982) Resistance response to *Phoma lingam* of plants regenerated from selected cell and embryogenic cultures of haploid *Brassica napus*. Theor Appl Genet 61:193–200.
56. Sacristan M and Melchers G (1969) The caryological analysis of plants regenerated form tumorous and other callus cultures of tobacco. Mol Gen Genet 105:317–333.

57. Scowcroft WR and Larkin PJ (1982) Somaclonal variation: A new option for plant improvement. In: Vasil IK, Scowcroft WR, and Frey KJ (Eds.) Plant Improvement and Somatic Cell Genetics. Academic Press. pp. 159–178.
58. Shamel AD and Pomeroy CS (1936) Bud mutations in horticultural crops. J Hered 27: 486–494.
59. Shephard JF, Bidney D, and Shahin E (1980) Potato protoplasts in crop improvement. Science 28:17–24.
60. Skirvin R and Janick J (1976a) 'Velvet Rose' Pelargonium, a scented geranium. HortScience 11:61–62.
61. Skirvin RM and Janick J (1976b) Tissue culture-derived variation in scented *Pelargonium* spp. J Amer Soc Hort Sci 101:281–290.
62. Sommer HE and Brown CL (1980) Embryogenesis in tissue cultures of sweetgum. For Sci 26:257–260.
63. Sree Ramula K, Dijkhuis P, and Roest S (1983) Phenotypic variation and ploidy level of plants regenerated from protoplasts of tetraploid potato (*Solanum tuberosum* L. cv. Bintje). Theor Appl Genet 65:329–338.
64. Takebe I, Labib G, and Melchers G (1971) Regeneration of whole plants from isolated mesophyll protoplasts of tobacco. Naturwissenschaften 58:318–320.
65. Valentine F, Baker S, Belanger R, Manion P, and Griffin D (1988) Screening for resistance to *Hypoxylon mammatum* in *Populus tremuloides* callus and micropropagated plantlets. In: Ahuja R (Ed.) Somatic Cell Genetics of Woody Plants, Dordrecht. Kluwer Acad Pub pp. 181.
66. Wersuhn G (1988) Selecting in higher plant cell cultures. A discussion. Biochem Physiol Pflanzen 183:361–369.
67. Wersuhn G (1989) Obtaining mutants from cell cultures. Plant Breeding 102:1–9.
68. Zimmerman RH (1981) Genetic stability of horticultural plants propagated by tissue culture. Intern Plant Prop Soc 31:118–122.

27. Conifer micropropagation: its function in tree improvement programs

Y. S. PARK and J. M. BONGA
Forestry Canada Maritimes Region, P.O. Box 4000, Fredericton, N.B. E3B 5P7, Canada

1. Introduction

In contrast to herbaceous crops, many of which are propagated *in vitro* on a commercial scale, micropropagation of many tree species, especially conifers, is still largely experimental. Among the hardwoods, a number of orchard and ornamental trees and a few forest species can now be micropropagated on a commercial scale. However, large-scale micropropagation is, at present, possible with only one or two conifers species. One reason for the lack of commercialization of conifer micropropagation is cost. To be of use for large-scale reforestation, the cost per micropropagule would have to be reduced to a value close to that of a seedling [19]. Furthermore, micropropagules sometimes show signs of early maturation and other abnormalities. Considering these problems, micropropagation generally does not yet compare favorably with traditional rooting of cutting techniques. In some species, the efficiency of traditional rooting of cuttings techniques can be further improved by spraying ortet seedlings with benzylaminopurine. This stimulates shoot formation from axillary buds and thus results in an increase in the number of shoots available for conventional rooting of cuttings [20,28,31]. In those species where such additional shoot production is possible, the competitiveness of micropropagation vis à vis rooting of cuttings is even further eroded.

In view of the above, it is obvious that more efficient micropropagation systems will have to be developed before micropropagation of conifers will become a viable technique in tree improvement and reforestation operations. The prospects that this will soon be achieved have improved considerably over the last few years. It is expected that recently developed techniques for somatic embryogenesis will soon replace the traditional micropropagation techniques based on rooting adventitious or axillary shoots. The objective of this paper is to consider how an efficient clonal propagation system for conifers could benefit deployment strategies in tree improvement programs.

2. Tree breeding problems

In comparison to herbaceous, agricultural crops, which have been bred for

M. R. Ahuja (Ed.), Micropropagation of Woody Plants, 457–470.

dozens or hundreds of generations, tree breeding generally starts with unimproved wild types of trees which have been selected for two or three generations. Therefore, genetic variation in conifer planting stock is, except for a few species (e.g. *Pinus resinosa* Ait. [13]), still high. Thus, the potential for genetic improvement is considerable. Since most conifer species are outbreeders, breeding methods that are used for herbaceous, cross-pollinating crop species, such as mass and recurrent selections, and hybridization, are generally also applicable in conifer breeding.

The effectiveness of recurrent selection, however, is limited by the fact that proper selection generally cannot be carried out until the specimens have attained a considerable size. For some traits in some species, selection is possible in seedlings as young as two to three years old, e.g. selection for height in Sitka spruce (*Picea sitchensis* (Bong) Carr) [15]. However, for some traits in some species, proper selection may require field testing for as long as half the length of the rotation [26]. Furthermore, flowering in coniferous species takes several years, resulting in long generation times. A single sexual generation may vary from four to 20 years [30]. Therefore, recurrent selection, is slow and expensive.

To develop hybrid varieties for tree species is nearly impossible, mainly because of our inability to obtain inbred lines for controlled hybridization. Most conifer species carry a high load of lethal recessive genes [5,22], and repeated selfing will likely lead to loss of vigor or death. Therefore, inbreeding through six to eight generations, followed by controlled hybridization, which is one of the most potent plant breeding techniques, is not practised with conifers. A more practical breeding practice is inter-specific or inter-racial hybridization, or specific combination of individuals within a breeding population, which can result in heterosis.

3. Cloning

3.1. *General aspects*

When initiating a conifer tissue culture program, one of the first choices that will have to be made is whether to use explants from juvenile or from mature specimens. Generally, true-to-type micropropagation is far more difficult with mature than with juvenile conifer specimens [7]. (Our definition of true-to-type implies that the clonal propagule has the same capacity for development as the ortet at the time it was a seedling.) In fact, reasonably true-to-type micropropagation at reasonable production rates has, except with *Sequoia sempervirens* [14], not been achieved with mature conifers. Consequently, most experimentation with conifers has been carried out with juvenile material. However, even though micropropagation of mature conifers is currently still difficult or impossible, there is a strong incentive to try to achieve such propagation. In the following, the merits and demerits of cloning of juvenile and mature conifers will be discussed. Furthermore, we will discuss some aspects of conventional breeding strategies, and the integration of these breeding strategies with cloning, to maximize genetic improvement (see also [18, 30 and 32] for information on this subject).

3.2. *Cloning in relation to breeding; cloning of juvenile versus mature*

A simplified flow chart of possible breeding-cloning strategies in a tree im-

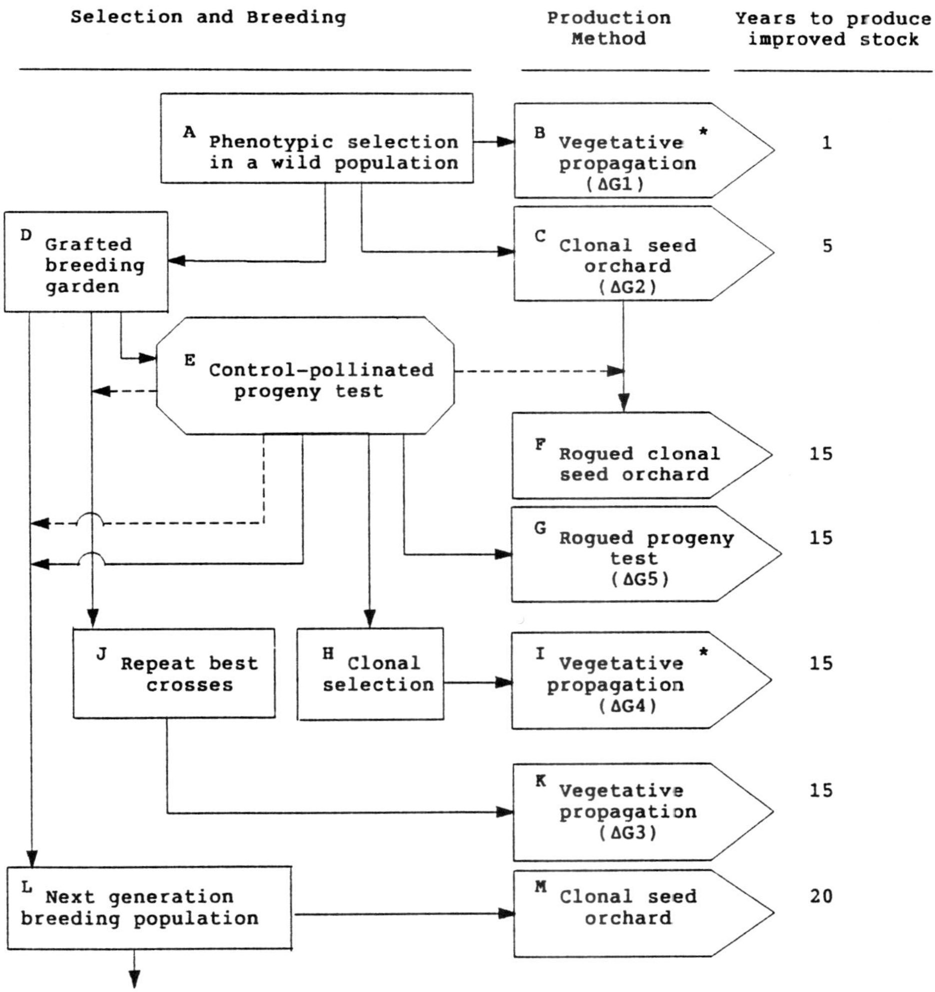

Figure 1. A simplified flow chart of possible breeding-cloning options in a tree improvement program. It is assumed that the time required for flowering and field testing is 4 and 10 years respectively.

* = It is assumed that clonal propagation is true-to-type.

---→ = Information flow.

provement program is presented in Figure 1. In the preparation of this flow chart, we decided arbitrarily that the time required for flowering is four years, and that field testing is carried out for ten years. Of course, modifications of the scheme in Figure 1 are possible. A detailed review of various combinations of breeding methods is presented by Shelbourne [30].

In Figure 1, a typical tree improvement program begins with phenotypic (plus tree) selection in the wild populaton (A). Bulk vegetative propagation of the selected trees can result in considerable genetic gain if the cloning can be done true-to-type, i.e. if there is no variation (c-effects) associated with cloning (B). Scions from the plus trees are grafted and planted in a clonal seed orchard to produce genetically improved seeds by intermating (C). Also, a portion of the grafted plus trees are placed in a breeding garden (D), where controlled-pollinations are performed. Subsequently, a progeny test is carried out (E). Based on the results obtained in this progeny test, inferior parents can be rogued from the clonal seed orchard (C), resulting in an improved clonal seed orchard (F). At this point, the progeny test (E) itself can be rogued of inferior families and used as a seed orchard (G). The best individuals from the progeny test (E) can be selected and cloned (H and I) for planting stock. However, cloning these individuals without c-effect may be difficult because the trees may no longer be juvenile after field testing. Alternatively, after the best cross combination has been identified in the field test (E), the best parents in the breeding garden (D) can be crossed again (J) and the embryos or seedlings obtained from the few seeds produced in these crosses can then be mass-cloned (K). Based on the results of the progeny test, the best individuals from the test plantation (E) and/or the breeding garden (D) can be selected to form the next generation breeding population (L). With this breeding population, steps (E) to (K) can be repeated to obtain further genetic gain, approximately within the same time frame indicated in Figure 1 for the first cycle of genetic improvement.

As mentioned, genetic variation in most conifers is high. How this variation is being made use of, and what degree of genetic gain one can expect from it in the breeding and cloning programs shown in Figure 1, is illustrated in Figure 2. For the sake of simplicity, only one trait, e.g. tree height, is displayed in Figure 2.

When we look at an evenly aged stand of a forest species on a uniform site, we generally find a tree height distribution curve that is bell-shaped and has a wide base (curve 1). This curve has a wide base because variation in tree height is large. Furthermore, the population distribution is symmetrical around the mean (dotted vertical line). Curve 1 shows total variation within the stand, i.e. it shows the sum of all genetically and environmentally caused variation. Symbolically, it can be expressed as

$$\sigma^2_P = \sigma^2_G + \sigma^2_E \tag{1}$$

where σ^2_P, σ^2_G, and σ^2_E are total phenotypic, genetic, and environmental variances, respectively. The genetic variance (σ^2_G) is the sum of three variances:

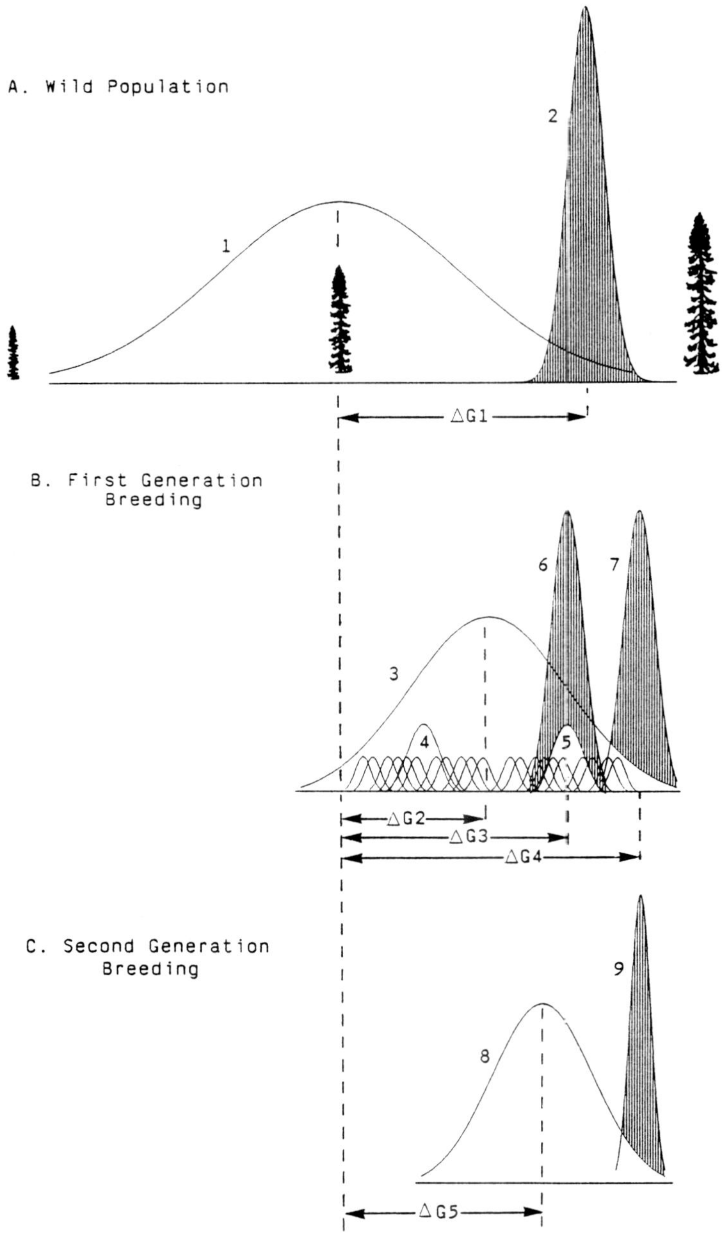

Figure 2. Genetic gain obtained by various breeding-cloning methods for one trait (height growth). Explanation see text.

additive variance due to the average effect of genes (σ^2_A), dominance variance due to interaction between alleles (σ^2_D), and epistatic variance (σ^2_I) due to interactions among alleles of different loci. Thus, Equation (1) can be written as

$$\sigma^2_P = \sigma^2_A + \sigma^2_D + \sigma^2_I + \sigma^2_E \tag{2}$$

To improve the genetic make-up of our planting stock, we have two options, i.e. sexual and asexual propagation. Dealing with the latter first, we can select the best trees from a genetically variable population (on the right hand side of curve 1, and in (A) in Figure 1), and clone them. The theoretical advantage of such cloning, if the cloning is true-to-type, is that it utilizes total genetic variance (σ^2_G), including dominance and epistatic (non-additive) variances that are difficult to capture in conventional breeding [10,25]. Thus, depending on the magnitude of non-additive variance, genetic gain for a particular trait can be substantial, especially if the trait has only a small additive variance. The magnitude of non-additive variance may vary, depending on the species and population structure. For example, in some tamarack (*Larix laricina* (Du Roi) K. Koch) populations, non-additive variance for 5-year height was about one-third to one-half the additive variance; in one population, it was larger than the additive variance [29].

The genetic gain ($\Delta G1$) expected to be obtained by 'phenotypic selection and mass-cloning', i.e. the difference between the mean of a wild population (curve 1) and that of a cloned population of selected trees within that wild population (curve 2) is

$$\Delta G1 = i \cdot H^2_I \cdot \sigma_P \tag{3}$$

where i is selection intensity, H^2_I is defined as the broad-sense heritability, which is the ratio of total genetic variance to total phenotypic variance, i.e. σ^2_G/σ^2_P, and σ_P is the phenotypic standard deviation ($\sqrt{\sigma^2_P}$) of the forest stand. One major problem currently is our inability either to clone the selected trees, or to clone them in true-to-type fashion, i.e. free of such negative cloning effects as plagiotropic growth and early maturation. In the latter case, negative propagation effects could cancel the genetic improvement that would have been gained if the propagation had been true-to-type (see Figure 1 (B)).

In conventional tree breeding, grafts of the selected plus trees are placed in a clonal seed orchard so that intermating among the trees produces genetically improved seeds. Expected genetic gain in the 'clonal seed orchard' option ($\Delta G2$) is

$$\Delta G2 = i \cdot h^2_I \cdot \sigma_P \tag{4}$$

where h^2_I is defined as the narrow-sense heritability, which is the ratio of additive variance to total phenotypic variance, i.e. σ^2_A/σ^2_P. Thus, the genetic gain under this option is mostly restricted to utilization of additive variance (σ^2_A) (see Figure 1 (C)). Since broad-sense heritability is generally larger than narrow-sense heritability, due to inclusion of non-additive variances, cloning (curve 2) will result in greater gain than sexual propagation (curve 3), provided that the clones are true-to-type.

The 'clonal seed orchard' procedure can be modified by making controlled crosses among the selected parents from the far right area of population curve 1. Thus, we obtain an improved offspring population (curve 3) consisting of several full-sib families. The short curves under curve 3 represent these full-sib families. These full-sib families can subsequently be field tested at several locations to identify superior, broadly adapted families (or parents).

In the figure, two groups of several full-sib families are shown, in which the field test (see Figure 1 (E)) either showed a minor genetic gain (curve 4) or a major genetic gain (curve 5). In a multi-generation breeding program, several superior families are selected (family selection) to produce an improved breeding population. Also, after roguing of inferior families from the test, the remaining selected families can be used to produce improved seeds for planting (see Figure 1 (G)). In both situations, the number of selected families should be sufficiently large to maintain a variable genetic base and to prevent the possibility of inbreeding. In this context, curve 5 represents several of the best families.

After the field testing has determined which crosses are superior, we can repeat the crosses (see Figure 1 (J)) and seedlings from the cross can be clonally propagated in large numbers for reforestation (see Figure 1 (K)). In this case, we combine 'family selection and mass clonal propagation'. If the initial, controlled pollination produces a sufficient number of seeds, e.g., in the order of several hundreds, half of the seeds can be stored during the field test period, to be used for mass-clonal propagation later, if the field test warrants this. In addition to mass selection gain ($\Delta G2$), the expected genetic gain in this procedure ($\Delta G3$) is

$$\Delta G3 = \Delta G2 + i \cdot H^2_{\mathrm{F}} \cdot \sigma_{\mathrm{PF}} \tag{5}$$

where H^2_{F} is known as the family mean heritability [33], which is the ratio of full-sib covariance (σ^2_{F}), i.e. $\frac{1}{2}\,\sigma^2_{\mathrm{A'}} + \frac{1}{4}\,\sigma^2_{\mathrm{D'}} + k\,\sigma^2_{\mathrm{I'}}$ where $k\,\sigma^2_{\mathrm{I'}}$ is the fraction of epistatic variance and where σ_{PF} is the phenotypic standard deviation of the full-sib family means. The primes on $\sigma^2_{\mathrm{A'}}$, $\sigma^2_{\mathrm{D'}}$, and $\sigma^2_{\mathrm{I'}}$ indicate that these variances represent the genetic variances in the selected portion of the original wild population. These are smaller than those in the original population [11].

Curve 6 represents the population obtained by mass cloning of the selected full-sib families (indicated by curve 5). Thus curve 6 has the same mean as the selected full-sib families, but its height is much greater, since it contains a large number of clonal propagules. This 'family selection and mass clonal propagation' procedure has three main advantages: firstly, the expected genetic gain is larger than the conventional clonal seed orchard procedure because it captures a portion of non-additive variance segregating among full-sib families; secondly, cloning can be done reasonably true-to-type because the ortets are young seedlings; thirdly, the cost of maintaining and operating seed orchards and of seed extraction can be reduced or eliminated. However, the greenhouse cost will be higher for producing rooted cuttings than for producing seedlings. Controlled pollinations are mostly carried out with selected grafts in breeding gardens (or clone banks), which are cheaper to maintain than seed orchards. Clonal multiplication of superior full-sib families is, for some conifers, achieved on a commercial scale through repeated cycles of rooted cuttings, initiating the process with juvenile seedlings [2,29]. A comparative economic analysis of clonal and seed orchard strategies in Norway spruce (*Picea abies* (L.) Karst) indicated that an additional genetic gain of 4% will make clonal propagation economical as long as interest rates are below 7% [27]. Cloning by *in vitro* methods has, so far, been mostly experimental (see the discussion below).

Further genetic gain can be achieved by 'clonal selection and mass vegetative propagation'. Instead of just mass cloning the population expressed by curve 5, we can restrict our cloning to those individuals that showed the best performance in the clonal field test (see Figure 1 (H)), i.e. those individuals belonging to the right-hand-side area of curve 3. Expected genetic gain in clonal selection ($\Delta G4$), in addition to $\Delta G2$ is

$$\Delta G4 = \Delta G2 + i \cdot H^2_{C} \cdot \sigma_{PC} \tag{6}$$

where H^2_C is defined as the heritability of clone means, which is the ratio of total genetic variance in the first generation breeding population ($\sigma^2_{G'}$) over phenotypic variance of clone means (σ^2_{PC}), in which the environmental variance is reduced by an increased number of ramets obtained by cloning [24]. By such clonal selection and mass vegetative propagation, we will obtain a population expressed by curve 7 (see Figure 1 (I)).

Alternatively, we can clone only the best performing clones within the best full-sib families. Expected genetic gain ($\Delta G4'$), in addition to family selection gain ($\Delta G3'$), is

$$\Delta G4' = \Delta G3' + i \cdot H^2_{C/F} \cdot \sigma_{C/F} \tag{7}$$

where $H^2_{C/F}$ is defined as the clone-within family heritability, which is the ratio of clone-within-family covariance, i.e. $\frac{1}{2}\sigma^2_{A'} + \frac{3}{4}\sigma^2_{D'} + (1-k)\,\sigma^2_{I'}$, over clone mean within family variance ($\sigma^2_{C/F}$), and where $\Delta G3'$ indicates family selection gain at a reduced selection intensity. This alternative scheme could be useful in maintaining a greater genetic diversity in planting stocks than is obtained by cloning the few best specimens.

The problem with this 'clonal selection and mass vegetative propagation' procedure is that: (1) after five or more years of field testing, the best individuals in this test have generally matured to the point where easy, true-to-type cloning is no longer possible; and (2) we cannot use seeds from a repeated cross because we do not know which of these seeds have the genetic potential expressed by the stippled area of curve 5. However, the following procedure circumvents these problems. Each individual of the best full-sib families is cloned as soon as possible, i.e. a clone, consisting of many propagules, is established from each individual prior to the field testing. A portion of the individuals of each clone is used for clonal testing and the rest are kept in a juvenile state during the clonal test (by hedging, if the clone consists of rooted cuttings; by cold storage [1], if the clone consists of micropropagules). After the clonal test has determined which of the clones are the best (which clones belong in the shaded part of curve 3), the half of these clones maintained in a juvenile state are available for mass cloning.

One problem with this approach is that hedges remain juvenile, and cold-stored micropropagated cultures remain viable, for only a limited number of years, thus limiting the length of time available for field testing. Recently, somatic embryogenesis has been induced in cryopreserved cultures of *Picea abies* and *Pinus taeda* [16], and of *Picea glauca* [21]. Because all metabolic activity ceases in cells stored in liquid nitrogen, it is expected that cryopreserved conifer cultures will remain in a physiologically unaltered state (i.e. juvenile) indefinitely in storage. If this assumption is correct, and if no genetic damage is incurred during storage, cryopreservation will allow us to field test for much longer periods of time than presently possible.

To provide a crude numeric example of the expected genetic gains obtained in the different breeding-cloning strategies discussed above, a hypothetical situation with a series of simplifying assumptions is considered. Under the additive-dominance genetic model, it is assumed that the amount of additive, dominance, and environmental variances in the wild population are in equal proportions, say 60 units each. Thus, the total phenotypic variation in the wild stand is 180 units, and narrow- and broad-sense heritabilities are 0.33 and 0.66, respectively. If we assume that the selection of plus trees is carried out at one in 100, then the selection intensity (i) is equivalent to 2.66 standard deviation units. If we clone these plus trees without c-effects [8], the expected genetic gain ($\Delta G1$) according to Equation [3], is 24. If these plus trees are used as parents in a grafted seed orchard, the expected genetic gain ($\Delta G2$) is 12. This is only half as much as that obtained by cloning, the reason being that only the additive variance is utilized in the sexual reproduction. Further, suppose that the next (first) generation is produced by a system of controlled pollinations such as a factorial mating design. For example, ten grafts from each of the plus trees are used as male and female parents producing 100 full-sib families. For each full-sib family, 10 seedlings are clonally propagated to produce 10 ramets each, amounting to a total of 10 000 propagules, which are subjected to a clonal field test. When this test is evaluated, it will provide estimates of genetic and environmental variances. Suppose the variance due to males, females, and male × female interactions is 10 units each, while that due to clones within family and within-ramets is 50 units each, then the total phenotypic variance is 130 units. After noting the covariance structure among full-sibs, the additive (σ^2_A) and dominance (σ^2_D) variances are estimated as 40 units each. Incidentally, the variation in this population is less than that in the original population as it is derived from a selected portion of the original population. If we select the ten best families from 100 full-sib families, or repeat controlled crosses that gave ten best families and clone them, the expected genetic gain, in addition to the initial plus tree selection gain, is 9, and thus the total gain in 'family selection and cloning' ($\Delta G3$) is 21. If we select the best 100 clones based on clone means and clone them without c-effects, the genetic gain, in addition to the initial plus tree selection gain, is 15. Thus the total gain in 'clonal test and mass vegetative propagation' ($\Delta G4$) is 27. Alternatively, if we select the 20 best families and clonally propagate, without c-effects, each of the five best clones within these best 20 families, cumulative genetic gain in this 'clonal selection within best families' ($\Delta G4$) is 25. If the ten best families were used as the parents of next (second) generation, expected cumulative gain ($\Delta G5$) is 18. This example is intended to illustrate that a substantial genetic gain is possible with cloning due to utilization of non-additive variance. Of course, in an operational breeding program, several alternative breeding-cloning schemes are possible. Note that the genetic gains (ΔG's) in Figure 2 are scaled approximately the same as in the above numeric example.

In many one-cycle breeding experiments, the mean of curve 5, and even of curve 7, may be to the left of the tail of curve 1, i.e. genetic gain will be less than if the best mature specimens are cloned (curve 2). However, in breeding experiments with some species, the mean of curve 7 may approach or even move to the right of that of curve 2. Whenever that is the case, there is little advantage in cloning selected mature trees over cloning selected juvenile populations. Furthermore, if more than one breeding cycle is used, the means of the population curves will tend to move to the right (curve 8 for the sexual, and curve 9 for the clonal population). Eventually this could result in a higher genetic gain, at least for the one characteristic expressed in the curves, than can obtained by cloning mature, dominant specimens.

Nevertheless, even though the combination of breeding and cloning of the juvenile, full-sib progeny is very effective in improving planting stock, there are still good reasons for trying to develop methods for cloning mature trees as well. First of all, the combination of breeding and cloning of juvenile material works well for additive genetic traits (high narrow-sense heritability), but is less effective when the traits are primarily non-additive (low heritability). Another problem is that field tests of juvenile material, lasting 5–10 years and sometimes more, are not always good indicators of what the trees will look like throughout their full rotation, i.e. perfect selections cannot be made in these tests. Finally, even though breeding is effective in improving one, or at best a few traits, it is less efficient in improving a larger number of traits simultaneously. On the other hand, where cloning of mature trees can be done effectively, mature trees with several good traits can be selected and all of these traits can then be transferred in unaltered form to the clonal propagules. For example, when selecting among mature trees, we may pick a tree which is not only tall, but also has little taper, good branching habit, good fiber length, is disease resistant, and has other good qualities. Of course, as outlined earlier, clones from mature trees will have to be field tested (preferably for many years) to determine if the desired traits are being passed on in true-to-type fashion. Furthermore, field testing is required to determine genotype-environment interactions and whether the clones are suitable for a variety of sites.

Where two species produce exceptional hybrids if crossed, e.g. when *Larix decidua* is crossed with *L. leptolepis* [12], there may be less justification for cloning superior parental trees (at least if we consider only those traits that show up well in the hybrid, ignoring those that are not or less improved by hybridization). In this situation, the mean of curve 7 in Figure 2, and possibly even the means of curves 5 and 6, may well be to the right of the mean of curve 2 of either parent species.

Cloning mature trees is also not very effective if the species we are dealing with has a narrow genetic base, e.g. in *Pinus resinosa*. In that case, curve 1 in Figure 2 is narrow at the bottom, and, consequently, the best plants (towards the right in the curve) do not have a genotype that is sufficiently better than that of the average plant in the population to warrant an expensive clonal propagation program. In this situation, sexually, and thus cheaply, obtained seedlings would, on average, be of almost the same quality as clones obtained from selected, mature trees.

4. Expected functions of micropropagation in breeding-cloning programs

As described in the foregoing, the combination of breeding and cloning, either by rooting of cuttings or micropropagation, can have an important role in tree improvement programs. Since several important new developments have occurred over the last few years in conifer tissue culture, the possible effects of these on future tree improvement schemes should be contemplated.

We can expect that recent advances in somatic embryogenesis [9] will soon lead to more efficient clonal propagation of juvenile material. This should be very helpful in providing, in conjunction with the breeding-cloning strategies described earlier, large numbers of genetically improved propagules for reforestation.

Protoplasts derived from embryogenic cultures obtained from juvenile material appear, at least in some species, to be capable of cell regeneration and morphogenesis [3,23]. Theoretically, haploid protoplasts could be fused to produce diploid hybrids that are not available through sexual propagation [6]. Protoplasts are also useful in genetic engineering. Transient genetic transformation has been achieved with protoplasts of *Picea glauca* [4].

Adding a new trait by genetic engineering to clones rather than by breeding has the advantage that we do not have to go through several breeding cycles to stabilize the new gene in the population. In addition, genetic engineering could add traits that are not available in the breeding populations, to clonal populations. However, it remains to be established how genetically stable transformed propagules are. This is of particular concern if the transformed propagules are to be used in subsequent breeding. To be useful in breeding the introduced gene has to remain stable until flowering and has to survive through meiosis and fertilization. If the transformed propagules are to be mass-propagated by cloning, it remains to be determined how well they will survive repeated cycles of clonal propagation.

The main justification for continued attempts to clone mature material is clonal propagation of specimens with superior genetic traits that have a low heritability. Once such specimens can be propagated true-to-type on a large scale, they may be further improved by careful genetic engineering. For example, a genotype superior for many traits but lacking one essential one, e.g. disease resistance, could, through genetic engineering, be supplemented with this desired trait. Research with that objective has already been initiated [17].

5. Summary

Cloning juvenile material, in conjunction with breeding programs, is an effective means of producing improved planting stock. Cloning juvenile material is relatively easy and plagiotropism and other cloning-related abnormalities are generally absent or minimal. A few cycles of traditional rooting of cuttings, for example, will rapidly increase superior full-sib family populations that were produced by breeding. In comparison, micropropagation by adventitious shoot induction and subsequent rooting of these shoots is often less effective because the propagation numbers are too low and labor costs are too high. However, somatic embryogenesis techniques, which are developing rapidly, promise to deliver large numbers of propagules much faster and more cheaply. This technique may eventually replace the traditional method of using successive cycles of rooted cuttings. By maintaining a portion of several full-sib family populations, or clones thereof, in a juvenile state by hedging or cryopreservation, while field testing the rest, it is possible to select the best clones within the best full-sib families for mass-propagation. This is a powerful tool to obtain genetically improved planting stock.

There are disadvantages in cloning juvenile material: (1) Breeding is most effective for traits with large additive variance, less so for those with large non-additive variance. Therefore, the genetic gain obtained by selection of the best full-sib families, and by subsequent cloning of these families, will only capture a fraction of non-additive variance: (2) Because of frequent unfavorable correlations among traits, the number of traits that can be improved simultaneously by breeding is often limited, at least if only a few breeding cycles are used.

Cloning selected mature trees is the preferred method if one wishes to capture non-additive variance and improve several traits simultaneously. However, methods for cloning mature conifers in a true-to-type fashion on a large scale are not yet available. Research in that direction is progressing only slowly and, consequently, it is expected that micropropagation as a tool in planting stock improvement will, for the next decade or so, remain restricted to juvenile material.

Acknowledgements

We wish to thank Drs. W. M. Cheliak and D. P. Fowler, Forestry Canada, for their critical review of this manuscript.

References

1. Aitken-Christie J and Singh AP (1987) Cold storage of tissue cultures. In: Bonga JM and Durzan DJ (Eds.) Cell and Tissue Culture in Forestry: Volume 2, Specific Principles and Methods: Growth and Developments, Martinus Nijhoff Publishers, Dordrecht, The Netherlands. pp. 285–304.
2. Armson KA, Fung M, and Bunting WR (1980) Operational rooting of black spruce cuttings. J For 78:341–343.
3. Attree SM, Bekkaoui F, Dunstan DI, and Fowke LC (1987) Regeneration of somatic embryos from protopolasts isolated from an embryogenic suspension culture of white spruce (*Picea glauca*). Plant Cell Rep 6:480–483.
4. Bekkaoui F, Pilon M, Laine E, Raju DSS, Crosby WL, and Dunstan DI (1988) Transient gene expression in electroporated *Picea glauca* protoplasts. Plant Cell Rep 7:481–484.
5. Bishir J and Namkoong G (1987) Unsound seeds in conifers: Estimation of numbers of lethal alleles and of magnitudes of effects associated with the maternal parent. Sivae Genet 36:180–185.
6. Bonga JM, von Aderkas P, and James D (1988) Potential application of haploid cultures of tree species. In: Hanover JW and Keathley DE (Eds.) Genetic Manipulation of Woody Plants, Plenum Publishing Corporation, New York. pp. 57–77.
7. Bonga JM and von Aderkas P (1992) Rejuvenation of tissues from mature conifers and its implications for propagation *in vitro*. In: Ahuja MR and Libby WJ (Eds.) Clonal Forestry: Genetics, Biotechnology, and Application, Springer Verlag (in press).
8. Cannell MGR, Sheppard LJ, and Cahalan CM (1988) c-Effects and second generation clone performance in *Picea sitchensis* and *Pinus contorta*. Silvae Genet 37:15–19.
9. Durzan DJ (1988) Somatic polyembryogenesis for the multiplication of tree crops. In: Russell GE (Ed.) Biotechnology & Genetic Engineering Reviews, Volume 6, Intercept, Wimborne, Dorset. pp. 341–378.
10. Farnum P, Timmis R, and Kulp JL (1983) Biotechnology of forest yield. Science 219: 694–702.
11. Finney DJ (1956) The consequences of selection for a variate subject to errors of measurement. Rev Inst Intern Stat 24:1–10.
12. Fowler DP (1986) Strategy for the genetic improvement of *Larix* species and hybrids in the Maritimes. In: 1986 Larch Workshop, The New Brunswick Forest Research Advisory Committee. pp. 73–78.
13. Fowler DP and Morris RW (1977) Genetic diversity in red pine: Evidence for low genic heterozygosity. Can J For Res 7:343–347.
14. Franclet A, Boulay M, Bekkaoui F, Fouret Y, Verschoore-Martouzet B, and Walker N (1987) Rejuvenation. In: Bonga JM and Durzan DJ (Eds.) Cell and Tissue Culture in Forestry: Volume 1, General Principles and Biotechnology, Martinus Nijhoff Publishers, Dordrecht, The Netherlands. pp. 232–248.

15. Gill JGS (1987) Juvenile-mature correlations and trends in genetic variances in Sitka spruce in Britain. Silvae Genet 36:189–194.
16. Gupta PK, Durzan DJ, and Finkle BJ (1987) Somatic polyembryogenesis in embryogenic cell masses of *Picea abies* (Norway spruce) and *Pinus taeda* (loblolly pine) after thawing from liquid nitrogen. Can J For Res 17:1130–1134.
17. Hammerschlag FA, Owens LD, and Smigocki AC (1989) *Agrobacterium*-mediated transformation of peach cells derived from mature plants that were propagated *in vitro*. 114:508–510.
18. Hasnain S and Cheliak W (1986) Tissue culture in forestry: Economic and genetic potential. For Chron 62:219–225.
19. Hasnain S, Pigeon R, and Overend RP (1986) Economic analysis of the use of tissue culture for rapid forest improvement. For Chron 62:240–245.
20. Inglis JE (1984) The effects of some growth substances on the promotion and rooting of interfascicular shoots in *Pinus caribaea* Morelet. Commonw For Rev 63:115–120.
21. Kartha KK, Fowke LC, Leung NL, Caswell KL, and Hakman I (1988) Induction of somatic embryos and plantlets from cryopreserved cell cultures of white spruce (*Picea glauca*). J Plant Physiol 132:529–539.
22. Klekowski EJ Jr (1988) Genetic load and its causes in long-lived plants. Trees 2:195–203.
23. Klimaszewska K (1989) Recovery of somatic embryos and plantlets from protoplast cultures of *Larix* × *eurolepis*. Plant Cell Rep 8:440–444.
24. Libby WJ (1964) Clonal selection and an alternative seed orchard scheme. Silvae Genet 13:32–40.
25. McKeand SE and Weir RJ (1984) Tissue culture and forest productivity. J For 82:212–218.
26. Namkoong G, Barnes RD, and Burley J (1980) Screening for yield in forest tree breeding. Commonw For Rev 59:61–68.
27. Nielsen C (1988) An economic analysis of tree improvement strategies for Norway spruce. Ontario Min Nat Res On line Vol. 5. No. 1.
28. Norton ME and Norton CR (1986) An alternative to *in vitro* propagation – Axillary shoot enhancement on whole plants. J Hortic Sci 61:423–428.
29. Park YS and Fowler DP (1987) Genetic variances among clonally propagated populations of tamarack and the implications for clonal forestry. Can J For Res 17:1175–1180.
30. Shelbourne CJA (1969) Tree breeding methods. NZ For Serv Tech Rep No 55, Wellington.
31. Stiff CM and Boe AA (1985) Effects of foliar-applied benzylaminopurine on fascicular bud development in mugo pine. HortScience 20:285–287.
32. Timmis R, Abo El-Nil MM, and Stonecypher RW (1987) Potential genetic gain through tissue culture. In: Bonga JM and Durzan DJ (Eds.) Cell and Tissue Culture in Forestry: Volume 1, General Principles and Biotechnology, Martinus Nijhoff Publishers, Dordrecht, The Netherlands. pp. 198–215.
33. Wright JW (1976) Introduction to Forest Genetiocs. Academic Press, New York. 463 pp.

28. Use of biotechnology for tree improvement in *Populus* model systems

M. E. OSTRY and C. H. MICHLER*

*USDA Forest Service, North Central Forest Experiment Station, St. Paul, MN 55108 and *Rhinelander, WI 54501, U.S.A.*

1. Introduction

Tree improvement involves the identification, recovery, and multiplication of unique, useful gene combinations that provide the desired growth, quality, and stress resistance traits. The long generation time of trees, lack of knowledge about juvenile-mature trait correlations, and difficulty in obtaining the transfer and expression of desirable genes at high frequencies seriously limit the success of the classical selection, breeding and testing cycle.

Biotechnology has several advantages over classical breeding in forest tree improvement. This technology may reduce the time necessary to introduce new traits into desirable species and can overcome long generation times typical with forest trees for production of improved genotypes [16].

The integral tools of the new biotechnologies in plant improvement include the manipulation of trees *in vitro* using cell and tissue culture techniques. *In vitro* techniques, if they can be applied to trees, provide an even greater potential advantage in trees than in other higher plants [9,17,43,52].

Widespread application of these new biotechnologies for tree improvement is not likely until successful model systems have been developed and some degree of achievement has been demonstrated. Many commercially important forest trees are presently difficult or impossible to regenerate efficiently *in vitro*. One of the major obstacles to biotechnological manipulations in woody plants is the lack of plant regeneration techniques. Although one to several *in vitro* culture and regeneration systems can be applied successfully to a given tree species, essentially all systems have been applied with varying degrees of success to members of the genus *Populus*.

Regeneration of poplars *in vitro* has been the subject of many studies since the 1930s. Black poplar (*P. nigra*) was among the first tree species from which cambial tissue was isolated and established under aseptic conditions in 1934 [15]. Thirty-five years later the first complete plantlet to be regenerated from unorganized callus of a woody plant species and then established under field conditions was obtained from quaking aspen (*P. tremuloides*) [53].

Populus in vitro regeneration techniques [1] are now readily available; the y-range from plant regeneration from protoplasts [45], callus [6], internode

M. R. Ahuja (Ed.), Micropropagation of Woody Plants, 471–483.

explants [8], leaf explants [5], shoot tips [47], bud meristems [1] root explants [54], nodules [30], micro-cross sections [27], to somatic embryos [32]. Poplar regeneration is reviewed in detail elsewhere in this volume.

Significant progress has been made in applying biotechnological techniques to *Populus* species. Most recently, through *Agrobacterium*-mediated gene insertion and somaclonal selection, hybrid poplars have been developed that resist damage due to herbicides and disease. Poplars have been used as models for genetic improvement of forest trees through biotechnology because of their excellent capacity for regeneration *in vitro*; relatively short sexual generation time; small genome size; our extensive knowledge of their genetics, physiology, and silvicultural management; and their world-wide utility and economic value.

In this chapter we focus on *Populus* as a model for tree improvement through biotechnology. Our objectives are to review how *in vitro* regeneration systems are being used in these approaches to tree improvement and to discuss the limitations and future prospects of *Populus* as a model in forest tree biotechnology.

2. Biotechnology systems

New methods for the introduction of foreign genes or selection of cells containing mutant genes in tree species are available as tools for tree improvement. Poplars have proven to be amenable to these new techniques, and rapid progress has been made in applying these new technologies. To date, every known method of genetic engineering and somaclonal biotechnology has been applied to poplars, including direct gene insertion and *Agrobacterium*-mediated gene insertion.

2.1. *Direct gene insertion*

2.1.1. *Electroporation of protoplasts*

Directed gene insertion using electric field-mediated pulses was attempted by Bednarek *et al.* [4]. Isolated protoplasts of *Populus alba* × *P. grandidentata* were electroporated with plasmid DNA containing a marker gene for selectable kanamycin resistance. Surviving protoplasts successfully formed colonies, but colonies could not be cultured on levels of kanamycin normally toxic to untransformed cells.

2.1.2. *Microprojectile bombardment*

Microprojectile bombardment or biolistics is a novel area of direct gene insertion that has only recently been applied to trees. Gold particles coated with plasmid DNA have been blasted into poplar, and stable incorporation has been confirmed in regenerated plants [11]. The plasmids contained a chimeric β-glucuronidase (GUS) gene linked to a 35s CaMV promoter, a *Bacillus thurengiensis* (BT) toxin gene linked to the 35s CaMV promoter, and a neomycin phosphotransferase II (NPT II) gene linked to a nos promoter. Expression of the GUS gene was detected in poplar protoplasts, nodules, and internode tissue and was expressed in protoplast-derived callus and plants as long as 4 months after insertion. Up to 6% of surviving protoplasts had transient GUS gene expression initially and 50% of these retained expression after 4 weeks.

2.2. *Agrobacterium-mediated gene insertion*

2.2.1. *Kanamycin resistance*

A readily available marker gene for determining plasmid gene insertion by *Agrobacterium* is the NPT II gene that confers resistance to the antibiotic, kanamycin. Pythoud *et al.* [41] were able to stably transform *in vitro* shoots of *Populus trichocarpa* × *P. deltoides* with *Agrobacterium rhizogenes* and have expression of kanamycin resistance. This poplar hybrid normally has low susceptibility to *A. rhizogenes*, but the virulence was increased after insertion of the vir region from a plasmid that had enhanced virulence. Shoots could not be regenerated from transformed adventitious roots, but were successfully regenerated from a transformed callus line.

Minocha *et al.* [36] used a similar transformation procedure with *A. tumefaciens* to insert NPT II into *P. tremuloides*. Leaf disks and stem segments were infected, and resultant transformed shoots were selected and showed stable expression of the gene for kanamycin resistance.

2.2.2. *Opine production*

Other marker genes commonly inserted to give evidence for transformation are for the production of either octopine, nopaline, or agropine. Parsons *et al.* [39] demonstrated transformation of *Populus trichocarpa* × *P. deltoides* stem sections and callus with *Agrobacterium tumefaciens*. Agropine and octopine were synthesized by the transformed tissues, and inserted DNA was confirmed by Southern blot analysis. Transformed plants were not recovered despite the successful insertion of foreign genes. In another example, Michel *et al.* [31] transformed hybrids of the Leuce section of *Populus* (*P. alba*, *P. tremula*, and *P. tremuloides*). Shoots were regenerated from galls, and nopaline was detected in the transgenic plants with T-DNA confirmed by Southern blot analysis.

2.2.3. *Glyphosate resistance*

Fillatti *et al.* [14] were the first researchers to insert an economically important gene into a forest tree. A binary oncogenic strain of *Agrobacterium tumefaciens* with a plasmid containing a proprietary gene (aroA) coding for bacterial 5-enolpyruvylshikimate-3-phosphate synthase was inserted into *Populus alba* × *P. grandidentata* leaf disks. The regenerated plants conferred kanamycin resistance, and evidence for insertion and expression of the *aroA* gene was confirmed by Southern and Western blot analysis. It has been reported that the same system could be applied to other *Populus* hybrids including *P. nigra betulifolia* × *P. trichocarpa, P. nigra* × *P. laurifolia,* and *P. maximowiczii* × *P. trichocarpa* [48]. Successful regeneration of transformed plantlets may have been aided by the *trans*-zeatin gene carried within the vir region or by the oncogenes.

Riemenschneider *et al.* [44] tested the transgenic plants produced by Fillatti [14] and confirmed an increased tolerance to the commercial formulation of glyphosate, Roundup (Monsanto Chemical Co.). Plants were sprayed with 0.07 and 0.28 kg/l active ingredient, and transformed plants grew significantly taller than untransformed plants. These rates of herbicide were not of commercial value, but the test demonstrated successful expression of the foreign *aroA* gene in a forest tree species. The lack of greater tolerance to Roundup was attributed to the genetic construct of the introduced plasmid. The mannopine synthase promoter did not allow for the product of the *aroA* gene to be carried into the chloroplasts where it is found in its native state. Thus, the EPSP synthase was only cytoplasmically active and probably less effective.

2.2.4. *Bialaphos resistance*

Bialaphos is an important non-selective herbicide whose active ingredient, phosphinothricin (PPT), inhibits glutamine synthase. Phosphinothricin acetyltransferase (PAT) from *Streptomyces hygroscopius* can deactivate PPT by acetylation of free NH_4 on PPT. The bar gene, which encodes for PAT, has been attached to the 35s promoter of CaMV and inserted into poplar by *Agrobacterium tumefaciens*. Plant Genetic Systems in Belgium is currently field testing these transgenic poplars that express the bar gene [3].

2.2.5. *Proteinase inhibitor*

Agrobacterium was used in a recent attempt to transform three hybrid *Populus* clones with the proteinase inhibitor gene pin-2 to increase pest resistance [22]. The proteinase inhibitor gene, containing a wound inducible promotor, the constitutive portion, and a terminator, from potato codes for a protein that inhibits insect and microorganism proteinases and is induced in solanaceous plants in response to wounding. The *Agrobacterium* binary vector system included the selectable marker gene encoding NPT II linked to pin-2.

Kanamycin was used to select putative transformed shoots from co-cultured leaves. NPT II expression has been obtained on 34 putative transformants of hybrid aspen clones Crandon or Hansen. A portion of these transformants contain the pin-2-CAT chimeric gene, confirmed by Southern blot assays. Other integration and expression tests are underway. Approval has been obtained for field planting a Hansen transformant for a 4-year growth and development study (H.S. McNabb, Jr., Personal Communication).

2.3. *Somaclonal variation*

Passage of plant cells through a tissue culture cycle can often result in increased spontaneous phenotypic and genetic variation in regenerated plants, referred to as somaclonal variation [24]. Somatic variation for several traits has been observed in many tissue culture-derived plants and is now considered to be a general phenomenon. Several authors have recently reviewed the examples of somaclonal variation, including the possible controlling factors responsible, and the potential applications of somaclonal selection for crop improvement [2,12,13,21,23,25,37].

Few examples of somaclonal variation in woody plants exist; however, research directed at using somaclonal selection for recovering useful variants in tree species is increasing. Conclusive demonstration of permanent genetic changes in trees via somaclonal variation is severely hampered by the difficulties of carrying out the necessary genetic analyses involving progeny tests.

Because of the long generation times of trees and the possibility of introducing desired traits more quickly, somaclonal variation may be an important adjunct technique to tree breeding. However, the factors influencing the frequency of somaclonal variation must be thoroughly understood to minimize cryptic genetic changes in recombinant DNA systems that use cell and tissue culture and also in micropropagation systems where clonal fidelity is critical.

Several examples involving tissue culture-associated variation in *Populus* species and hybrids have recently been reported. These examples will be elaborated on in Michler's chapter on somaclonal variation elsewhere in this volume.

2.3.1. *Morphological variance*

Lester and Berbee [28] were first to discover somaclonal variation in poplar when they tested tissue-cultured trees derived from callus culture. Trees from *P.* × *euramericana* hybrid poplar clones displayed changes in height, leaf morphology, and branching habit. Causes of variation could not be distinguished because the plants were heat treated for virus elimination and also cultured on 2,4-dichlorophenoxyacetic acid, which is mildly mutagenic. Virus elimination could partly or wholly account for increased vigor of the clones. The authors had strong evidence for chromosome instability as a partial cause for the phenotypic variation.

Variation has also been noted for protoplast-derived plants of *P. alba* × *P. grandidentata* [29]. (B. McCown, Personal Communication). These plants were evaluated for 13 morphological traits including leaf, stipule, and stem characteristics. Seventy-five of 192 plants had variant phenotypes with at least one notable variant characteristic after shoot regeneration. Thirty-five of these plants with the highest number of variant characteristics were field planted along with appropriate controls. Fifteen of these trees still had stable variation by the second growing season.

Suggested causes were given for somaclonal variation seen in this work. These causes included minor point mutations, polyploidy, transposable elements, aneuploidy, and epigenetic changes. Types of stable variation included increased leaf length, altered leaf pubescence, and dwarfism. The possibility of transposable elements existed in plants that contained sectored areas of more or less leaf pubescence surrounding tissue with normal amounts of that characteristic.

Leaf variegation is a common mutation that arises after adventitious shoot regeneration *in vitro* (C. Michler, unpublished). Six examples of poplars within the hybrid clones *P. nigra* × *P. laurifolia* and *P. nigra betulifolia* × *P. trichocarpa* with leaf variegation have been studied in greenhouse evaluations, including examples of periclinal and mericlinal chimeras. The periclinal chimeras may have commercial value as ornamental trees.

2.3.2. *Glyphosate and sulfometuron methyl tolerance*

Somaclonal selection to increase chemical tolerance in poplars was first achieved by using glyphosate and sulfometuron methyl as selective herbicidal agents [33–35]. Shoots and somatic embryos were regenerated from leaf disks and cell suspension callus in the presence of herbicide-containing regeneration medium. Variant plants produced from leaf disks were tested in the greenhouse and several plants were selected that survived field-rate dosages of herbicide. Enzymatic analyses of several somaclones found evidence for overproduction and less sensitivity of the target enzymes to the herbicides. These somaclonal variants have not yet been field tested.

2.3.3. *Septoria leaf spot resistance*

Septoria leaf spot, caused by the fungus *Septoria musiva*, is a damaging disease of many hybrid poplars. Clones susceptible to *Septoria* cannot be successfully grown for fiber or energy production.

Putative resistance in hybrid poplars previously susceptible to Septoria leaf spot was obtained using somaclonal selection [38]. Adventitious bud formation and shoot proliferation were induced on callus derived from internodal stem segments. Elongated shoots were excised, rooted under high humidity, and transferred to the greenhouse. Variant plants with increased levels of resistance to Septoria leaf spot were selected among the regenerants by using a leaf disk bioassay. The incidence of somaclonal variation in resistance differed among the five *Populus* genotypes tested. Plants with increased resistance were recovered from two of the three susceptible clones. These variant plants are now being field tested.

2.3.4. *Melampsora leaf rust resistance*

Leaf rust of poplars caused by *Melampsora medusae* is a serious disease that limits fiber production throughout the world. Several physiologic races of the fungus have been indentified. A recent study explored the occurrence and extent of phenotypic variation in leaf rust resistance in *P. deltoides* somaclones regenerated from leaf callus [40].

Adventitious shoots from callus of one *P. deltoides* genotype were excised and rooted under mist in the greenhouse. Two races of *M. medusae* were inoculated onto leaf disks from each of the somaclones. The parental genotype was susceptible to one of the rust races and moderately resistant to the other. Most of the somaclones reacted to disease in the same way as the parent, but some were either more or less resistant than the parent. A race-specific interaction with somaclones was found for the latent period and the number of uredia produced per leaf disk.

2.4. In vitro *screening and selection*

2.4.1. *Hypoxylon canker resistance screening*

Tissue culture has been used as a bioassay to screen poplar clones for disease resistance. A tissue culture selection procedure was developed for screening aspen plantlets resistant to the toxin from *Hypoxylon mammatum* by regenerating plantlets from cotyledon explants on toxin-containing media [10]. Valentine *et al.* [51] screened clones of *P. tremuloides* for resistance to *H. mammatum*. Callus cultures were exposed to fungal culture filtrates, and tissue-cultured plantlets were inoculated with ascospores. *In vitro* callus growth inhibition was positively correlated with callus inhibition from cankers of infected 10-year-old field-test trees, and moisture stress was found to be necessary for infection of plantlets by ascospores.

2.4.2. *Salt tolerance screening*

Lee *et al.* [26] used *in vitro* methods to screen 11 hybrid poplar clones for salt tolerance. Artificial seawater was added to culture medium, and shoot tips were grown on various concentrations of the salt. Clones were rated for tolerance after 6 weeks in culture. Two clones (*P. alba* × *P. glandulosa* and *P. nigra* × *P. maximowiczii*). showed successful growth under high salt concentrations, but were not tested *ex vitro*.

2.5. *Protoplast fusion*

Jang *et al.* [19] obtained fusion of *Populus glandulosa* protoplasts with high protoplast and fusion yields. Isolation techniques were standard, and viable protoplasts were osmotically maintained on WPM with 20 ml/l coconut water. Fusion yield was approximately 30%, using polyethylene glycol. Further cell division and regeneration were not reported.

3. Limitations of biotechnology for *Populus* improvement

3.1. *Plant regeneration*

In vitro-based biotechnology requires high frequency regeneration from cells, protoplasts, organs, and tissues with or without the intervention of callus. Although poplars are generally responsive to *in vitro* regeneration, large differences in regenerative capacity exist among species and clones. Because some clones are much more recalcitrant than others, culture techniques may have to be modified for selected genotypes. Further research on developmental biology and *in vitro* processes [42] is needed before biotechnological approaches can be widely applied.

3.2. *Knowledge of stress tolerance*

Throughout their lives, trees are susceptible to stress from a complex of environmental and biological factors which can result in reduced growth, yield, or quality, or tree death. Incorporating stress tolerance into poplars is hindered by our incomplete understanding of the basic physiological and biochemical processess involved in host defense. The recent successes achieved using biotechnology to improve trees have been limited to traits that are probably controlled by single or few genes. However, multiple-gene interactions are probably involved in the response of trees to stress, making it difficult or impossible to incorporate tolerance to selected stress agents. Finally, genetic engineering is, hampered by our need for research in the molecular genetics of forest trees, including identification and recovery of genes controlling stress tolerance.

3.3. *Screening and selection*

Many of the limitations of *in vitro* screening and selection for disease resistance [7,18] also apply to resistance to other stresses. Using a cellular-based selection system is better than screening large numbers of regenerated plants, but a correlation must exist between the resistance expressed at the cellular level and that at the whole plant level. Our lack of understanding of the physiology of stressed plants and the biochemical mechanisms in host-parasite interactions complicates the development of efficient selection systems.

3.4. *Genetic stability*

The objective of genetic manipulation using recombinant DNA techniques is the stable integration of desirable genes, expression of those genes in regenerated plants, and the transmission of those genes during the breeding cycle. Introduced genes may not be expressed immediately or may be lost. An inherent limitation to determining genetic stability of traits in trees is their long generation cycles. The relatively high frequency of random genetic changes that have been associated with a tissue culture cycle of poplars raises questions about the control of unwanted variation in plants produced by *in vitro*-based biotechnology systems. Many of the somatic variations that can occur are not desirable, and they cannot be minimized until their origin and cause are determined. It has been documented in other tissue culture-derived plants that multiple traits are often changed. Thus, presence of the desirable traits existing before *in vitro* manipulations will also have to be confirmed in the regenerated plants.

3.5. *Field testing*

Biotechnology may increase the efficiency of tree improvement, but it does not replace the classical selection, breeding, and field testing cycle. Field testing will remain a critical link between the *in vitro* genetic manipulation and selection phase and the release of the improved tree for further breeding. Adequate field testing is especially important for evaluating pest resistance [49]. Because of the long lives of trees, sexual sterility might need to be used to prevent the accidental introduction of modified genes into the native tree population. In the United States, guidelines are being developed for the safe testing and release of genetically modified plants through recombinant DNA techniques. At this time, plants produced through somaclonal selection are not regulated.

4. Future applications

In addition to the above applications that are still being developed in some cases, other areas of poplar biotechnology are emerging.

4.1. *Wound resistance*

Wounding is a detrimental stress imposed upon poplars from mechanical means or insect attack. Research is ongoing to identify genes induced by mechanical leaf wounding (M. Gordon, Personal Communication). cDNAs have been cloned, and corresponding mRNAs have been identified in leaves proximal to damaged leaves on the lower portion of *P. trichocarpa* × *P. deltoides* hybrids. These cDNA clones were similar to multigene families that encode chitinases in bean, tobacco, and barley.

If this technology is used, genes could be isolated for production of glucanases, chitinases, proteinase inhibitors, and other pathogen-related proteins, and then inserted into poplar to enhance its natural defense mechanisms against pathogen invasion.

4.2. *Environmental stress tolerance*

Biotechnology techniques combined with classical breeding offer the possibility of introducing valuable additional traits into otherwise superior poplar clones. Tolerance to environmental stresses such as drought and cold will increase the adaptability of clones to a wider range of sites and environments.

4.3. *Wood quality and secondary products*

Poplars are a source of biomass for fuel and for the manufacture of various wood and paper products. Biotechnology may offer approaches to improve traits for manufacturing applications such as fiber length, wood brightness, and lignin content [20]. Production and utilization of secondary products, such as flavonoids, may also be realized by biotechnological approaches.

4.4. *Molecular genetics research*

Poplars have a small genome size, making them ideal candidates for molecular genetic analyses. Application of somatic cell genetic research techniques such as RFLP mapping [46] to *Populus* will enhance our understanding of the genome organization of a forest tree and increase the likelihood of success in using recombinant DNA technology for tree improvement.

4.5. *Commercial propagation*

Traditional means of planting poplar by hardwood cuttings have not always succeeded because of poor rooting and establishment. Hybrid poplar exhibits clonal variability for rooting, and many of the commercially valuable clones cannot be used because of this limitation. Using *Agrobacterium rhizogenes* to infect hardwood cuttings may increase rooting of these important clones and may enhance drought tolerance. Strobel and Nachmias [50] have demonstrated this application with almond trees.

5. Summary

Application of biotechnology to tree improvement offers the possibility of developing useful new variability without hybridization. Important traits such as rapid growth and disease resistance may be introduced into selected genotypes much more quickly than through classical breeding. One of the major obstacles to using many of the biotechnological approaches with forest trees is the lack of *in vitro* plant regeneration techniques. *Populus* species and hybrids have been successfully manipulated *in vitro* using several regeneration systems. Largely on account of their capacity for high frequency of competency, *Populus* species are being used in model systems to demonstrate the potential of biotechnology in tree improvement. The recent successes using the new biotechnological tools with *Populus* give hope that these techniques will be effective with other tree species as well. Many traits are desired in forest trees, including such attributes as rapid growth, stress resistance, adaptability to a wide variety of sites, and specific fiber characteristics. Many of these are multigenic traits or traits for which we have not isolated the specific genes of interest. The new molecular genetic techniques hold great promise for research into the fundamental genetic basis of these traits.

References

1. Ahuja MR (1987) *In vitro* propagation of poplar and aspen. In: Bonga JM and Durzan DJ (Eds.) Cell and Tissue Culture in Forestry. Martinus Nijhoff Publishers. pp. 207–223.
2. Ahuja MR (1987) Somaclonal variation. In: Bonga JM and Durzan DJ (Eds.) Cell and Tissue Culture in Forestry. Martinus Nijhoff Publishers. Vol. 1, pp. 272–285.
3. Anonymous (1988) Genetic Engineering News. 15
4. Bednarek SY, Russell JA, Sellmer JC, and McCown BH (1987) Stable transformation of a woody plant by electric field-mediated gene transfer. *In vitro* Cellular Developmental Biology 23(3) Part II: 66A.
5. Bychenkova EA and David A (1978) Callus formation and organogenesis in tissues of leaves of *Populus balsamifera* L. Fiziol Rast 25(2):274–282.
6. Chalupa V (1974) Control of root and shoot formation and production of trees from poplar callus. Biol Plant 16:316–320.
7. Daub ME (1986) Tissue culture and the selection of resistance to pathogens. Annu Rev Phytopathol 24:159–186.
8. Douglas GC (1984) Formation of adventitious buds in stem internodes of *Populus* spp. cultured *in vitro* on basal medium: influence of endogenous properties of explants. J Plant Physiol 116:313–321.
9. Durzan DJ (1985) Tissue culture and improvement of woody perennials: an overview. In: Henke RR, Hughes KW, Constantin MJ, and Hollaender A (Ed.) Tissue Culture in Forestry and Agriculture. New York, NY. Plenum Press. pp. 233–256.

10. Einspahr DW and Wann SR (1985) Use of tissue culture techniques in a hardwood tree improvement program. In: Schmidtling RC and Griggs M (Eds.) Proceedings, 18th Southern Forest Tree Improvement Conference. Long Beach, MS. pp. 33–41.
11. Ellis DD, McCown BH, Russell JA, Sellmer JC, and McCabe DE (1989) Genetic transformation of trees: the introduction of foreign genes via electrical discharge particle acceleration. XIX Stadler Genetics Symposium on Gene Manipulation in Plant Improvement II, University of Missouri-Columbia, Columbia, Missouri. (Poster presentation).
12. Evans DA, Sharp WR, and Medina-Filho HP (1984) Somaclonal and gametoclonal variation. Amer J Bot 71:759–774.
13. Evans DA and Sharp WR (1986) Applications of somaclonal variation. Bio/Technology 4:528–532.
14. Fillatti JJ, Sellmer J, McCown B, Haissig B, and Comai L (1987) *Agrobacterium* mediated transformation and regeneration of *Populus*. Mol Gen Genet 206:192–199.
15. Gautheret RJ (1934) Culture du tissu cambial. CR Acad Sci Paris 198:2195–2196.
16. Haissig BE (1986) Tissue culture-based biotechnology for *Populus* clones. In: Klass DL (Ed.) Energy from Biomass and Wastes X. Elsevier Publishers. pp. 155–175.
17. Haissig BE, Nelson ND, and Kidd GH (1987) Trends in the use of tissue culture in forest improvement. Bio/Technology 5:52–59.
18. Hammerschlag FA (1984) *In vitro* approaches to disease resistance. In: Collins GB and Petoline JG (Eds.) Applications of Genetic Engineering to Crop Improvement. Martinus Nijhoff/ Dr W. Junk Publishers. pp. 453–470.
19. Jang SS, Lee JJ, Lee JS, and Lee SK (1987) Isolation, culture, and fusion of protoplasts of *Populus glandulosa* Uyeki. Res Rep Inst For Gen Korea 23:137–142.
20. Jurasek L (1989) Biotechnological approaches to wood utilization. In Vitro Cell & Dev Biol 25(3) Part II:13A.
21. Karp A and Bright SWJ (1985) On the causes and origins of somaclonal variation. In: Miflin BJ (Ed.) Oxford Surveys of Plant Molecular and Cell Biology 2:199–234.
22. Klopfenstein NB, Heuchelin SA, McNabb Jr HS, Thornburg RW, Hall RB, and Hart ER (1989) Transformation of hybrid *Populus* with the proteinase inhibitor II gene. Phytopathology 79:1004.
23. Larkin PJ (1987) Somaclonal variation: history, method, and meaning. Iowa State J Res 61:393–434.
24. Larkin PJ and Scowcroft WR (1981) Somaclonal variation – a novel source of variability from cell cultures for plant improvement. Theor Appl Genet 60:197–214.
25. Lee M and Phillips RL (1988) The chromosomal basis of somaclonal variation. Ann Rev Plant Physiol Mol Biol 39:413–437.
26. Lee BS, Youn Y, and Kim YJ (1986) Variation in salt tolerance of hybrid poplars through *in vitro* culture. Res Rep Inst For Gen Korea 22:139–144.
27. Lee-Stadelmann OY, Lee SW, Hackett WP, and Read PE (1989) The formation of adventitious buds *in vitro* on micro-cross sections of hybrid *Populus* leaf midveins. Plant Science 61:263–272.
28. Lester DT and Berbee JG (1977) Within-clone variation among black poplar trees derived from callus culture. Forest Sci 23:122–131.
29. Lindgren DT, Russell JA, and McCown BH (1988) Greenhouse and field evaluation of protoplast-derived *Populus* trees. HortScience 23(3):759.
30. McCown BH, Zeldin EL, Pinkalla HA, and Dedolph RR (1988) Nodule culture: A developmental pathway with high potential for regeneration automated micropropagation, and plant metabolite production from woody plants. In: Hanover JW and Keathley DE (Eds.) Genetic Manipulation of Woody Plants. Plenum Press. pp. 149–166.
31. Michel MF, Delmotte F, and Depierreux C (1988) Transformation of hybrids *Populus tremula* × *P. alba* by *Agrobacterium tumefaciens*. In: Somatic Cell Genetics of Woody Plants, Groshansdorf, FRG. Kluwer Acad Publ p. 81.
32. Michler CH and Bauer E (1987) Somatic embryogenesis in plant cell cultures of *Populus*. In Vitro 23(3) Part II: 46A.

33. Michler CH and Bauer E (1988) Selection for herbicide tolerance in *Populus* to sulfometuron methyl. In: Hanover JW and Keathley DE (Eds.) Genetic Manipulation of Woody Plants. Plenum Press. pp. 479–480.
34. Michler CH and Haissig BE (1988) Increased herbicide tolerance of *in vitro* selected hybrid poplar. In: Ahuja MR (Ed.) Somatic Cell Genetics of Woody Plants. Kluwer Acad Pub pp. 183–189.
35. Michler CH, Voelker TM, and Bauer EO (1989) Production of putative tolerant plants from somatic embryos regenerated from *in vitro* selected callus. In Vitro Cell Dev Biol 25(3) Part II:62A.
36. Minocha SC, Noh EW, and Kausch AP (1986) Tissue culture and genetic transformation in *Betula papyrifera* and *Populus tremuloides*. TAPPI J 68:116–119.
37. Orton TJ (1984) Genetic variation in somatic tissues: method or madness? Adv Plant Pathol 2:153–189.
38. Ostry ME and Skilling DD (1988) Somatic variation in resistance of *Populus* to *Septoria musiva*. Plant Disease 72:724–727.
39. Parsons TJ, Sinkar VP, Stettler RF, Nester EW, and Gordon MP (1986) Transformation of poplar by *Agrobacterium tumefaciens*. Bio/Technology. 4:533–536.
40. Prakash CS and Thielges BA (1989) Somaclonal variation in eastern cottonwood for race-specific partial resistance to leaf rust disease. Phytopathology 79:805–808.
41. Pythoud F, Sinkar VP, Nester EW, and Gordon M (1987) Increased virulence of *Agrobacterium rhizogenes* conferred by the vir region of pTiBo542:application to genetic engineering of poplar. Bio/Technology 5:1323–1327.
42. Ranch J and Pace GM (1988) Science in the art of plant regeneration from cultured cells: An essay and a proposal for a conceptual framework. Iowa State J Res 62:537–569.
43. Riemenschneider DE, Haissig BE, and Bingham ET (1987) Integrating biotechnology into woody plant breeding programs. In: Hanover JW and Keathley DE (Eds.) Genetic Manipulation of Woody Plants. New York, NY. Plenum Press. pp. 433–449.
44. Riemenschneider DE, Haissig BE, Sellmer J, and Fillatti JJ (1988) Expression of an herbicide tolerance gene in young plants of a transgenic hybrid poplar clone. In: Somatic Cell Genetics of Woody Plants. Groshansdorf. FRG. Kluwer Acad Pub pp. 73–80.
45. Russell JA and McCown BH (1986) Culture and regeneration of *Populus* leaf protoplasts isolated from non-seedling tissue. Plant Science 46:133–142.
46. Schlarbaum SE (1988) Somatic cell genetic research in forestry: Integration of cytogenetics, tissue culture, and molecular genetics. In: Ahuja MR (Ed.) Somatic Cell Genetics of Woody Plants. Kluwer Acad Pub pp. 103–118.
47. Sellmer JC, Russell JS, Zeldin EL, and McCown BH (1985) Utilization of cytokinin response curves in tissue evaluation of *Populus* for biotechnology research. HortScience 20(3):593.
48. Sellmer JC and McCown BH (1989) Transformation in *Populus* species. In: Bajaj YPS (Ed.) Biotechnology in Agriculture and Forestry, Plant Protoplasts and Genetic Engineering II. Springer Verlag, Berlin 9:155–172.
49. Stack RW (1987) Long generation times in breeding trees – a pest management blessing in disguise. In: Proceedings 5th North Central Tree Improvement Conference. August 10–12, Fargo, N.D. pp. 72–81.
50. Strobel GA and Nachmias A (1985) *Agrobacterium rhizogenes* promotes the initial growth of bare root stock almond. J Gen Microbiol 131:1245–1249.
51. Valentine F, Baker S, Belanger R, Manion P, and Griffin D (1988) Screening for resistance to *Hypoxylon mammatum* in *Populus tremuloides* callus and micropropagated plantlets. In: Somatic Cell Genetics of Woody Plants, Grosshansdorf, FRG, Kluwer Acad Pub p. 181.
52. Wenzel G (1985) Strategies in unconventional breeding for disease resistance. Ann Rev Phytopathol 23:149–172.
53. Winton LL (1970) Shoot and tree production from aspen tissue cultures. Amer J Bot 57:904–909.
54. Zeldin EL and McCown B (1986) The dynamics of poplar root culture and the differentiation of shoots from cultured roots. HortScience 21(3):815.

29. Strategic planning for applying biotechnology to woody plant genetics and breeding

BRUCE E. HAISSIG and DON E. RIEMENSCHNEIDER
USDA Forest Service, North Central Forest Experiment Station, Forestry Sciences Laboratory, P.O. Box 898, Rhinelander, WI 54501, U.S.A.

1. Introduction

Biotechnology can be used to produce products or knowledge, or both (e.g. [22]). But biotechnology means different things to different people. There are the so-called old biotechnologies of, for instance, vegetative propagation, and genetic selection and breeding. Indeed, 'casual' vegetative propagation and breeding are about as old as agriculture, and have been systematized for centuries [10,44]. There are also what are collectively termed the new biotechnologies, for example, recombinant DNA (e.g., genetic engineering, gene mapping); somaclonal *in vitro* genetic selection; in vitro embryo rescue; *in vitro* sexual hybridization; *in vitro* somatic cellular hybridization; *in vitro* vegetative propagation (i.e., micropropagation, micrografting); and *in vitro* production of haploid, aneuploid, or polyploid plants. However, like other technologies, the new biotechnologies have past histories. For example, protoplast fusion was first achieved 56 years ago [33] and is the basis of somatic cellular hybridization.

The repertoire of new biotechnologies is limitless because each composition has many themes and important variations, each of which can be orchestrated differently. And, scientists throughout the world are discovering new compositions, themes, variations, and orchestrations. Many of the new biotechnologies can be equally applied to animals, and higher and lower plants. Within a given species, biotechnology may be used to change readily obvious morphological features, or macroscopically or even microscopically invisible features such as biosynthesis of a single chemical or altered physiological potential for a morphological expression such as flower production.

Our dicussion of biotechnology in this chapter will be limited to the new biotechnologies (which we have termed 'biotechnology' for simplicity). The emphasis will be on those technologies that modify heritable (genetic) characteristics of the plant or that support such modifications (i.e., most biotechnologies). As indicated above, many of the new biotechnologies are based on *in vitro* methods that are also fundamental to micropropagation. Overall, we will discuss strategies to improve and smooth the application of forest tree biotechnology in genetics and breeding in order to genetically improve

M. R. Ahuja (Ed.), Micropropagation of Woody Plants, 485–500.

forest tree crops. As the previous sentence indicates, our presentation will use forestry, forest trees, and forest tree biotechnology as examples, but our discussion will apply about equally to horticulture and many woody plants [32]. We will exclude the application of biotechnologies to genetic modification of microbes, except as essential to our discussion. Most of what has been written about biotechnology (apart from refereed journal articles about actual results) is learned speculation. The present chapter is not an exception, of which we warn the reader.

2. Background

There has not been much time to formulate and test hypotheses in higher plant biotechnology, which would have provided a sounder understanding of the correct placement of biotechnology in the overall scheme of research and applied plant biology – and how to place biotechnology there. The major interest in biotechnology has arisen in forestry mostly within the last eight years, although biotechnology-like or biotechnology-related research has been conducted for much longer. Indeed, biotechnology is relatively new to all biological scientists [47]. Some exceptions are microbiologists and molecular biologists, who were largely responsible for developing the new biotechnologies.

Opinions differ about the potentials of forest biotechnology, as Hanover [22] stated: 'Viewpoints among foresters range from skepticism through unguarded optimism to apparent fantasy'. The potentials of biotechnology in tree genetics and breeding have become important subjects for discussion for two main reasons. First, tree genetics has been studied for at most 75 years, under conditions of minimal scientific infrastructure, so more knowledge is needed (cf. [5,28,36]). Second, human needs and expectations have progressed faster than product development through tree breeding, so more products are quickly needed [6,12,22,26]. Application of biotechnology may alleviate the two forementioned deficiencies.

However, we perceive the need for more discussion of how biotechnology can be generally joined to forest biology research and specifically to tree genetics and breeding. Forest biotechnologists are mostly optimistic and generally believe that biotechnology will be a means to effectively increase genetic understanding, and to speed genetic improvement of trees. We also believe that biotechnology can and will be successfully joined to tree genetics and breeding [44]. For example, there have been ample technical presentations of theoretical and technological linkages between biotechnology and tree genetics and breeding [8,11,22,44]. One optimistic strategic vision (mission) for tree biotechnology might be that better genetic knowledge and more robust genetic technologies will initiate a chain reaction of genetic-based improvement in productivity, fueled by ever increasing biological knowledge. We willingly accept such a grand strategic vision. Nonetheless, it is unclear to us how biotechnology should or could be cohesively applied to trees to efficiently achieve that strategic vision or any other strategic vision.

Scientific uncertainty within forest biology has not fostered effective communication about biotechnology among all interested and affected parties. The current situation is confusing because biotechnology is being rapidly incorporated, or at least talked about, across fields within forest biology such as genetics, breeding, physiology, biochemistry, and ecology. Even economics should have reasons for interest (e.g. [29]). But, even apart from biotechnology, it is not generally understood how knowledge and pre-existing technologies from these separate fields in forest biology should be combined to enhance our understanding of forest tree biology. Uncertainty has increased with the advent of biotechnology, especially because recognizing biotechnology has increased research interest in definitive genetic approaches to problem solving (e.g. [18,20]). Such genetic approaches were not previously available to tree molecular biologists, physiologists, and biochemists due to the slowness of tree breeding (cf. [18,20,44]).

The literature indicates that the higher plant biotechnology community agrees on several points. Papers have been written about future applications of biotechnology to genetics and breeding of crops for agronomy [7,9,16,17, 31,38,43] and forestry ([4,8,11,22,25,26,42]; additional refs. in [1,2,3,23]). Forestry and agronomy papers have differed little because most have dealt with the array of technologies that are or soon will be available for vegetative propagation or genetic modification of all higher plants [1,19,22,23,41,44,48]. Thus, many scientists agree that most of the technologies can be used – perhaps with differing efficiency – with most higher plants. Some technological obstacles are nearly always noted, but then largely dismissed. Secondly, most scientists agree that biotechnology *and* breeding are necessary to fulfill certain objectives, ranging from the purely scientific to the purely practical. Regarding the second point, most forest biotechnology papers refer to a connection or linkage with breeding or, more generally, to 'conventional' tree genetic improvement; indeed agricultural biotechnology always starts with selected genetic lines that will be specially modified and continued (incorporated) in the breeding program (previous refs.). Thirdly, most scientists (and financiers) agree that few of the remarkable promises of biotechnology have been realized with either agronomic or forestry crops. Short time-frames of research, and research and development are often cited or alluded to as the reasons that the potentials have not been realized.

Most forestry biotechnology papers suggest that biotechnology is a more effective substitute for conventional genetics and breeding; problems are not categorized by species, except perhaps between angiosperms and gymnosperms (e.g. [4,11]); and technological operation or planning of operation have been emphasized. In comparison, agronomic papers are more species-oriented and tend to better explain how biotechnology fits or should fit into existing strategies for gaining knowledge to help genetically improve crops (e.g. ([9,16,43]).

The foregoing may indicate that forest biotechnology is proceeding properly. However, analysis of what has been written and said about the future of forest biotechnology also provides reasons for concern: a certain euphoric blindness pervades the scientific literature and meetings about biotechnology. Perhaps forest biotechnologists have been necessarily distracted in an initial exploration of the newly available technologies, but the exploratory phase is rapidly approaching its end.

3. Taking the agronomic lead

A possible pitfall in forest biotechnology would be to copy exactly the major leads of agronomy in strategic planning, for two main reasons. First, there are few, if any, instances where forestry has succeeded with pure agronomic strategies. Most likely, biotechnology will not be the exception, and there are good bases for that assumption: agronomy worldwide deals with few, mostly annual (herbaceous) species; existing genotypes are already highly tailored to purpose; substantial genetic knowledge is available; entire crops are replanted frequently; and huge genetic improvement programs are in place, along with strong supporting programs in soils, entomology, and pathology (to name but a few).

For each of the foregoing, almost the exact opposite is true for forest biotechnology. Second, a major (perhaps the major) strength of agronomy has been breeding. Breeding has not been a strength in forestry, mainly because of insufficient scientific infrastructure (as stated above) and because flowering of trees is a late-life trait of many species. Indeed there is an insidious 'Catch-22' for forestry: biotechnology is supposed to hasten development of knowledge or products that tree breeding has been unable to accomplish because of various constraints. But, forest biotechnology will almost certainly require substantial breeding to verify new knowledge and for genetic manipulations to develop products. The late-life flowering characteristic of many important tree species causes a dilemma common to breeding and biotechnology.

There is no firm basis for categorically excluding use of agronomic biotechnological strategies in forestry. However, neither can agronomic strategies be accepted until three basic theorems are fully evaluated: (1) important types of biological knowledge fully transfer between agronomy and forestry; if so, important forestry knowledge and products can be produced by directly using agronomic biotechnology; (2) agronomic knowledge does not apply to the most important forestry targets; if so, more basic biological knowledge about trees must be obtained before attempting to produce forestry products; or (3) a combination of (1) and (2); if so, forestry targets must be skillfully chosen.

Theorem 1 is sometimes true. For example, herbicide tolerance was first identified as an agronomic biotechnology target [15,38] with possible application in forestry by using essentially agronomic strategies [37]. Subsequently, herbicide tolerances in hybrid poplars were achieved both by *Agrobacterium*-mediated insertion of a mutant bacterial gene [13,45] and by tissue culture somaclonal selection [34,35]. However, this is a unique forestry situation because the hybrid poplar tree crop is grown like an herbaceous perennial crop, and the trait of herbicide tolerance is only important during the first two years of growth (i.e., the period of potentially lethal weed competition for poplars). Similar uses of agronomic biotechnology in forestry are difficult to imagine; the biology of tree development and/or heterogeneity and complexity of forest ecosystems invariably complicates use. In particular, forestry problems are seldom associated with ideal species (i.e., those easy to work with) or simple genetic traits, and fundamental knowledge is usually minimal.

Theorem 2 is also sometimes true because agronomic knowledge may not apply to important forestry targets. Indeed, the necessary knowledge may not even be a concern in agronomy because of basic biological differences between agronomic and forestry crops. Even though agronomic and forestry crops are flowering, vascular species, agronomic species are relatively short lived and do not form the secondary vascular system characteristic of trees. Maturation and secondary vascularization are closely linked with oft-mentioned forestry genetic improvement targets such as sexual and vegetative regenerative capabilities, wound healing, tolerances to damaging biotic agents, lignin characteristics, and wood characteristics. More knowledge of maturation and secondary vascularization will probably not come from agronomy.

Theorem 3 is the best alternative for current planning. We suggest that, except for certain opportunities, agronomic biotechnology strategies will not be directly useful for forestry or, if they are useful, will be impossible to make operational either in the laboratory or in the field, or both. Therefore, forest biotechnology needs to create its own strategic plans to foster the applications of biotechnology in forest biology and, specifically, in tree genetics and breeding.

Our summary view is that agronomy will supply forestry with useful examples but not with exact – or even nearly exact – templates for solving major problems. We suggest that unless the previously discussed theorems are intelligently evaluated, there is the persistent danger of transferring biotechnologies developed for agronomy to forestry *along with the agronomic research goals that spawned the techniques.* Such linked transfer would, by-and-large, be a mistake because the availability of techniques is only one consideration in developing research strategy. Existing technologies are only tools for operating within a strategy. Indeed, the fulfillment of strategy may require development of new tools.

4. Strategic planning for forestry

Strategic planning may seem trite and unnecessary to scientists. However, only clearly expressible strategic plans will efficiently guide forest biotechnologists in their quest for new knowledge. That new knowledge is generally perceived as essential to keep forestry competitive within agricultural and, therefore, global economics. Concrete achievements based on a vision, such as the strategic vision stated above, require vision-based strategies that operations support. However, in forest tree biotechnology there are no well-critiqued, commonly accepted strategic plans (hypotheses) upon which to base operations. Forestry is mostly copying – or maybe hoping to copy – agronomy, which has developed some major strategic plans (e.g. [30]). The consequence of lack of planning would be greater for forestry than agronomy because human and financial resources in forestry are substantially less, which means that resource-consuming errors could not be offset by human and financial efforts, and would be relatively more damaging to progress in forestry. Engaging in any operation, including forest biotechnology, requires planning, organizing, directing, leading, and controlling – equally in the strategic and operational (tactics and execution) arenas.

Strategic plans are needed to move from inaction or reaction to action and to eliminate or reduce surprises, and for greater efficiency and productivity. Strategic planning focuses on long-term co-alignment with the future global environment, in order to maintain or improve competitive position. Good strategic plans are simple and short, flexible and current, and credible and accepted by those involved. Strategic plans state destinations, hope of accomplishment, situational conditions, provisions to reach targets, and approaches for future reference.

Perhaps no single strategic plan can or even should be created for forest biotechnology, nor is our purpose the creation of a strategic plan *per se*. Our purpose is to point out important elements of strategic planning for efficiently applying forest biotechnology to genetics and breeding. However, even if there is no universally applicable strategic plan for forest biotechnology, planning by individual researchers or research groups is presently possible and desirable for the individual and collective good.

5. Elements of the forestry plan

Strategic planning usually involves the consideration of five elements: strategic vision (or mission), goals, rationale for selection of goals, changes needed to reach goals, and program integration. The following subsections offer some thoughts that should be considered in the five elements of strategic planning for forest biotechnology.

5.1. *Strategic vision*

A clear strategic vision, the underpinning of mission, for forest biotechnology is needed to consolidate support within the forestry community and to gather external support from related areas of science and, very importantly, from the public. In our view, there are very few major forestry problems that can be effectively solved right now using biotechnologies. Nonetheless, forest biotechnology will not develop efficiently and productively unless its vision includes obtaining knowledge for problem arenas – such as diseases, insects, wood characteristics, growth characteristics, etc. To a great extent, there has been a dichotomy of vision in forest biotechnology. At one extreme, the need for forest biotechnology products (i.e., something practical) has prompted us to apply existing knowledge but not necessarily generate new knowledge; at the other extreme, the intrinsic desire for more 'pure' knowledge has caused us to justify or at least discuss biotechnology research, and research and development based on illusory products, those which could only be achieved in some ill-defined future, if at all. The forest biotechnology vision needs to link the development of knowledge (research) and products (research and development) so they are more logical and understandable by scientists and the public alike.

Time-frame needs to be considered. Probably anything can be accomplished if time is unlimited. But in research the time-scale for accomplishment has usually been a prime consideration and projects are terminated if not completed in a 'reasonable' period of time. We presume that there is a finite amount of time in which to demonstrate the efficacy of biotechnology in forest biology, whether for the generation of unique knowledge or products, or both. Even now the newness of biotechnology is disappearing, before much progress has been reported in scientific journals, before products have been marketed.

We should consider how biotechnology can be flexibly applied to tree genetics and breeding over a 10-year period; a longer period would be unrealistic. Ten years is well within the usual length of scientific careers, is adequate for obtaining important new knowledge, and is about as long as public and organizational support can be expected for any unproven research program. Finally, biotechnologies and knowledge of tree biology will probably change substantially over time, which may impact planning. Indeed, the best strategy is to allow modification during the planned period, to compensate for unforeseeable change and to allow for innovation.

5.2. *Goals*

We suggest developing goals with high expectations but without concomitant, ridiculously great risk and probability of failure. The public will not long accept the promise of products, including knowledge, that are not forthcoming or products whose added value is disappointingly less than anticipated. A major problem in goal setting for forest biotechnology is that the junctures of technology and specific problems have been chosen or made the focus of discussion before there is adequate supporting knowledge. For example, it is difficult at present to unerringly select a specific major disease or insect problem for forest *tree* biotechnology to solve, even though there is substantial evidence to suggest that tree biotechnologies will help solve insect and disease problems. We just do not know which problems and when. In this instance, biotechnology and breeding (e.g. [28]) should be used to obtain more knowledge of how to solve insect and disease problems directly at the tree level of biology, if possible, without the preconceived goal of solving the problem of a specific insect or disease. Overall, we suggest developing knowledge through applying biotechnology in important problem arenas, communicating this to interested parties, but without having preconceived, highly specific problems targeted.

5.3. *Rationale for selection of goals*

An acceptable strategic plan should clearly show that goals are commensurate with resources, i.e., that the goals are achievable. Historically, forest biology research has not had resources commensurate with the complexity of goals. The inadequacy of resources has continued and most probably will continue with forest biotechnology.

An acceptable rationale for selecting biotechnology goals must also consider humans and forests, in that order. As Gifford Pinchot said: 'To start with I had to know something about the people, the country, and the trees. And of the three the first was the most important.' The biotechnologies that prevail will be the ones that most efficiently meet human needs. Regarding forests, replacement forests and existing forests deserve equal consideration. There are ample opportunities for deploying genetically modified plants because reforestation and afforestation are needed in most nations, and these needs will increase due to misuse and destruction of forests due to settlement, urbanization, and industrialization. However, most of the world's present forests will be in place for a long time – at least relative to a 10-year base for strategic planning – so they must also be considered. A rationale for helping existing forests is paricularly important for those regions that are not faced with any near-term timber famine or where forests are threatened because of changing environment.

5.4. *Changes needed to reach goals*

5.4.1. Communication

Forestry biotechnologists must actively seek opportunities to communicate the goals of forest biotechnology to the public. For example, it is uncertain whether the goal of field distribution of recombinant DNA trees will be achieved in the near term. Field tests of recombinant DNA trees appear to be nearly achievable worldwide; however, whether recombinant DNA trees can be commercialized is unknown because there are no mechanisms in place for 'certification' of such trees for general distribution, if the public wants them distributed. Public opinion about widespread release of perennial, recombinant DNA higher plants may be neutral or favorable, but it is essentially unknown because there have not been test cases or public opinion surveys. We anticipate that biotechnology goals will best be achieved when scientists are sensitive to public perceptions and willing to educate the public where misconceptions exist (e.g. [46]). Thus, progress is needed in explaining the potentials, limitations, and progress of forest biotechnology to the non-scientific public in 'understandable' language. These days, understandable language is defined as that which effectively communicates ideas to about a sixth grade student.

Forest biotechnologists also need to improve internal communications. Effective use of biotechnology in tree genetic improvement requires much fundamental information, including related terminology. Terminology links that which has been done with that which might be done. However, there are no meaningful linkages in terminology between breeding and biotechnology: biotechnology is inherently deterministic (qualitative) while tree breeding is most often statistical (quantitative). Tree breeding has focused on quantitative techniques because the multigeneration studies that are prerequisite to qualitative knowledge, such as linkage analyses, have not been done. Thus, genetic characterization of valuable traits has been generally limited to estimation of their heritabilities, i.e., ratios of genetic variance to total phenotypic variance; heritability estimates the importance of genetic effects relative to environmental effects [20]. Heritability and other quantitative parameters describe the importance of genetic effects but provide little, if any, knowledge of their mechanics. Indeed, proper use of such quantitative parameters is continually disputed (e.g. [6]). In contrast to breeding, biotechnology is more deterministic. For example, full knowledge of a genetic system, from gene size to gross structure, is required to insert useful genes into plants. Such knowledge is expressed in terms of physical quantities (gene size, location, and morphology) and rates (gene product activity).

It is difficult to envision how descriptive *quantitative* information developed through experimental breeding can be used to foster biotechnology. The two fields do not necessarily need common knowledge or, therefore, terminology for sharing knowledge. Examples of effective integration of breeding and biotechnology have been mostly limited to application of *qualitative* knowledge, and only for agronomic crops (e.g. [39]). However, recent advances in the mapping of quantitative trait loci (QTLs) [27,40] suggest that it may be possible to more fully integrate *quantitative* and *qualitative* information. However, creation of the detailed genetic map that is required to locate QTLs may be inconsistent with scientific and technical resources, especially if mapping is envisioned for many different conifers.

5.4.2. *Tree species*

We suggest that whether or not economics is initially made the primary basis for species selection, in the end the species selected will be the most economically valuable because they are the ones that will be planted most or need the most protection from damaging agents (as in agronomy). But it is questionable whether biotechnology can produce plants cheap enough, regardless of improved character, even for the commercially most important species (cf. [21,24]). Forestry is a consistent, low financial return, long-term investment. Successful biotechnology will only enhance the investment opportunity in forestry if close attention is given to selection of species. At the state of present knowledge, biotechnology cannot be applied to all commercial forest tree, agroforestry, or multipurpose species because there are too many worldwide. Indeed, the worldwide capacity for forest biotechnology would be overwhelmed if biotechnology were to address the collective commercial forest tree species of the US alone. Therefore, global, national, and regional timber and nontimber values must be considered in selecting forest biotechnology goals. The potential value of tree biotechnology can only be achieved (as opposed to demonstrated) if the genetically better clone or population is much planted or better protected, or both. It is not going to be economically viable to genetically tailor or protect forest trees for niche markets because the cost per plant would have to be unacceptably large to amortize research and development costs (cf. [29]).

5.4.3. *Traits and organisms*

Selection of target traits is critical to the success of forest biotechnology, yet there is little consensus as to how that selection should be made. Nonetheless, two gross biological aspects must be considered; the objective for modifying the trait and the logical target organism.

In terms of objectives, there are two gross classes of traits, those whose improvement would directly improve applying biotechnology to forest trees, and those whose improvement would directly improve trees. Vegetative and sexual regeneration abilities are examples of traits whose improvement would directly and immediately benefit plant genetic improvement through biotechnology and breeding, and the gathering of genetic knowledge. Late and inadequate flowering, and poor vegetative regeneration *in vitro* and *ex vitro* presently limit forest tree genetics and genetic improvement [18,19]. Regeneration traits have proven difficult to study without biotechnology but perhaps they can be efficiently studied with biotechnology (e.g. [18,20]).

Attention is also due for traits whose changed characteristics would improve trees by eliminating deleterious responses to stresses or by enhancing innate tree productivity. However, such problems may or may not be best solved by applying biotechnology to trees themselves. Trees do not exist in isolation, but as members of complex ecological communities. Thus, the first objective in problem solving should be to select the optimum target organism in the community. For example, tree resistance to insects is a complex problem that involves the host tree, the host tree's physical and microbial environment, the damaging insect, and agents that damage the insect. We suggest that the most appropriate biotechnology targets in such systems are the lower organisms, i.e., tree microflora, soil microflora, and parasites that may control insect predations indirectly or directly (cf. [14]). Our suggestion is based on the fact that microorganisms are more amenable to genetic manipulation than higher plants. Thus, biotechnological modification of the tree should be considered only after it has been determined that a problem cannot be solved by manipulation of other organisms. Furthermore, modified, nontree components of the forest ecosystem such as microbes can be deployed in forests that are not replanted or not frequently replanted, as is the case in much of the world. As an extension of the concept of focusing biotechnology on nontree components where possible, diagnostic tools are one potentially important forest biotechnology product that could be deployed in existing forests, new forests, and forest tree nurseries.

When the tree is targeted, progress in genetic modification will be most rapid when trait knowledge can be transferred from agronomy to forestry without modification and when substantial species knowledge exists (e.g., the foregoing example of herbicide tolerant hybrid poplars). Much fundamental knowledge is needed to modify higher plants using, for example, direct gene insertion. That knowledge can be divided into trait components and species components. Trait components such as molecular biology and biochemistry are essentially invariant among species, whereas species components such as *in vitro* regeneration ability and earliness to flower are highly variable among species. Unskilled selection of either trait or species will limit the rate of progress because of insufficient fundamental knowledge to meet or exceed the critical threshold for progress (Figure 1).

Overall, we suggest that appropriate target traits for trees should be those

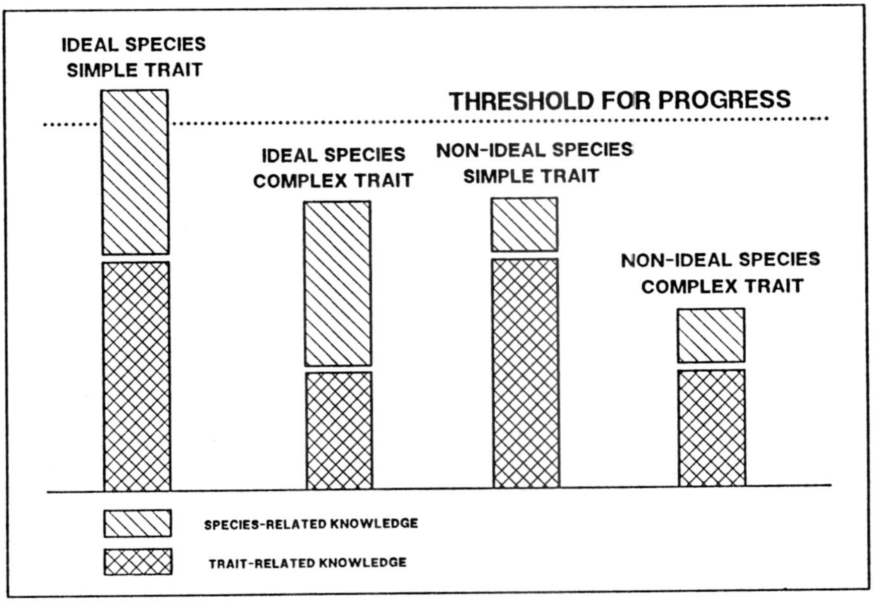

Figure 1. Interdisciplinary research may gain knowledge that is generally applicable to tree-damaging agent interactions, whereas monodisciplinary research may not.

whose characteristics cannot be equalled or exceeded by manipulation of other organisms in the ecosystem and/or that result in such high added value that economic opportunity drives deployment of the genetically different trees. In addition, we suggest that the most suitable target traits will have four characteristics: (1) definition based on biochemistry rather than morphology; (2) commonness to various plant species, thus having a substantial knowledge base; (3) importance in either maturation or secondary vascularization (i.e., wood formation); and (4) association with tree response/resistance to biotic or abiotic damaging agents (stresses). Assertion of (1) and (2) should limit euphoric discussions of traits for which adequate molecular and biochemical knowledge does not exist (i.e., yield). Nevertheless, assertion of (1) and (2) should not be meant to limit research to gain needed biochemical and molecular knowledge for future application. Assertion of (3) ensures a high probability of significant value-added. Assertion of (4) links biochemical/molecular processes to stress resistance or reaction, which is prerequisite to development of molecular diagnostic tools.

5.4.4. *Program integration*

We suggest that research that incorporates biotechnology across scientific fields will prove most efficient in generating the new knowledge needed to solve problems. It has, however, proven difficult within forest biology to make multi-field approaches operational because of the idiosyncrasies of individual scientists, groups of scientists, and scientific institutions. Hence, fields within forest biology have tended to operate disciplinarily rather than multidisciplinarily. But program integration in forest biotechnology means that scientific fields need to be integrated among themselves and integrated among diverse and seemingly ever-expanding technologies.

As an example, monodisciplinary study of individual damaging agents such as insects and diseases may produce results that apply only to the damaging agent under study. On the surface that seems to be expected and reasonable. However, there is always the possibility, perhaps a great possibility, that a single discipline will prove to be ill-equipped to obtain critical missing knowledge. Such possibilities exist because it is presently unclear which scientific field-problem intersections (Figure 2) are most important to obtaining critical new knowledge for problem solution; picking the optimum intersection *a priori* is about impossible. Given such uncertainty, multidisciplinary approaches have a better probability of solving the problem of a specific damaging agent and also other damaging agents because more complete knowledge is likely to be obtained (Figure 1). Study of the molecular, genetic, biochemical, and physiological response of trees to damaging agents may lead to knowledge that is broadly applicable to major forestry problems. Biotechnology affords a prime setting within which to construct multidisciplinary research approaches because the technologies, taken collectively, are multidisciplinary.

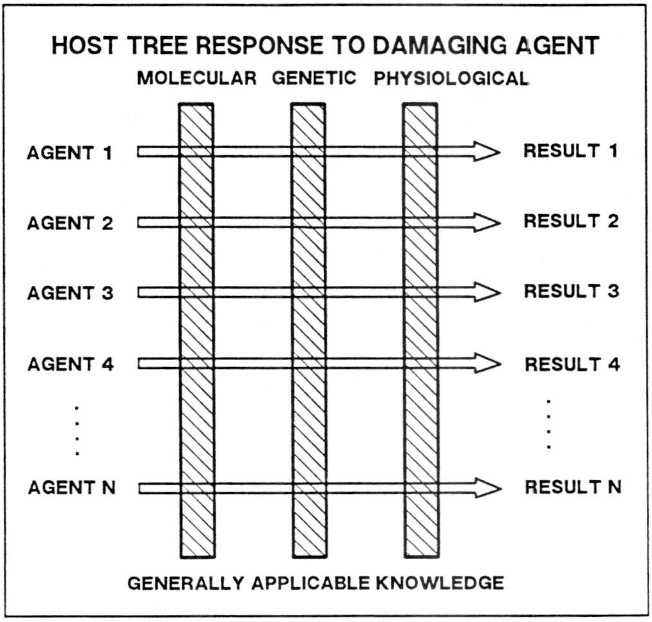

Figure 2. The magnitudes of species- and trait-related knowledge determine the probability of successfully modifying genetic traits.

6. Summary

Forestry is substantially different than agronomy in many comparisons. A significant difference is between the biology of forestry *vs.* agronomic crops. The sum of critical differences between forestry and agronomy suggests that forestry needs its own unique strategies for applying biotechnology in tree genetics and breeding. Such strategies do not exist in forestry or are deficient. However, the ill consequences of any poor planning in forestry will be great. Human and financial resources are inadequate to responsibilities in forest biology research, which means that any resource-consuming errors will be very damaging to forestry. Forestry plans, like all strategic plans, need to consider five main elements: strategic vision and mission, goals, rationale for selection of goals, changes needed to reach goals, and program integration. The most effective strategic plans for forestry will build on differences, as opposed to similarities, with agronomy because the plans must fully consider the unique biological characteristics of trees and forest ecosystems, which make trees such valuable crops to meet human needs worldwide.

Acknowledgements

We thank Dr C.F. Bey for ideas and insights concerning strategic planning, and Dr L.W. Handley and Dr S.L. Krugman for their critical review of the manuscript.

References

1. Ahuja MR (Ed.) (1988) Somatic Cell Genetics of Woody Plants. Kluwer Academic Pubs Dordrecht/Boston/London. 225 pp.
2. Bajaj YP (Ed.) (1988) Trees I. Biotechnology in Agriculture and Forestry. Vol 1. Springer-Verlag, New York, Berlin, Heidelberg, London, Paris, Tokyo. 515 pp.
3. Bajaj YP (Ed.) (1988) Trees II. Biotechnology in Agriculture and Forestry. Vol 5. Springer-Verlag, New York, Berlin, Heidelberg, London, Paris, Tokyo. 595 pp.
4. Berlyn GP, Beck RC, and Renfroe MH (1986) Tissue culture and the propagation and genetic improvement of conifers: problems and possibilities. Tree Phsyiol 1:227–240.
5. Bey CF, Houston DB, and Dinus RJ (Eds.) (1986) II. The business of tree improvement. Tree genetics and improvement. J For 84:45–55.
6. Birks JS and Kanowski PJ (1988) Interpretation of the composition of coniferous resin. Silvae Genetica 37:29–39.
7. Bliss FA (1984) The application of new plant biotechnology to crop improvement. HortScience 19:43–48.
8. Chatarpaul L and Carlisle A (1983) Nitrogen fixation: A biotechnological opportunity for Canadian forestry. For Chron 59:249–259.
9. Cohen JI (1988) Models for integrating biotech into crop improvement programs. Bio/Technology 6:387–392.
10. Davis TD, Haissig BE, and Sankhla N (Eds.) (1988) Adventitious Root Formation by Cuttings. Advances in Plant Sciences Series, Vol. 2, Dioscorides Press, Portland, OR. 315 pp.
11. Dunstan DI (1988) Prospects and progress in conifer biotechnology. Can J For Res 18:1497–1506.
12. Farnum P, Timmis R and Kulp JL (1983) Biotechnology of forest yield. Science 219:694–702.
13. Fillatti JJ, McCown BH, Sellmer J, Haissig B, and Comai L (1987) *Agrobacterium* mediated transformation and regeneration of poplar. Mol Gen Genet 206:192–199.
14. Glenister C (1989) The promise of biocontrol. Amer Nurseryman, Feb. 15, pp. 41–47.
15. Gressel J (1986) The challenges of conferring herbicide resistance on crops. In: Augustine PC, Danforth HD, and Bakst (Eds.) Biotechnology for Solving Agricultural Problems. Beltsville Symp in Agric Res, Nijhoff Pubs, Dordrecht, Boston, Lancaster. pp. 369–383.
16. Goodman RM, Hauptli H, Crossway A, and Knauf VC (1987) Gene transfer in crop improvement. Science 236:48–54.
17. Green CE (1977) Prospects for crop improvement in the field of cell culture. HortScience 12:131–134.
18. Haissig BE (1988) Future directions in adventitious rooting research. In: Davis TD, Haissig BE, and Sankhla N (Eds.) Adventitious Root Formation by Cuttings. Advances in Plant Sciences Series, Vol 2, Dioscorides Press, Portland, OR. pp. 303–810.
19. Haissig BE (1989) Status of forest tree vegetative regeneration for biotechnology. Am Biotech Lab 7(1):48–51.
20. Haissig BE and Riemenschneider DE (1988) Genetics of adventitious rooting. In: Davis TD, Haissig BE, and Sankhla N (Eds.) Adventitious Root Formation by Cuttings. Advances in Plant Sciences Series, Vol 2, Dioscorides Press, Portland, OR. pp. 47–60.
21. Haissig BE, Nelson ND, and Kidd GH (1987) Trends in the use of tissue culture in forest improvement. Bio/Technology 5:52–59.
22. Hanover JW (1987) Applications of biotechnology in forest tree improvement. In: Proc 19th Southern For Tree Improv Conf Pub No 41, Southern For Tree Improv Cttee College Station, TX. pp. 59–70.

23. Hanover JW and Keathley DE (Eds.) (1987) Genetic Manipulation of Woody Plants. Basic Life Sciences, Vol 44, Plenum Press, NY and London. 519 pp.
24. Hasnain S, Pigeon R, and Overend RP (1986) Economic analysis of the use of tissue culture for rapid forest improvement. For Chron 62:240–245.
25. Karnosky DF (1981) Potential for forest tree improvement via tissue culture. BioScience 31:114–120.
26. Lakshmi Sita G, Vaidyanathan CS, and Ramakrishnan T (1982) Applied aspects of plant tissue culture with special reference to tree improvement. Current Sci 51:88–92.
27. Lander ES and Botstein D (1989) Mapping Mendelian factors underlying quantitative traits using RFLP linkage maps. Genetics 121:185–199.
28. Loehle C and Namkoong G (1987) Constraints on tree breeding: Growth tradeoffs, growth strategies, and defensive investments. For Sci 33:1089–1097.
29. Löfgren KG (1988) On the economic value of genetic progress in forestry. For Sci 34:708–723.
30. MacKenzie DR, Faust RM, and Russell KW (Eds.) (1988) Plant Genome Conference Report. Science and Education, US Dept Agric Dec 12–14, 1988, Washington, DC. 29 pp.
31. McCormick D (1988) How biotech is dealing with its nitrogen fixation. Bio/Technology 6:383–385.
32. McCown BH (1989) The biotechnology of urban trees. J Arboriculture 15(4):77–83.
33. Michel W (1937) Über die experimentelle Fusion pflanzlicher Protoplasten. Arch Exp Zellforsch 20:230–252.
34. Michler CH and Haissig BE (1988) Increased herbicide tolerance of *in vitro* selected hybrid poplar. In: Ahuja MR (Ed.) Somatic Cell Genetics of Woody Plants. Kluwer Academic Pubs, Dordrecht/Boston/London. pp. 183–188.
35. Michler CH, Voelker TM, and Bauer EO (1989) Production of putative herbicide tolerant plants from somatic embryos regenerated from *in vitro* selected callus. In Vitro 25(3):207 (Abstract).
36. Namkoong G, Kang HC, and Brouard JS (1988) Tree Breeding: Principles and Strategies. Monographs on theoretical and applied genetics. 11. Springer-Verlag, New York, Berlin, Heidelberg, London, Paris, Tokyo. 177 pp.
37. Nelson ND and Haissig BE (1986) Herbicide stress: Use of biotechnology to confer herbicide resistance to selected woody plants. In: Hennessey TC, Dougherty PM, Kossuth SV, and Johnson JD (Eds.) Stress Physiology and Forest Productivity. Martinus Nijhoff/Dr. W. Junk Publishers, The Hague, The Netherlands. pp. 191–215.
38. Netzer WJ (1984) Engineering herbicide tolerance: When is it worthwhile? Bio/Technology 2:939–944.
39. Nienhuis J, Helentjaris T, Slocum M, Ruggero B, and Schaefer A (1987) Restriction fragment length polymorphism analysis of loci associated with insect resistance in tomato. Crop Sci 27:797–803.
40. Paterson AH, Lander ES, Hewitt JD, Peterson S, Lincoln SE, and Tanksley SD (1988) Resolution of quantitative traits into Mendelian factors by using a complete linkage map of restriction fragment length polymorphisms. Nature 335:721–726.
41. Perani L, Radke S, Wilke-Douglas M, and Bossert M (1986) Gene transfer methods for crop improvement: Introduction of foreign DNA into plants. Physiol Plant 68:566–570.
42. Powledge TM (1984) Biotechnology touches the forest. Bio/Technology 2:763–773.
43. Reid WJ (1987) Biotechnology and breeding team up in agriculture. Bio/Technology 5: 899–906.
44. Riemenschneider DE, Haissig BE, and Bingham ET (1987) In: Hanover JW and Keathley DE (Eds.) Genetic Manipulation of Woody Plants. Basic Life Sciences, Vol. 44, Plenum Press, NY and London. pp. 433–449.
45. Riemenschneider DE, Haissig BE, Sellmer J, and Fillatti JJ (1988) Expression of an herbicide tolerance gene in young plants of a transgenic hybrid poplar clone. In: Ahuja MR (Ed.) Somatic Cell Genetics of Woody Plants. Kluwer Academic Pubs, Dordrecht/Boston/London. pp. 73–80.

46. Tiedje JM, Colwell Rk, Grossman YL, Hodson RE, Lenski RE, Mack RN, and Regal PJ (1989) The planned introduction of genetically engineered organisms: Ecological considerations and recommendations. Ecology 70:298–315.
47. Torrey JG (1985) The development of plant biotechnology. Amer Sci 73:354–363.
48. Wilke-Douglas M, Peranai L, Radke S, and Bossert M (1986) The application of recombinant DNA technology toward crop improvement. Physiol Plant 68:560–565.

Index

M. R. Ahuja (Ed.), Micropropagation of Woody Plants, 501–507.

FORESTRY SCIENCES

1. P. Baas (ed.): *New Perspectives in Wood Anatomy.* Published on the Occasion of the 50th Anniversary of the International Association of Wood Anatomists. 1982 ISBN 90-247-2526-7
2. C.F.L. Prins (ed.): *Production, Marketing and Use of Finger-Jointed Sawnwood.* Proceedings of an International Seminar Organized by the Timber Committee of the UNECE (Halmar, Norway, 1980). 1982 ISBN 90-247-2569-0
3. R.A.A. Oldeman (ed.): *Tropical Hardwood Utilization.* Practice and Prospects. 1982 ISBN 90-247-2581-X
4. P. den Ouden (in collaboration with B.K. Boom): *Manual of Cultivated Conifers.* Hardy in the Cold- and Warm-Temperate Zone. 3rd ed., 1982 ISBN Hb 90-247-2148-2; Pb 90-247-2644-1
5. J.M. Bonga and D.J. Durzan (eds.): *Tissue Culture in Forestry.* 1982 ISBN 90-247-2660-3
6. T. Satoo: *Forest Biomass.* Rev. ed. by H.A.I. Madgwick. 1982 ISBN 90-247-2710-3
7. Tran Van Nao (ed.): *Forest Fire Prevention and Control.* Proceedings of an International Seminar Organized by the Timber Committee of the UNECE (Warsaw, Poland, 1981). 1982 ISBN 90-247-3050-3
8. J.J. Douglas: *A Re-Appraisal of Forestry Development in Developing Countries.* 1983 ISBN 90-247-2830-4
9. J.C. Gordon and C.T. Wheeler (eds.): *Biological Nitrogen Fixation in Forest Ecosystems.* Foundations and Applications. 1983 ISBN 90-247-2849-5
10. M. Németh: *Virus, Mycoplasma and Rickettsia Diseases of Fruit Trees.* Rev. (English) ed., 1986 ISBN 90-247-2868-1
11. M.L. Duryea and T.D. Landis (eds.): *Forest Nursery Manual.* Production of Bareroot Seedlings. 1984; 2nd printing 1987 ISBN Hb 90-247-2913-0; Pb 90-247-2914-9
12. F.C. Hummel: *Forest Policy.* A Contribution to Resource Development. 1984 ISBN 90-247-2883-5
13. P.D. Manion (ed.): *Scleroderris Canker of Conifers.* Proceedings of an International Symposium on Scleroderris Canker of Conifers (Syracuse, USA, 1983). 1984 ISBN 90-247-2912-2
14. M.L. Duryea and G.N. Brown (eds.): *Seedling Physiology and Reforestation Success.* Proceedings of the Physiology Working Group, Technical Session, Society of American Foresters National Convention (Portland, Oregon, USA, 1983). 1984 ISBN 90-247-2949-1
15. K.A.G. Staaf and N.A. Wiksten (eds.): *Tree Harvesting Techniques.* 1984 ISBN 90-247-2994-7
16. J.D. Boyd: *Biophysical Control of Microfibril Orientation in Plant Cell Walls.* Aquatic and Terrestrial Plants Including Trees. 1985 ISBN 90-247-3101-1
17. W.P.K. Findlay (ed.): *Preservation of Timber in the Tropics.* 1985 ISBN 90-247-3112-7
18. I. Samset: *Winch and Cable Systems.* 1985 ISBN 90-247-3205-0

FORESTRY SCIENCES

19. R.A. Leary: *Interaction Theory in Forest Ecology and Management*. 1985 ISBN 90-247-3220-4
20. S.P. Gessel (ed.): *Forest Site and Productivity*. 1986 ISBN 90-247-3284-0
21. T.C. Hennessey, P.M. Dougherty, S.V. Kossuth and J.D. Johnson (eds.): *Stress Physiology and Forest Productivity*. Proceedings of the Physiology Working Group, Technical Session, Society of American Foresters National Convention (Fort Collins, Colorado, USA, 1985). 1986 ISBN 90-247-3359-6
22. K.R. Shepherd: *Plantation Silviculture*. 1986 ISBN 90-247-3379-0
23. S. Sohlberg and V.E. Sokolov (eds.): *Practical Application of Remote Sensing in Forestry*. Proceedings of a Seminar on the Practical Application of Remote Sensing in Forestry (Jönköping, Sweden, 1985). 1986 ISBN 90-247-3392-8
24. J.M. Bonga and D.J. Durzan (eds.): *Cell and Tissue Culure in Forestry*. Volume 1: General Principles and Biotechnology. 1987 ISBN 90-247-3430-4
25. J.M. Bonga and D.J. Durzan (eds.): *Cell and Tissue Culure in Forestry*. Volume 2: Specific Principles and Methods: Growth and Development. 1987 ISBN 90-247-3431-2
26. J.M. Bonga and D.J. Durzan (eds.): *Cell and Tissue Culure in Forestry*. Volume 3: Case Histories: Gymnosperms, Angiosperms and Palms. 1987 ISBN 90-247-3432-0
    Set ISBN (Volumes 24-26) 90-247-3433-9
27. E.G. Richards (ed.): *Forestry and the Forest Industries: Past and Future*. Major Developments in the Forest and Forest Industries Sector Since 1947 in Europe, the USSR and North America. In Commemoration of the 40th Anniversary of the Timber Committee of the UNECE. 1987 ISBN 90-247-3592-0
28. S.V. Kossuth and S.D. Ross (eds.): *Hormonal Control of Tree Growth*. Proceedings of the Physiology Working Group, Technical Session, Society of American Foresters National Convention (Birmingham, Alabama, USA, 1986). 1987 ISBN 90-247-3621-8
29. U. Sundberg and C.R. Silversides: *Operational Efficiency in Forestry*. Vol. 1: Analysis. 1988 ISBN 90-247-3683-8
30. M.R. Ahuja (ed.): *Somatic Cell Genetics of Woody Plants*. Proceedings of the IUFRO Working Party S2.04-07 Somatic Cell Genetics (Grosshansdorf, Germany, 1987). 1988 . ISBN 90-247-3728-1
31. P.K.R. Nair (ed.): *Agroforestry Systems in the Tropics*. 1989 ISBN 90-247-3790-7
32. C.R. Silversides and U. Sundberg: *Operational Efficiency in Forestry*. Vol. 2: Practice. 1989 ISBN 0-7923-0063-7
    Set ISBN (Volumes 29 and 32) 90-247-3684-6
33. T.L. White and G.R. Hodge (eds.): *Predicting Breeding Values with Applications in Forest Tree Improvement*. 1989 ISBN 0-7923-0460-8
34. H.J. Welch: *The Conifer Manual*. Volume 1. 1991 ISBN 0-7923-0616-3
35. P.K.R. Nair, H.L. Gholz, M.L. Duryea (eds.): *Agroforestry Education and Training*. Present and Future. 1990 ISBN 0-7923-0864-6
36. M.L. Duryea and P.M. Dougherty (eds.): *Forest Regeneration Manual*. 1991 ISBN 0-7923-0960-X

FORESTRY SCIENCES

37. J.J.A. Janssen: *Mechanical Properties of Bamboo.* 1991 ISBN 0-7923-1260-0
38. J.M. Bonga and P. Von Aderkas: *In Vitro Culture of Trees.* 1992 ISBN 0-7923-1540-5
39. L. Fins, S.T. Friedman and J.V. Brotschol (eds.): *Handbook of Quantitative Forest Genetics.* 1992 ISBN 0-7923-1568-5
40. M.J. Kelty, B.C. Larson and C.D. Oliver (eds.): *The Ecology and Silviculture of Mixed-Species Forests.* A Festschrift for David M. Smith. 1992 ISBN 0-7923-1643-6
41. M.R. Ahuja (ed.): *Micropropagation of Woody Plants.* 1992 ISBN 0-7923-1807-2
42. W.T. Adams, S.H. Strauss, D.L. Copes and A.R. Griffin (eds.): *Population Genetics of Forest Trees* (Corvallis, Oregon, USA, 1990) 1992 ISBN 0-7923-1857-9

KLUWER ACADEMIC PUBLISHERS – DORDRECHT / BOSTON / LONDON